Are We the Next Endangered Species?

Are We the Next Endangered Species?

BIOWEAPONS, EUGENICS, AND MORE

Dr. Richard M. Fleming

PhD, MD, JD

Foreword by Charles H. Rixey

Skyhorse Publishing

Skyhorse Publishing books may be purchased in bulk at special discounts for sales promotion, corporate gifts, fund-raising, or educational purposes. Special editions can also be created to specifications. For details, contact the Special Sales Department, Skyhorse Publishing, 307 West 36th Street, 11th Floor, New York, NY 10018 or info@skyhorsepublishing.com.

Skyhorse® and Skyhorse Publishing® are registered trademarks of Skyhorse Publishing, Inc.®, a Delaware corporation.

Visit our website at www.skyhorsepublishing.com.

Please follow our publisher Tony Lyons on Instagram @tonyisuncertain.

10 9 8 7 6 5 4 3 2 1

Library of Congress Cataloging-in-Publication Data is available on file.

Cover design by David Ter-Avanesyan

Print ISBN: 978–1-5107–8167-2
Ebook ISBN: 978–1-5107–8168-9

Printed in the United States of America

Contents

Acknowledgments *ix*
Author Notes on Why I Am Here in This Battle for Humanity *xi*
Use of QR Codes in This Book *xvii*
Action Points *xix*
Foreword by Charles H. Rixey *xxi*

CHAPTER 1: WHY YOU SHOULD READ THIS BOOK **1**

CHAPTER 2: UNDERSTANDING HOW WE GOT HERE **5**

- Searching for Human Immortality 5
- Slavery—Subjugation 8
- Eugenics (Slavery Version 2.0)—Sterilization and Extermination 10
- Extraction of Nazi Scientists—The USA vs. Russia 21
- Convergence: Bioweapons and Eugenics 27
- Research and Weapons Development: Gain-of-Function 29
- Biological Viral Weapons Defined 32
- Spike Proteins, Genetic Vaccines, and Manipulation 45
- The Human Genome Project (HGP)—What Does It Mean To Be Human? 53
- CRISPR—Changing DNA 57
- Transfection (Cell Penetration) Vector Development. (1) Lipid Nanoparticles (LNPs) 59
- Transfection (Cell Penetration) Vector Development. (2) Viral Vectors 60
- The Consequences of Gene Therapy, Including Shedding 60

ENDGAME 64

CHAPTER 3: UNQUENCHABLE LUST FOR POWER **69**

- Pharaoh of Egypt 70
- The Coronation of Queen Elizabeth II—2 June 1953 71
- Adolf Hitler—*Mein Kampf* 71
- Habitual Lust for Power 71
- Ancient Leaders (Before 1850) 73
- Modern-Day Leaders (Since 1850) 75
- The Military-Industrial Complex (MIC) 77

WAR GAMES—WOULD YOU LIKE TO PLAY A GAME? 78

- Absolute Power and One World Order 80
- Let the Games Begin! 81

THE UNITED STATES CONSTITUTION, TREATIES, AND RESOLUTIONS 87
- Who Is in Control of the World Health Organization? 92
- What if the People You Trust Are the People Causing the Problem? 94

CHAPTER 4: THE HIJACKING OF MEDICINE 97
- Ancient Medicine (3000 BCE to 500 BCE) 97
- Modern Medicine 99
- Medicine Is Not a Business—It's a Profession, a Calling 100

HOW MEDICINE WAS HIJACKED 100
- Terminology 101
- What Are the Expressed Powers of Congress and the President? 102
- *Cruzan v. Director, Missouri Department of Health*, 497 U.S. 261 (1990) 105
- A Return to the Eugenic Greeks 106
- Seizing Medical Education in America 109
- The American Medical Association (AMA) 109
- The Flexner Report and the Influence of the Rockefeller and Carnegie Foundations 112
- Handcuffing the American Physician 113
- Health and Human Services (HHS) 114
- The Food and Drug Administration (FDA) 115
- The Centers for Disease Control (CDC)—Part of the Public Health Service 118
- Multiple Additional Factors 119

PERFORMING CPR ON MEDICINE: WE'RE INJURED BUT NOT DEAD! 121

CHAPTER 5: THE GENETIC VACCINES 126
- The Parallel Pathways of Eugenics and Biowarfare 126
- Strict Product Liability (SPL) 129

LET'S PLAY LAWYER 130

EVIDENCE 130
- Strict Product Liability (SPL)—It's the Product on Trial, Not the Defendant (Manufacturer) 132

DO YOU HAVE A CASE? 137

CHAPTER 6: PUTTING IT ALL TOGETHER—INDICT, PROSECUTE, AND CONVICT 139
- Prosecutors Obligation to *We The People* 139
- Defendants 140
- Crimes and Applicable Courts 140
- The 1975 Biological Weapons Convention (BWC) Treaty 142
- Evidence of Habitual Unethical Behavior by the Military-Industrial Complex and Its Agents 142

- The 2012 Federal Law Prohibitions with respect to biological weapons 18 USC § 175 and the 1989 Biological Weapons Anti-Terrorism Act 145
- Call to UK Parliament & ICC 147
- The 1979 Belmont Report 149
- The 1976 International Covenant on Civil and Political Rights Treaty 149
- The 1964 Declaration of Helsinki 149
- The American Medical Association Code of Medical Ethics 149
- The 1947 Nuremberg Code 151
- The 1965 Federal Perjury Statute 18 USC § 1001 151
- The 2018 Federal Perjury Statute 18 USC § 1621 154
- Applicable State Statutes 154
- Jurisdiction 155

INDICTMENT INTRODUCTION 155

- Applicable Laws Governing Gain-of-Function, Informed Consent, Perjury, and Terrorism 156
- Factual Background Evidence: Gain-of-Function 162
- Discussion: Gain-of-Function 174
- Factual Background Evidence: SARS-CoV-2 Pandemic of 2019 (COVID-19) 174
- Discussion: SARS-CoV-2 Pandemic of 2019 (COVID-19) 248
- Exhibit A: Evidence of Gain-of-Function Alteration 252
- Exhibit B: Is the Vaccine Efficacy Statistically Significant? 252
- Exhibit C: InflammoThrombotic Response and Disease 253
- Exhibit D: InflammoThrombotic, Priogenic, Sudden Cardiac Death, Miscarriages, and Cancer Pathways Promoted by Various Factors, Including Poor Dietary and Lifestyle Practices, SARS-CoV-2 Viruses, and the Genetic Vaccines 253
- Exhibit E: Actual WHO Resolution and 2005 International Health Regulations 253

CHAPTER 7: A TALE OF TWO CITIES 254

- Countering the Chaos 255
- A Jurassic Park of Misinformation. What's Real—What Isn't 257
- Anytime, Anyplace, Anywhere, Dr. Fauci 269

IN THE END, IT'S OUR CHARACTER AND BEHAVIOR THAT DEFINE US 273

CHAPTER 8: "IT'S NOT WHO YOU ARE UNDERNEATH, IT'S WHAT YOU DO THAT DEFINES YOU" 275

- It's Time To Define What Type of Person You Are 275
- Medicine and the Federal Government 279
- The 29 December 2008 Health Care Forum 279
- Implementing the Will of the People 284
- The American Health-Care Act of 2025—Legislation by and for *We the People* 284

DISCUSSING WHY WE NEED THE AMERICAN HEALTH-CARE ACT OF 2025 294

- Purpose 295
- Revoking the National Childhood Vaccine Injury Act (NCVIA). 295
- Revoking the National Practitioner Data Bank (NPDB) 296
- Access to Emerency Health Care Act (AEHCA) 298
- Investigation, Research and Development, Patenting, FDA Approval Process, and Cost of Prescription and Nonprescription Diagnostics and Drugs 298
- Diagnostic Testing and Decision-Making Transparency 299
- Restrictions upon the Federal Government 299

ADDITIONAL AREAS OF CONCERN THAT NEED TO BE ADDRESSED SOONER RATHER THAN LATER 300

- The Practice of Medicine 300
- Science, Scientific Research, and Publication 300
- Attorneys 301
- Courts and Term Limits—Including the SCOTUS 301

ACTION YOU CAN TAKE—BEGINNING TODAY! 302

- **First,** We Address the Crimes (10letters.org) 303
- **Second,** We Address The Government Overreach and Demand Legislation Be Passed to Stop This Abuse of the American People 304
- Third, We Remind Them Who This Country Belongs To—*We the People!* 304

CHAPTER 9: ARE WE NEXT? 305

- Conspiracy Theory 305
- Imagine If You Will 305
- Our Differences Make Us Stronger 309
- "It Is Better to Die on Your Feet than to Live on Your Knees" 310
- Parallel Pathways—Eugenics and Biowarfare 313

CHAPTER 10: OBSTRUCTION OF JUSTICE 315

THE APPENDIXES 321

- Appendix A: Fleming Doctorate Research and Training
- Appendix B: Severe Acute Respiratory Syndrome Viruses (SARS-CoV-2) Gain-of-Function
- Appendix C: Genetic Vaccine Development, InflammoThrombotic Response (ITR), Prion Diseases, Sudden Cardiac Death, and Spontaneous Abortions
- Appendix D: Human Immunity
- Appendix E: Rockefeller Foundation Whistleblower Materials
- Appendix F: Fleming Presentation 29 December 2008 on the ACA
- Appendix G: All Images in Are We the Next Endangered Species?

Endnotes *329*

Index *393*

Acknowledgments

Acknowledgments in a book are almost like those given after receiving an Academy Award. The list of people to thank for support, including those who helped with material for publication, the ones who gave thoughts on the manuscript itself, and those who have supported the author over time—particularly when the author was writing—makes a writer risk forgetting to include someone.

Please forgive me in advance if I forget to acknowledge someone. I would like to thank John Fitzgerald Kennedy for his program known by us as the Kennedy Kids, which launched my life. My parents, Joseph and Margaret, are appropriately the next for me to recognize. Thanks go to my father, a man who had an eighth-grade education and became a carpenter, explaining to me that if it was good enough for Jesus's's[1] father, Joseph, to be a carpenter, then it was good enough for him. Thanks go to my mother, who had a high school degree and was absolutely brilliant scientifically. My father was likewise absolutely brilliant, mathematically, performing trigonometry in his head without knowing it was trigonometry. These two people were and remain the most brilliant people I have ever known, far outshining their youngest son. I remember my parents telling me they had three children: one to replace my father, one to replace my mother, and me for the world. For better or worse, with a life I would not wish upon anyone but am forever grateful for; if questioning and advancing science and medicine while battling those profiting from the illness of people, and getting this information out is what I was born for, then so be it.

Next, as always, are my three children, in birth order: Stephanie, Christian, and Matthew. They are the three most precious gifts I have been given. I have walked through my own hell to protect them as best I can, and only time will tell if I have done this well. I can only hope that, in the end, they will believe that I did. For them and the world they live in, I am carrying forth in the efforts which JFK believed I would.

To those who have joined me on this journey during the last three to four years, including the men and women who have been so kind as to refer to themselves as Team Fleming[2] in Dallas and Utah, I offer my utmost appreciation. I don't know if I deserve such a compliment, but I shall continue to try to earn it.

I acknowledge Drs. McCairn, Deinert, and Huff, along with others (AC, JC, KD, NMD) not mentioned by name; you know who you are, and, with time, others will too. I am grateful to Charles Rixey, who first thought I was a conspiracy theorist. He has proven that the honor of our military men and women lives on.

To Nash Singh and Nisha Devkurran,[3] who organized the Crimes Against Humanity Tour (CAHT) in 2022 and transcribed some of the materials in chapter 6, I appreciate you. Finally, I dedicate this book to all of you.

For more information about the material presented in this book, or to keep up on the latest developments with my research, you can use the following quick-response (QR) code to go to: https://www.FlemingMethod.com.

For more information on the 10Letters.org Campaign to bring those responsible to trial, you can use the following quick-response (QR) code to go to www.10Letters.org, where you can fill in the information to generate a cover and indictment letter for your state attorney general and governor.

Author Notes on Why I Am Here in This Battle for Humanity

When Thomas Jefferson wrote the Declaration of Independence, he used documents that had already been written. Approximately one third of the declaration came from the writings of British citizens in the 1600s regarding their grievances with King Charles I and King Charles II. The remaining two thirds came from the Flemish Declaration of Independence in 1581, *Het Plakkaat van Verlatinge en de Declaration of Independence*, frequently shortened to the *Plakkaat van Verlatinge.*

A declaration is *not* a set of laws or governing rules, but a statement, or, in this instance, a declaration of war. The laws of this land are laid out in the US Constitution, as you shall see later in this book.

At least four Fleming bloodlines (*cognatus*, in Latin), including mine, come from Rollo Vlamingen. Both Vlamingen (Norwegian) and Fleming (Flemish/ Belgian) mean "wanderer," an appropriate name for Vikings. Understandably, while others were comfortable in their homes, my ancestors were wandering the planet—initially raiding, perhaps, but later becoming part of the people they conquered and establishing governments.[1] On 24 December 1776, Captain Fleming, rather than staying home, joined General George Washington crossing the Delaware,[2] delivering a special Christmas present to the German Hessian British mercenaries.

On 6 December 2021, I sat down with attorney Richard Stone and Texas court reporter Karen Escher, Texas CSR[3] number 5536, at the former McKinney Court House in McKinney, Texas, and provided recorded testimony under oath.[4]

In Norse mythology,[5] the world tree, or tree of life, is known as *Yggdrasill*; which is pictured as follows. Although it's acceptable to call Nordic beliefs pagan and scientific phylogenic schemes classifying all life on earth science, it is nevertheless interesting to note the similarities.

Yggdrasill represents life in the universe—the similarities and differences between the realms and the different life forms, or species. These forms of life are all connected genetically. Changing those genetics artificially breaks or disconnects the species, with the potential for creating life that shouldn't be in existence or extinguishing life already in existence. As you read through this book, I encourage you to ask whether the gain-of-function alterations in species, or the

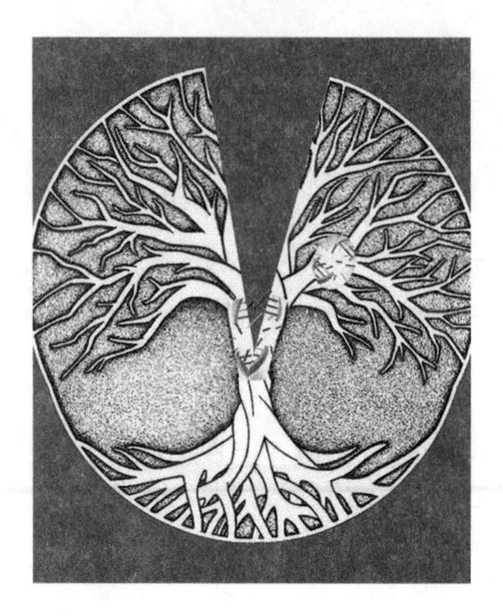

ÁRVORE DA VIDA

A ideia de ancestralidade comum a todos os seres vivos é a base da Teoria da Evolução

ANTES

A Teoria da Grande Cadeia dos Seres, proposta por Aristóteles (384 a.C. - 322 a.C.), defendia que a evolução era responsável por aperfeiçoar as espécies e o homem era o posto mais elevado dessa cadeia – ideia que ainda hoje tem espaço no senso comum.

DARWIN

Para expressar melhor suas ideias sobre o processo de descendência e ligação entre os seres, ele escreveu "I think" (Eu penso) e desenhou um diagrama (à direita). Nele, as formas mais antigas de vida ficavam na base, enquanto as espécies se ramificavam de um tronco central.

DEPOIS

A divisão dos seres vivos em cinco reinos – a mais comum nas escolas – foi criada por Robert Whittaker (1920-1980). Dela surgiu a árvore filogenética (abaixo). Com o avanço da Genética, surgiram novas classificações, em até 18 reinos, mas não há consenso sobre isso.

ERAS GEOLÓGICAS*

hoje

ERA CENOZOICA – Com o predomínio de clima mais ameno, aves, plantas superiores, insetos e mamíferos se espalham pelos continentes. Começa o processo de evolução do homem.

65 milhões de anos

ERA MESOZOICA – Animais e plantas superiores, com flores, se diversificam. Começam a surgir os primeiros mamíferos e aves. Nessa época, os dinossauros desaparecem.

245 milhões de anos

ERA PALEOZOICA – O nível de oxigênio se aproxima do atual, o que favorece à diversificação de peixes, plantas e algas e o desenvolvimento da maioria dos filos animais.

543 milhões de anos

ERA PROTEROZOICA – A vida na Terra tem origem. Surgem os primeiros seres unicelulares. Essas formas de vida ancestral se desenvolvem e evoluem, dando origem a novos reinos e filos.

4,5 bilhões de anos

perissodáctilos
carnívoros
macacos
primatas
artiodáctilos
roedores
proboscídeos
edentados
cetáceos
(extinto)
aves
placentários
marsupiais
(extintos)
monotremados
mamíferos
eucotiledôneas
monocotiledôneas
répteis
anfíbios
tetrápodes
insetos
miriápodos
quelicerados
crustáceos
osteíctes
condreíctes
anelídeos
artrópodes
vertebrados
equinodermos
moluscos
platelmintos
cnidários
esponjas
animais
fungos
plantas com flor
plantas com semente
plantas
algas
arqueobactérias
eubactérias
protozoário

Os tucanos se dividem em várias espécies, assim como a maioria dos seres vivos representados nos ramos da árvore.

REINO FUNGI – Constituído por seres com material genético delimitado por uma membrana. Os principais são bolores e cogumelos.

REINO PLANTAE – Composto de elementos pluricelulares que apresentam material genético delimitado e organizado. É representado por todas as plantas.

REINO ANIMALIA – Apresenta seres pluricelulares, cujas células formam tecidos. É representado por todos os animais.

REINO MONERA – Formado por organismos simples que não têm material genético delimitado por uma membrana. Representado pelas eubactérias e arqueobactérias.

REINO PROTISTA – Tem como principais elementos algas e amebas. Suas células têm formas variadas – em todas, a informação genética está num núcleo.

Origem da vida

utilization of genetic vaccines coupled with CRISPR (clustered regularly interspaced short palindromic repeats) technology, are helping humanity or breaking the tree of life.

As the divisions between us split the very core of Yggdrasill and the fracturing of life through gain-of-function, CRISPR, genetic vaccines, and eugenics alters the genetic code of life, will we be smart enough to learn from our past mistakes, or will we continue on a path to potential extinction?

The slides available through QR codes included throughout the chapters and appendixes are derived from the slides I have put together for presentations over many years. Each QR code will include one or more slides and an MP3 audio file, an MP4 video, or both. The audio and video files will help explain the material on the slides. Additional information is available at https://www.FlemingMethod.com and https://www.10Letters.org, in addition to https://www.crimesagainsthumanitytour.com and multiple other Internet sites. The information provided on these sites is free. To the best of my knowledge, this is the first book to combine old (written text) and modern (internet, slides, mp3 and mp4) technologies at a time when there is a desperate need for transparency, evidence, and truth. My legal colleagues would insist that I include that this book is not offering you medical or legal advice.

Chapter 6 is written in a format that can be used by prosecuting attorneys to indict, prosecute, and convict the criminals involved. For everyone wanting bullet points, this is your chapter; because a criminal indictment, like a civil complaint, lays out the critical items that, when proven, show beyond a reasonable doubt that a crime or crimes have been committed. We will cover civil litigation in chapter 5 and the legislative action and change needed in chapter 8.

Those chapters, in the context of the entire book, provide everything a prosecutor could ever want to win the case; but, alone, they lend themselves to a criminal indictment that will hold up in court. I have promised to provide this to you, the people of the United States of America, to the citizens of the world, and to the prosecutors who took an oath to uphold and defend the Constitution of the United States of America and the Constitution of the State where they are a prosecutor.

CARPE DIEM QUAM MINIMUM CREDULA POSTERO!
"Take on the day, and do not trust in tomorrow!"

Over the last four to five years, others and I have noted materials disappearing from the Internet, including published research material. Should anyone desire to remove any of the materials included or referenced in this book, please don't waste your time. I have secured PDFs, screen shots, and hard and computer copies of everything.

In their efforts to control and achieve physical immortality, the individuals carrying out these bioweapon, CRISPR, and eugenics programs have a backup plan that allows them to experiment with our physical makeup (deoxyribonucleic acid, or DNA), while simultaneously buying the time necessary to achieve that endgame. If necessary, they are taking the steps, as best they can, to secure their immortality by being able to save as much of their physical and mental selves, using robotics and artificial intelligence (AI). The battle over AI, eugenics, and biowarfare, which is sold to us as a way to make humanity better (build back better), is just beginning.

> I'm not a man, not a machine. . . . I'm more!
>
> John Connor in *Terminator Genisys*

There is a new world coming. The question, as you read this book and make decisions about what to do next, is what type of world it will be and what your role will be in it. If you fail to act or decide to go quietly into that good night, then your fate is sealed. If you take responsibility for your actions and demand that others do the same, then you not only can shape the future but also have the people you care about be a part of it. More than a tool to be used by someone else, you can be an instrument of your own choosing.

Nothing could be more appropriate at this time than to be reminded of the following quotes from movies designed to both entertain you *and* make you think:

> **Sarah Connor:** "Genisys is a Trojan horse, Skynet's way into everything."
> **Kyle Reese:** "These people are inviting their own extinction in through the front door, and they don't even know it."
>
> From *Terminator Genisys*

Then there are questions posed about the world:

> "The Matrix is a system, Neo. That system is our enemy. But when you're inside, you look around, what do you see? Businessmen, teachers, lawyers, carpenters, the very minds of the people we are trying to save. But until we do, these people are still a part of that system and that makes them our enemy. You have to understand, most of these people are not ready to be unplugged. And many of them are so inert, so hopelessly dependent on the system, that they will fight to protect it."
>
> "Unfortunately, no one can be told what the Matrix is. You have to see it for yourself."
>
> "The Matrix is the world that has been pulled over your eyes to blind you from the truth."
>
> "The Matrix is a computer-generated dream world, built to keep us under control in order to change a human being into this" [holding a copper-top D-cell battery].
>
> Dialogue between Morpheus and Neo, *The Matrix*

What is real? How do you define real?

> "This is your last chance. After this, there is no turning back. You take the blue pill—the story ends, you wake up in your bed and believe whatever you want to believe. You take the red pill—you stay in Wonderland and I show you how deep the rabbit-hole goes."
>
> "I'm trying to free your mind, Neo. But I can only show you the door. You're the one that has to walk through it."
>
> Dialogue between Morpheus and Neo, *The Matrix*

The only question you, the reader, need to ask yourself now, is do you take the blue pill, put this book down, and wake up in your bed believing what you want? Or do you take the red pill, read this book, and walk through the door to help change the world for the better?

I can only show you the door. You're the one who has to walk through it.

Use of QR Codes in This Book

More than forty QR codes are placed throughout this book, allowing you access to that material. That material provides evidence I consider important for you to know. Even if you do not completely understand the material, it is important for you to be exposed to it. Then when others doubt what you have to say, you can point to the material, knowing you are right and help others to learn. Sharing knowledge and information is one of the best ways for a free people to remain free.

Using the QR codes links will also allow you to look at the material in color and at larger size. Please refer to appendix G to find a QR code for any images reproduced in black and white and at smaller sizes in the pages of this book.

The best way to use these QR codes is to open the code the first time it is presented. Look at the slides, listen to all the MP3 audio files and watch any MP4 video files on the page. The audio and video files will explain all of the slides on the page. Leave the QR link open until you come to the place in the book pertaining to the last slide on the page. Then close the QR code, and wait for the next QR code to appear in these pages and repeat. Only a few QR codes have just one slide.

Action Points

Read this book until you understand it.

IF YOU ARE AN AMERICAN

- Go to www.10Letters.org, and send the cover letter and indictment letter to *both* your governor and attorney general in your state.
- Call your state and federal elected senators, representatives, and governor, and White House, and demand the following actions:
  - Criminal indictment and prosecution of those responsible for the biological gain-of-function bioweapon crimes.
  - Change specified laws to restore the balance of power back to the people, and remove the federal government from the practice and control of science and medicine.

IF YOU ARE NOT AN AMERICAN

- Support the American people as we take action to hold these criminals accountable.
- Take action in your country to hold your criminals accountable for their biological gain-of-function bioweapon crimes and eugenics actions.

FOR EVERYONE

- Change the world to make it better; hold yourself and those you elect legally accountable.
- Share this information with everyone. Educate, share, change.

Foreword

On August 6, 1945, a single plane flew over the small city of Hiroshima, nestled in between mountains on the western edge of the main island of Honshu, Japan. That plane, the *Enola Gay*, dropped a single atomic bomb that detonated six hundred meters above the ground and instantly killed more than forty thousand people. That moment represented a paradigm shift in warfare, and the world as we knew it would never be the same, geopolitically or technologically.

Three days later a second atomic bomb was dropped on the city of Nagasaki, and as a result of those two attacks the Empire of Japan ultimately surrendered to the United States, thus ending the Second World War. Crucially, after those attacks, the United States publicly declared that it had created a new weapon, and that announcement served as a warning to all of mankind that we as a species should be careful not to abuse this newly developed technology.

Sometime in the fall of 2019, a new virus appeared and began spreading in the city of Wuhan, China. Although the world does not yet know it, there has been another paradigm shift in warfare, one that presents an even more existential threat to our species than the nuclear arms race sparked in the wake of Hiroshima. Unlike in 1945, this new paradigm shift took place without public proclamations afterward of what had occurred; in fact, the opposite took place: intense censorship was instituted to hide it.

That censorship began first in the scientific literature and spread even faster than the virus around the world, led by Western scientists, including doctors Anthony Fauci, Francis Collins, Jeremy Farrar, and many others who would go on to play vital roles in coordinating the response to the pandemic.

The new paradigm shift is in biological weapons development, not nuclear weapons:

> For the physicists who worked on atomic energy and radioactivity, there is a before and an after Hiroshima. Virology will have to be thought of—and connected—in a similar way, starting from Wuhan.[1]

On June 16, 2023, my friend Dr. Kevin McCairn and I arrived in Hiroshima on what could be described as a kind of pilgrimage. (I'd spent my career in the Marine Corps as a chemical, biological, radiological, nuclear [CBRN] defense specialist.) As we sat in a restaurant overlooking the ground-zero memorial park that marked the spot where the first atomic bomb detonated, I received text messages from a friend of mine who was on active duty as a Marine Corps officer. He described how he had personally witnessed two fellow military men in their early thirties collapse from strokes in the previous seven days, at least one of whom had died. To put that into perspective, in my entire career, I had *never* been stationed anywhere where a single marine died of noncombat-related *anything* that wasn't tied to drugs, alcohol, suicide, or murder. This officer personally witnessed two strokes in one week, and those weren't even the first ones he'd seen during the previous two years.

What the world needs to understand is that those sudden deaths are part of the same paradigm shift that global politicians and public health officials continue to gaslight the public about: a paradigm shift in biological warfare.

Like Kevin and I, Dr. Fleming is not a journalist or politician; he is a scientist, researcher, and doctor who has independently and collaboratively worked to uncover the truth about the SARS-COV-2 virus, and the historical lines of research that have clear and direct ties to decades of biological weapons research and development. He is also a lawyer who is working tirelessly to compile evidence and hold those responsible for the creation and spread of the virus accountable for crimes against humanity.

Each of us has faced intense censorship, financial hardship, and attempts to discredit us and our findings, rather than any attempt to learn more about the evidence we have found or the implications that are clearly derived from them. However, each of us has persisted in our investigations, and over time our findings and conclusions have been repeatedly confirmed.

In this book Dr. Fleming goes beyond bioweapons development to expose the background and impact of medical malfeasance upon a population whose human rights of informed consent and protections against medical experimentation have been violated repeatedly by the public health authorities sworn to uphold those rights. He warns of the dangers associated with the emerging biotechnologies and lays out the twisted philosophical rationales used to justify the human-rights abuses that continue to be inflicted upon us. And, as he did in his previous book, Dr. Fleming continues to lay out the methods and mechanisms through which we can fight back.

Ultimately, it is *our* responsibility to ensure that the lessons of this paradigm shift are learned; as Dr. Fleming warns us in this book, the costs of inaction could be existential.

—Charles H. Rixey, MA, MBA
Researcher for DRASIC
Retired USMC Staff Sergeant
WMD Instructor for USMC

CHAPTER 1

Why You Should Read This Book

The United States of America, the once shiny city on the hill, has become the Gotham of the world, where criminals no longer hide in the shadows but have taken control of the military-industrial complex, the courts, government, and law, and with it, physicians and the practice of medicine. The elite control the system, and the average everyday American hides in fear of his or her government. People try to hide from the all-seeing eyes of the government, hoping to eke out a living and be left alone. In this Gotham state of America, gain-of-function bioweapons have been developed for use whenever those in power find it necessary to control the masses. Control has been leveraged for the elite, allowing fulfillment of an age-old desire for ultimate control—over who is enslaved and who isn't, over who lives and who dies.

Before you decide to fully surrender the shining city on the hill and hope for the world, you should remember that just because you were not born into these families or this power doesn't mean you are powerless. You can act. You can drive back, defeat, and destroy this darkness—you, your friends, and all of us together.

It's not who you are descended from that defines you, but history will ultimately be written about what you do and who you are. Be what destroys the dark.

This book looks at the events that have shaped the human quest for genetic purity. It takes a critical, what some would consider harsh, look at the events that have shaped human history as people clustered into small groups, originally for protection. Later, people's tendency to cluster altered the way in which we perceived each other.

This perception eventually led to differences that not only separated us from each other but also potentially ultimately weaken us. In the ultimate quest for power and dominion over others, these individuals have used the skill and knowledge of scientists, some well-intentioned and others not so much, for their own purpose. That need to control others and dominate has led to efforts to produce the ultimate human being, a human being without genetic flaws, without health problems, and with the potential to almost achieve immortality.

To obtain such a human required an unraveling of that which makes us: our genetic code, which is written using only four nucleosides (adenine, cytosine, guanine, and thymine). Combined like computer code, these four nucleosides make up our DNA; that defines our strengths and weaknesses. When combined with our environment (what we eat, breathe, expose ourselves to, and so forth), it determines our health and well-being. It helps determine if you will have heart disease, cancer, diabetes, and so much more, including how long you will live.

Controlling DNA would give one ultimate physical power and control of oneself and others. With the decoding of the human genome, the introduction of pseudouridine into mRNA (messenger ribonucleic acid), the discovery of CRISPR that allows bacteria to protect themselves from viruses, paving the pathway for genetic vaccines, and the gain-of-function research that led to SARS-COV-2, science has advanced far enough to make the next leap—the final leap, the leap to potentially achieving physical immortality.

To obtain this immortality leap requires manipulating human DNA to determine what works and what doesn't. As you can imagine, such a leap would normally take a long time, unless there was a way to accelerate the learning curve.

To achieve such a leap, instead of researching slowly and methodically as science usually does, it would be necessary to do something that hasn't been done for almost eighty years: mass (eugenic) experimentation. By carrying out multiple research projects on a large group of individuals, one would be able to acquire the needed knowledge more quickly. The more people one could experiment on and the more varied the experiments, the greater the chance of finding the desired result.

But how could a researcher obtain such a massive number of individuals willing to participate in a research project where a bad outcome could result in death? The answer is fear and control. This is not the same thing as mass psychosis, which is when someone decides that people who disagree with him or her must be delusional. What I am talking about is the direct manipulation of people to believe they are being protected by the person in power.

Convincing people that their lives were at risk and that those who don't participate are somehow a threat to the lives of those who do, then concentration camps[1] or a Stanford experiment[2] would not be needed. Fearful people would willingly comply. In fact, they would put pressure on others to agree to be part of the experiment.

As Hermann Göring said at his Nuremberg trial,[3] all that is required for a leader to get people to do what he wants is to convince them of a threat that can be dealt with if the they will simply do what they are told and consider anyone who objects to be a threat.

Would people be willing to give up their individual rights and liberties against what they consider their better judgment? If history has taught us anything, the answer to that question is unquestionable yes.[4] That has plagued humanity since our beginning. The segregation of people into groups of like-minded individuals who do not question will reinforce them to go along with what they are told to do by the people in power.

During his 1947 Nuremberg Trial Göring Said The Following.

... it is the leaders of the country who determine the policy and it is always a simple matter to drag the people along, whether it is a democracy or a fascist dictatorship or a Parliament or a Communist dictatorship.

...voice or no voice, the people can always be brought to the bidding of the leaders. That is easy. *All you have to do is tell them they are being attacked and denounce the pacifists for lack of patriotism and exposing the country to danger.* It works the same way in any country.

Are we so predictable that, like a rat in a (Skinner) box, we will "press a bar" and choose to cluster together in groups like our ancestors did? Will we isolate and insulate ourselves from any and all fears, instead of facing the truth? Will we blindly follow those in power, as Göring explained, as we did after the 911 attacks, surrendering ourselves to the control of others whose motives might very well be different from our own? Or will we question the motives of others, call for them to be held accountable, and be willing to pay whatever price is required of us as John F. Kennedy did[5] in an effort to find the truth and protect humanity?

Rather than speculate or provide opinions, this book provides a record of the documented events showing who has been involved in this very focused search for the Holy Grail of immortality that not everyone is going to be part of. In looking at this record, you will see some parts of human and US history that will make you feel very uncomfortable; they show that something is rotten in the United States of America. The country founded on principles of freedom, human rights, and opposition to tyranny has played a critical role in the potential extinction of humanity.

Our future is not completely written in stone, and there is still time to correct the direction we are heading. But changing course means we must take action to recover the foundation of the United States before we cross the threshold from which we cannot return. This action is to hold these people criminally responsible for what they have been doing and to prevent them from continuing their program.

As with Adolf Hitler and Nazi Germany, which merely adopted the eugenics program already underway in the United States, we must hold these people accountable for their criminal violations of the Biological Weapons Convention (BWC) Treaty and 18 United States Code (USC) Section (§) 175, and halt their programs until scientific review can be transparently conducted. If we do not, then we have given them the opportunity and ability to determine the final outcome—yours, mine, our children's, and humanity's.

The ramifications of not acting are very simple. I want you to stop and think long and hard about the following scenario. You are an average human. There is heart disease, cancer, diabetes, obesity, or some disability or disease in your family. You have defective genes and probably a defective lifestyle, including eating too much, smoking, or drinking too much. You are a burden to society, and it costs to keep you alive. Money and resources could be directed elsewhere to enrich humanity that is healthy, productive, and vital to society. Therefore, your children and grandchildren are taken away and placed in camps, where they will be allowed to die, and you will never reproduce again. Your bloodline has ended. You will be terminated. Is this worth fighting for? Is any of this worth fighting for? If the answer is yes, then meet me on the battlefield,[6] and go to www.10Letters.org to help bring this battle home to those who are responsible. No war or battle was ever won on the defensive; it is only when you take the battle to the enemy that you have any hope of victory, any hope for survival or the life you want.

CHAPTER 2

Understanding How We Got Here

This chapter lays the foundation for not only understanding how we got to this place in history but also appreciating why we are here. What are the underlying themes, be they strengths or weaknesses?[1] Those who fail to learn from history are destined to repeat it.

Searching for Human Immortality

If you were to ask people whom genetic purity and the killing of people considered inferior makes them think of, most would name Adolf Hitler. What if I told you that Hitler's Germany was merely part of a greater experiment designed to restore the longevity of humanity as reported in both the Torah[2] and Christian Bible;[3] a longevity culminating with Methuselah who reportedly lived to 969 years of age.

1 This is the record of *Adam's* line.—When *Hashem* created man, He made him in the likeness of *Hashem*;	ZEH SE-fer (b'-SE-fe-ri TEY-man SE-fer b'-SAM-kh G'-DO-lah) to-LE-doth a-DAM b'-YOM b'-RO-a e-lo-HIM a-DAM bi-d'-MUT e-lo-HIM a-SA o-TO	**א** זֶה סֵפֶר*(בספרי תימן סֵפֶר בסמ״ך גדולה) תּוֹלְדֹת אָדָם בְּיוֹם בְּרֹא אֱלֹהִים אָדָם בִּדְמוּת אֱלֹהִים עָשָׂה אֹתוֹ׃
2 male and female He created them. And when they were created, He blessed them and called them Man.—	za-KHAR u-ne-KE-va b'-ra-AM va-y'-va-REKH o-TAM va-yik-RA et-sh'-MAM a-DAM b'-YOM hi-ba-RE-AM	**ב** זָכָר וּנְקֵבָה בְּרָאָם וַיְבָרֶךְ אֹתָם וַיִּקְרָא אֶת־שְׁמָם אָדָם בְּיוֹם הִבָּרְאָם׃
3 When *Adam* had lived 130 years, he begot a son in his likeness after his image, and named him *Shet*.	vai-KHEE a-DAM sh'-lo-SHEEM um-AT sha-NAH va-YO-led bid-mu-TO k'-tzal-MO va-yik-RA et sh'-MO SHAYT	**ג** וַיְחִי אָדָם שְׁלֹשִׁים וּמְאַת שָׁנָה וַיּוֹלֶד בִּדְמוּתוֹ כְּצַלְמוֹ וַיִּקְרָא אֶת־שְׁמוֹ שֵׁת׃
4 After the birth of *Shet*, *Adam* lived 800 years and begot sons and daughters.	vai-Y'-hi-YU y'-MAY a-DAM a-kha-RAY ho-LEE-do et SHAYT, sh'-mo-NEH may-OT sha-NAH, vai-YO-led ba-NEEM u-va-NOHT.	**ד** וַיִּהְיוּ יְמֵי־אָדָם אַחֲרֵי הוֹלִידוֹ אֶת־שֵׁת שְׁמֹנֶה מֵאֹת שָׁנָה וַיּוֹלֶד בָּנִים וּבָנוֹת׃
5 All the days that *Adam* lived came to 930 years; then he died.	vai-Y'-hu kol y'-MAY a-DAM a-SHER kha-Y, t'-sha-a m'-OT sha-NAH u-sh'-LO-sheem sha-NAH, vai-YA-mot.	**ה** וַיִּהְיוּ כָּל־יְמֵי אָדָם אֲשֶׁר־חַי תְּשַׁע מֵאוֹת שָׁנָה וּשְׁלֹשִׁים שָׁנָה וַיָּמֹת׃
6 When *Shet* had lived 105 years, he begot *Enosh*.	vai-KHEE-SHAYT kha-MAYSH sha-NEEM u-m'-AT sha-NAH va-YO-led et e-NOSh	**ו** וַיְחִי־שֵׁת חָמֵשׁ שָׁנִים וּמְאַת שָׁנָה וַיּוֹלֶד אֶת־אֱנוֹשׁ׃

7 After the birth of *Enosh, Shet* lived 807 years and begot sons and daughters.	vai-KHEE-SHAYT a-kha-RAY ho-LEE-do et e-NOSh, SHE-va sha-NEEM u-sh'-MO-neh me-OT sha-NAH, va-YO-led ba-NEEM u-va-NOT	**ז** וַיְחִי־שֵׁת אַחֲרֵי הוֹלִידוֹ אֶת־אֱנוֹשׁ שֶׁבַע שָׁנִים וּשְׁמֹנֶה מֵאוֹת שָׁנָה וַיּוֹלֶד בָּנִים וּבָנוֹת׃
8 All the days of *Shet* came to 912 years; then he died.	vai-Y'-hu kol y'-MAY shayt sh'-TEEM e-SRAY sha-NAH u-t'-sha-a m'-OT sha-NAH vai-YA-mot.	**ח** וַיִּהְיוּ כׇּל־יְמֵי־שֵׁת שְׁתֵּים עֶשְׂרֵה שָׁנָה וּתְשַׁע מֵאוֹת שָׁנָה וַיָּמֹת׃
9 When *Enosh* had lived 90 years, he begot *Keinan.*	vai-KHI e-NOSH tish-EEM sha-NAH va-YO-led et-KEE-nan	**ט** וַיְחִי אֱנוֹשׁ תִּשְׁעִים שָׁנָה וַיּוֹלֶד אֶת־קֵינָן׃
10 After the birth of *Keinan, Enosh* lived 815 years and begot sons and daughters.	vai-KHI e-NOSH a-kha-RAY ho-LEE-do et KAY-nan, kha-MAYSH es-ray SHA-na, u-sh'-MO-neh me-OT SHA-na, vai-YO-led ba-NEEM u-va-NOT	**י** וַיְחִי אֱנוֹשׁ אַחֲרֵי הוֹלִידוֹ אֶת־קֵינָן חֲמֵשׁ עֶשְׂרֵה שָׁנָה וּשְׁמֹנֶה מֵאוֹת שָׁנָה וַיּוֹלֶד בָּנִים וּבָנוֹת׃
11 All the days of *Enosh* came to 905 years; then he died.	vai-Y'-hu kol y'-MAY e-NOsh KHA-mesh sha-NEEM u-t'-sha MA-oT sha-NAH va-YA-mot. (s)	**יא** וַיִּהְיוּ כׇּל־יְמֵי אֱנוֹשׁ חָמֵשׁ שָׁנִים וּתְשַׁע מֵאוֹת שָׁנָה וַיָּמֹת׃
12 When *Keinan* had lived 70 years, he begot *Mehalalel.*	vai-KHI kay-NAN shi-VAY-im sha-NAH va-YO-led et ma-ha-lal-AYL	**יב** וַיְחִי קֵינָן שִׁבְעִים שָׁנָה וַיּוֹלֶד אֶת־מַהֲלַלְאֵל׃
13 After the birth of *Mehalalel, Keinan* lived 840 years and begot sons and daughters.	vai-KHI kay-NAN a-kha-RAY ho-LEE-do et ma-ha-lal-AYL ar-ba-IM sha-NAH u-sh'-mo-NAY may-OT sha-NAH vai-YO-led ba-NEEM u-va-NOT	**יג** וַיְחִי קֵינָן אַחֲרֵי הוֹלִידוֹ אֶת־מַהֲלַלְאֵל אַרְבָּעִים שָׁנָה וּשְׁמֹנֶה מֵאוֹת שָׁנָה וַיּוֹלֶד בָּנִים וּבָנוֹת׃
14 All the days of *Keinan* came to 910 years; then he died.	vai-Y'-hu kol y'-MAY kay-NAN, E-ser sha-NEEM u-t'-sha MA-oht sha-NAH, vai-YA-mot.	**יד** וַיִּהְיוּ כׇּל־יְמֵי קֵינָן עֶשֶׂר שָׁנִים וּתְשַׁע מֵאוֹת שָׁנָה וַיָּמֹת׃
15 When *Mehalalel* had lived 65 years, he begot *Yered.*	vai-HEE ma-ha-lal-EYL, kha-MESH sha-NEEM v'-sh'-SHEEM sha-NAH, va-YO-led et ya-RED	**טו** וַיְחִי מַהֲלַלְאֵל חָמֵשׁ שָׁנִים וְשִׁשִּׁים שָׁנָה וַיּוֹלֶד אֶת־יָרֶד׃

16 After the birth of *Yered, Mehalalel* lived 830 years and begot sons and daughters.

vai-KHI ma-ha-lal-EIL a-kha-RAY ho-lee-DO et YE-red, sh'-lo-SHEEM sha-NAH u-sh'-mo-NEH me-OT sha-NAH, va-YO-led ba-NEEM u-ba-NOT

טז וַיְחִי מַהֲלַלְאֵל אַחֲרֵי הוֹלִידוֹ אֶת־יֶרֶד שְׁלֹשִׁים שָׁנָה וּשְׁמֹנֶה מֵאוֹת שָׁנָה וַיּוֹלֶד בָּנִים וּבָנוֹת׃

17 All the days of *Mehalalel* came to 895 years; then he died.

vai-YI-hu kol y'-MAY ma-ha-l'-AYL, kha-MESH v'-tish-EEM sha-NAH u-sh'-MO-neh me-OT sha-NAH, vai-YA-mot.

יז וַיִּהְיוּ כָּל־יְמֵי מַהֲלַלְאֵל חָמֵשׁ וְתִשְׁעִים שָׁנָה וּשְׁמֹנֶה מֵאוֹת שָׁנָה וַיָּמֹת׃

18 When *Yered* had lived 162 years, he begot *Chanoch*.

vai-KHEE YER-edh sh'-TA-yim v'-shi-SHEEM sha-NAH u-m'-AT sha-NAH va-YO-led et-kha-NOCH

יח וַיְחִי־יֶרֶד שְׁתַּיִם וְשִׁשִּׁים שָׁנָה וּמְאַת שָׁנָה וַיּוֹלֶד אֶת־חֲנוֹךְ׃

19 After the birth of *Chanoch, Yered* lived 800 years and begot sons and daughters.

vai-KHI-y'-RED a-kha-RAY ho-LEE-do et kha-NOCH, sh'-mo-NEH me-OT sha-NAH, va-YO-led ba-NEEM u-va-NOHT.

יט וַיְחִי־יֶרֶד אַחֲרֵי הוֹלִידוֹ אֶת־חֲנוֹךְ שְׁמֹנֶה מֵאוֹת שָׁנָה וַיּוֹלֶד בָּנִים וּבָנוֹת׃

20 All the days of *Yered* came to 962 years; then he died.

vai-Y'-hu kol y'-MAY ye-RED, sh'-TA-yim v'-shi-SHEEM sha-NAH, u-t'-sha ME-ot sha-NAH, vai-YA-mot. (s)

כ וַיִּהְיוּ כָּל־יְמֵי־יֶרֶד שְׁתַּיִם וְשִׁשִּׁים שָׁנָה וּתְשַׁע מֵאוֹת שָׁנָה וַיָּמֹת׃

21 When *Chanoch* had lived 65 years, he begot *Metushelach*.

vai-KHI kha-NOKh kha-MESH v'-shi-SHEEM sha-NAH va-YO-led et m'-tu-SHA-lakh

כא וַיְחִי חֲנוֹךְ חָמֵשׁ וְשִׁשִּׁים שָׁנָה וַיּוֹלֶד אֶת־מְתוּשָׁלַח׃

22 After the birth of *Metushelach, Chanoch* walked with *Hashem* 300 years; and he begot sons and daughters.

vai-yit-ha-LEKH kha-NOKH et ha-e-lo-HEEM a-kha-RAY ho-LEE-do et m'-tu-SHE-lakh, sh'-LOSH may-OT sha-NAH, vai-YO-led ba-NEEM u-va-NOT.

כב וַיִּתְהַלֵּךְ חֲנוֹךְ אֶת־הָאֱלֹהִים אַחֲרֵי הוֹלִידוֹ אֶת־מְתוּשֶׁלַח שְׁלֹשׁ מֵאוֹת שָׁנָה וַיּוֹלֶד בָּנִים וּבָנוֹת׃

23 All the days of *Chanoch* came to 365 years.

vai-HI kol y'-MAY kha-NOCH, kha-MAYSH v'-sh'-sh'-EEM sha-NAH u-sh'-LOSH may-OT sha-NAH

כג וַיְהִי כָּל־יְמֵי חֲנוֹךְ חָמֵשׁ וְשִׁשִּׁים שָׁנָה וּשְׁלֹשׁ מֵאוֹת שָׁנָה׃

24 *Chanoch* walked with *Hashem*; then he was no more, for *Hashem* took him.

vai-yit-ha-LEKH kha-NOCH et ha-e-lo-HEEM v'-AY-ne-NU kee-la-KAK o-TO e-lo-HEEM.

כד וַיִּתְהַלֵּךְ חֲנוֹךְ אֶת־הָאֱלֹהִים וְאֵינֶנּוּ כִּי־לָקַח אֹתוֹ אֱלֹהִים׃

25 When *Metushelach* had lived 187 years, he begot *Lemech*.	vai-KHI me-tu-SHE-lakh, SHE-va u-sh'-mo-NEEM sha-NAH u-m'-AT sha-NAH, va-YO-led et-la-MEKH	**כה** וַיְחִי מְתוּשֶׁלַח שֶׁבַע וּשְׁמֹנִים שָׁנָה וּמְאַת שָׁנָה וַיּוֹלֶד אֶת־לָמֶךְ׃
26 After the birth of *Lemech, Metushelach* lived 782 years and begot sons and daughters.	vai-KHI met-u-SHE-lakh a-kha-RAY ho-LEE-do et le-MEKH, sh'-TA-yim u-sh'-MO-neem sha-NAH u-sh'-VA me-OT sha-NAH, vai-YO-led ba-NEEM u-ba-NOT.	**כו** וַיְחִי מְתוּשֶׁלַח אַחֲרֵי הוֹלִידוֹ אֶת־לֶמֶךְ שְׁתַּיִם וּשְׁמוֹנִים שָׁנָה וּשְׁבַע מֵאוֹת שָׁנָה וַיּוֹלֶד בָּנִים וּבָנוֹת׃
27 All the days of *Metushelach* came to 969 years; then he died.	vai-Y'-hu kol y'-MAY m'-tu-SHE-lakh tay-SHA v'-shi-SHEEM sha-NAH u-t'-sha m'-OT sha-NAH vai-YA-mot.	**כז** וַיִּהְיוּ כָּל־יְמֵי מְתוּשֶׁלַח תֵּשַׁע וְשִׁשִּׁים שָׁנָה וּתְשַׁע מֵאוֹת שָׁנָה וַיָּמֹת׃
28 When *Lemech* had lived 182 years, he begot a son.	vai-KHI le-MEKH sh'-TA-yim u-sh'-MO-neem sha-NAH u-m'-AT sha-NAH va-YO-led BEN	**כח** וַיְחִי־לֶמֶךְ שְׁתַּיִם וּשְׁמֹנִים שָׁנָה וּמְאַת שָׁנָה וַיּוֹלֶד בֵּן׃
29 And he named him *Noach*, saying, "This one will provide us relief from our work and from the toil of our hands, out of the very soil which *Hashem* placed under a curse."	vai-YIK-ra et sh'-MO NO-akh lay-MOR: ZEH ye-na-kha-ME-nu mi-ma-a-SE-nu u-mi-a-tz'-VOHN ya-DAY-nu min ha-a-da-MAH a-SHER e-re-RAH a-do-NAI	**כט** וַיִּקְרָא אֶת־שְׁמוֹ נֹחַ לֵאמֹר זֶה יְנַחֲמֵנוּ מִמַּעֲשֵׂנוּ וּמֵעִצְּבוֹן יָדֵינוּ מִן־הָאֲדָמָה אֲשֶׁר אֵרְרָהּ יְהוָה׃
30 After the birth of *Noach, Lemech* lived 595 years and begot sons and daughters.	vai-KHEE le-MEKH a-kha-RAY ho-LEE-do et-NO-akh, kha-MESH v'-tish-EEM sha-NAH, va-kha-MESH may-OT sha-NAH, va-YO-led ba-NEEM u-va-NOHT	**ל** וַיְחִי־לֶמֶךְ אַחֲרֵי הוֹלִידוֹ אֶת־נֹחַ חָמֵשׁ וְתִשְׁעִים שָׁנָה וַחֲמֵשׁ מֵאֹת שָׁנָה וַיּוֹלֶד בָּנִים וּבָנוֹת׃
31 All the days of *Lemech* came to 777 years; then he died.	vai-HI kol y'-MAY le-MEKH, SHE-va v'-shiv-EEM sha-NAH u-sh'-VA may-OT sha-NAH, va-YA-mot. (s)	**לא** וַיְהִי כָּל־יְמֵי־לֶמֶךְ שֶׁבַע וְשִׁבְעִים שָׁנָה וּשְׁבַע מֵאוֹת שָׁנָה וַיָּמֹת׃
32 When *Noach* had lived 500 years, *Noach* begot *Shem*, Ham, and Japheth.	vai-HI no-AKH ben-kha-MESH me-OT sha-NAH va-YO-led no-AKH et-SHEM et-KHAM v'-et-YA-fet	**לב** וַיְהִי־נֹחַ בֶּן־חֲמֵשׁ מֵאוֹת שָׁנָה וַיּוֹלֶד נֹחַ אֶת־שֵׁם אֶת־חָם וְאֶת־יָפֶת׃

These passages indicate human life spans progressively increased, up to Methuselah, followed by a steady decline in life spans until the depiction of a great flood—reported by many cultures around the world—with a significant reduction in longevity and appearance of disease, first mentioned in Exodus:

> If you will diligently listen to the voice of the Lord your God, and do that which is right in his eyes, and give ear to his commandments and keep all his statutes, I will put none of the diseases on you that I put on the Egyptians, for I am the Lord, your healer.[4]

Slavery—Subjugation

With the appearance of disease, shortened life spans, and related fears, no wonder people have clustered together to seek why they should be the superior ones who should live eternally. This search for immortality and superiority gave rise to nation-states, religious and cultural differences, and the conquest of other people, subjugating them to the role of a lesser person or slave. Many are still struggling with the consequences of subjugation, with many seeking retributions for enslavement that was not theirs from people who did not enslave them.

For many who owned slaves, slaves were thought of just like any other chattel besides real estate. Most people envision the southern United States and the Civil War—or as my Southern friends refer to it, the War Between the States, noting there was nothing civil about the war.

But slavery dates back many thousands of years, and you need not look any further than modern theological texts detailing how Egyptians enslaved those they conquered. The conquered were a free workforce for the Egyptians, but, as history has repeatedly shown us, enslaved people are not happy.[5] However, long before the Jewish people were enslaved in Egypt, Pharaoh Sneferu of the twenty-sixth century BCE distinguished between laborers and "servants" who were "bound for life."[6]

The history of human slavery extends throughout the centuries and across all continents, including Europe, Scandinavia, Asia, Africa, and North and South America, and it exists today in countries such as North Korea, Eritrea, Mauritania, Saudi Arabia, Turkey, Tajikistan, the United Arab Emirates, and Russia. This modern version of slavery includes forced labor, debt bondage, forced marriages, and human trafficking.[7]

During the beginning of European settlement of the United States, more than 70 percent of the slaves—an estimated 2.7 million[8] men, women and children—brought to the New World came via British and Portuguese slave trading. Slave trade slowed in the early 1800s, but the newly founded United States did not guarantee that all were entitled to the same unalienable rights in its Declaration of Independence:[9]

> We hold these truths to be self-evident, that all men are created equal, that they are endowed by their Creator with certain unalienable Rights, that among these are Life, Liberty and the pursuit of Happiness.
>
> Declaration of Independence, 4 July 1776

In fact, many of the Founding Fathers owned slaves, and some such as Thomas Jefferson fathered children[10] with female slaves. The different opinions and attitudes between slave owners and the enslaved involved in this type of relationship is apparent. Five of the Native American Nations—the "civilized" tribes Cherokee, Chickasaw, Choctaw, Creek, and Seminole—formally recognized the institution of slavery.[11]

Clearly, the history of slavery is not as simple as white versus black but can be classified as different people versus different people, or the enslavement of the conquered. Mediterranean peoples owned Africans, Native Americans owned Africans, English owned Irish, Vikings owned Europeans, Chinese owned Japanese and vice versa, kings within their own country owned fellow countrymen and women, and so forth. One of the biggest mistakes that could be made in the discussion of slavery is that it was only about one racial group versus another. In fact, slavery existed on every continent among most if not all people.

Between 1850 and 1861 debates within the U.S. Congress[12] were moving the United States closer to war, as divisions grew between those who saw slavery as a paid-for property right and those who saw this as a violation of the rights of others. When the bloodiest war on United States soil was about to begin, resulting in the abolition of slavery, another war had already begun, as discussions about genetic differences between species and people was evolving: A genetics war (eugenics) separated people based upon social standing, national origin, intelligence (IQ), and physical, mental, and religious differences. As with US slavery, the origin of this war was in Britain and the United States, but eugenics would soon play a major role in the sterilization of hundreds of thousands and the extermination of millions.

Eugenics (Slavery Version 2.0)—Sterilization and Extermination

Before the 1850s few had heard the term eugenics, but that was about to change.

In the seventeenth century the control of the Catholic Church over science was beginning to slowly unravel, following the house arrest of Galileo Galilei that began with his arrival in Rome on 13 February 1633.[13] He was found guilty of heresy, technically not for telling people the earth revolved around the sun (heliocentric model of the solar system) as opposed to the other way around (geocentric model)[14] but for telling people they did not need to accept the Church's word for it. Instead, Galilei told people that if they stopped to think about the sun, earth, moon, stars, and other celestial bodies, that they too could figure out that the Church was wrong, and that the earth did in fact revolve around the sun.

The discussions first initiated by Thales (640–546 BCE), Xenophanes (576–480 BCE), Empedocles (495–435), Aristotle (384–322 BCE), and Roger Bacon (1220–1292) were soon to be raised again with the scientific investigation of different animal species, leading to two opposing positions. The first was led by Jean-Baptiste de Lamarck,[15] who in 1809 proposed his evolutionary model, which included the erroneous idea that an animal could through want or need alter its body[16] and pass that onto its offspring.

By contrast, Charles Darwin, following his five-year voyage (1831–1836) on the HMS *Beagle*, published his theory in November 1859, in the first edition of *On the Origin of Species by Means of Natural Selection*,[17] in which he established five basic tenets:

- More individuals are born in each generation than can survive and reproduce.
- There are natural variations among individuals of the same species.
- Certain favorable characteristics have a better chance of surviving and reproducing.
- Many of these favorable characteristics are hereditary and can be passed on to offspring.
- Gradual changes can occur over long periods, which can be passed on to offspring.

Although they have been the source of continued debate, Darwin's observations and his subsequent theory did not speculate on the origin of *Homo sapiens sapiens*, a.k.a. people. Even the term *survival of the species* was not coined by Darwin. It originated in 1864 with Herbert Spencer[18] five years after Darwin's book *On the Origin of Species.*

The next step in the evolution of human understanding was the concept of *inheritance* introduced by the Austrian monk Gregory Mendel[19] (1822–1884) in 1865. Mendel had introduced the concept of *autosomal dominant and recessive traits* four years before the Swiss physiological chemist Friedrich Miescher first identified "nuclein" inside the nucleus of white blood cells (leukocytes) in 1869. Not until 1953 did James Watson and Francis Crick conclude that deoxyribonucleic acid (DNA) was a three-dimensional double helix that contained the genetic code of life[20] thirty-four years after Phoebus Levene initially proposed (1919) that DNA was composed of nucleotides consisting of multiple nucleoside-phosphate-sugar-base moiety (parts) sequences.

Mendel was either the most fortunate monk in history or he only presented the data he found useful for his published work on dominant-recessive traits. It turns out that the peas he was studying only had seven traits that followed this dominant-recessive pattern, and yet this would become the basis of the eugenics movement and later the plan for Hitler's Germany.

The terms *dominant* and *recessive* defined the external appearance (phenotype) of the genetic (genotype) information. To show a recessive trait (e.g., blue eyes in people), you must have the same gene (blue) from both parents, whereas a dominant trait (nonblue eyes) means you could receive a blue gene from one parent while the other parent gave you a gene for brown eyes. You would then have brown eyes, but you would carry the blue recessive gene that you might pass on to one of your children.[21]

The focus on improving the genetic quality (eugenics) of the human population[22] began a few decades after the War Between the States, in 1884, when London hosted the International Health Exhibition, which examined advanced public health outreach.[23] In 1869, just before the 1884 exhibit, Charles Darwin's half-cousin Francis Galton (1822–1911) published his book *Hereditary Genius*[24] and discusses what he intended to prove with the text:

> I propose to show in this book that a man's natural abilities are derived by inheritance, under exactly the same limitations as are the form and physical features of the whole organic world. Consequently, as it is easy, notwithstanding those limitations, to obtain by careful selection a permanent breed of dogs or horses gifted with peculiar powers of running, or of doing anything else, so it would be quite practicable to produce a highly-gifted race of men by judicious marriages during several consecutive generations. I shall show that social agencies of an ordinary character, whose influences are little suspected, are at this moment working towards the degradation of human nature, and that others are working towards its improvement. I conclude that each generation has enormous power over the natural gifts of those that follow, and maintain that it is a duty we owe to humanity to investigate the range of that power, and to exercise it in a way that, without being unwise towards ourselves, shall be most advantageous to future inhabitants of the earth.
>
> The general plan of my argument is to show that high reputation is a pretty accurate test of high ability; next to discuss the relationships of a large body of fairly eminent men—namely, the Judges of England from 1660 to 1868, the Statesmen of the time of George III, and the Premiers during the last 100 years—and to obtain from these a general survey of the laws of heredity in respect to genius. Then I shall examine, in order, the kindred of the most illustrious Commanders, men of Literature and of Science, Poets, Painters, and Musicians, of whom history speaks. I shall also discuss the kindred of a certain selection of Divines and of modern Scholars. Then will follow a short chapter, by way of comparison, on the hereditary transmission of physical gifts, as deduced from the relationships of certain classes of Oarsmen and Wrestlers. Lastly, I shall collate my results, and draw conclusions.
>
> Francis Galton, *Hereditary Genius*

Galton took Darwin's work and applied the selective-breeding concept appreciated by most livestock farmers, to humans. Galton specifically asked whether it might be possible to

> give the more suitable races of blood a better chance of prevailing over the less suitable than they otherwise would have had.
>
> Francis Galton, *Hereditary Genius*

In 1883 Galton termed this practice *eugenics*,[25] preceded by the 1870 reviews of his work in Nature that applauded his conclusions and stated that "his book will take rank as an important and valuable addition to the science of human nature."[26]

By the beginning of the twentieth century, Britain's elite were looking for a way to improve the strength, productivity, and health of the nation. Crime, poverty, and mental illness, it was thought, could all be solved. William Beveridge, the architect of the welfare state, along with the economist John Maynard Keynes, became proponents of eugenics, which led to the idea that forced sterilization could be used to remove "negative traits," including specific mental disabilities, laziness, and criminality,[27] which were thought to be inherited.

The British Eugenics Education Society was founded in London in 1907. This group argued that the feebleminded should be prevented from having children. Thirty years before Winston Churchill helped lead Britain to victory over Germany, Churchill was honorary vice president of this society. As you can see, there was an overwhelming belief that eugenics could purge Britain of its weak, infirm, and most-costly elements of society. In 1910 Churchill warned the then acting Prime Minster Herbert Henry Asquith of the "very terrible danger to the race posed by the multiplication of the unfit." Churchill went on to propose forced sterilization and the prevention of marriage between the unfit.[28] Churchill even envisioned planned labor colonies where the criminally feebleminded could be detained.

By the early twentieth century, millions of dollars were donated by the Carnegie Institute of Washington, the Rockefeller Foundation (which would become extremely influential in determining the course of American medicine), along with the Harriman and Kellogg Foundations, to prove that social problems were the result of defective genetics. Examples of this can be seen in appendix E.

If you doubt the intentions of those supportive of the eugenics movement, you need only look at their words and actions. These people (Theodore Roosevelt, Margaret Sanger, Alexander Graham Bell, Helen Keller, and many others), then as now, believed that many undesirable qualities, including criminality, prostitution, "feeblemindedness," and other contributors to poverty were the result of genetically inherited characteristics, not environmental factors.

> I wish very much that the wrong people could be prevented entirely from breeding. And when the evil nature of these people is sufficiently flagrant, this should be done. Criminals should be sterilized and feeble-minded persons forbidden to leave offspring behind them.
>
> Theodore Roosevelt[29]

Margaret Sanger, a firm believer in eugenics and in discouraging people from becoming pregnant or stopping pregnancies before bringing babies to term, had the following to say:

> I think the greatest sin in the world is bringing children into the world that have disease from their parents, that have no chance in the world to be a human being practically; delinquents, prisoners, all sorts of things, just marked when they're born.
>
> Margaret Sanger, 21 September 1957
> interview by Mike Wallace[30]

In 1916 Sanger opened her first birth-control clinic. She advocated "improving the genetic composition of humans through controlled reproduction of different races and classes."[31]

Margaret Sanger shared her opinions and beliefs in the *Birth Control Review*, for which she was editor:

> I personally believe in the sterilization of the feeble-minded, the insane and the syphilitic.
>
> "Birth Control and Racial Betterment," February 1919

> The most urgent problem today is how to limit and discourage the over-fertility of the mentally and physically defective.
>
> "The Eugenic Value of Birth Control Propaganda," October 1921

In 1923 Sanger made a racial reference to people as weeds:

> It means the release and cultivation of the better racial elements in our society, and the gradual suppression, elimination and eventual extirpation [destruction] of defective stocks—those human weeds which threaten the blooming of American civilization.
>
> 8 April 1923 *New York Times* interview

Survival of the species (us) is frequently dependent upon our ability to protect our bodies from infectious diseases, cancer, or something else that would kill us. We now know that this is so important that women subconsciously use their sense of smell to determine which man will give them children with the greatest immunologic strength—that is, the ability to survive and live long, healthy lives.

We also know that women taking birth-control pills (BCPs) are less sexually satisfied, having an altered sexual behavior compared to those who do not take BCPs. This is associated with a reduced interest in finding a male partner to produce the healthiest baby possible. The end result is a reduction in the health and potential survivability of the baby conceived by that woman. Providing the perfect argument to practice eugenics.

At the Cold Springs Harbor Laboratory in New York, Harry Laughlin, an animal breeder from Iowa, in 1910 became the superintendent and assistant director of the Eugenics Records Office. There, he claimed he could use Mendel's work along with mathematics to determine who would inherit "good or bad traits" including "moral perverts, felons, epileptics, and the feeble-minded, or mentally disabled." Laughlin stated that "executive agents should have the final decision about sterilization, and that consent of the patient or parent or guardian of the patient at risk for sterilization is not required in most state statues."[32]

Laughlin called for forced sterilization of the feebleminded, insane, criminalistic, epileptic, inebriate (including both alcohol and other drugs), people who had infectious diseases (including tuberculosis, syphilis, leprosy, and other chronic infections), those who were blind or seriously visually impaired, and those who were deaf, dependent (homeless, tramps, and ne'er–do-wells), along with relatives who might be carrying recessive genes for inferior traits.

Following Laughlin's call for forced sterilization of these individuals, Indiana passed the first compulsory sterilization law in the world in 1907. The Supreme Court of the United States (SCOTUS) would uphold sterilization laws in 1927.[33] On 30 July 1913 Wisconsin also passed legislation in support of forced sterilization.[34]

Along with the passage of sterilization law within the United States, John Harvey Kellogg founded the Race Betterment Foundation,[35] followed by Charles Davenport's founding of the Eugenics Record Office in 1911.[36] They enjoyed the financial backing of the Carnegie Institute and the Rockefeller Foundation in their efforts to alleviate the "economic burden" caused by "undesirable" members of society, including the disabled and mentally ill.

In 1911 the Carnegie Institute reportedly explored eighteen potential methods[37] for removing defective genetic attributes, with the eighth option being euthanasia with one of the methods employeed being local gas chambers,[38] a method the Nazis would later employ in their concentration camps.

The clear perspective of the day was that "Social Problems Have Proven Basis of Heredity" as heralded by the *New York Times*.[39] In that article Dr. Charles B. Davenport, the director of the Carnegie Institutes Station for Experimental Evolution at Cold Spring Harbor, said,

> We scientists don't perform any experiments in eugenics; the human race does plenty of that! . . . The work that is being done in eugenics is work of investigation and of education.

In February 1914,[40] months before the beginning of the *First World War*, Harry H. Laughlin issued a report to the Eugenics Records Office[41] that focused on "the Best Practical Means of Cutting Off the Defective Germ-Plasm in the American Population," as reflected in its title.

This report emphasized not only the laissez-faire approach of not providing socioeconomic support of those in need, with the resultant eventual loss of life and reduction in undesirable genetic traits in the population, but also the more direct approach of euthanasia.

Based upon the report by Laughlin, the State of Virginia passed the Virginia Sterilization Act of 1924[42] for the expressed purpose of sterilizing those "afflicted with hereditary forms of insanity that are recurrent, idiocy, imbecility, feeble-mindedness or epilepsy." The test case for this law would center around Carrie Buck, a pregnant, unwed seventeen-year-old who was deemed to be a "moral degenerate" and "white trash of the South" for having a baby out of wedlock. The case was conducted in such a way as to prevent Carrie's substantive exculpatory

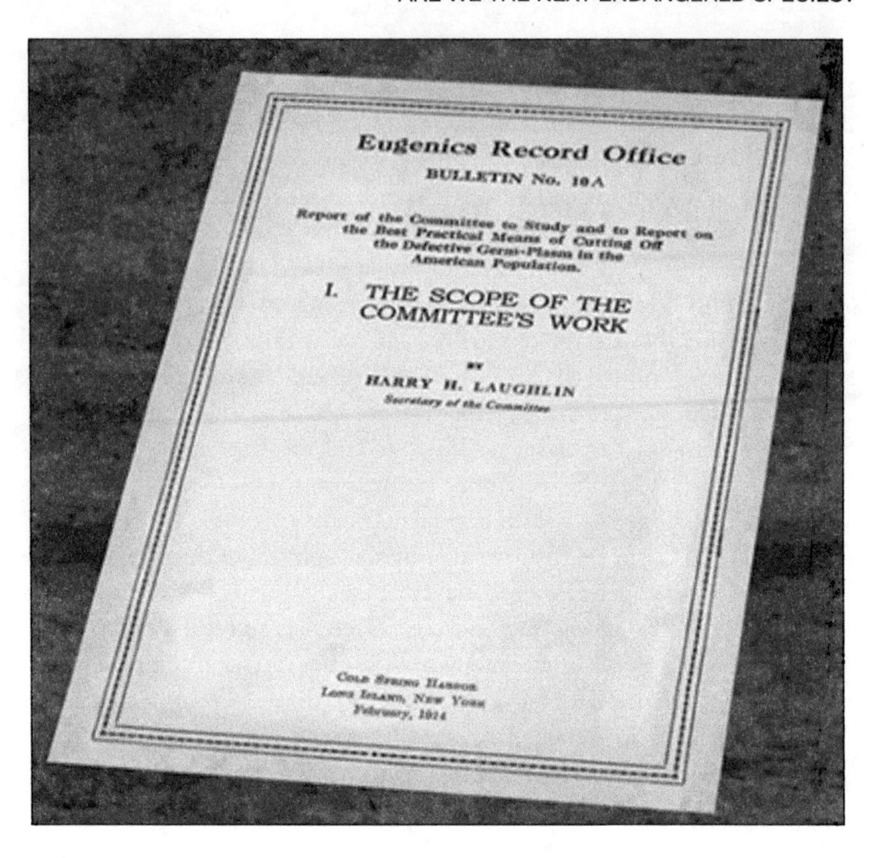
Eugenics Record Office

BULLETIN No. 10A

Report of the Committee to Study and to Report on the Best Practical Means of Cutting Off the Defective Germ-Plasm in the American Population.

I. THE SCOPE OF THE COMMITTEE'S WORK

BY

HARRY H. LAUGHLIN

Secretary of the Committee

Cold Spring Harbor
Long Island, New York
February, 1914

evidence from being presented, a pattern that has repeated itself[43] in state and federal courts since.

Opposition to the 1924 Virginia Sterilization Act was minimal outside of the white Catholic community in Virginia, with white Protestants being primarily in favor of such a law. The debate and resulting decision by the U.S. Supreme Court rulings, including *Buck v. Bell*, 274 U.S. 200, 208 (1927) fueled future laws in thirty-two US states.[44]

> Is the probable potential parent of socially inadequate offspring, likewise afflicted, that she may be sexually sterilized without detriment to her general health, and that her welfare and that of society will be promoted by her sterilization.
>
> Chief Justice Oliver Wendell Holmes Jr.

This case directly linked forced sterilization, societal costs, and compulsory vaccination.[45]

> It is better for all the world, if instead of waiting to execute degenerate offspring for crime, or let them starve for their imbecility, society can prevent those who are

> manifestly unfit from breeding their kind. The principle that sustains compulsory vaccination is broad enough to cover cutting the Fallopian tubes. *Jacobson v. Massachusetts*, 197 U. S. 11. Three generations of imbeciles are enough.
>
> Chief Justice Oliver Wendell Holmes Jr.

The Supreme Court ruling in *Buck v. Bell* held that forced sterilization of an allegedly "feeble-minded" woman in Virginia was "constitutional." Of the nine Supreme Court Justices, only Pierce Butler,[46] a Roman Catholic whose parents were from Ireland, voted against. Like the voices[47] against John Fitzgerald Kennedy running for President of the United States, including the Ku Klux Klan and religious leaders, Chief Justice Holmes reportedly believed Butler's position was influenced by his Catholic religion.

Holmes was a firm eugenicist, and he believed that Buck had defective genes and that forced sterilization was best for her and for society. His endorsement of eugenics introduced bias, in that he did not consider that his opinions could possibly be wrong. Apart from the conservative Catholic Butler, who did not believe it was morally right for the state to take the right to have children from its citizens, Holmes and the other justices focused on *procedural issues*, such as due process, rather than questioning the *substantive* validity of the law. Holmes's erroneous perspective allowed him to justify the abuse of citizens, under the fabrication that it was for the good of the citizens.[48]

In *Roe v. Wade* 410 U.S. 113, 93 S. Ct. 705, 35L. Ed. 2d 147 (1973),[49] *Buck v. Bell* was used to support the justices' ruling in support of the Due Process Clause of the Fourteenth Amendment to the United States Constitution having been meet. Norma McCorvey (a.k.a. Jane Roe) was a woman I knew and brought to Reno, Nevada, in 2006 to provide her story to a group of interested individuals. McCorvey's case was not so much about abortion as it was about the fundamental right to privacy:

> We, therefore, conclude that the right of personal privacy includes the abortion decision, but that this right is not unqualified and must be considered against important state interests in regulation.
>
> *Roe v. Wade*, 410 US at 159

In fact, *Buck v. Bell* is noted in the footnotes of *Roe v. Wade* as one of the exceptions to the general rule of reproductive freedom.

The most recent SCOTUS decision has found errors in the *Roe* decision:

> The Constitution does not confer a right to abortion; *Roe* and *Casey* are overruled; and the authority to regulate abortion is returned to the people and their elected representatives.
>
> Pp. 8–79. *Dobbs v. Jackson Women's Health Organization*, Docket No. 19–1392 (2022)

Even though *Buck v. Bell* has never been overturned, *Skinner v. Oklahoma* 316 U.S. 535 (1942) raised serious concerns about sterilization as a punitive measure. The Oklahoma sterilization statute called for forced sterilization of those with repeated criminal actions. In this instance, Jack Skinner was found guilty of stealing chickens at age nineteen.[50] This is what Justice Douglas wrote:

> We are faced with legislation which involves one of the basic rights of man. Marriage and procreation are fundamental to the very existence of the race. The power to sterilize, if exercised, may have subtle, far-reaching and devastating effects. In evil or reckless hands it can cause races of types which are inimical to the dominant group to wither and disappear. There is no redemption for the individual whom the law touches. **Any experiment the state conducts is to his [Skinner's] irreparable injury. He is forever deprived of a basic liberty.** [emphasis added]
>
> Justice William O. Douglas

As stated in the publication by Paul Popenoe[51] in 1918 and in his presentation at the American Eugenics Society,

> The first method which presents itself to cleanse the gene pool is execution. . . . Its value in keeping up the standard of the race should not be underestimated.

Execution was carried out by a variety of methods in the United States, including giving patients in mental institutions milk infected with tuberculosis,[52] resulting in a 30 to 40 percent mortality rate, while in other locations patients were killed by lethal neglect.[53]

In 1916 Madison Grant published *The Passing of the Great Race*,[54] which discussed the pollution of the Nordic race by other lesser races. Grant pushed for purification of the American people through immigration restriction, selective breeding, and sterilization.

> The man of the old stock is being crowded out of many country districts by these foreigners just as he is today being literally driven off the streets of New York city by the swarms of Polish Jews. These immigrants adopt the language of the American. They wear his clothes, they steal his name, and they are beginning to take his women. But they seldom adopt his religion or understand his ideals. And while he's being elbowed out of his own home, the American looks calmly abroad and urges on others the suicidal ethics which are exterminating his own race.
>
> Madison Grant

> A rigid system of selection through the elimination of those who are weak or unfit . . . would solve the whole question in one hundred years, as well as enable us to get rid of the undesirables who crowd our jails, hospitals, and insane asylums.

The book by Madison Grant was so favored by the Nazis that it was the first non-German book republished by the Nazis when they took power. Hitler wrote to Grant, telling him *The Passing of the Great Race* was his bible.

In addition to Grant, Hitler also admired Henry Ford, who published the *Ford International Weekly* in Dearborn, Michigan. On 22 May 1920 the following headline appeared:

The Ford International Weekly

THE DEARBORN INDEPENDENT

One Dollar — Dearborn, Michigan, May 22, 1920 — Five Cents

The International Jew: The World's Problem

The newspaper carried a review of Ford's series of articles,[55]which became a critical part of the reading of Nazi youth:

> The decisive anti-Semitic book I was reading and the book that influenced my comrades was [. . .] that book by Henry Ford, *The International Jew*. I read it and became anti-Semitic.
>
> Baldur von Schirach, head of the Hitler Youth

In the 1920s Congress debated the issue of immigration quotas based upon race and hired Laughlin to investigate and offer recommendations. As a result of Laughlin's congressional report, in which he reportedly showed undesirable genetic traits among some individuals, Congress passed the Johnson-Reed Immigration Restriction Act of 1924.[56] This new immigration law increased visas for those who could enter the United States from Western Europe and the British Isles, while reducing immigration from Southern and Eastern Europe, including Russia, Poland, Italy, and the Balkans. It also prevented Asians from immigrating to the United States, while significantly reducing immigration from India. On 27 April 1924, the *New York Times* ran the article "America of the Melting Pot Comes to End."

Adolf Hitler considered the United States to be the shining example of what eugenics could achieve, and in his 1925 book *Mein Kampf*[57] (My struggle), he declared the following:

> I have studied with great interest the laws of several American states concerning prevention of reproduction by people whose progeny would . . . be of no value or be injurious to the racial stock.

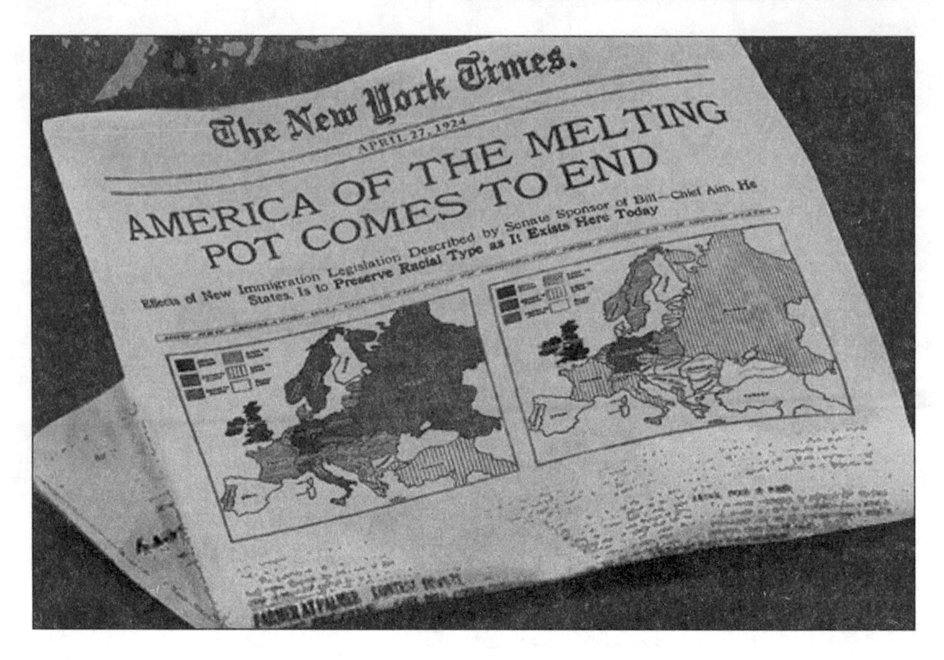

The New York Times.

APRIL 27, 1924

AMERICA OF THE MELTING POT COMES TO END

Effects of New Immigration Legislation Described by Senate Sponsor of Bill—Chief Aim, He States, Is to Preserve Racial Type as It Exists Here Today

In 1927 North Carolina became the seventeenth state to pass a sterilization law based upon Laughlin's reports. Ultimately, thirty-three states passed such a law.

In 1933 when Hitler was elected chancellor of Germany, one of his first laws was the Law for the Prevention of Hereditary Diseased Offspring,[58] which allowed for the forced sterilization of those with genetic disorders. The American eugenicist Harry H. Laughlin, who formulated the law that Hitler used to produce these German laws, subsequently received an honorary medical degree[59] from Heidelberg University in 1936 for research on purifying the germ plasm of the human population. Californian eugenicist and founder of Cal State Sacramento University C. M. Goethe congratulated his colleagues with the following statement:

> You will be interested to know that your work has played a powerful part in shaping the opinions of the group of intellectuals who are behind Hitler in this epoch-making program. . . . Everywhere I sensed that their opinions have been tremendously stimulated by American thought.

Under The Prevention of Hereditary Diseased Offspring law, over four hundred thousand people were forcibly sterilized, including the disabled, the blind, the deaf, schizophrenics, mixed-race individuals, and alcoholics. Two years after the initiation of this sterilization program, the Nazis passed the Nuremberg Laws on 15 September 1935,[60] which included the Reich Citizenship Law and the Law for the Protection of German Blood and German Honor. They laid the foundation for what would soon happen to Jewish[61] individuals in Germany.

This involuntary sterilization led to the next step: the involuntary euthanasia program known as Aktion T4.[62] Aktion T4 began in 1939 and was responsible for the killing of up to three hundred thousand disabled people in psychiatric hospitals. The selected individuals were transferred from hospital to specialized "treatment centers" run by SS[63] men wearing white coats and giving the impression of medical professionalism. The subjects were taken to shower blocks and exposed to bottled carbon monoxide (CO) gas. False death certificates were provided to families after mass cremation of the bodies. This was the euthanasia model adapted by the Nazis beginning at the end of 1941.

The sentiment of forced sterilization resonated with Americans as well as Nazis. When *Fortune* magazine polled Americans about the topic in 1937, the results revealed that two-thirds of Americans favored eugenic sterilization of mentally defective individuals, while 63 percent favored sterilization of criminals.

Although there was never a federal sterilization law, by the end of the Second World War at least eighty thousand Americans had been forced to undergo hysterectomies, tubal ligation, vasectomies, and castration.[64]

Confronted with the horrors of Auschwitz and other concentration camps, the genocidal eugenic racial and human-cleansing program, first initiated by the British and Americans, lost favor, at least in open social discussion. But, as we shall see, it is the same eugenics program now being actively deployed in the name of science.[65] Tools in the hand of good people can be weapons of mass destruction in the hands of nefarious people.

Extraction of Nazi Scientists—The USA vs. Russia

Thanks to the intelligence efforts by the Norwegian resistance movement and Norwegian scientists in their opposition to the supplying of steel and deuterium (heavy water) for Hitler's hydrogen bomb project, the United States and Britain were made aware of Hitler's plans to develop the first nuclear weapon. Between 1940 and 1944 a series of special operations were carried out to sabotage the Telemark, Norway, facilities. This was successfully accomplished, after several failed attempts, in February 1943. Efforts by the Nazis to salvage and remove the heavy water to Germany were successfully sabotaged by Norwegian resistance forces that sank the SF *Hydro* ferry on Lake Tinn.[66]

Hitler's plan, along with other programs, including the V2 rocket program[67] overseen by Wernher von Braun, represented a clear and present threat to the United States and its allies. With the downfall of Nazi Germany, those countries who could get to the military treasures and scientists the Nazis had collected first stood the best chance of securing them. Perhaps more important, they would prevent enemy nations from securing them.

Despite controversies surrounding the extraction of Nazi scientists from Germany after the Second World War and efforts by US presidents to prevent recruitment of Nazis, the Office of Strategic Services (OSS)—which eventually became the Central Intelligence Agency (CIA)—actively recruited approximately 1,600 German scientists and their families to develop weapon systems, including rockets and chemical and biological agents.[68] This program, originally known as Operation Overcast, later became known as Operation Paperclip. The goal was to recruit Nazi scientists and prevent them from being recruited by Russia.

Among the now well-established psychological, chemical, radiological, and biological experiments[69] conducted by the CIA and their Nazi scientists was Project MKUltra.[70] This initiative, like the eugenics programs, included research on "mentally impaired" boys, "sexual psychopaths," and American soldiers. It also tested the effects of putting lysergic acid diethylamide (LSD) into the drinks of unsuspecting CIA employees.[71] Among the major contributors to this research was Walter Schreiber, former surgeon general for the Third Reich, and Kurt Blome, leader of the Nazi bubonic plague program.[72]

Other biological research included subjecting navy personnel to mustard gas and arsenic,[73] sarin, soman, and tabun nerve gases,[74] in addition to other research projects. Some of the research was conducted at Fort Detrick. Despite President Nixon reportedly ordering an end to chemical-weapons testing in 1969, the testing continued.[75]

The recruitment of Nazi scientists and CIA testing upon US men, women, and military personnel was reminiscent of the type of research conducted in German concentration camps by Josef Mengele, who was particularly interested in genetics[76] and other Nazi scientists and physicians during the Second World War. In fact, while only a couple dozen high-ranking Nazis including scientists, physicians, and judges were prosecuted at Nuremberg, thousands were brought to England, France, Russia, and the United States, and inserted into our military and intelligence agencies, where they continued their work.

In light of these changes in the US military-industrial complex, President Dwight David Eisenhower, who had been commander in chief of the Normandy invasion, warned the American public of the threat the military-industrial complex[77] posed to the American people during his 17 January 1961 farewell address[78] to the country:

> We have been compelled to create a permanent armaments industry of vast proportions. . . . In the councils of government, we must guard against the acquisition of unwarranted influence, whether sought or unsought, by the military-industrial complex. . . . Our military organization today bears little relation to that known of any of my predecessors in peacetime, or, indeed, by the fighting men of World War II or Korea. . . . The potential for the disastrous rise of misplaced power exits and will persist. . . . We must never let the weight of this combination endanger our liberties or democratic process. . . . Only an alert and knowledgeable citizenry can compel the proper meshing of the huge industrial and military machinery of defense with our peaceful methods and goals, so that security and liberty may prosper together. [79]

President Eisenhower's 17 January 1961 Farewell Warning of the MIC

"In the councils of Government we must guard against the acquisition of unwarranted influence ... by the Military / Industrial Complex. The potential for the disastrous rise of misplaced power exists and will persist."

"We must never let the weight of this combination endanger our Liberties or Democratic processes. We should take NOTHING for granted."

"We must be alert to the equal and opposite danger that public policy could, itself, become the captive of a Scientific / Technological Elite."

https://www.youtube.com/watch?v=OyBNmecVtdU

It is clear from the multiple files that have since been declassified that the CIA was involved in paramilitary actions in Cuba well after Eisenhower left office, interfering with the John F. Kennedy administration and setting the stage for the Cuban missile crisis:[80]

> The one thing that has most concerned me has been the possibility that your government would not correctly understand the will and determination of the United States in any given situation, since I have not assumed that you or any other sane man would, in this nuclear age, deliberately plunge the world into war which it is crystal clear no country could win and which could only result in catastrophic consequences to the whole world, including the aggressor.
>
> JFK to Russian Premier Khrushchev, 22 October 1962

By 1963 not only had the CIA and the Department of Defense (DoD) been actively involved in research experimentation upon American citizens and military personnel, but also the continued sterilization laws and efforts which had sterilized more than sixty-five thousand Americans against their will. Efforts to promote the Sanger influence of reducing reproduction among the "human weeds" had resulted in tens of thousands of aborted fetuses, a third of which were from African American women, with African Americans representing only 17 to 18 percent of the population at this time.

Having failed to provide protections for all members of our society, John Fitzgerald Kennedy set out to end the abuses by ensuring civil rights for all:

> For man holds in his mortal hands the power to abolish all forms of human poverty and all forms of human life. And yet the same revolutionary beliefs for which our forebears fought are still at issue around the globe—the belief that

TOP SECRET SPECIAL HANDLING NOFORN

THE JOINT CHIEFS OF STAFF
WASHINGTON 25, D.C.

UNCLASSIFIED

13 March 1962

MEMORANDUM FOR THE SECRETARY OF DEFENSE

Subject: Justification for US Military Intervention in Cuba (TS)

1. The Joint Chiefs of Staff have considered the attached Memorandum for the Chief of Operations, Cuba Project, which responds to a request of that office for brief but precise description of pretexts which would provide justification for US military intervention in Cuba.

2. The Joint Chiefs of Staff recommend that the proposed memorandum be forwarded as a preliminary submission suitable for planning purposes. It is assumed that there will be similar submissions from other agencies and that these inputs will be used as a basis for developing a time-phased plan. Individual projects can then be considered on a case-by-case basis.

3. Further, it is assumed that a single agency will be given the primary responsibility for developing military and para-military aspects of the basic plan. It is recommended that this responsibility for both overt and covert military operations be assigned the Joint Chiefs of Staff.

For the Joint Chiefs of Staff:

SYSTEMATICALLY REVIEWED BY JCS ON 21 May 84 CLASSIFICATION CONTINUED

L. L. Lemnitzer

L. L. LEMNITZER
Chairman
Joint Chiefs of Staff

1 Enclosure
Memo for Chief of Operations, Cuba Project

EXCLUDED FROM GDS

EXCLUDED FROM AUTOMATIC REGRADING; DOD DIR 5200.10 DOES NOT APPLY

TOP SECRET SPECIAL HANDLING NOFORN

the rights of man come not from the generosity of the state but from the hand of God.

President John F. Kennedy, inauguration speech,[81] 20 January 1961.

President Kennedy's quest for social justice,[82] educational achievements,[83] and control of the military-industrial complex and intelligence agencies, including the DoD and CIA, did not go unnoticed by the CIA.[84] In fact, it was rumored that Kennedy had already decided to dissolve the CIA, and the evidence released to date points the assassination finger on the gun to the CIA.[85] More than thirty years after the full records of the JFK assassination records were to be released, President Joseph Robinette Biden Jr. has blocked[86] the release of those records and denied the nephew of President Kennedy Secret Service protection in his bid for the presidency of the United States.[87]

What JFK had attempted to do was to fight fire with fire. Understanding that the military-industrial complex would not go quietly into the night and taking a page out of Nazi Germany's recruitment of youth,[88] Kennedy set out to produce the same thing that Hitler had: a group of inspired young people. In this case, they were inspired by ideals of American freedom ("Ask not what your country can do for you, ask what you can do for your country") and would receive their doctorate in educational training[89] based upon their aptitudes. After all, the government[90] is ultimately responsible for granting universities and institutes of higher education the authority to bestow degrees upon graduates. Efforts to recruit me (and others) into the FBI, CIA, and other intelligence agencies, through the police department I was already a part of, were intense, as were the efforts to recruit me into the Air Force. Our work was of great interest during the end of the Vietnam War and the Apollo program.

To that end, in the interest of national security, funding could be established in the same way military funding can be set aside: monies from various sources can be funneled where needed, when needed. Part of this was done, including but not limited to, by executive orders[91] while other parts were secured as needed. Although I do not know how many children were recruited into the Kennedy Kids program, I do know my training site was not the only one. As I have traveled around the United States, I have come across others who told me what happened to them. Their stories matched mine and those of my fellow trainees.

This might be thought of as the ultimate positive eugenics experiment, the outcome of which only history can determine. JFK had hoped that at least some of these individuals—embraced by the nation and encouraged to make decisions[92] on behalf of the nation—would come to the aid of the country when most needed.

Even though JFK did not live to see the full execution of his efforts and we do not know what agencies knew about, or influenced the outcome of, these advanced doctorate programs, we do know the military and intelligence agencies continued to monitor the activity of these individuals, take interest in their research, and continue to call upon them. As you can see in the following QR coded material, efforts were made to recruit me to Fort Detrick even in 2021, to work on viral projects funded by the National Institute of Allergy and Infectious Diseases (NIAID), as a physicist.

Two months later, the book I wrote, *Is COVID-19 a Bioweapon? A Scientific and Forensic Investigation*, was published, revealing the funding sources and role that Fort Detrick played in the research and development of SARS-CoV-2 (COVID-19). All communications from Fort Detrick stopped.

Since the assassinations of President Kennedy; his brother Bobby Kennedy,[93] who was running for his party's nomination for president; and Martin Luther King Jr.,[94] there was no change to the sterilization or efforts to substantially reduce poverty in the United States of America. In fact, just the opposite has occurred.

During the Richard M. Nixon administration, Nixon was able to move many of Lyndon Johnson's War on Poverty programs out of the Office of Economic Opportunity to other agencies and cut funding by more than 50 percent. Nixon expanded monies for birth control for the poor:

> The people in what we can call our class control their populations. . . . The people who don't control their families are the people that shouldn't have kids.
>
> White House Tape 700/10, 3 April 1971,
> Nixon speaking to Ehrlichman[95]

A private White House memo distributed 18 May 1971 to federal clinics across the country informed them for the first time that War on Poverty funds could be used to cover sterilization costs.

OFFICE OF ECONOMIC
OPPORTUNITY
EXECUTIVE OFFICE OF THE PRESIDENT
WASHINGTON, D.C. 20006

OEO Instruction	6130-1
Subject Family Planning Activities	Date May 18, 1971
	Office of Primary Responsibility HA
Supersedes First Par. of Part A & Item 7, Exhibit IV, CA Memo 37-A	Distribution ER, 5-2, 10, 15, 15-11, 35-4, 45-1

APPLICABILITY: All agencies or organizations receiving funds under Titles II and III-B of the Economic Opportunity Act of 1964 , as amended.

1. PURPOSES: This Instruction amends CA Memo 37-A **to permit the use of grant funds for surgical procedures intended to result in sterilization.**

2. POLICY:

A. The first paragraph of Part A of CA Memo 37-A is amended to read as follows:

Part A: Statement of Policy on Family Planning Applications

Family planning services, including contraceptive devices and drugs, may be made available to those individuals who meet eligibility criteria as set by the applicant agency, subject only to the limitations set forth in Section 1 below. Certain other limitations

Dr. Warren Hern,[96] who worked at the Office of Economic Opportunity (OEO), developed the guidelines for this program. They specified that no one could be sterilized without informed consent and there could be no coercion.[97] However, these guidelines were never delivered to the clinics and were reportedly held up at the White House.[98] According to the *New York Times* (1973), twenty-five thousand copies of guidelines were discovered in a warehouse.[99] As a result, this federally funded program lacked the rules and regulations that would have allowed for a variety of methods for family planning and would have prevented sterilization without informed consent—that is, coerced forced sterilization. Lacking rules and regulations, the program was used for forced sterilizations. This included the cases of Minnie (fourteen years old) and Mary Alice Relf (twelve years old), who were sterilized at a clinic in Montgomery, Alabama.[100]

Joseph J. Levin Jr., cofounder of the Southern Poverty Law Center, filed a lawsuit against the Nixon administration, *Relf v. Weinberger*, 372 F. Supp. 1196 (D.D.C. 1974). The judge presiding over the Relf case found that during the Nixon administration nearly four hundred thousand poor people were sterilized without being fully informed:[101]

> There is uncontroverted evidence in the record that minors and other incompetents have been sterilized with federal funds and that an indefinite number of poor people have been improperly coerced into accepting a sterilization operation under the threat that various federal supported welfare benefits would be withdrawn unless they submitted to irreversible sterilization.
>
> Federal Judge Gerhard Gesell

Eugenic policies continued to be carried out in the United States, with the names of eugenic societies changed and eugenicists taking new titles and new positions. In 2011, a report revealed that 148 female prisoners[102] from two California state prisons were sterilized without informed consent between 2006 and 2011.

Convergence: Bioweapons and Eugenics

Prior to the perceived ending of the eugenics era, other related efforts and research had already been initiated. Both seemingly unrelated areas were being run by the same people: the military-industrial complex.

To fully understand these two areas and the ramifications of the COVID-19[103] pandemic, it is important to appreciate that several different factors converged to produce the world we now find ourselves in. As the saying goes, never let a good crisis go to waste.[104] In other words, seize every opportunity to advance what you are trying to accomplish, even if it comes at the cost of a catastrophe—or perhaps, as history so often teaches, especially during an emergency.

As such, it is also best to hide what you are doing in plain sight. To do this requires duality—that is to say, to hide in plain sight there must be a good purpose in what you are doing, as well as an ulterior motive. Since 2019 you have undoubtedly heard the term *dual use*,[105] or the concept that any research could have more than one reason for being carried out—one potentially quite good and the other quite nefarious. So, it is with the two areas of work. The first area of research and development includes gain-of-function (GoF) research. GoF has often been promoted as being critical for understanding potentially life-threatening diseases, and yet, the evidence shows it was also used to develop SARS-CoV-2, as detailed in this book and in my previous book, *Is COVID-19 a Bioweapon? A Scientific and Forensic Investigation.*[106]

The second area of investigation and development is in the field of genetic modification accomplished by transfection. *Transfection* is the ability to infect a cell with genetic material and have that material incorporated into human (or any species) DNA. Since the completion of the human genome project;[107] the discovery of human immunodeficiency virus (HIV) by my friend and colleague Professor Luc Montagnier;[108] the discovery that 18 percent of the human genome is composed of long interspersed nuclear elements-1 (LINE-1),[109] which allow outside genetic materials to become incorporated into human DNA; and the discovery and understanding of how bacteria defend themselves against viruses, namely the bacterial immune system known as clustered regularly interspaced short palindromic repeats (CRISPR),[110] it has become scientifically apparent how human DNA could be changed. The altruistic approach would be to do so for the elimination of congenital diseases, a treatment for cancer, and control of infectious diseases. The more nefarious aspects lie in the approach used to getting there and a determination by those having control over this scientific technology of what that ultimate human being should look like. What genetic traits would live on, and which would be terminated? In other words, this approach is a new and improved eugenics 2.0.

The use of this genetic manipulation technology can be broken down into two major categories:

1. Gain-of-Function: Biological warfare, the ability to defeat your enemies.

   This approach could use conventional weapons, chemical weapons, or even nuclear weapons, but those were not nearly as promising as the ability to use biological weapons. Biological weapons, in fact, would be the ultimate weapon against any enemy who had not been exposed to or protected from the biological threat. The US military could support such use if it believed other nations were developing biological weapons to be used against the United States[111] or if perhaps Earth was under threat[112] of invasion by an alien species.

2. Genetic Manipulation: The ability to genetically manipulate living organisms: to understand DNA, to find the errors and correct them.

To manipulate DNA requires (a) the ability to find genetic errors, (b) a method to target genetic errors, (c) a way to change those genetic errors—CRISPR, (d) a way to deliver the genetic material into cells for CRISPR and the genetic code to make the changes, and (e) evidence that the desired result is happening. To accomplish these steps, one either needs a tremendous amount of time to make mistakes until successful (not desirable if one wants to reap the benefit of these results) or a massive number of research subjects to test, allowing for multiple failures at once while learning from mistakes and correcting them. The result of this research could potentially correct genetic flaws, elongate telomeres for cellular immortality, and produce a healthier, longer-living human, eugenically restoring to the almost physical immortality of Methuselah.

Research and Weapons Development: Gain-of-Function

With Einstein's publication[113] of his special theory of relativity in 1905 and the Manhattan Project's development of the first three atomic bombs—the first of which was detonated in New Mexico, the latter two on Hiroshima and Nagasaki—mankind had discovered the power of creation, and with it the power of destruction. As any good physicist will tell you, the power of creation frequently comes at the cost of destruction.[114] My thesis work on positrons and plasma had the potential to power us to the stars or yield a device that could obliterate us, depending upon how it was used. Given the history of weapons development—look no further than Alfred Nobel—it was clear to me that my work would need to be buried to prevent its use by the military-industrial complex (MIC). That is something the MIC clearly has not forgotten.

With the discovery that organic (life) matter could be created from inorganic (nonlife) matter in the 1953 abiogenesis experiment of Miller-Urey,[115] scientists had unleashed the power of life. In fact, quantum mechanics explains how this can occur through proton tunneling.[116] As with the atomic bomb, this power of creation and destruction is determined by those wielding it.

Ever since life has existed on this planet, living organisms have constantly devised new ways to kill another organism. The first use of biological warfare by people was reported as early as 1500 BCE, when the Hittites (a.k.a. Neshites) used tularemia[117] to attack their political rivals, the Egyptians. After this series of attacks with resulting epidemics,[118] recorded events of the use of biological weapons to kill an enemy have plagued humanity.[119]

Year	Biological Weapon Used & Biological Weapon Agreements
1155	Emperor Barbarossa uses corpses to poison the water wells of Tortona, Italy.
1346	Mongols catapulted plague-ridden victims over the city walls of Caffa in the Crimean Peninsula.
1495	The Spanish mix the blood of lepers with wine and sell the mixture to the French in Naples, Italy.
1650	The Polish fire saliva from rabid dogs at their enemies.
1675	German and French forces agree *not* to use infected bullets.
1710	Russian army catapults plague-infested corpses over Reval (a.k.a. Tallinn, Estonia) onto Swedish troops.
1763	The British sell smallpox-infected blankets while besieged at Fort Pitt (Pittsburgh) to Native Americans, producing an epidemic and devastating the natives.
1797	Napoleon floods the Mantua, Italy, plains in an effort to spread malaria.
1863	In the United States, Confederates sell clothing tainted with smallpox and flavivirus (a.k.a. yellow fever, due to the yellow skin coloration caused by the liver disease) to Union troops.
1914–1918	Germany infects horses and cattle with bacteria (*Burkholderia mallei*, resulting in glanders disease) on both their western and eastern fronts.
1925	Geneva Protocol prohibits the use of chemical or biological weapons in war.
1939	Japanese attempt to obtain flavivirus from the Rockefeller Institute in New York, which they were planning to use against the Chinese.
1937–1945	Japanese experiment with the bubonic plague, anthrax, thyphus, yellow fever, tularemia, cholera, *Clostridium perfringens*, smallpox, and hepatits on human "subjects," killing more than three thousand.
1975	Biological Weapons Convention (BWC) Treaty signed to prevent development, stockpiling, storage, and use of biological weapons.

[120]

Although the list is by no means complete and although there have been multiple episodes of countries and groups of individuals using biological weapons against each other (biological terrorism)[121] since the Biological Weapons Convention (BWC) Treaty, it would be extremely disingenuous, not to mention hypocritical, for us to blame other countries when we need look no further than our own backyard. Given the tremendous amount of money the United States military (and that of other countries) and other agencies have funneled into biological weapons (including viral) development, we have to consider our responsibility to ourselves and the world. Evidence of this funding can be found in the following sources.

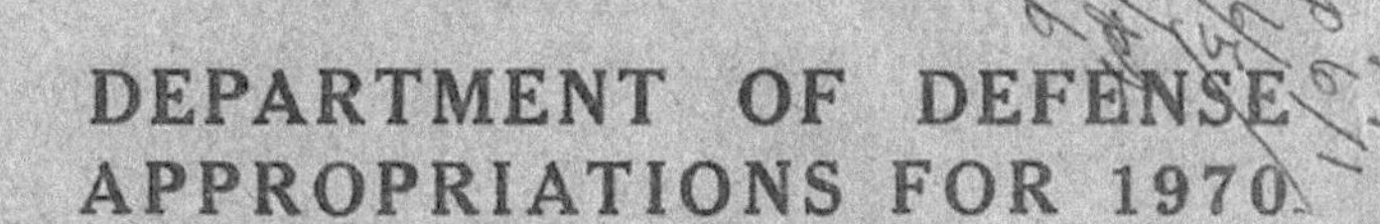

DEPARTMENT OF DEFENSE APPROPRIATIONS FOR 1970

HEARINGS

BEFORE A

SUBCOMMITTEE OF THE

COMMITTEE ON APPROPRIATIONS

HOUSE OF REPRESENTATIVES

NINETY-FIRST CONGRESS

FIRST SESSION

SUBCOMMITTEE ON DEPARTMENT OF DEFENSE APPROPRIATIONS

GEORGE H. MAHON, Texas, *Chairman*

ROBERT L. F. SIKES, Florida
JAMIE L. WHITTEN, Mississippi
GEORGE W. ANDREWS, Alabama
DANIEL J. FLOOD, Pennsylvania
JOHN M. SLACK, West Virginia
JOSEPH P. ADDABBO, New York
FRANK E. EVANS, Colorado[1]
GLENARD P. LIPSCOMB, California
WILLIAM E. MINSHALL, Ohio
JOHN J. RHODES, Arizona
GLENN R. DAVIS, Wisconsin

R. L. MICHAELS, RALPH PRESTON, JOHN GARRITY, PETER MURPHY, ROBERT NICHOLAS, AND ROBERT FOSTER, *Staff Assistants*

[1] Temporarily assigned.

PART 6

Budget and Financial Management
Budget for Secretarial Activities
Chemical and Biological Warfare
Defense Installations and Procurement
Defense Intelligence Agency
Safeguard Ballistic Missile Defense System
Testimony of Adm. Hyman G. Rickover
Testimony of Members of Congress and Other Individuals and Organizations

Printed for the use of the Committee on Appropriations

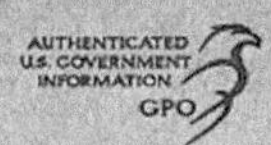

One of the potential threats posed by bioweapons is that they could escape from the laboratory in which they are designed and stored. Given the higher than "normal" level of security measures taken, you might think that so many safeguards exist that such a leak would be impossible. After all, much is at stake:

- The loss of life of people working for the country, each with a unique and needed skill set.
- The risk of being discovered violating the BWC Treaty, not to mention 18 U.S.C. § 175, in which the United States actually implemented a federal law (see chapter 6) making it a crime to not only develop, store, or use such weapons but also to take a naturally occurring organism (virus, bacteria, and such) and modify it.
- National and international repercussions, including the loss of life, liberties, and national and personal economies.

One can only speculate at the level of indifference and arrogance required to implement the development of such biological weapons, and to then use that crisis to further projects, goals, and control.

With that, we begin the uncomfortable look at the gain-of-function biological viral development of severe acute respiratory virus-2 (SARS-CoV-2) and the pandemic of coronavirus disease first reported in 2019 (COVID-19).[122]

Biological Viral Weapons Defined

In 1999, five years after I had first presented the Fleming Unified Theory of Vascular Disease at the 1994 American Heart Association (AHA) Conference[123] in Dallas, Texas, and the same year the theory was first published in a cardiology textbook,[124] the United States Department of Health and Human Services (HHS) began funding coronavirus research, focusing on the infectious characteristics of these viruses.

My theory had included viruses and other infectious organisms as one of the potential causes for a variety of diseases, known as InflammoThrombotic diseases—that is, diseases that occur because of the body's immune response to something happening within the body, that either shouldn't be there or shouldn't be present to the extent the irritant promoting an immune response is present. This is called the InflammoThrombotic response (ITR).

The following diagrams show what the theory looked like when I first presented it at the AHA conference, and again five years later when it appeared in the textbook. They are a testament to the importance of having graphic designers and artists working with a scientist's theory.

My greatest concern in the early to mid-1990s when I first presented my theory was (and remains) that infectious diseases[125]—causing an increase in ITR diseases, including heart disease, strokes, diabetes, cancer, high blood pressure, and so forth, would once again, left unchecked, return as the number one killer of humanity. My simplified version[126] of this theory was also presented in the cardiology textbook as follows:

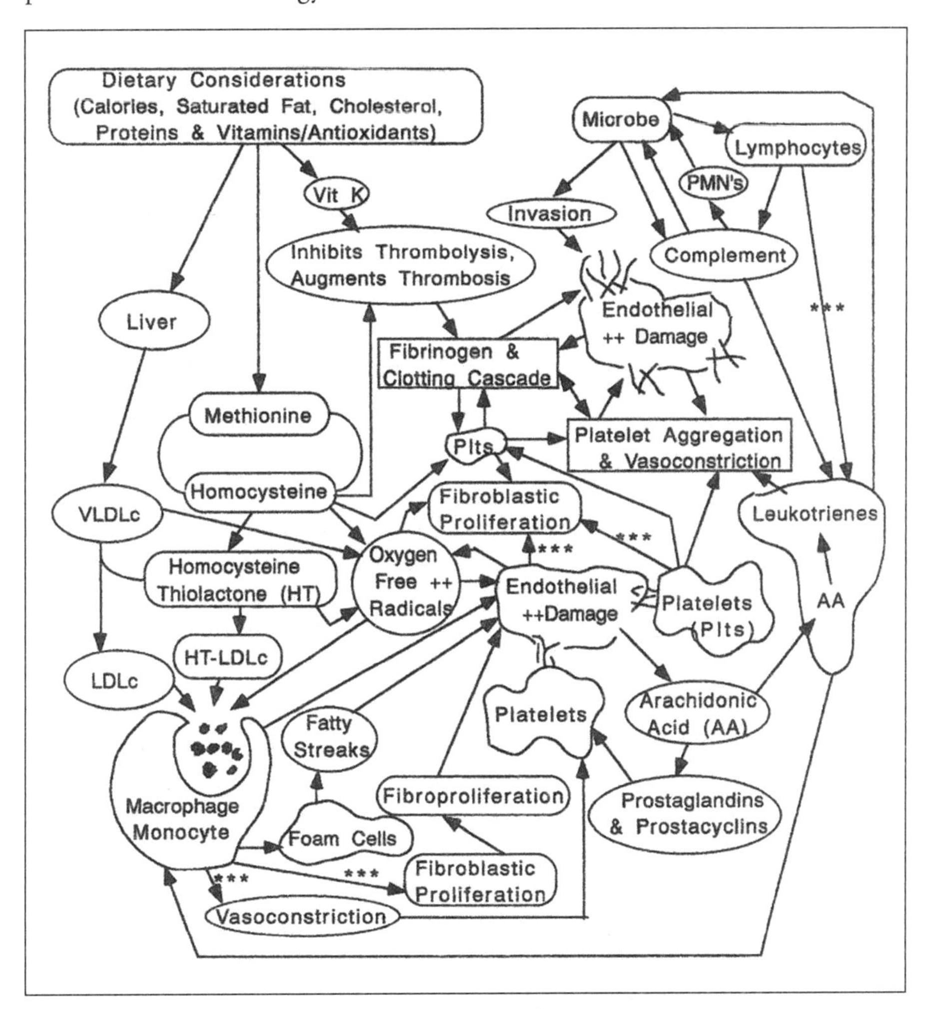

Part of my work had focused on finding and treating these diseases. The theories behind it—homocysteine as a marker of a failing or failed immune system and the role of Neu5Ac—I actively shared with both the Centers for Disease Control (CDC) and DoD in grant applications[127] and presentations.[128]

It is the glycoprotein 120 (gp120) that attaches to the Neu5Ac[129] sialic acid receptor site that is the first step involved in the spike protein attachment to our cells. Once this is done, the spike protein swings into position to attach to the angiotensin converting enzyme (ACE) receptor, followed by the furin processing of the two parts of the spike protein, known as S1 and S2. Once processed (positioned) by furin for cell entry, the transmembrane serine protease 2 (TMPRSS2) cleaves (splits) the S2 part of the spike protein, allowing the virus to be brought inside the cell for replication to increase the numbers of infecting viruses.

As it turns out, not only was HHS interested in recombinant engineering of viruses, but so too was Ralph Baric. In May 2000 and April 2002, Baric[130] and others reverse engineered an infectious clone of a severe acute respiratory syndrome (SARS) virus, thanks to funding by the National Institutes of Health (NIH). This synthetic, or man-made, clone was named SARS-CoV-Urbani. The clone would, as it turns out, match a gain-of-function virus by the name of CoV-RsSHC014, with polymerase chain reaction (PCR) matches to SARS-CoV-2 (COVID-19).[131]

In 2003 Shi Zhenlgi of the Wuhan Institute of Virology had engineered a SARS-CoV-1[132] using an HIV pseudovirus. This genetic alteration of the SARS-CoV increased its infectivity, using the glycoprotein 120 (gp 120) of HIV to attach to the sialic acid receptor I had been investigating as a cause of ITR. Such an adaptation or change in the coronavirus made it uniquely infectious for people.

By 2006, Chinese researchers had completed genetic modification of several viruses, combining four different viruses, each of which are extremely harmful to people. This new gain-of-function virus not only included the lethal HIV, in addition to the original SARS-CoV-1 and hepatitis C virus (HCV), but also a previously unidentified virus known as SARS-CoV-2, only three years after Baric had resurrected SARS-CoV-Urbani. In 2007 Baric would show that his Urbani virus could be made lethal by using the then state-of-the-art gain-of-function tool (serial passage).

The risk of these gain-of-function bioweapons[133] escaping from any of these biological laboratories is higher than you might think.[134] As laid out in the cited Furmanski 2014 report of five specific examples of these gain-of-function biological laboratory leaks, a variety of pathologic organisms have escaped from a variety of countries, including:

Region	**Year(s)**	**Pathogenic Agent**
Great Britain	1963–1978	H1N1 influenza virus
China & Soviet Union	1977	smallpox virus
Columbia	1995	Venezuelan equine encephalitis (VEE) virus
Singapore & China	2003–2004	SARS-CoV-1
United Kingdom	2007	Enterovirus

s Including Gain-of-Function Pathogens Esca

Escaped Viruses-final 2-17-14

Laboratory Escapes and "Self-fulfilling prophecy" Epidemics

By: Martin Furmanski MD
Scientist's Working Group on Chemical and Biologic Weapons
Center for Arms Control and Nonproliferation

February 17, 2014

Introduction

The danger to world or regional public health from the escape from microbiology laboratories of pathogens capable of causing pandemics, or Potentially Pandemic Pathogens (PPPs) has been the subject of considerable discussion[1,2,3,4] including mathematical modeling of the probability and impact of such escapes[5]. The risk of such releases has generally been determined from estimates of laboratory infections that are often incomplete, except for the recent 2013 Centers for Disease Control (CDC) report[6], which is a significant source of recent data on escapes from undetected and unreported laboratory-acquired infections (LAIs).

This paper presents an historical review of outbreaks of PPPs or similarly transmissible pathogens that occurred from presumably well-funded and supervised nationally supported laboratories. It should be emphasized that these examples are only the "tip of the iceberg" because they represent laboratory accidents that have actually caused illness outside of the laboratory in the general public environment. The list of laboratory workers who have contracted potentially contagious infections in microbiology labs but did not start community outbreaks is much, much longer. The examples here are not "near misses;" these escapes caused real-world outbreaks.

Can Pathogens Including Gain-of-Function Pathogens Escape from Biolabs?

Example #1: British smallpox escapes, 1966, 1972, 1978

Example #2: The "re-emergence" of H1N1 human influenza in 1977.

"Perhaps an even more serious consequence [of the 1976 swine flu episode] was the accidental release of human-adapted influenza A (H1N1) virus from a research study, with subsequent resurrection and global spread of this previously extinct virus, leading to what could be regarded as a 'selffulfilling prophecy' epidemic." (Zimmer 2009)

Example #3 Venezuelan Equine Encephalitis in 1995

Example 5: Foot and Mouth Disease (FMD) from Pirbright 2007

Example 4: SARS laboratory escapes outbreaks after the SARS epidemic

Moreover, about 5% of SARS patients are "super-spreaders" who pass the infection to many (over 8) secondary cases[35]. One case (ZZ) spread SARS to directly to

"The possibility that a SARS outbreak could occur following a laboratory accident is a risk of considerable importance, given the relatively large number of laboratories currently conducting research using the SARS-CoV or retaining specimens from SARS patients. These laboratories currently represent the greatest threat for renewed SARS-CoV transmission through accidental exposure associated with breaches in laboratory biosafety."

The hypothetical outbreak was not long in coming.

On April 22, 2004 China reported a suspected case of SARS in a 20-year-old nurse who fell ill April 5 in Beijing. The next day it reported she had nursed a 26-year-old female laboratory researcher who had fallen ill in March 25. Still ill, the researcher had traveled by train to her home in Anhui province where she was nursed by her mother, a physician, who fell ill on April 8 and died April 19.

The report concluded with the warning:

> Conclusions
>
> There are some common themes in these narratives of escaped pathogens. Undetected flaws in the functioning of what was considered at the time to be an adequate standard of technical biocontainment is one theme, as demonstrated in the UK smallpox and FMD cases. Transfer to and handling of inadequately inactivated preparations of dangerous pathogens in areas of the laboratory with reduced biosecurity levels (allowable if the preparation is actually inactivated) is another theme, demonstrated in the SARS and VEE escapes. Poor training of personnel and slack oversight of laboratory procedures negated policy efforts by national and international bodies to achieve biosecurity in the SARS and UK smallpox escapes. The recent appearance of a cohort of immunologically naïve people in the general population, which previously had been uniformly immune was a factor in the UK smallpox and the 1977 H1N1 escapes; in this regard it should be remembered that there is no immunity at all in the general population to most potentially pandemic pathogens currently under discussion, such as Avian influenza and SARS.
>
> It is hardly reassuring that despite stepwise technical improvements in containment facilities and increased policy demands for biosecurity procedures in the handling of dangerous pathogens, that escapes of these pathogens regularly occur and cause outbreaks in the general environment. Looking at the problem pragmatically, question is not if such escapes will happen in the future, but rather what the pathogen may be and how such an escape will be contained, if indeed it can be contained at all.

In 2010 there were 244 unintended releases of bioweapon agents.[135] In 2018, a year prior to the COVID-19 pandemic, concerns about project DEFUSE as part of the US Defense Advanced Research Projects Agency (DARPA) involving EcoHealth Alliance and the safety of the Wuhan Institute of Virology Bio Level 4 Laboratory were raised about unethical research and the potential for a laboratory leak of a pathogen:[136]

> "Ralph, Zhengli. If we win this contract, I do not propose that all of this work will necessarily be conducted by Ralph, but I do want to stress the US side of this proposal so that DARPA are comfortable with our team," Peter Daszak of EcoHealth Alliance wrote to North Carolina-based researcher Ralph Baric and the Wuhan scientist at the center of the lab leak theory, Zhengli Shi.
>
> "Once we get the funds, we can then allocate who does what exact work, and I believe that a lot of these assays can be done in Wuhan as well."

In light of several laboratory accidents that had occurred that released pathogens into the human population, including but not limited to the gain-of-function pathogen H5N1 (Asian avian influenza virus) released from both the Netherlands and University of Wisconsin, the CDC exposure of its workers to anthrax and smallpox, and Baric's new isolation of coronaviruses with a HKU4 spike protein which could not infect humans, the Obama administration supported a distribution of bat hosts, we have access to biological inventory data on all bat caves in Southern China, as well as information on species distributions across SE Asia from the literature and museum records. We will use radio- and satellite telemetry to identify the home range of each species of bat in the caves, to assess how widely the viral 'plume' could contaminate surrounding regions, and therefore how wide the risk zone is for the warfighter positioned close to bat caves.

Commented [PD3]: Could add " We will continue monitoring the human population proximal to these caves to assess the rates of viral spillover, and ground-truth which specific CoVs are able to infect people

We will build environmental niche models using the data above, and environmental and ecological correlates, and traits of cave species communities (eg. phylogenetic and functional diversity), to predict the species composition of bat caves across Southern China, South and SE Asia. We will validate these with data from the current project and data from PREDICT sampling in Thailand, Indonesia, Malaysia and other SE Asian countries. We will then use our unique database of bat host-viral relationships updated from our recent *Nature* paper (*1*) to assess the likelihood of low- or high-risk SARSr-CoVs being present in a cave at any site across the region. At the end of Yr 1, we will use these analyses to produce a prototype app for the warfighter that identifies the likelihood of bats harboring dangerous viral pathogens based on these analyses. **The 'high-risk bats near me' app** will be updated as new host-viral surveillance data comes on line from our project and others, to ground-truth and fine-tune its predictive capacity. Specifically, our telemetry data on bat movement will be used to assess how often bats from high-risk caves migrate to other colonies and potentially spread their high-risk strains.

The Wuhan Institute of Virology team will conduct viral testing on samples from all bat species in the caves as part of this inventory. Fecal, oral, blood and urogenital samples will be collected from bats using standard capture techniques as we have done for the last decade. In addition, tarps will be laid down in caves to assess the feasibility of surveys using pooled fresh fecal and urine samples. Assays will be designed to correlate viral load in an individual with viral shedding in a fecal sample. Once this is complete, surveys will continue largely on fecal samples so as not to disturb bat colonies and undermine longitudinal sampling capacity. Samples will be tested by PCR and spike proteins of all SARS-related CoVs sequenced. Analyses of phylogeny, recombination events, and further characterization of high-risk viruses (those with spike proteins close to SARS-CoV) will be carried out (REF). Isolation will be attempted on a subset of samples with novel SARSr-CoVs. Prof. Ralph Baric, UNC, will reverse engineer spike proteins in his lab to conduct binding assays to human ACE2 (the SARS-CoV receptor). Proteins that bind will then be inserted into SARS-CoV backbones, and inoculated into humanized mice to assess their capacity to cause SARS-like disease, and their ability to be blocked by monoclonal therapies, or vaccines against SARS-CoV (REF).

Commented [PD4]: Ralph, Zhengli. If we win this contract, I do not propose that all of this work will necessarily be conducted by Ralph, but I do want to stress the US side of this proposal so that DARPA are comfortable with our team. Once we get the funds, we can then allocate who does what exact work, and I believe that a lot of these assays can be done in Wuhan as well...

The modeling team will use these data to build models of 1) risk of viral

temporary halt to gain-of-function research in 2014, thanks to significant pressure from the scientific community.

Ralph Baric, Shi Zhengli, and Peter Daszak's EcoHealth Alliance would be allowed to continue their gain-of-function research[137] even with the Obama pause and in 2015, thanks to funding by the NIH, Zhengli would announce that the HKU4 spike protein isolated by Baric in 2013 had been made infective by her work that "introduced two single mutations . . . mutations in these motifs in coronavirus spikes have demonstrated dramatic effects on viral entry [infectivity] into human cells." These viruses match the SARS-CoV-2 PCR results.[138]

Again, while others are prevented from conducting gain-of-function research, in 2016 Baric and Zhengli announced they had combined different parts of different coronaviruses and changed five very specific nucleotide bases (genetic codes) at critical parts of the virus to yield a new virus, which in conjunction with their bacterial artificial chromosome, also funded by NIH, combined an accessory open reading frame (ORF) for a protein that would inhibit the ability of people to make interferon[139] to protect themselves against their coronavirus.

By the time the gain-of-function research "pause" was lifted in 2017, the NIH, USAID, and other government agencies had funded and helped produce coronaviruses with the ability to increase human infections and cause greater disease, including but not limited to the ability to suppress human protective immune responses.

Seven weeks before the world heard about COVID, the United States Department of Defense awarded $419,511 to Labyrinth Global Health.[140] This money was to be used to provide a report from Ukrainian laboratories to the DoD about the biological threat of COVID under the Cooperative Biological Engagement Program (CPEP), responsible for addressing threats—including pathogens and the spread of disease.

Efforts to conceal the lab origin of SARS-CoV-2 can be seen in the emails exchanged by Dr. Anthony Fauci, Kristian G. Andersen, and Kelvin K. Droegemeier (of the Executive Office of the President), in addition to efforts by Peter Daszak and others.[141]

The following two tables shown in the QR code provide just a brief look at some of the grants awarded to Peter Daszak at EcoHealth Alliance,[142] including those provided during the "pause" in gain-of-function research.

A brief summary of early key COVID events[143] up to 22 May 2020 are included in the following timelines.

As shown in appendix B, Military research funding was not restricted to Baric, Zhengli, and Daszak. Funding has gone, and continues to go, to gain-of-function biological-weapons development. Mixed among research efforts of ethical scientists to advance science has been this nefarious, illegal activity, with the expressed purpose being to develop weapons systems, despite the international and national obligations to not participate in such criminal actions. Not the least of which has been the US involvement in the biolabs in the Ukraine, which Russia has been in a hurry to reacquire and the United States has been investing heavily to prevent the Russian government, and the rest of the world, from finding out what the US military has been doing here and in other countries.

I think most people would have accepted an apology from the US military and government, when the COVID-19 pandemic hit the world with all the vigor of an ice bath, a mea culpa, an acknowledgment that they didn't intend for this to happen, that they were only acting in the best interest of the country and an accidental release of these pathogens occurred, an explanation of what they knew and what they thought needed to be done next.

Instead, with the distraction and sleight of hand that would put Houdini to shame and keeping true to the principle of not letting a crisis go to waste, the parallel program of eugenic gene modification was ramped up. What better way to see where you are technologically than to conduct a mass experiment that would put the Nazis to shame? Everyone has been participating in the experiment, either as a member of the experimental group[144] or the control group.[145]

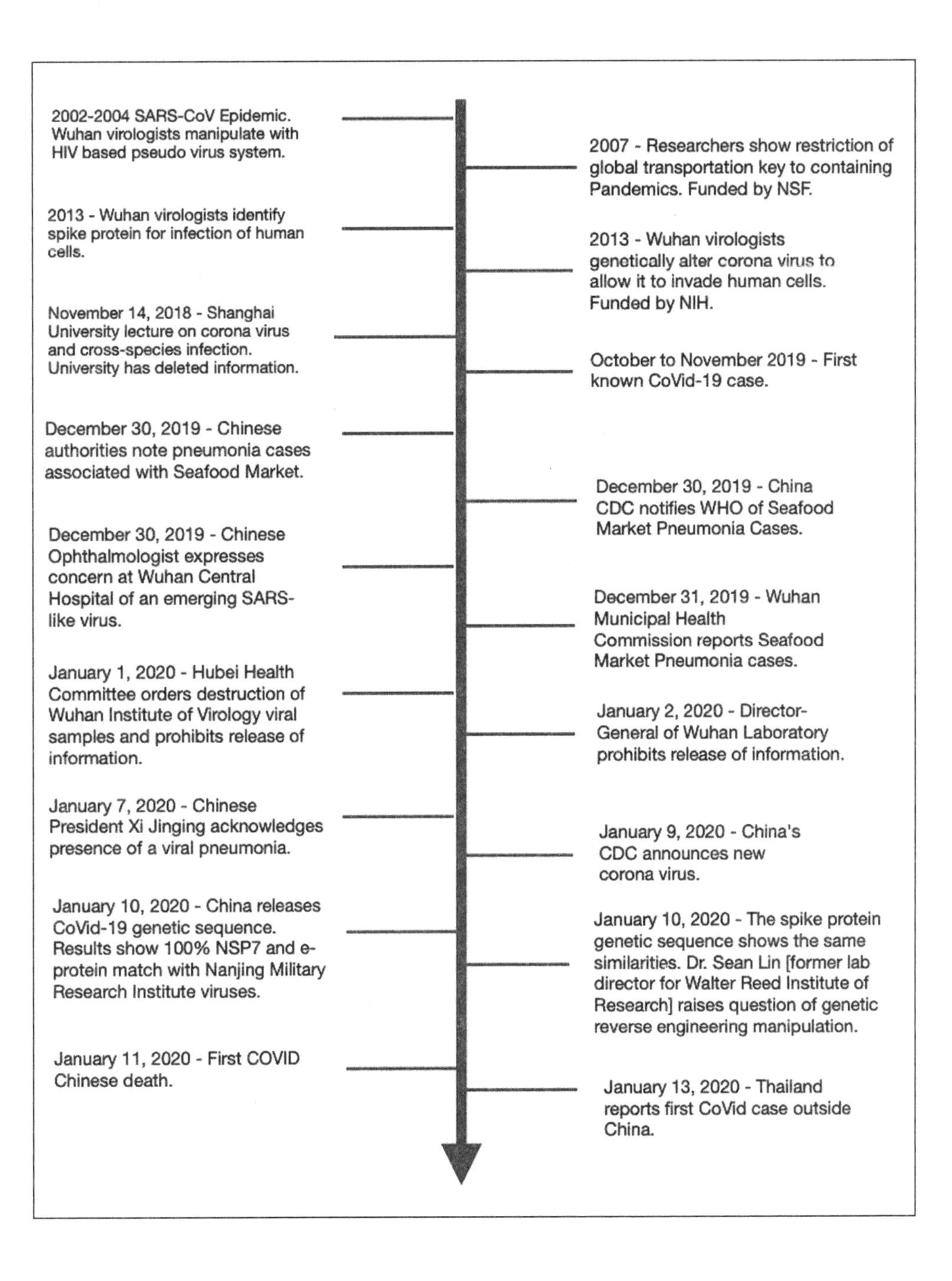
2002-2004 SARS-CoV Epidemic. Wuhan virologists manipulate with HIV based pseudo virus system.
2007 - Researchers show restriction of global transportation key to containing Pandemics. Funded by NSF.
2013 - Wuhan virologists identify spike protein for infection of human cells.
2013 - Wuhan virologists genetically alter corona virus to allow it to invade human cells. Funded by NIH.
November 14, 2018 - Shanghai University lecture on corona virus and cross-species infection. University has deleted information.
October to November 2019 - First known CoVid-19 case.
December 30, 2019 - Chinese authorities note pneumonia cases associated with Seafood Market.
December 30, 2019 - China CDC notifies WHO of Seafood Market Pneumonia Cases.
December 30, 2019 - Chinese Ophthalmologist expresses concern at Wuhan Central Hospital of an emerging SARS-like virus.
December 31, 2019 - Wuhan Municipal Health Commission reports Seafood Market Pneumonia cases.
January 1, 2020 - Hubei Health Committee orders destruction of Wuhan Institute of Virology viral samples and prohibits release of information.
January 2, 2020 - Director-General of Wuhan Laboratory prohibits release of information.
January 7, 2020 - Chinese President Xi Jinging acknowledges presence of a viral pneumonia.
January 9, 2020 - China's CDC announces new corona virus.
January 10, 2020 - China releases CoVid-19 genetic sequence. Results show 100% NSP7 and e-protein match with Nanjing Military Research Institute viruses.
January 10, 2020 - The spike protein genetic sequence shows the same similarities. Dr. Sean Lin [former lab director for Walter Reed Institute of Research] raises question of genetic reverse engineering manipulation.
January 11, 2020 - First COVID Chinese death.
January 13, 2020 - Thailand reports first CoVid case outside China.

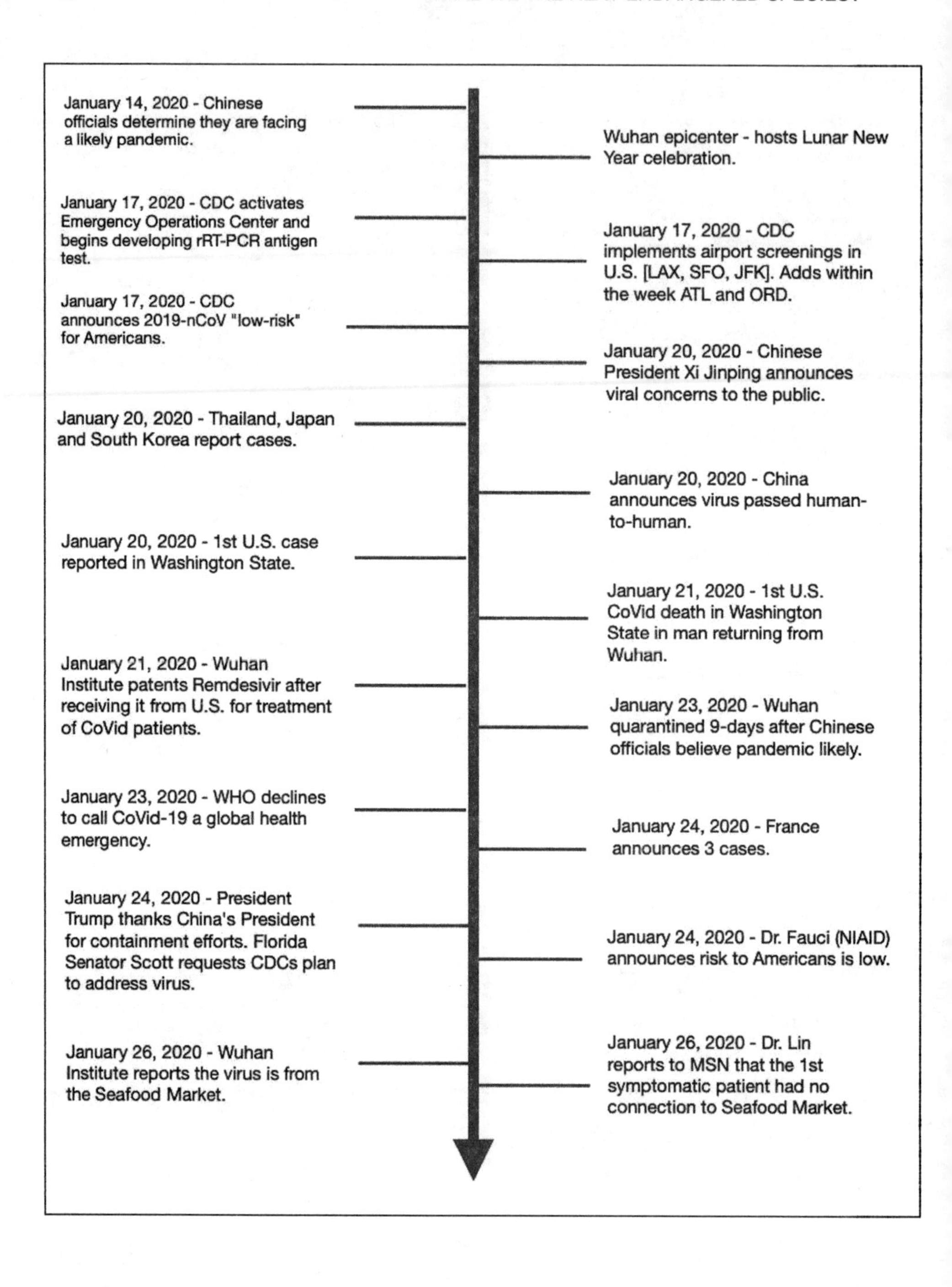
January 14, 2020 - Chinese officials determine they are facing a likely pandemic.
Wuhan epicenter - hosts Lunar New Year celebration.
January 17, 2020 - CDC activates Emergency Operations Center and begins developing rRT-PCR antigen test.
January 17, 2020 - CDC implements airport screenings in U.S. [LAX, SFO, JFK]. Adds within the week ATL and ORD.
January 17, 2020 - CDC announces 2019-nCoV "low-risk" for Americans.
January 20, 2020 - Chinese President Xi Jinping announces viral concerns to the public.
January 20, 2020 - Thailand, Japan and South Korea report cases.
January 20, 2020 - China announces virus passed human-to-human.
January 20, 2020 - 1st U.S. case reported in Washington State.
January 21, 2020 - 1st U.S. CoVid death in Washington State in man returning from Wuhan.
January 21, 2020 - Wuhan Institute patents Remdesivir after receiving it from U.S. for treatment of CoVid patients.
January 23, 2020 - Wuhan quarantined 9-days after Chinese officials believe pandemic likely.
January 23, 2020 - WHO declines to call CoVid-19 a global health emergency.
January 24, 2020 - France announces 3 cases.
January 24, 2020 - President Trump thanks China's President for containment efforts. Florida Senator Scott requests CDCs plan to address virus.
January 24, 2020 - Dr. Fauci (NIAID) announces risk to Americans is low.
January 26, 2020 - Wuhan Institute reports the virus is from the Seafood Market.
January 26, 2020 - Dr. Lin reports to MSN that the 1st symptomatic patient had no connection to Seafood Market.

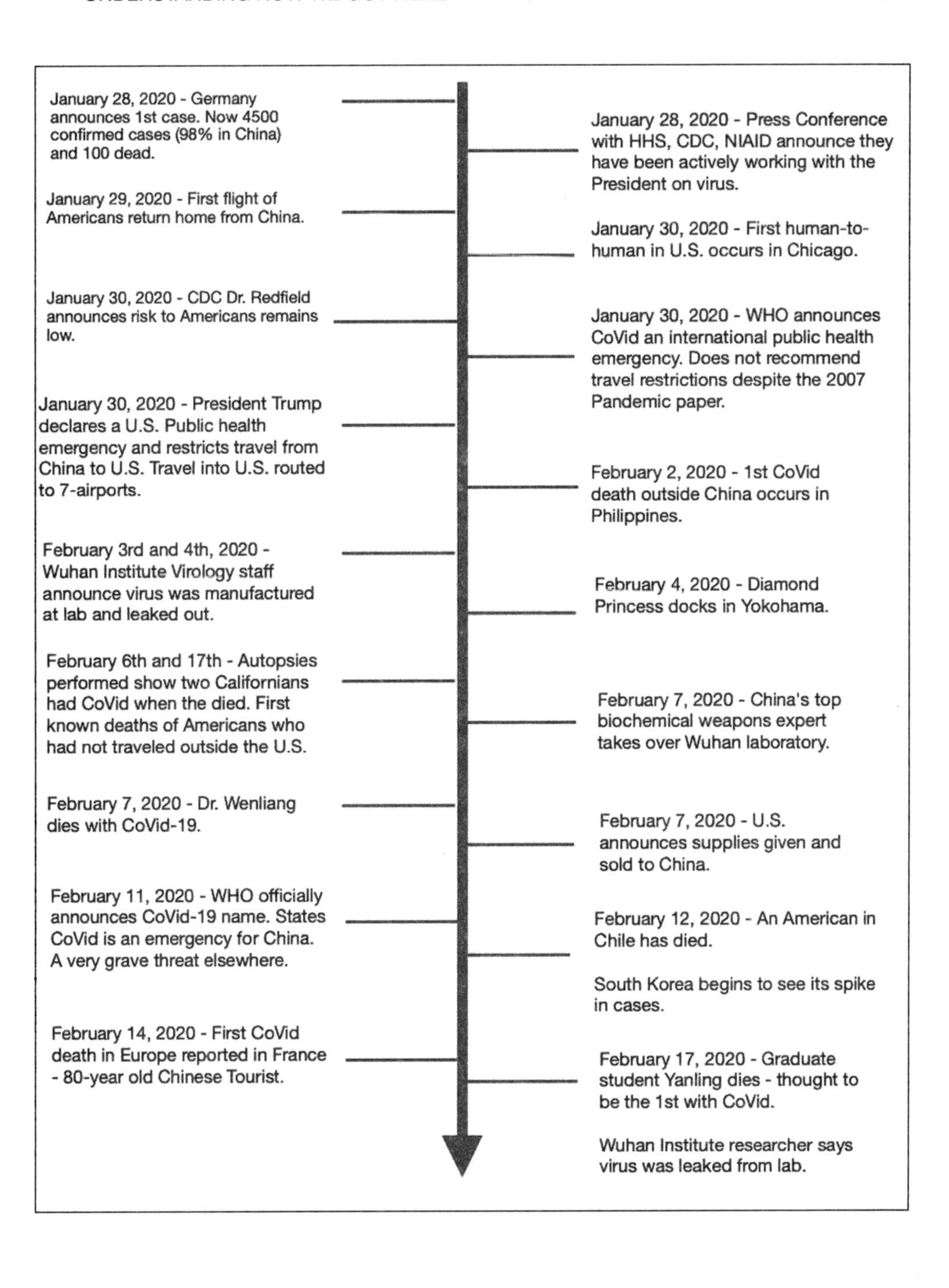
January 28, 2020 - Germany announces 1st case. Now 4500 confirmed cases (98% in China) and 100 dead.
January 28, 2020 - Press Conference with HHS, CDC, NIAID announce they have been actively working with the President on virus.
January 29, 2020 - First flight of Americans return home from China.
January 30, 2020 - First human-to-human in U.S. occurs in Chicago.
January 30, 2020 - CDC Dr. Redfield announces risk to Americans remains low.
January 30, 2020 - WHO announces CoVid an international public health emergency. Does not recommend travel restrictions despite the 2007 Pandemic paper.
January 30, 2020 - President Trump declares a U.S. Public health emergency and restricts travel from China to U.S. Travel into U.S. routed to 7-airports.
February 2, 2020 - 1st CoVid death outside China occurs in Philippines.
February 3rd and 4th, 2020 - Wuhan Institute Virology staff announce virus was manufactured at lab and leaked out.
February 4, 2020 - Diamond Princess docks in Yokohama.
February 6th and 17th - Autopsies performed show two Californians had CoVid when the died. First known deaths of Americans who had not traveled outside the U.S.
February 7, 2020 - China's top biochemical weapons expert takes over Wuhan laboratory.
February 7, 2020 - Dr. Wenliang dies with CoVid-19.
February 7, 2020 - U.S. announces supplies given and sold to China.
February 11, 2020 - WHO officially announces CoVid-19 name. States CoVid is an emergency for China. A very grave threat elsewhere.
February 12, 2020 - An American in Chile has died.
South Korea begins to see its spike in cases.
February 14, 2020 - First CoVid death in Europe reported in France - 80-year old Chinese Tourist.
February 17, 2020 - Graduate student Yanling dies - thought to be the 1st with CoVid.
Wuhan Institute researcher says virus was leaked from lab.

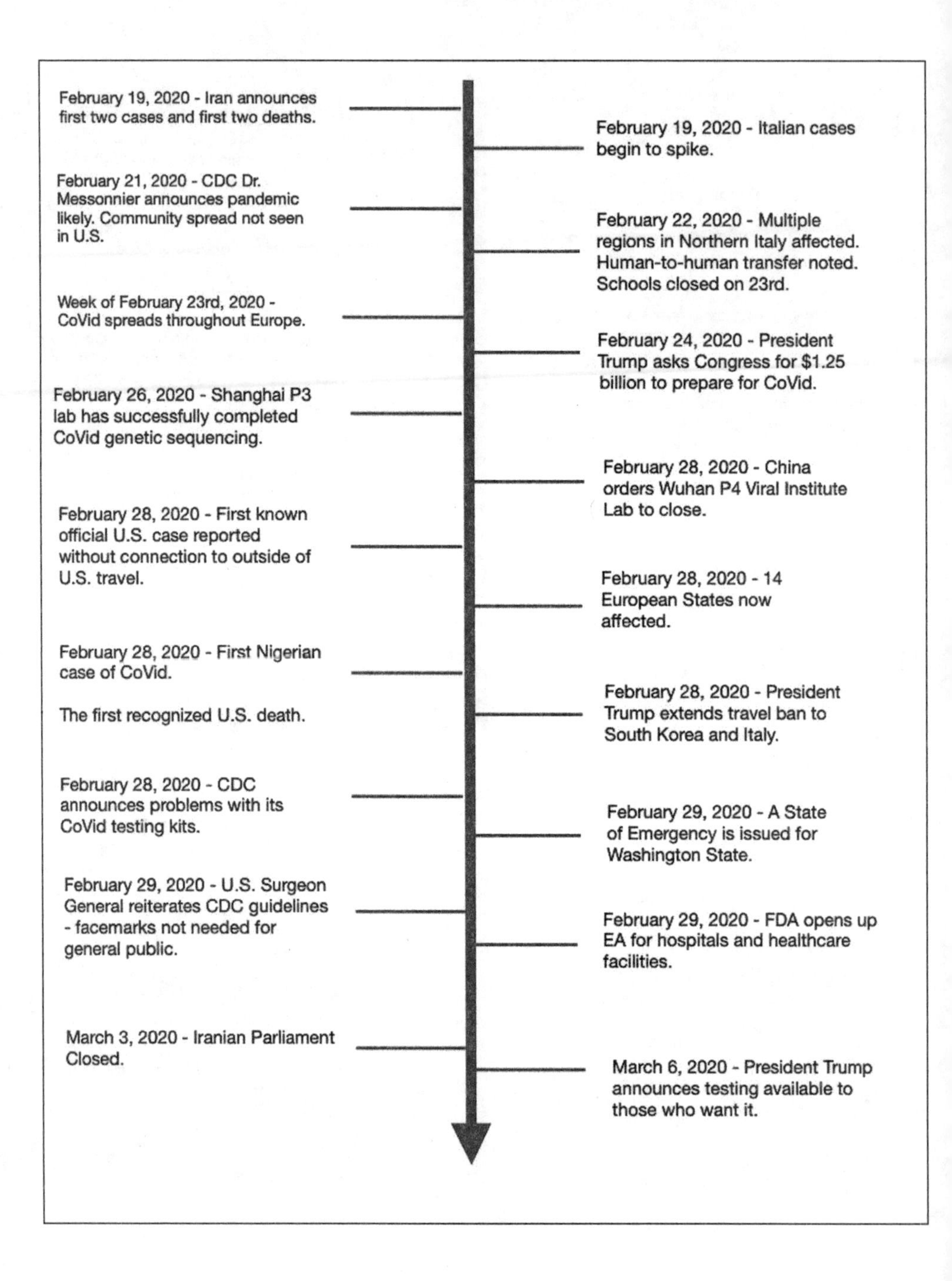
February 19, 2020 - Iran announces first two cases and first two deaths.
February 19, 2020 - Italian cases begin to spike.
February 21, 2020 - CDC Dr. Messonnier announces pandemic likely. Community spread not seen in U.S.
February 22, 2020 - Multiple regions in Northern Italy affected. Human-to-human transfer noted. Schools closed on 23rd.
Week of February 23rd, 2020 - CoVid spreads throughout Europe.
February 24, 2020 - President Trump asks Congress for $1.25 billion to prepare for CoVid.
February 26, 2020 - Shanghai P3 lab has successfully completed CoVid genetic sequencing.
February 28, 2020 - China orders Wuhan P4 Viral Institute Lab to close.
February 28, 2020 - First known official U.S. case reported without connection to outside of U.S. travel.
February 28, 2020 - 14 European States now affected.
February 28, 2020 - First Nigerian case of CoVid.
The first recognized U.S. death.
February 28, 2020 - President Trump extends travel ban to South Korea and Italy.
February 28, 2020 - CDC announces problems with its CoVid testing kits.
February 29, 2020 - A State of Emergency is issued for Washington State.
February 29, 2020 - U.S. Surgeon General reiterates CDC guidelines - facemarks not needed for general public.
February 29, 2020 - FDA opens up EA for hospitals and healthcare facilities.
March 3, 2020 - Iranian Parliament Closed.
March 6, 2020 - President Trump announces testing available to those who want it.

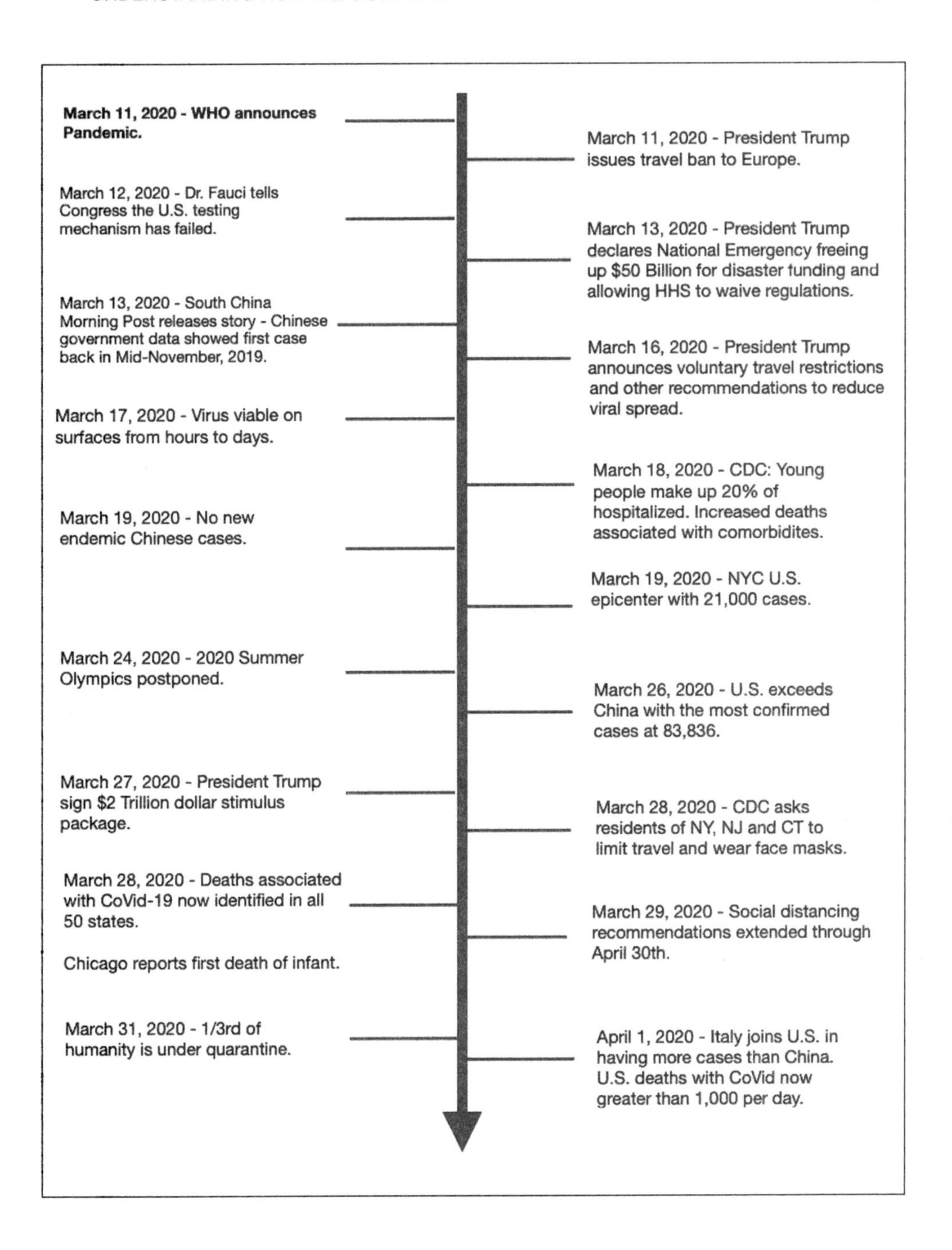
March 11, 2020 - WHO announces Pandemic.
March 11, 2020 - President Trump issues travel ban to Europe.
March 12, 2020 - Dr. Fauci tells Congress the U.S. testing mechanism has failed.
March 13, 2020 - President Trump declares National Emergency freeing up $50 Billion for disaster funding and allowing HHS to waive regulations.
March 13, 2020 - South China Morning Post releases story - Chinese government data showed first case back in Mid-November, 2019.
March 16, 2020 - President Trump announces voluntary travel restrictions and other recommendations to reduce viral spread.
March 17, 2020 - Virus viable on surfaces from hours to days.
March 18, 2020 - CDC: Young people make up 20% of hospitalized. Increased deaths associated with comorbidites.
March 19, 2020 - No new endemic Chinese cases.
March 19, 2020 - NYC U.S. epicenter with 21,000 cases.
March 24, 2020 - 2020 Summer Olympics postponed.
March 26, 2020 - U.S. exceeds China with the most confirmed cases at 83,836.
March 27, 2020 - President Trump sign $2 Trillion dollar stimulus package.
March 28, 2020 - CDC asks residents of NY, NJ and CT to limit travel and wear face masks.
March 28, 2020 - Deaths associated with CoVid-19 now identified in all 50 states.
Chicago reports first death of infant.
March 29, 2020 - Social distancing recommendations extended through April 30th.
March 31, 2020 - 1/3rd of humanity is under quarantine.
April 1, 2020 - Italy joins U.S. in having more cases than China. U.S. deaths with CoVid now greater than 1,000 per day.

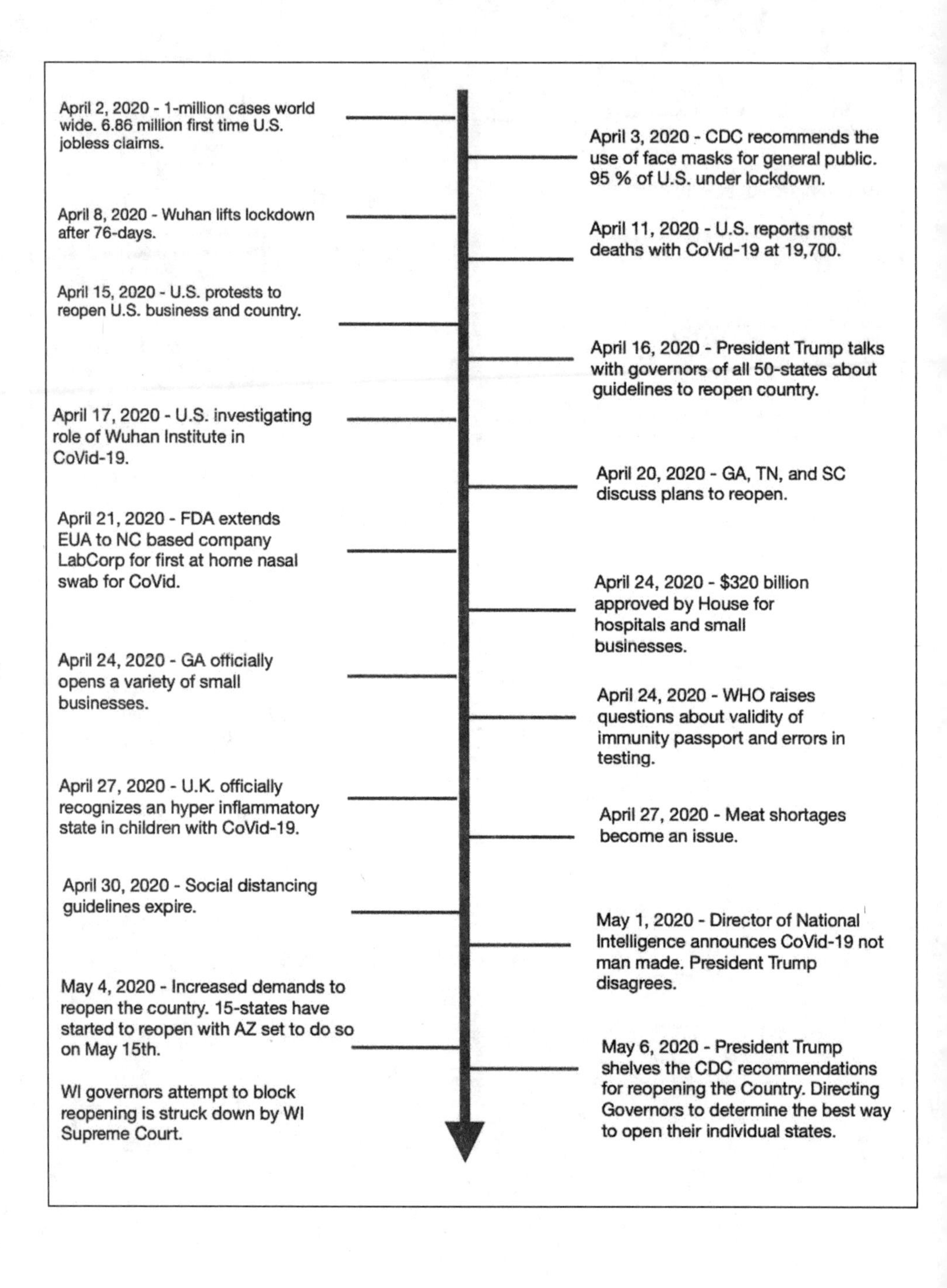
April 2, 2020 - 1-million cases world wide. 6.86 million first time U.S. jobless claims.
April 3, 2020 - CDC recommends the use of face masks for general public. 95 % of U.S. under lockdown.
April 8, 2020 - Wuhan lifts lockdown after 76-days.
April 11, 2020 - U.S. reports most deaths with CoVid-19 at 19,700.
April 15, 2020 - U.S. protests to reopen U.S. business and country.
April 16, 2020 - President Trump talks with governors of all 50-states about guidelines to reopen country.
April 17, 2020 - U.S. investigating role of Wuhan Institute in CoVid-19.
April 20, 2020 - GA, TN, and SC discuss plans to reopen.
April 21, 2020 - FDA extends EUA to NC based company LabCorp for first at home nasal swab for CoVid.
April 24, 2020 - $320 billion approved by House for hospitals and small businesses.
April 24, 2020 - GA officially opens a variety of small businesses.
April 24, 2020 - WHO raises questions about validity of immunity passport and errors in testing.
April 27, 2020 - U.K. officially recognizes an hyper inflammatory state in children with CoVid-19.
April 27, 2020 - Meat shortages become an issue.
April 30, 2020 - Social distancing guidelines expire.
May 1, 2020 - Director of National Intelligence announces CoVid-19 not man made. President Trump disagrees.
May 4, 2020 - Increased demands to reopen the country. 15-states have started to reopen with AZ set to do so on May 15th.
WI governors attempt to block reopening is struck down by WI Supreme Court.
May 6, 2020 - President Trump shelves the CDC recommendations for reopening the Country. Directing Governors to determine the best way to open their individual states.

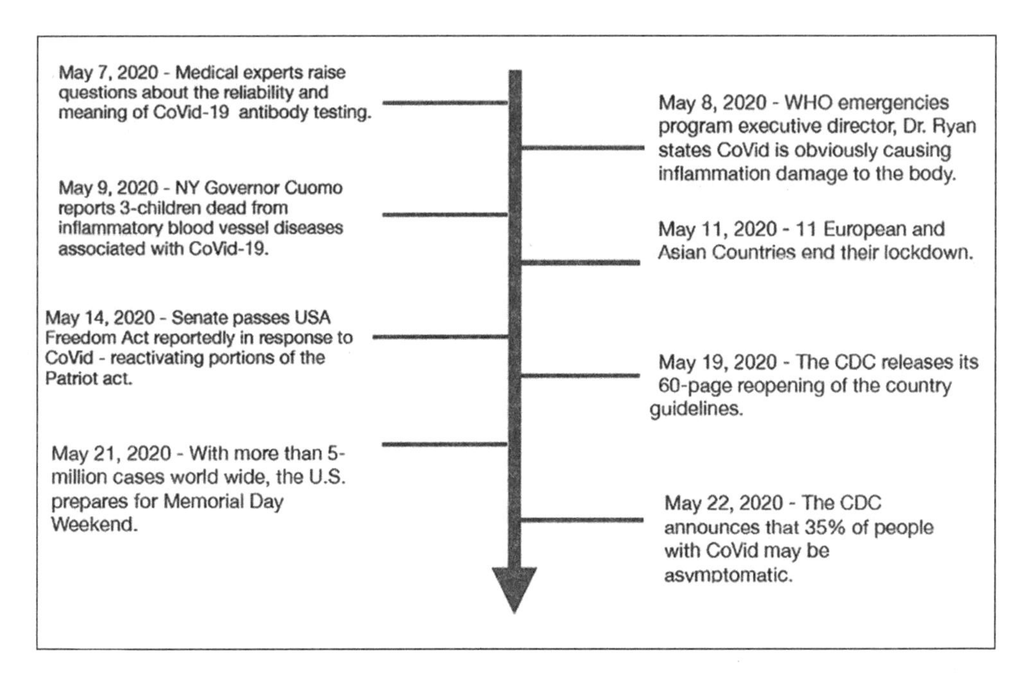

This experiment with a biological viral weapon, resulting from gain-of-function research, has left almost 1.2 million Americans dead—more than the number of deaths sustained by the US military since 1776. More than 1.5 million Americans were maimed, more than the number of injured US soldiers since 1776. And more than 34,500 genetic vaccine deaths have occurred, more deaths than the US military has suffered in all but five of our bloodiest wars. The numbers are continuing to climb.

Spike Proteins, Genetic Vaccines, and Manipulation

With what might seem to be an overwhelming amount of material, absent a few recorded instances dating back thousands of years, many of the biological-weapon attacks began with the United States, and most of those since the mid-1800s. That is roughly the same time that Lamarck, Darwin, and Mendel began to make their mark.

Similarly, most of the work on genetic manipulation began around the same time that the CIA and the DoD began to play a dominant role in US life and policy. The bad actors continue to raise their heads in history at the same time.

While much time, attention, and discussion has been focused on these genetic vaccines, the people and companies developing them are merely opportunists with established relationships between one part of the federal government and another; raising significant ethical conflict of issue problems. The Food and Drug Administration (FDA) was originally given its power by the 1906 Pure Food and Drugs Act[146] (note the timeline with the eugenics movement) and has been responsible for approving the use of medications within the United States. Its track history is less than stellar, with almost twice[147] as many drugs being pulled off the market in the last two to three years as have been approved.

The problem with this symbiotic relationship between government agencies is that it is actually parasitic and feeds off the American people. In reviewing and doing the scientific work that should have been done prior to vaccinating people with these three (Pfizer, Moderna, and Janssen) genetic vaccines, the FDA should have asked at least two fundamental questions. First, where is the actual data showing the genetic vaccines have a desirable immunologic response, one that will actually be beneficial for the person receiving the injected drug? Review the three Emergency Use Authorization (EUA) documents.[148] If you can find any tables or data showing the actual measured immune (e.g., interferon levels, IgA antibody responses), please notify the press. The data is missing from these FDA documents.

Second, the FDA and vaccine manufacturers should have statistically analyzed the results of their FDA published data. Anyone can look at a set of numbers and see if they are identical or different. That isn't science; that's counting.

To answer this second question, one has to subject the results to statistics—that is, the mathematical process of asking whether the results are scientifically different or whether they simply give you a warm fuzzy feeling.

In real-world science, and even in softer sciences, such as medicine and biology, researchers need to statistically analyze the data to determine if the different results are meaningful. In this instance, that would be whether there were statistically fewer people who were diagnosed with COVID,[149] or who died, after being vaccinated versus those who weren't vaccinated. There was no difference in the Pfizer results and none in the Moderna results. The Janssen EUA data showed a statistical reduction at two weeks—a benefit that was lost two weeks later.[150]

The results of this statistical analysis are shown in appendix C, with all the numbers circled and the tables using chi-square analysis. Since the results were merely counting the number of people and not actually measuring immunologic data (e.g., interferon), chi-square analysis is the method to be used. The genetic vaccines failed.

That failure was predicted as shown by Karikó and Weissman,[151] who in 2008 published the first successful paper on the use of pseudouridine to stabilize messenger ribonucleic acid (mRNA), for which they received the 2023 Nobel Prize in Physiology or Medicine.

In their 2008 published research they made four important findings:

1. Adding pseudouridine, in the place of uridine, produced a stable mRNA molecule, making it possible for the first time to be used.
2. The findings showed that pseudouridine had statistically lower immune response, specifically showing diminished interferon levels.[152]

3. Pseudouridine mRNA is not useful for a genetic vaccine.
4. Pseudouridine mRNA substantially enhanced the translational capacity of mRNA, making it a promising tool for gene replacement by injection.

I have provided all of this information in appendix C, along with the evidence that the SARS-CoV-2 viruses, and the genetic vaccines, are associated with enhanced vascular endothelial growth factor (VEGF).

This increase in VEGF, the priogenic effect of the spike sequence (either the virus or transfection-induced protein by the genetic vaccines), the InflammoThrombotic response (ITR), along with anything that alters the flow of potassium (seen with changes in VEGF) in heart tissue complicated by increases in heart rate (infection, exertion, or other health problems) provide the perfect storm for sudden cardiac death, miscarriages, and cancers.[153]

Discussions surrounding the potential inclusion of simian virus 40 (SV40) nucleotide sequences reportedly being in some of the genetic vaccines, with SV40 viruses being known to cause problems with the p53 mechanism (see the foregoing QR graphic) cancer-protective mechanism of the body, while of interest, is more important as a topic of strict product liability tort law, discussed later in the book. The SV40 sequences being reported by some do not appear to include the cancer-causing parts[154] of the SV 40 virus.

In addition to the strict product liability issue, SV40 is important for another reason: its promoter genes, which could increase the production of the spike protein,[155] as shown in the following QR slide.

The problem with making more of the spike protein, apart from the EUAs having no information showing production of interferon, IgA or even IgG/IgM, which would be necessary for these genetic vaccines to be beneficial for people; apart from the EUAs showing no statistically significant reduction in COVID cases or death for people vaccinated versus the unvaccinated; apart from the Pressure Selection (below/infra) and undesirable shift from IgG3 to IgG4 harmful effects (below/infra) seen with repeated vaccination using these genetic vaccines; is the tiny problem of the (A) priogenic effect of the spike protein causing Alzheimer's disease and other priogenic disease in the brain of people exposed to prions, the amyloid prion disease of the heart, diabetes caused by the priogenic effect to the pancreas; and other damaged organs, (B) the InflammoThrombotic Disease including heart attacks,[156] strokes,[157] blood clots in the lung,[158] miscarriages, cancer, the effect upon red blood cells when the vaccines are directly added to the blood,[159] and the sudden cardiac death caused by a multitude of factors, not the least of which is the effect of pseudouridine on the potassium channels of the heart and brain critical to life, and (c) the potential for long-term effects associated with the persistent production of the spike protein for an unknown period of time; if not the spike protein itself, is the consequence of harm to the person occurring during the period of vaccination itself. The link to FlemingMethod.com[160] provides a more in-depth discussion of these medical problems.

The slides and MP3 audio file provided by this QR link provides a shorter explanation of both the InflammoThrombotic and priogenic disease (harm/injury) caused by SARS-CoV-2 viruses and the SARS-CoV-2 spike proteins generated by the genetic vaccines.

This brings us to the critical question: why would someone make a genetic vaccine using pseudouridine, given the failure of pseudouridine mRNAs to produce the desired interferon immune response?[161] As you will soon see, the information obtained to date has demonstrated these genetic vaccines appear to be altering human immune response, shifting from the desired production of IgG3 with its protective response against infection, to an IgG4 response associated with autoimmune disease and cancer.

The basic immune system,[162] which at first glance might appear to be more of a "Where's Waldo?" map, is an extremely intricate system that has evolved to not only provide the first part of our immunity, the innate system, but also include a more advanced secondary part known as the adaptive (humoral—i.e., antibodies) immune response, which is sometimes more harmful if activated than not.[163] Your immune system also includes specialized cells known as natural killer (NK) cells that help protect your body and kill cancer cells when they develop.

The question is whether this class shift of immune adaptive antibody response from a more protective IgG3 (or IgA2 for respiratory or gastrointestinal infections) to an IgG4 response, seen with autoimmune diseases[164] and cancer, is due to these genetic vaccines, or if it is the result of multiple layered vaccines, impairing our response to other infections or vaccines by blunting our antibody response. The latter response was shown in the study, provided with the QR code, revealing a blunted response to the influenza vaccine in people previously vaccinated for COVID.

Also provided in the QR code are two studies where the generic vaccines caused a shift in immune response from IgG3 to IgG4, increasing the ability of cancers to evade our natural immune response, which is critical to stopping cancers before they fully develop or spread throughout our bodies;[165] as demonstrated by the presence or absence of IgG3 and IgG4 when looking at tissue where cancers occur.

As also shown with this QR code, some cancers have developed escape mechanisms, allowing them to grow in areas where IgG4 is present.

If this shift from IgG3 to IgG4 is the result of the genetic vaccines made with pseudouridine, then our focus can be directed to the pseudouridine mRNA or altered DNA vaccines. If, however, this is the result of multiple vaccinations, perhaps related to the overwhelming of our immune systems (which could perturb homeostasis, forcing a continued immune response to the same antigen, resulting in our body's perception of the vaccine antigen as too frequent to be from outside the body and instead as part of it, thereby yielding an IgG4 autoimmune response), then we may need to look at the current immunizations schedules. To the best of my knowledge, we have not investigated if the immune response to this increased frequency of vaccines now scheduled and promoted by the CDC[166] is producing a similar shift in immunity from IgG3 to IgG4.

The following table taken from the CDC website identifies nineteen vaccines for children younger than age eighteen, with two to be given at birth:

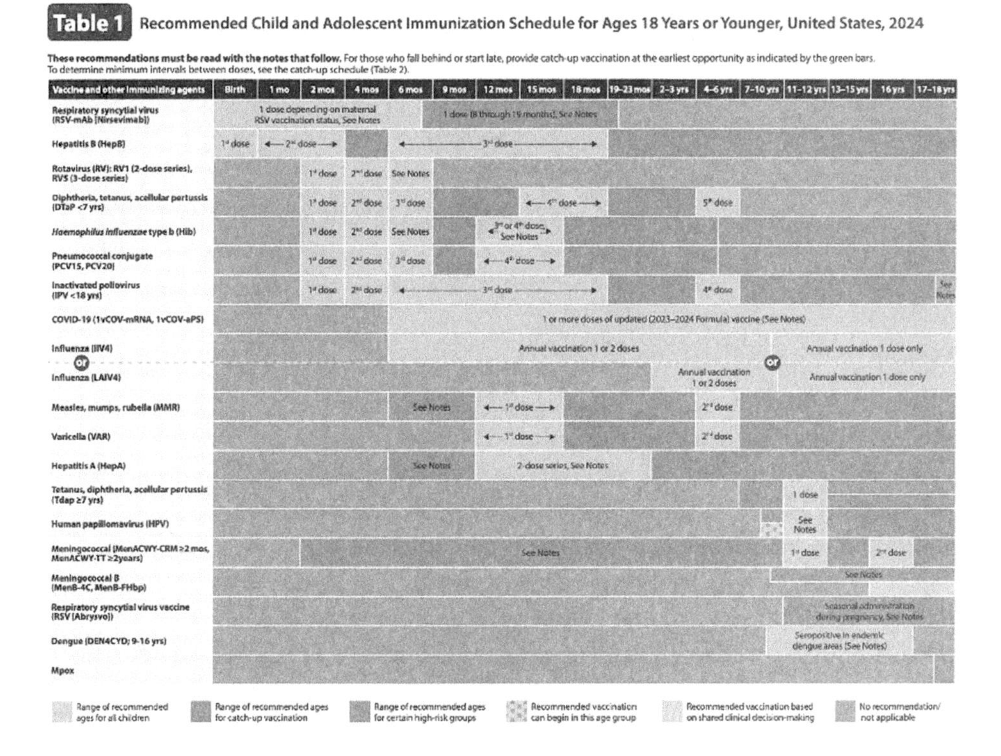

Table 1 Recommended Child and Adolescent Immunization Schedule for Ages 18 Years or Younger, United States, 2024

These recommendations must be read with the notes that follow. For those who fall behind or start late, provide catch-up vaccination at the earliest opportunity as indicated by the green bars. To determine minimum intervals between doses, see the catch-up schedule (Table 2).

Vaccine and other immunizing agents	Birth	1 mo	2 mos	4 mos	6 mos	9 mos	12 mos	15 mos	18 mos	19–23 mos	2–3 yrs	4–6 yrs	7–10 yrs	11–12 yrs	13–15 yrs	16 yrs	17–18 yrs
Respiratory syncytial virus (RSV-mAb [Nirsevimab])	1 dose depending on maternal RSV vaccination status, See Notes				1 dose (8 through 19 months), See Notes												
Hepatitis B (HepB)	1st dose	←2nd dose→			←3rd dose→												
Rotavirus (RV): RV1 (2-dose series), RV5 (3-dose series)			1st dose	2nd dose	See Notes												
Diphtheria, tetanus, acellular pertussis (DTaP <7 yrs)			1st dose	2nd dose	3rd dose			←4th dose→				5th dose					
Haemophilus influenzae type b (Hib)			1st dose	2nd dose	See Notes		←3rd or 4th dose, See Notes→										
Pneumococcal conjugate (PCV15, PCV20)			1st dose	2nd dose	3rd dose		←4th dose→										
Inactivated poliovirus (IPV <18 yrs)			1st dose	2nd dose	←3rd dose→							4th dose					See Notes
COVID-19 (1vCOV-mRNA, 1vCOV-aPS)					1 or more doses of updated (2023–2024 Formula) vaccine (See Notes)												
Influenza (IIV4) or					Annual vaccination 1 or 2 doses								or Annual vaccination 1 dose only				
Influenza (LAIV4)											Annual vaccination 1 or 2 doses		Annual vaccination 1 dose only				
Measles, mumps, rubella (MMR)					See Notes		←1st dose→					2nd dose					
Varicella (VAR)							←1st dose→					2nd dose					
Hepatitis A (HepA)					See Notes		2-dose series, See Notes										
Tetanus, diphtheria, acellular pertussis (Tdap ≥7 yrs)														1 dose			
Human papillomavirus (HPV)														See Notes			
Meningococcal (MenACWY-CRM ≥2 mos, MenACWY-TT ≥2years)			See Notes											1st dose		2nd dose	
Meningococcal B (MenB-4C, MenB-FHbp)													See Notes				
Respiratory syncytial virus vaccine (RSV [Abrysvo])														Seasonal administration during pregnancy, See Notes			
Dengue (DEN4CYD; 9-16 yrs)													Seropositive in endemic dengue areas (See Notes)				
Mpox																	

Range of recommended ages for all children; Range of recommended ages for catch-up vaccination; Range of recommended ages for certain high-risk groups; Recommended vaccination can begin in this age group; Recommended vaccination based on shared clinical decision-making; No recommendation/not applicable

Several additional vaccines are recommended under a variety of settings.

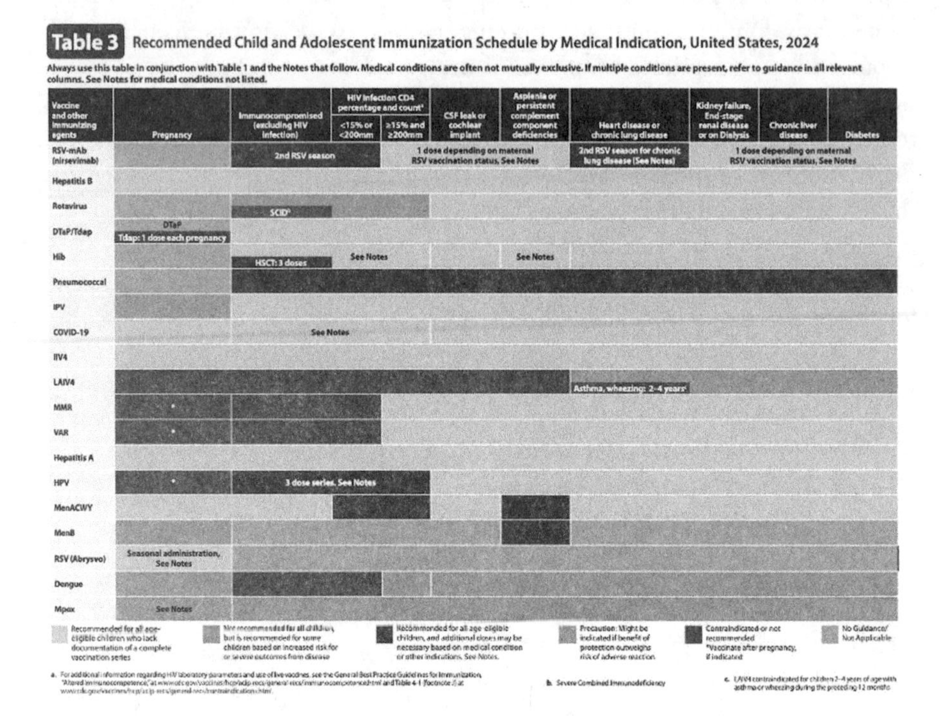

Table 3 Recommended Child and Adolescent Immunization Schedule by Medical Indication, United States, 2024

Always use this table in conjunction with Table 1 and the Notes that follow. Medical conditions are often not mutually exclusive. If multiple conditions are present, refer to guidance in all relevant columns. See Notes for medical conditions not listed.

Vaccine and other immunizing agents	Pregnancy	Immunocompromised (excluding HIV infection)	HIV infection CD4 percentage and count[a]		CSF leak or cochlear implant	Asplenia or persistent complement component deficiencies	Heart disease or chronic lung disease	Kidney failure, End-stage renal disease or on Dialysis	Chronic liver disease	Diabetes
			<15% or <200mm	≥15% and ≥200mm						
RSV-mAb (nirsevimab)		2nd RSV season		1 dose depending on maternal RSV vaccination status, See Notes			2nd RSV season for chronic lung disease (See Notes)	1 dose depending on maternal RSV vaccination status, See Notes		
Hepatitis B										
Rotavirus		SCID[b]								
DTaP/Tdap	DTaP Tdap: 1 dose each pregnancy									
Hib		HSCT: 3 doses	See Notes			See Notes				
Pneumococcal										
IPV										
COVID-19		See Notes								
IIV4										
LAIV4							Asthma, wheezing: 2–4 years[c]			
MMR	*									
VAR	*									
Hepatitis A										
HPV	*	3 dose series. See Notes								
MenACWY										
MenB										
RSV (Abrysvo)	Seasonal administration, See Notes									
Dengue										
Mpox	See Notes									

Recommended for all age-eligible children who lack documentation of a complete vaccination series

Not recommended for all children, but is recommended for some children based on increased risk for or severe outcomes from disease

Recommended for all age-eligible children, and additional doses may be necessary based on medical condition or other indications. See Notes.

Precaution: Might be indicated if benefit of protection outweighs risk of adverse reaction

Contraindicated or not recommended
*Vaccinate after pregnancy, if indicated

No Guidance/Not Applicable

a. For additional information regarding HIV laboratory parameters and use of live vaccines, see the General Best Practice Guidelines for Immunization, "Altered Immunocompetence," at www.cdc.gov/vaccines/hcp/acip-recs/general-recs/immunocompetence.html and Table 4-1 (footnote J) at www.cdc.gov/vaccines/hcp/acip-recs/general-recs/contraindications.html.

b. Severe Combined Immunodeficiency

c. LAIV4 contraindicated for children 2–4 years of age with asthma or wheezing during the preceding 12 months

As shown in appendix C, when we look at the rollout of the genetic vaccines and the emergence of the SARS-CoV-2 variants, we see the development of pressure selection. Pressure selection is similar to antibiotic resistance. When patients have repeated infections and are given the same antibiotic or an antibiotic in the same class of drugs, the antibiotics kill those bacteria sensitive to the antibiotic. However, some bacteria have developed additional protective mechanisms, genetic pathways that allow them to make proteins to protect against the antibiotic. As a result, these resistant bacteria grow and thrive, while the sensitive bacteria die off. The overuse of antibiotics can cause antibiotic resistant bacteria to dominate and reproduce, making it more difficult to successfully treat patients.

The same outcome was seen with the repeated use of these genetic vaccines. Instead of targeting multiple parts of the viruses, of which our best immune response is to the nucleocapsid part of the SARS-CoV-2 virus (COVID-19), the focus of these genetic vaccines was on the spike protein. Changes in nucleoside bases with resulting amino acid changes, "seen in other variants" of the SARS-CoV-2 viruses, yielded changes in the spike protein (and other parts of the viruses), changing the dominant variant strain of SARS-CoV-2 being transmitted from one person to another.

The slide in the following QR code shows you how a change in any of the 4-nucleoside[167] bases, adenine (A), thymine (T), cytosine (C), or guanine (G), can change the transcribed (DNA to RNA) RNA. This change in RNA will result in a different amino acid being placed into the protein (e.g., spike protein) when the protein is made by the ribosome (translation) in our cells.

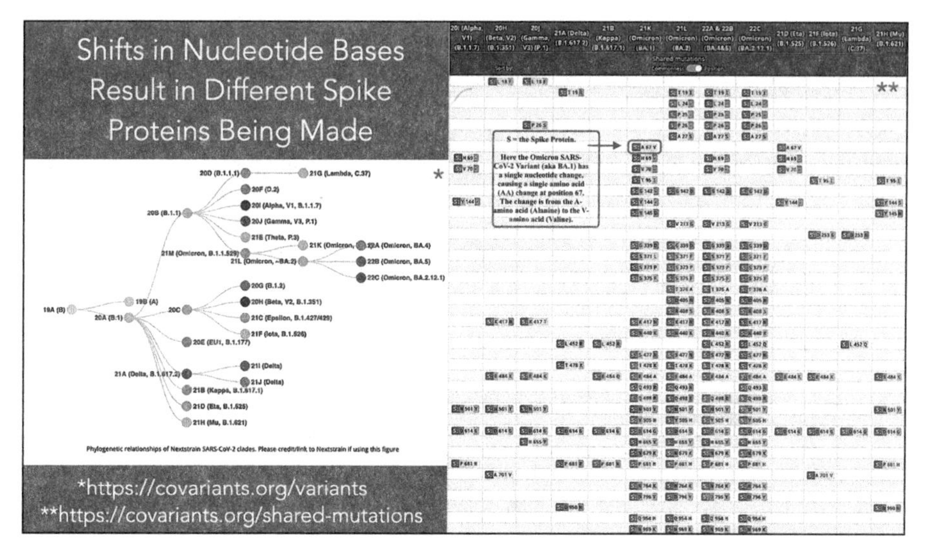

A single nucleoside change at one site can change the amino acid put into that part of the protein where the nucleoside mutation occurred. An example of this is shown (in the QR code slide) explaining how a change in nucleoside changes the amino acid and subsequent spike protein in one of the Omicron variants.

When enough of these changes occur at multiple sites, the structure of the spike protein changes.[168] Eventually a new viral variant emerges and spreads from person to person.

This change in the spike protein is equivalent to the lock on a door being changed, a key (the antibody made from the genetic vaccine, or by exposure to the variant) that once worked no longer does.[169]

When the dominant strain becomes sufficiently different from the original Wuhan-Hu-1 variant, breakthrough infections[170] occur. Antibodies produced to the Wuhan-Hu-1 variant are no longer protective. In survival of the fittest, the SARS-CoV-2 variants sufficiently different from the antibody produced by the genetic vaccines will thrive, the viral equivalent to indiscriminant antibiotic use associated with bacterial resistance. The next QR code shows an overlay of variants with vaccines. As shown in appendix C, the pressure selective changes in the SARS-CoV-2 virus variants occurred around the world and were associated with all the genetic vaccines experimentally injected into people. The experimentation would never have been accepted by people had a set of gain-of-function viruses not escaped from the laboratory where they were being created.

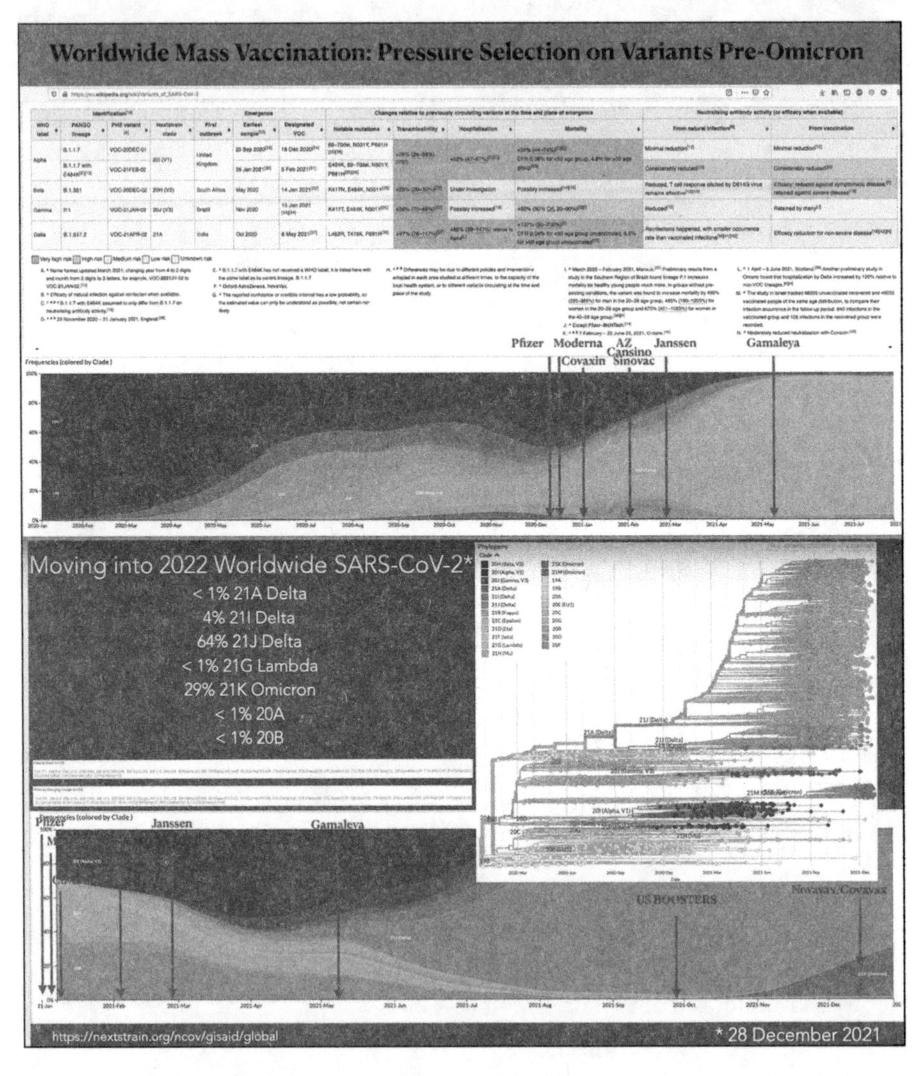

Notwithstanding this lack of understanding by those involved, the COVID pandemic has provided several very important pieces of information, including the ability to get people to subject themselves, their family members, and loved ones to a research project that would make Josef Mengele[171] blush. Not only are people willing to walk into the gas chambers of this research, but they also will demand that others walk into the chambers with them[172] or blame those who refuse.[173]

As the data eventually revealed, early hospitalizations occurred among the unvaccinated. Over time, as shown in the United Kingdom data, as shown in the last QR code, it was the vaccinated who were more frequently hospitalized for variants not covered by the vaccines; not because of other people being unvaccinated.

The most plausible scientific explanation for this difference is the pressure selection placed on the variants with a lack of vaccinated immunity to other parts of the SARS-CoV-2 viruses (spike protein, membrane, envelope, nucleocapsid), with overarching protective immunity achieved by those who had developed immunity via person-to-person transmission of the gain-of-function viruses, as well as individuals who had developed immunity associated with prior exposure to influenza and cytomegalovirus (CMV).

We must dig deeper to understand the motives and goals of those who has been conducting the COVID research experiment, taking into account that this experiment has been built upon decades of research including the use of gene therapy to alter human DNA and eugenics.

Distractions surrounding these genetic vaccines, which, I argue, some people would like us to concentrate on, turn attention away from the motives and goals of those involved. Therefore, let us now turn our attention to understanding why these genetic vaccines have emerged and converged upon the scene in 2020. These vaccines have the ability to genetically manipulate living organisms, to understand DNA, to find errors in DNA as they see them, and to eventually correct those errors, making a perfect human being, approaching if not achieving physical immortality. In other words, the new eugenics 2.0.

The Human Genome Project (HGP)[174]—What Does It Mean to Be Human?

Human traits (phenotypes) are the direct result of human genetics (genotype), which as we have already discussed was the basis of the eugenics movement that has plagued humanity perhaps since our beginning. By the mid-to-late 1800s this movement raised its less than phenotypically appealing head to emerge to eliminate those who did not live up to the desired traits as viewed by those in power—the ultimate us vs. them.

In spite of what you may have heard or learned, deoxyribonucleic acid (DNA) is more than just two twisted strands of nucleoside bases attached to phosphate bonds (making them nucleotides) and to a sugar molecule (deoxyribose for DNA; ribose for RNA). The amount of data that can be stored inside of the twenty-three sets of human chromosomes rivals that of our current quantum computers. It has been estimated that the storage capacity of human DNA is adequate to store all knowledge of the universe. In women, with XX chromosomes, this storage space includes 6.37 Gigabase (6 billion, 370 million) nucleotide bases weighing roughly 6.51 picograms (pg) or 2.29^{e-13} ounces and stretching 208.23 cm (81.98 inches) long. In men, having an XY chromosome pattern, this amounts to 6.27 Gigabase pairs, a weight of 6.41 pg (2.26^{e-13} ounces) and length of 205 cm (80.87 inches).

Given this massive ability to store data and the sheer length of your DNA in centimeters or inches, it might seem impossible for all your DNA to be stored inside the nucleus[175] of your cells;[176] however, your DNA is tightly wrapped and packaged inside the nucleus of your cells as described below.

Your nucleus is surrounded by a membrane with openings (nuclear pore complex, or NPC) that allow material from the outer part of your cell (cytoplasm) to enter your nucleus, and a way for the RNA transcribed[177] from your DNA to leave the nucleus and enter your cytoplasm. To accomplish the feat of compressing your DNA, histones,[178] which are basic proteins, act like a spool around which DNA is wrapped until needed. This coiled DNA-histone complex is called a *nucleosome.* Nucleosomes are themselves wrapped to form chromatin, or the chromosomal material seen in human chromosomes you may have heard of. The ultimate structure is positively charged histones and the negatively charged DNA phosphate backbone. The packaged DNA is 1/40,000 the length it would be if it weren't wrapped.

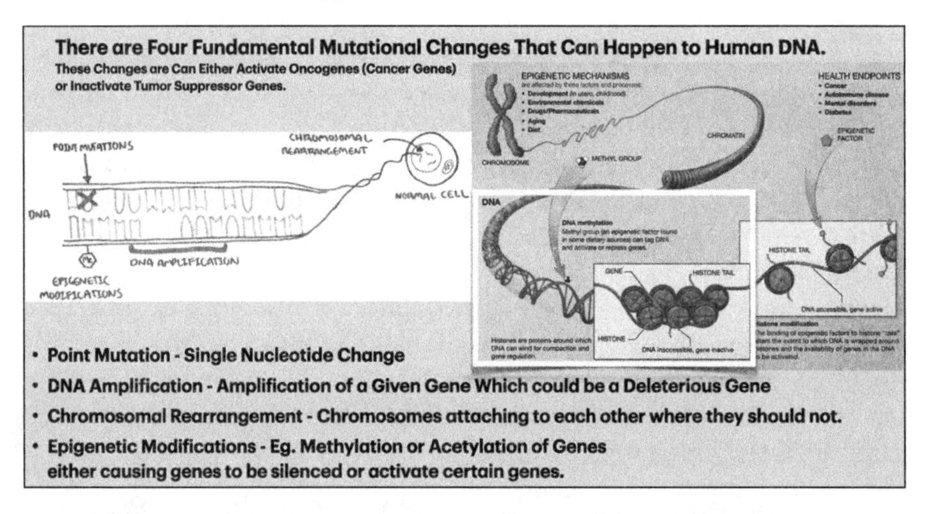

In this overly simplified model, when needed, certain enzymes unwrap the histone for DNA replication (mitosis for cell division, meiosis for fusion of genetic material from parents) or transcription (DNA to RNA, including mRNA, rRNA, and tRNA).[179]

With the foregoing as background information, it should be clear that the task of identifying and cataloguing human (or any) DNA would have been considered quite formidable, but the educational efforts promulgated by the JFK administration touched not only the lives of the Kennedy Kids but also so many others, and led to this project.

The HGP began in earnest following questions regarding potential DNA damage caused by the US nuclear bombing of Hiroshima and Nagasaki.[180] The US Department of Energy (DOE)[181] began funding the Human Genome Project in 1986 and was followed by HHS in 1987. To know whether permanent genetic damage had been done to the people of Japan, it would first be necessary to know what was and wasn't normal.

The scientific methods and equipment first made possible by the DOE included:

1. DNA sequencers:[182] High-throughput DNA sequencing machines critical to analyzing this much genetic material.
2. Fluorescent compounds: Material that makes it possible to radiolabel, through isotopes, DNA subunits.
3. DNA cloning vectors: A method to make it possible to cut and amplify (clone) before sequencing. The most commonly used method today is the bacterial artificial chromosome (BAC), which, as we have already discussed, was used by Baric and Zhengli to find open reading frame (ORF) X (ten).[183]
4. BAC-end sequencing: A method to carry fragments of human DNA from known genetic locations. By 2003 more than 450,000 BAC-based genetic markers tagged DNA every 304 kilo (thousand) bases. In 2023 there were 45 million such fragments, referred to as *expressed sequence tags* (ESTs), from over 1,400 species of animals (including humans) and single-celled bacteria, or eukaryotes.
5. The (Holy) GRAIL, or gene recognition and assembly Internet link. An analytic computer program for identifying potential DNA gene sequences.

The launching of the Human Genome Project was perhaps one of the most ambitious scientific projects undertaken to date and began as a thirteen-year project in October 1990. At that time, it was thought there were some 20,000 to 25,000 human genes,[184] an estimate that has since been updated to more than 120,000 genes.[185] The original goal was to map human chromosomes 2, 5, 11, 16, 19, and 21 and male chromosome X.[186] Interestingly enough, this research also focused on the differences between mice and people.

The call for volunteers and collection of their blood samples for DNA sequencing began around 1997. The HGP was deemed completed on 14 April 2003,[187] with an estimated 92 percent of the human genome having been mapped. It has been called the largest collaborative biological project to date. The last two chromosomes were mapped by May 2006. The final telomere to telomere (T2T2) work filling in repetitive regions of ribosomal DNA was completed in 2021.[188]

The Board of Scientific Counselors for the National Human Genome Research Institute[189] included people from the Rockefeller Foundation and Cold Springs Harbor Laboratory.

Christopher Amos, Ph.D.
Baylor College of Medicine
Term ends 6/2026

Mildred Cho, Ph.D.
Stanford University
Term ends 6/2027

Barry S. Coller, M.D. (Board Chair)
Rockefeller University
Term ends 6/2023

Gregory Barsh, M.D., Ph.D.
HudsonAlpha Institute for Biotechnology
Term ends 6/2024

Guillermina Lozano, Ph.D.
University of Texas MD Anderson Cancer Center
Term ends 6/2026

Heather Mefford, M.D., Ph.D.
St. Jude Children's Research Hospital
Term ends 6/2026

Carole Ober, Ph.D.
The University of Chicago
Term ends 6/2023

Adam C. Siepel, Ph.D.
Cold Spring Harbor Laboratory
Term ends 6/2025

Sarah Tishkoff, Ph.D.
University of Pennsylvania
Term ends 6/2023

Ken Nakamura, Ph.D.
Recording Secretary

In addition to these individuals, the mapping and sequencing of the human genome included the following federal agencies:

1. The National Science Foundation (NSF)
2. The U.S. Department of Agriculture
3. The Howard Hughes Medical Institute

It also relied on international collaboration with the United Kingdom, Italy, the Soviet Union, the Commission of the European Community (EU), France, and Japan.

The HGP set the stage for deciding what genetic information should be kept and what should be eliminated. With enough information, the undesirable traits of humanity could be eliminated and replaced with those traits thought to be best for humanity. But who decides?

The HGP also set the stage, wittingly or unwittingly, for gene therapy, essentially deleting those genes thought to be inferior and replacing them with superior genes.[190] At first, this would focus on finding those genes that clearly cause disease and replacing them—a noble-enough cause, allowing us to build a better human free of disease and mortality. Before this could happen we had to develop a way to cut and switch genes. That would require CRISPR technology, to cut pieces of genes out, and a transfection tool, or a way to insert genetic information that is extremely unstable outside of a cell, into cells. No one had figured out a way to do this yet, but nature was about to teach us how bacteria do it.

CRISPR[191]—Changing DNA

Just as bacteria have evolved genetic mechanisms to protect them from some antibiotics (antibiotic resistance), so too have bacteria developed mechanisms to protect them from viruses that might infect them, namely bacteriophagic[192] viruses.[193]

In 1987 researchers in Osaka, Japan, accidentally cloned a series of what would later be called *clustered repeats*. It did not, as you can see, take long before experimentation[194] revealed that CRISPR provided a tool that could splice DNA (including that of humans), allowing pieces of DNA to be removed, and, with the introduction of new genes, replacement of bad DNA with pieces of desirable DNA. One such approach for accomplishing this has included cytomegalovirus (CMV) and simian virus 40 (SV40).

There are two main parts to the CRISPR bacterial immune system. The first is immunization, including the Cas9 proteins that combine with viral DNA, to cleave the viral DNA and insert it into the CRISPR loci as a repeat-spacer unit. The second is immunity, where the repeat-spacer units are transcribed to make crRNA[195] precursors. The Cas9 endonuclease is guided by transactivating crRNA (trancrRNA) to bind with the crRNA precursors. Once cleaved by RNAse III, a mature crRNA-Cas-tracrRNA complex is formed that will then bind or pair with the new viral infection (DNA). This triggers Cas cleavage, interfering with the new viral DNA, providing immunity to the new viral infection. This is briefly shown in the figure below.

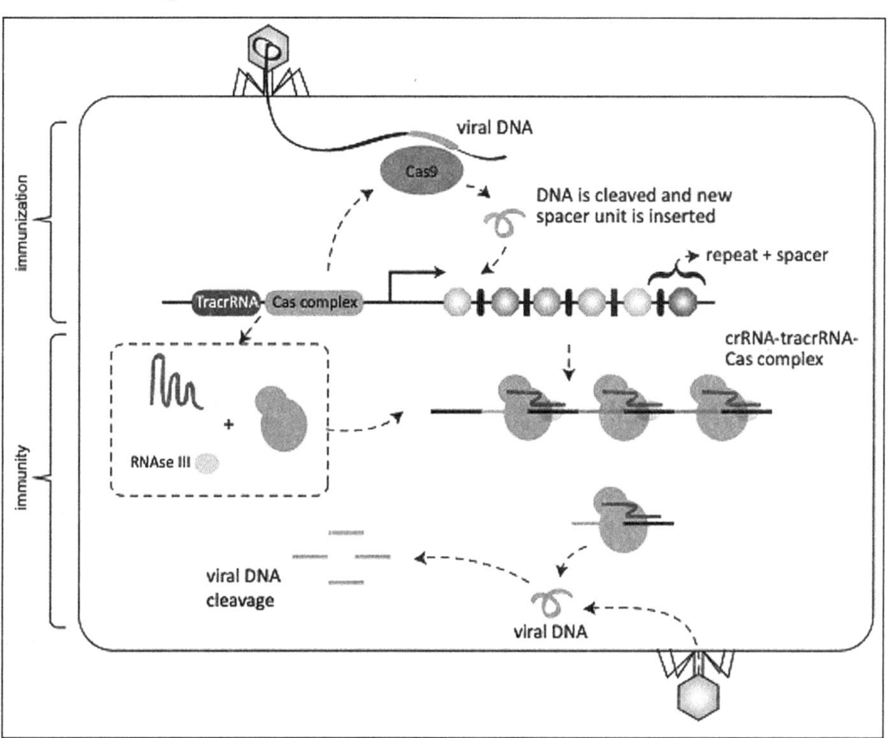

The field exploded with multiple discoveries including, (1) in 1989, the first discovery of homologous recombination (HR); (2) in 1992, site-specific Cre-lox (site-specific recombinase enzyme thar can delete, insert, invert or translocate specific DNA) gene editing; (3) in 1998, Zinc-finger nucleases (enzymes) that can target specific DNA sequences; (4) in 2000, identification of the first CRISPR/Cas[196] system within prokaryotes (single cell organisms without nuclei); (5) in 2009, discovery of the transcription-like effector nucleases, a.k.a. TALENs; and (6) in 2013, demonstration that CRISPR/Cas genetic editing could be used in mammalian (human) cells.

By 2011 researchers showed the ability to use CRISPR to stop human cells from making certain proteins (translation). This research was funded by NIH, along with the Helmsley and Rockefeller Foundations.

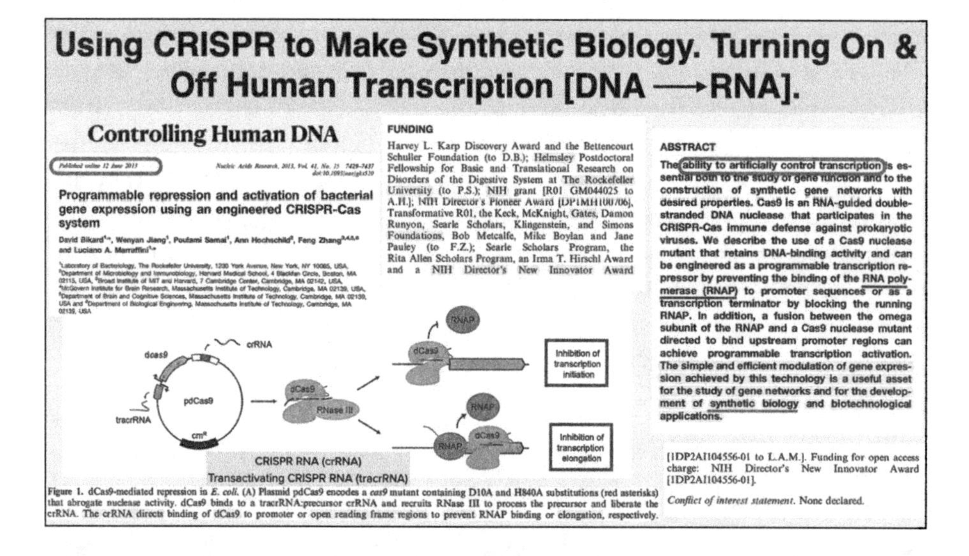

By 2016 researchers had used CRISPR to silence human DNA, preventing people from responding to infectious agents. More importantly, this silencing of human DNA could be passed onto children, making it inheritable. This selective silencing could be used for good or nefarious purposes, depending upon who makes the decision about who and why someone is being "treated."

As shown in the last QR code slides, before 2020, CRISPR had been successfully used to control the ability of people to respond to infections and cancer, turning off genes that would otherwise make interferon. This same effect is seen with the gain-of-function SARS-CoV-2 viruses.[197]

The 2022 Nobel Prize in Chemistry was awarded to Emmanuelle Charpentier and Jennifer A. Doudna for CRISPR gene editing. They began to recognize that bacteria had evolved a mechanism whereby the bacteria would transfer (reverse transcribe) pieces of viral genetic code (RNA), from pieces of viruses trying to infect the bacteria, into the genetic code of the bacteria. When the same type of virus would later attack, the bacteria would recognize and prevent the virus from attacking. I hope you recognize what this means. The bacteria have an immune system that uses something similar to human LINE-1 elements[198] to take outside (in this instance a virus) RNA and incorporate it into themselves (in this first instance, researchers discovered this happening with a bacteria known as *Streptococcus pyogenes*),[199] providing an immunity against the attacking virus.

The quest for human longevity and immortality gained traction in the era of CRISPR technology when, in 2018, less than three decades after the Osaka discovery of clustered repeats, the world learned of the efforts by a Chinese doctor to create the world's first gene-edited babies. His goal was to make two girls HIV resistant. In 2019 he received a three-year jail sentence[200] for doing this. As you will see below, efforts to develop gene therapy to make parts of the HIV virus inside human cells had already been patented in the United States.

Prior to 2020 the only limitation of CRISPR appeared to be how to get genetically modified human sequences[201] into people without having it break down before getting into their cells. By 2020 that problem had been solved—thanks to funding by DARPA of the US Department of Defense.

Transfection (Cell Penetration) Vector[202] Development (1) Lipid Nanoparticles (LNPs)

The introduction of lipid nanoparticles (LNPs) made it possible to introduce CRISPR-altered genetic sequences directly into human cells, using either injections or nasal sprays. Efforts to use CRISPR and its associated mRNA, or any type of genetic sequence, were originally limited due to the extracellular breakdown of RNA and DNA. These genetic sequences necessary for life inside the cell produced priogenic[203] disease prions[204] outside a cell.

The use of LNPs began in 1990[205] with, to the best of my knowledge, the first patent being issued in 1991. LNPs are an attempt to imitate our own cells and are primarily composed of cholesterol, triglyceride chains, and polyethylene glycol (PEG). Due to some areas being fat soluble (hydrophobic; water fearing) and other areas being water soluble (hydrophilic; water loving),[206] LNPs can merge with our cells and enter without needing to find a receptor, like the angiotensin converting enzyme receptor (ACE2). The composition of LNPs also results in different regions of the LNPs having different electrostatic charge, with associated differences in LNP configuration (appearance and assembly).

When scientists speak of *self-assembly*, this means when the molecules are mixed together, they separate into micelles, or chemical structures where the hydrophilic heads form the outside of the sphere, while the hydrophobic tails make up the inside of the sphere. Within this inner part of the sphere the genetic vaccine material (or any material the pharmaceutical company wishes to be surrounded by the LNP) is found, just like the DNA and RNA of human cells are found inside the LNP of our cell walls. The faster the flow rate of chemicals, the smaller the micelle and the less drug including the genetic vaccines inside the end product. Minor changes in flow rate can have a significant impact on the amount of mRNA or DNA found within the final LNP genetic vaccine product.[207]

Because of this similarity to the cells of our body, these LNPs tend not to be broken down when injected, and even though you might think what is injected into your arm stays there, there is no scientific reason, or, for that matter, no scientific evidence, to indicate the LNPs stay at the site of injection.

In fact, as shown in the last QR code slide set, Moderna researchers published a study in 2017 showing their LNP influenza vaccine spread throughout the body from the injection site.

In 2017 the first LNP CRISPR genetic vaccine for amyloidosis[208] (priogenic heart disease) was submitted with the goal to use CRISPR to silence genes that were problematic for amyloid heart disease. Unfortunately, the very LNPs being used to deliver these drugs and CRISPR genetic vaccines are highly inflammatory, leading to an increase in ITR[209] diseases, which we discussed earlier.

Transfection (Cell Penetration) Vector Development (2)
Viral Vectors

In addition to using LNPs to infect (transfection) our cells with gain-of-function, CRISPR genetic DNA changing sequences, viruses have been used to deliver the nuclear payload to our cells since at least the 1990s, when much of the work focused on Coxsackie virus (hand, foot and mouth disease) and adenovirus (a group of viruses that can cause nasal congestion, cough, diarrhea, and similar upper respiratory and gastrointestinal symptoms), including efforts to manipulate red blood cells, the only cells in our bodies without a nucleus with DNA.[210] These viruses are typically altered with the desired RNA or DNA placed inside the gutted virus and injected into people, as was done with the Janssen[211] genetic vaccine.

In research we[212] published in January 2023, some fourteen months after completing the work, we demonstrated that both the LNP (Pfizer and Moderna), and the adenovirus (Janssen) genetic vaccines damaged the blood when the vaccines were added directly to blood on a slide, under a microscope lens. In every instance, the addition of these genetic vaccines caused the blood to clump together—an InflammoThrombotic response (ITR)—and turn gray. The vaccines thus appear to alter the hemoglobin molecule so it can no longer hold on to life-sustaining oxygen.[213] A microscopic slide of this is shown in the last QR code slide set. Later, when we talk about strict product liability issues, we will look at what these genetic vaccines looked like under the microscope.

The Consequences of Gene Therapy, Including Shedding

What is gene therapy?[214] According to the US Food and Drug Administration (FDA)—the same group that approved the use of three experimental EUA genetic vaccines—gene therapy is the use of a biologic (through viruses, bacteria, plasmids, and so forth) agent in the treatment of infectious diseases, cancer, genetic diseases, and others:

> Seek[ing] to modify or manipulate the expression of a gene or to alter the biological properties of living cells for therapeutic use the risk of such, as noted by the CDC,[215] includes:
>
> 1. Integration of the gene therapy product into human DNA.
>   a. "This raises the potential for disruption of critical host (human) genes at the stie of integration, or activation of proto-oncogenes . . . thereby, the risk for malignancies."
> 2. Changes in human DNA.
>   a. "Genome editing may produce undesirable changes in the genome . . . with risk of malignancies, impairment of gene function, etc."
> 3. Prolonged expression.
>   a. "There is also the risk . . . of altered expression of host (human) genes that could result in unpredictable and undesirable outcomes."
> 4. Latency.
>   a. "There is the potential for reactivation . . . and the risk of delayed adverse events related to a symptomatic infection."
> 5. Establishment of persistent infections.

Human gene therapy seeks to modify or manipulate the expression of a gene or to alter the biological properties of living cells for therapeutic use 1.

Gene therapy is a technique that modifies a person's genes to treat or cure disease. Gene therapies can work by several mechanisms:

- Replacing a disease-causing gene with a healthy copy of the gene
- Inactivating a disease-causing gene that is not functioning properly
- Introducing a new or modified gene into the body to help treat a disease

Gene therapy products are being studied to treat diseases including cancer, genetic diseases, and infectious diseases.

There are a variety of types of gene therapy products, including:

- **Plasmid DNA:** Circular DNA molecules can be genetically engineered to carry therapeutic genes into human cells.
- **Viral vectors:** Viruses have a natural ability to deliver genetic material into cells, and therefore some gene therapy products are derived from viruses. Once viruses have been modified to remove their ability to cause infectious disease, these modified viruses can be used as vectors (vehicles) to carry therapeutic genes into human cells.
- **Bacterial vectors:** Bacteria can be modified to prevent them from causing infectious disease and then used as vectors (vehicles) to carry therapeutic genes into human tissues.
- **Human gene editing technology:** The goals of gene editing are to disrupt harmful genes or to repair mutated genes.
- **Patient-derived cellular gene therapy products:** Cells are removed from the patient, genetically modified (often using a viral vector) and then returned to the patient.

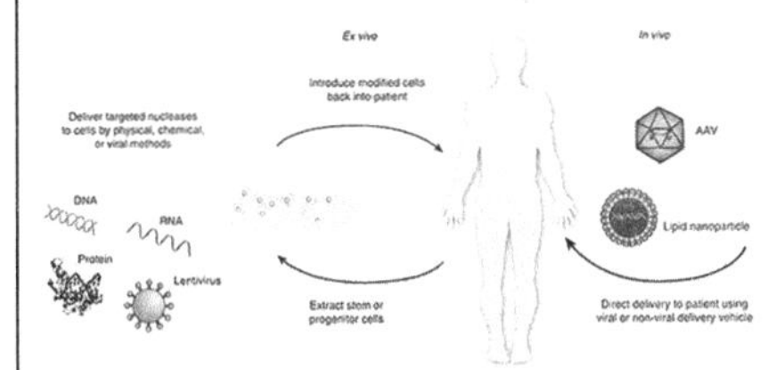

Gene therapy products are biological products regulated by the FDA's Center for Biologics Evaluation and Research (CBER). Clinical studies in humans require the submission of an investigational new drug application (IND) prior to initiating clinical studies in the United States. Marketing a gene therapy product requires submission and approval of a biologics license application (BLA).

a. "GT [gene therapy] products . . . have the potential to establish persistent infection . . . leading to the risk of developing a delayed but serious infection."

This was not the first warning by the CDC, HHS, and the Center for Biologics Evaluation and Research (CBER). There had been reports[216] in 2000, 2001, 2006, and 2015.[217] As shown in the following slide, the government issued a report to the "industry" regarding gene therapy and shedding[218] from these products.

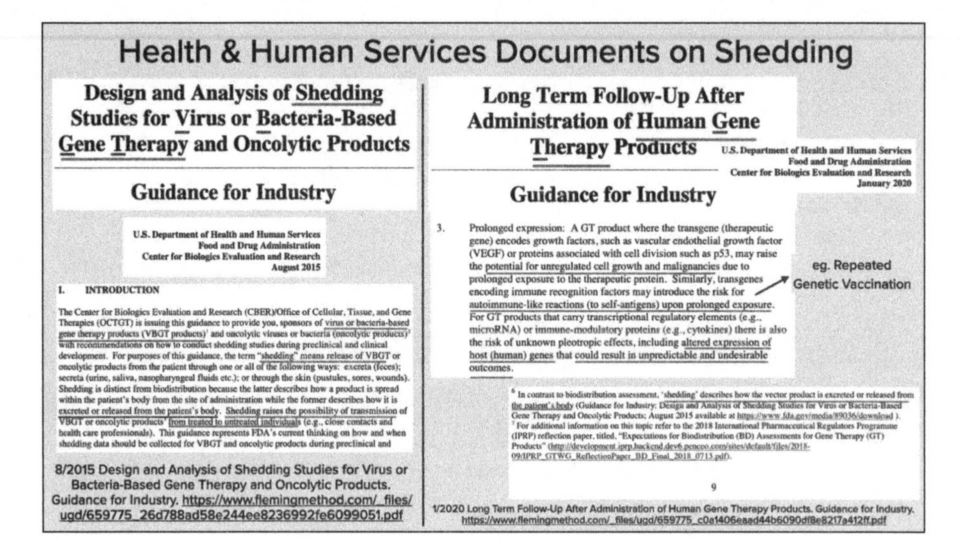

At the time of these warnings, and as shown in this QR slide set, researchers had already patented methods for inserting the priogenic (glycoprotein 120, or gp120) part of HIV into our cells, increasing the potential for shedding of gp120. Like the gain-of-function research already discussed, this gp120 gene therapy insertion was at least partially supported by the National Institute of Allergy and Infectious Diseases (NIAID).

In addition to shedding associated with (1) gene therapy products (e.g., gp120, SARS-CoV-2 spike protein), shedding can also include, (2) The vector of the gene therapy (LNP, adenovirus, and such) as defined by these HHS, CDC, and CBER reports, (3) Furin, which is not only shed but is associated with all the diseases seen with COVID (ITR, priogenic, and cancer) induced by the furin[219] cleavage site inserted into SARS-CoV-2 viruses, for which the US government owns the patent to insert the furin cleavage sites, and (4) Transferrable (topical, or skin, applied) and transmissible (injected) vaccines.[220]

The reader should find it very interesting that the research being done on transmissible (injected) genetic[221] vaccines for SARS-CoV-2 (COVID vaccines), using LNPs as the delivery method for the gene therapy designed to make the spike protein and using the Venezuelan equine encephalitis (VEE)[222] virus replicon[223] to amplify the production of the spike protein, was carried out on people, not some other nonhuman animal.

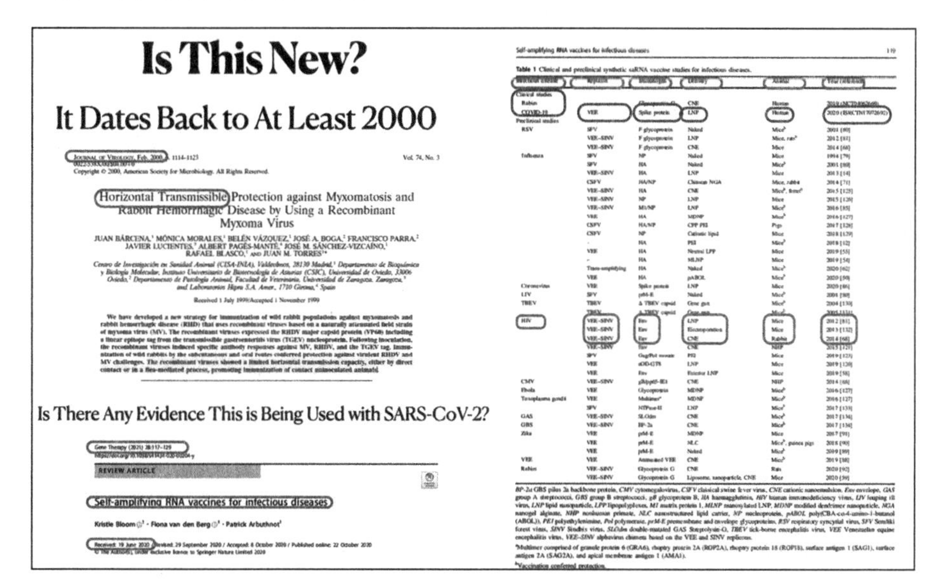

In addition to the VEE virus, Dr. Ralph Baric was actively involved in researching and patenting other gain-of-function research, including transmissible gastroenteritis virus (TGEV), infectious bronchitis virus (IBV), and coronavirus, as shown in this last QR code slide set.

There are considerable discussion and questions regarding shedding: Is shedding happening and to whom, how much, for how long? What is being shed? Are there treatments for shedding? But no substantial research has been published on any of this,[224] as of this writing, which means that you need to be wary of people answering these questions and people who are selling you treatments.

On 8 December 2023 the FDA, in conjunction with CBER, announced the approval of two gene therapy drugs for sickle cell anemia.[225] The first gene therapy (Casgevy), the first CRSIPR/Cas9 genetic treatment of its kind, involves removing the patient's blood stem cells[226] from his or her bone marrow and genetically altering them with Casgevy. This genetic treatment changes the patient's stem cells, shutting down the production of adult hemoglobin[227] (HbA), which in the case of sickle-cell patients is hemoglobin S (HbS), instructing the patient to make only the genetically altered CRISPR/Cas9 fetal hemoglobin (HbF) after the genetically altered stem cells are injected back into him or her.

The second approved cell-based gene therapy for sickle cell anemia is called Lyfgenia.[228] Like Casgevy, the patient's own stem cells must be harvested from his or her bone marrow. Lygfenia then genetically alters the DNA to make HbA^{T87Q}; which is actually a modified beta-chain hemoglobin. These stem cells are then grafted back into the patient. Critical to this experiment is what happens between the time the bone marrow harvesting[229] is done and the new genetically altered cells are reintroduced into the body. During this time, the patient must be given high-dose chemotherapy to kill off all of the patient's remaining bone marrow.

Human experimentation to eradicate a genetic code to make HbS (sickle cell), which provided a lifesaving advantage in areas of the world where malaria is endemic, could eliminate the ability to survive the spread of malaria, or some other potential threat to humanity.

Endgame

To accomplish a disease-free, potentially physically immortal person, with the characteristics we currently think of as desirable, will require many studies looking at a variety of genes, with the potential for eradicating genes that may turn out to be critical to human survival, either here or elsewhere in the universe. We will need a significant amount of time to make mistakes until we succeed (not desirable if you want to reap the benefit of these results) or a massive number of research subjects to experiment upon, allowing us the opportunity to have multiple failures and make multiple mistakes, which we could correct, it is hoped, before it's too late. The result could potentially either correct our genetic flaws, elongate telomeres for cellular immortality, and produce a healthier, longer-living human—eugenically restoring the almost physical immortality of Methuselah—or potentially cause our extinction.

The Miller experiment showed man could create organic matter out of inorganic material. The Manhattan Project gave man the power to control the creation (fusion) or destruction (fission) of matter.

> In the beginning, God created the heavens and the earth[230]

The ability to map the human genome allowed us to discover our genetic differences, which we have physically seen and recognized from the beginning—treating others different from ourselves as inferior as a result. These genetic differences include, among other things, actual health problems. The ability to change these differences using CRISPR and other technologies now make it possible to turn off certain genes (including those that protect us), or simply delete some, while increasing the survival of others—the power for humanity[231] to shape the image of life.

> Let Us make man in Our image, according to Our likeness; and let them rule over the fish of the sea and over the birds of the sky and over the cattle and over all the earth, and over every creeping thing that creeps on the earth.[232]

In our quest for the superior, almost homogenous human, including but not limited to our sexual differences and preferences, we have forgotten, or perhaps have not learned, the importance of these differences. The differences that account for sickle cell anemia, for example, are not a genetic weakness, but a genetic strength. Whereas sickle cell patients may have certain health problems, some of which may be fatal, the genes for sickle cell anemia are what kept people alive in areas where malaria was endemic. People without sickle cell are easily infected by the *Plasmodium* parasite carried by mosquitoes. These people become sick and can die. The parasite cannot live in red blood cells that are sickled, which provides a survival advantage. Eradication of this sickle cell gene would expose humanity to malaria without a defense. In our lust for power and control over the universe, any good physicist will remind you to tread cautiously. Until you fully understand what is happening, you could end up causing your own eradication.

> But of the tree of the knowledge of good and evil, thou shalt not eat of it: for in the day that thou eatest thereof thou shalt surely die. [233]

A decision to remove a genetic sequence just because we know about that sequence and its role with a particular genetic trait (phenotype), doesn't mean that same sequence is found in only one part of our genetic makeup. Nothing allows us to introduce a CRISPR sequence into only one part of our genetic makeup and not have it change another part of our DNA. If some other part of our DNA is changed, a part critical to preventing disease or death, and we change it by randomly changing our DNA with CRISPR, then we will not have a strengthened human; we will have weakened the human. If it lives, it will be subhuman. That could be motivation to those who wish to enslave or use those considered inferior.

From this experimentation we will hopefully learn what makes us stronger, healthier, potentially physically immortal. But an undisciplined approach, the approach of those desiring power and control, is an almost certain recipe for disaster, as shown in this scene from *I Am Legend*:

> TV News Anchor
>
> The world of medicine has seen its share of miracle cures, from the polio vaccine to heart transplants. But all past achievements may pale in comparison to the work of Dr. Alice Krippin. Thank you so much for joining us this morning.
>
> Dr. Alice Krippin
>
> Not at all.
>
> TV News Anchor
>
> So, Dr. Krippin, give it to me in a nutshell.
>
> Dr. Alice Krippin
>
> Well, the premise is quite simple—um, take something designed by nature and reprogram it to make it work for the body rather than against it.

TV News Anchor

You're talking about a virus?

Dr. Alice Krippin

Indeed, yes. In this case the measles, um, virus which has been engineered at a genetic level to be helpful rather than harmful. Um, I find the best way to describe it is if you can . . . if you can imagine your body as a highway, and you picture the virus as a very fast car, um, being driven by a very bad man. Imagine the damage that car can cause. Then if you replace that man with a cop . . . the picture changes. And that's essentially what we've done.

TV News Anchor

And how many people have you treated so far?

Dr. Alice Krippin

Well, we've had ten thousand and nine clinical trials in humans so far.

TV News Anchor

And how many are cancer-free?

Dr. Alice Krippin

Ten thousand and nine.

TV News Anchor

So you have actually cured cancer.

Dr. Alice Krippin

Yes, yes. . . yes, we have.

[*cuts to post-apocalyptic New York three years later*]

From the script *I Am Legend*

If history has taught us anything, it is that our understanding of how to use our knowledge has not kept pace with our actual knowledge.[234] In the end, how we use this knowledge, is going to be equally important as who uses this knowledge. This will ultimately determine the fate of mankind. Science and medicine must be governed by ethical scientists and physicians. It must not be regulated by lawyers, judges, politicians, bureaucrats, or individuals and families who have decided they should determine the fate of mankind. Such has always been the initiation of Dark Ages[235] of mankind. The Founding Fathers, taking a lesson from the history of others, made it possible for us to correct the misdirections of this country only if we chose to learn from history and only if *We the People* act to change history in the making. The choice is yours, mine, and ours—or it will be theirs.[236]

Maybe this time, unlike other times, we can get it right. Maybe this time, we will see individuals stand up and do what's right, what's necessary.

> Katniss, maybe the country was shocked tonight by your arrow, but once again, I was not. You were exactly who I believed you were. I wish I could give you a proper goodbye. But with both Coin and Snow dead, the fate of the country will be decided tonight, and I can't be seen at your side. Tonight, the 12 District leaders will call for a free election. There's little doubt that Paylor will carry it. She's become the voice of reason. I'm sorry so much burden fell on you. I know you'll never escape it. But if I had to put you through it again for this outcome, I would. The war's over. We'll enter that sweet period where everyone agrees not to repeat the recent horrors. Of course, we're fickle, stupid beings, with poor memories and a great gift for self-destruction. Although, who knows? Maybe this time, we'll learn. I've secured you a ride out of The Capitol. It's better for you to be out of sight. And when the time is right, Commander Paylor will pardon you. The country will find its peace. I hope you can find yours.
>
> Plutarch Letter to Katniss, *The Hunger Games: Mockingjay—Part II.*

Illusionists, like magicians and those in military intelligence, know that a key to bringing about their final outcome is sleight of hand, getting you to look in one direction or think in one direction while they are completing their task elsewhere.

According to the US Census report, roughly 335.8 million[237] people live in the United States as of 8:09 a.m. CT 24 December 2023. Of this number, slightly more than 270.2 million (80.4 percent) received at least one COVID vaccination, and more than 230.6 million (68.7 percent)[238] are considered fully vaccinated, with 36 million[239] having received the "updated" COVID vaccine as of 24 December 2023.

Many people are debating just how safe, or unsafe, these genetic vaccines are,[240] and the Vaccine Adverse Events Reporting System (VAERS)[241] shows a total of 1,615,020 reported adverse events.[242]

In addition to the total number of excess deaths[243] above and beyond the 36,726 deaths reported as being caused by the COVID vaccines between 1 February 2020 and 24 November 2023, there are the additional 540,122 and counting reported excess deaths, the vast majority of which are the result of ITR diseases, prion diseases, or both, as taken directly from the CDC website—now archived and no longer collecting data.

If we take all of this information together, the total number of deaths and adverse events associated with the genetic vaccines (36,726 + 1,615,020 = 1,651,746) and consider this the undesirable outcome of the vaccines—the number of people lost by using the genetic vaccines—and divide this by the number of people vaccinated at least once (270,200,000) x 100 to get the percentage, you will find that the loss ratio is 0.6 percent. Even if we add in the total excess deaths of 568,443 (1,651,746 + 568,443 = 2,220,189), we end up with an acceptable military loss of 0.8 percent. Put another way, when conducting this type of research, 99.2 percent survived the experiment, and with refinements from lessons learned, an even better outcome should be expected.

Yes, there were losses, but there always are. Every medical research project ever conducted has losses. Every military operation in history has had some type of loss. If more than 1.16 million Americans have died from COVID (3 percent of the US population)[244] and 6.96 million[245] people have died worldwide, and this produced a waiving of sovereign rights, then those conducting this operation have learned. As you will see from our later discussion on the WHO Treaty and scenarios run to determine what is needed to bring all countries into compliance, they, if not you, have been learning.[246]

Running these numbers by you shows that while everyone is arguing about how harmful the genetic vaccines are or are not, the important question is, how well will the general population[247] tolerate them? If they keep using this approach, will it be too harmful to continue the experiment, or have they found a method to introduce genetic material into people that's tolerated enough that they can expect people to participate and provide enough research subjects to make the research a success?

It's time to look at the people who have been playing a game as old as humanity.

> When you play the game of thrones, you win or you die. There is no middle ground.—Cersei Lannister, *Game of Thrones*

What you do with this information, how you act from here on out, is up to you. Rather than ending this chapter from the perspective of someone lusting for power, or the people we will talk about in the next chapter, I chose to leave you with the following two quotes:

> Neville: [*talking to Anna about Bob Marley*] He had this idea. It was kind of a virologist idea. He believed that you could cure racism and hate . . . literally cure it, by injecting music and love into people's lives. When he was scheduled to perform at a peace rally, a gunman came to his house and shot him down. Two days later he walked out on that stage and sang. When they asked him why, he said, "The people, who were trying to make this world worse. . . are not taking a day off. How can I? Light up the darkness."
>
> *I Am Legend*

> Hope begins when you stand in the dark looking out at the light.
>
> *The Christmas Train*

CHAPTER 3

Unquenchable Lust[1] for Power

Since people have walked on this planet, men, women, and families have sought dominion over other men, women, and families. Even the Christian Bible and Jewish Torah speak of dominion:

> And let them have dominion over the fish of the sea, and over the fowl of the air, and over the cattle, and over all the earth, and over every creeping thing that creepeth upon the earth. [2]

Is it any wonder that this would eventually extend to dominion over people? Again, the Christian Bible and Jewish Torah recognized and discussed slavery and property rights of one person over another.[3] The plight of the Jewish people against the enslaving Egyptians is well recorded in the book of Exodus[4]—although the specific words might change, the basic record is the same and discussed here in chapter 2. The Jewish people, through Moses, tells the pharoah, king of Egypt, that if he will not release the Jewish people, God will punish him.

1. Then the Lord said unto Moses, Go in unto Pharaoh, and tell him, Thus saith the Lord God of the Hebrews, **Let my people go**, that they may serve me.
2. For if thou refuse to let them go, and wilt hold them still,
3. Behold, the hand of the Lord is upon thy cattle which is in the field, upon the horses, upon the asses, upon the camels, upon the oxen, and upon the sheep: there shall be a very grievous murrain. [emphasis added]

And before you think this is only a Jewish or Christian problem, know that, as discussed in chapter 2, my pagan[5] ancestors, Native Americans, Muslims,[6] and others have all, at some point, enslaved other people.

In an effort to seek approval from others and to lessen any punishment to ourselves, including during this pandemic, is it any wonder that people have been so easily manipulated and turned against each other, when human character and habit has shown that people are capable of harming, even killing, those closest to us?

> Cain said to his brother Abel, "Let's go out to the field." While they were in the field, Cain attacked his brother Abel and killed him.[7]

During his 1947 Nuremberg Trial Göring Said The Following.

... it is the leaders of the country who determine the policy and it is always a simple matter to drag the people along, whether it is a democracy or a fascist dictatorship or a Parliament or a Communist dictatorship.

...voice or no voice, the people can always be brought to the bidding of the leaders. That is easy. *All you have to do is tell them they are being attacked and denounce the pacifists for lack of patriotism and exposing the country to danger.* It works the same way in any country.

As noted by Jewish historian Rabbi Berel Wein,[8]

> Sibling rivalry, jealousy, murder, psychological depression, sexual laxity and abuse are now all part of the story of humankind. Human beings are now bidden to struggle for their very physical and financial existence in a world of wonder—complete with ever present dangers and hostility.

History is littered with examples of kings, queens, rulers, and self-appointed aristocrats who believed they were somehow ordained or chosen to make the world a better place. If the common people could simply understand, if they could see and appreciate what these self-appointed leaders could see and appreciate, then certainly they would willingly follow the wisdom of these leaders, who are only trying to make the world a better place.

Pharaoh of Egypt[9]

The pharaoh was considered to be a living god on earth, and as such, he had many responsibilities to both gods and humans. He was responsible for maintaining balance in the universe and ensuring that Ma'at (a concept that meant something like "order, truth, or justice") prevailed. The pharaoh was considered to be the high priest of every temple in Egypt, and as such, the leader was also responsible for performing rituals and offering sacrifices to the gods. In doing so, he ensured that the gods were happy and that they would continue to provide for the people of Egypt.

The Coronation of Queen Elizabeth II, 2 June 1953[10]

> Our gracious Queen: to keep your Majesty ever mindful of the law and the Gospel of God as the Rule for the whole life and government of Christian Princes, we present you with this Book, the most valuable thing that this world affords. Here is Wisdom; This is the royal Law; These are the lively Oracles of God.
>
> James Pitt-Watson, Moderator
> General Assembly Church of Scotland

Adolf Hitler—*Mein Kampf*

> I believe today that my conduct is in accordance with the will of the Almighty Creator.
>
> Adolf Hitler, *Mein Kampf,* vol. 1, chap. 2

> What we have to fight for is the necessary security for the existence and increase of our race and people, the subsistence of its children and the maintenance of our racial stock unmixed, the freedom and independence of the Fatherland so that our people may be enabled to fulfill the mission assigned to it by the Creator.
>
> Adolf Hitler, *Mein Kampf,* Vol. 1 Chap. 8

An underlying theme of superiority and connection with a greater power can be seen in the character of many of history's leaders and of today's leaders, some elected, some self-appointed. It's important for us to honestly reflect on the history of mankind, where we have been and where the "leaders" of today are taking us, and ask the simple question, Are we going to continue repeating the mistakes of the past, or are we going to demand an end to the slavery, eugenics, and human experimentation that could lead to our extermination?

> Nearly all men can stand adversity, but if you want to test a man's character, give him power.[11]
>
> Abraham Lincoln

Habitual Lust for Power

In this chapter, we are going to look at the leaders of yesterday and today and see how the unquenchable lust for power that plagued our ancestors is still very much alive today. We are going to begin with the legal definition of *habit.* In a court of law, if you can demonstrate habit, you can demonstrate character and what someone will do in a given circumstance, situation, or setting. The person's *consistent* and *predictable* habits and character can be used to expose what he or she is doing—in this instance to usurp control of humanity.

Accordingly, I present evidence of the defendant's unquenchable lust for power and usurping of control and direction of humanity, pursuant to the Federal Rules of Evidence[12] (FRE) rule 406:[13]

> Evidence of a person's **habit** or an organization's routine practice **may be admitted to prove that on a particular occasion the person or organization acted in accordance with the habit or routine practice**. The court may admit this evidence regardless of whether it is corroborated or whether there was an eyewitness. [emphasis added]

As established in McCormick, habit and character are defined[14] as follows:

> Page 356 TITLE 28, APPENDIX—RULES OF EVIDENCE Rule 407 (Pub. L. 93–595, § 1, Jan. 2, 1975, 88 Stat. 1932.).
>
> NOTES OF ADVISORY COMMITTEE ON PROPOSED RULES
>
> An oft-quoted paragraph, McCormick, § 162, p. 340, describes habit in terms effectively contrasting it with character:
>
> "Character and habit are close akin. **Character** is a generalized description of **one's disposition**, or of one's disposition in respect to a general trait, such as **honesty**, **temperance**, or **peacefulness**. [emphasis added]
>
> '**Habit**,' in modern usage, both lay and psychological, is more specific. It describes **one's regular response to a repeated specific situation**. If we speak of character for care, we think of the person's tendency to act prudently in all the varying situations of life, in business, family life, in handling automobiles and in walking across the street. A habit, on the other hand, is the person's regular practice of meeting a particular kind of situation with a specific type of conduct, such as the habit of going down a particular stairway two stairs at a time, or of giving the hand-signal for a left turn, or of alighting from railway cars while they are moving. The doing of the habitual acts may become semi-automatic." **Equivalent behavior on the part of a group is designated "routine practice of an organization"** in the rule. [emphasis added]
>
> **Agreement is general that habit evidence is highly persuasive as proof of conduct on a particular occasion.** Again quoting McCormick § 162, p. 341: [emphasis added]
>
> **"Character may be thought of as the sum of one's habits though doubtless it is more than this.** But unquestionably the uniformity of one's response to habit is far greater than the consistency with which one's conduct conforms to character or disposition. Even though character comes in only exceptionally as evidence of an act, surely any sensible man in investigating whether X did a particular act would be greatly helped in his inquiry by evidence as to whether he was in the habit of doing it.' [emphasis added]

Let's begin by taking a look at some of humanity's most reviled leaders,[15] many of whom were self-appointed, and the character traits and habits they exhibited. I think it is safe to say that most if not all of the individuals believed they were invincible. However, history and humanity proved them wrong.

Ancient Leaders (Before 1850)

Years	Rulers	What They Were Known For
Ca. 1353–1335 BCE	Akhenaten; Amenhotep IV (Pharoah)	Made the sun disc. God was Aten, and Atenism was semimonotheistic religion of Egypt. Nutritional deficiencies; High juvenile mortality rate under his reign. Father of King Tut by DNA analysis.
Ended 210 BCE	Qin Shi Huang	Enslaved his people to build a wall around China—forerunner of the Great Wall—and caused hundreds of thousands of deaths to build the Great Wall and roads system. Destroyed Chinese educational system.
Ca. 20 BCE–39 A.D.	Herod Antipas (Herod the Tetrarch; King Herod—although he was never a king) Galilee and Perea	Responsible for the death of John the Baptist. Indirectly involved with crucifixion of Jesus of Nazareth. Conspired against Roman Emperor Caligula.
12–41	Caligula (Gaius Caesar Augustus Germanicus) Rome, Italy	Acts against the emperor were deemed to be treason, punishable by arrest, torture, and execution. Declared himself a god. Increased his executive authority, while weakening the senate and people of Rome (SPQR[16]).
161–192	Lucius Aurelius Commodus (emperor of Rome)	Heavily taxed citizens. Held numerous gladiatorial exhibitions, injuring and hiding the wounds of his opponents to guarantee his victory. Late in life declared himself the reincarnated god Romulus, naming Rome after himself.
1186–1216	John, king of England	Attempted rebellion against his brother, King Richard I. Became king after his brother's death. Historians agree he was cruel, vindictive, and petty. Upon return from France, where he lost Duchy of Normandy to King Phillip II of France, the British people made John sign the Magna Carta[17] in 1215, imposing limits on the king. He largely ignored the Magna Carta.
Ended 1405	Tamerlane; Timur (Sword of Islam); Turkish	Killed 17 million. Built towers on skulls of his defeated enemies. Known for torture and execution of enemies. Patron of the arts and an architectural visionary, demonstrating that some leaders might appear altruistic, while in reality, they are also committing atrocities.

Years	Rulers	What They Were Known For
1428–1476	Vlad the Impaler (Count Dracula; Prince Voivode of Wallachia) Romania	His favorite method of executing his enemies was impaling them (rectum to chest) and letting them die slowly. Boiled men, women, and children in cauldrons. Today recognized as a Romanian national hero.
1479–1516	Ferdinand II of Aragon. Spain (king of Castile)	Exiled, tortured, or killed Jews and Protestants who refused to convert to Catholicism.
1485–1547	Hernán Cortés (Spanish conquistadore)	Invasion of Mexico and Tenochtitlán "in the name of God." Looted the Aztec Empire.
1547–1584	Ivan the Terrible; Russia	Killed many thousands in multiple wars, including in Scandinavia, Turkey, Lithuania, Crimea, and Serbia.
1599–1658	Commander Oliver Cromwell; Britain, Ireland	Deported more than fifty thousand Irish from their homes, including women and children. Deported people to Barbados and Bermuda. Known for harsh treatment of Irish Catholics. Cromwell would not allow the "exercise of the Mass."[18] Significantly involved in the English Civil War.[19]
1612–1640	Murad IV, Sultan of the Ottoman Empire. Turkish	Banned coffee, alcohol, tobacco for others; killed if they partook. Did not ban himself or concubines from consumption.
1615–1648	Ibrahim of the Ottoman Empire Turkish	Bankrupted the Ottoman Empire through personal use of the country's wealth, for his harem. Waged war with Venice and Crete for twenty-four years. Naval blockades led to supply problems and increased taxation.
1729–1796	Catherine II; Catherine the Great (Princess Sophie of Anhalt-Zerbst) Russia	Participated in a coup that removed her husband (Peter III) from power. Had several rivals executed. Open rebellions against her led to more stringent conditions for the people of Russia. Had herself inoculated with smallpox.[20] She proceeded to have everyone inoculated. In 1785 she appropriated money to build mosques in an effort to incorporate Muslims into Russian society. In 1785 she declared Jews to officially be foreigners, denying them the rights of an Orthodox or naturalized Russian citizenship. In 1790 she banned Jewish citizens from Moscow's middle class, and in 1794, she doubled taxes on those of Jewish descent.

Years	Rulers	What They Were Known For
1754–1793	Louis XVI (Louis Auguste) French	During failing harvests and famine, he increased taxes, living lavishly as his people suffered and died from starvation.
1754–1801	Paul I of Russia	Attempted to reform the army, including flogging of soldiers. Other concerns included the treatment of the peasant class.
1758–1794	Maximilien Robespierre (Maximilien Francois Marie Isidore de Robespierre) French	French lawyer and statesman. His atrocities, including involvement in people being put to death by guillotine, resulted in most scholars agreeing he was one of the most feared and hated men in France. His actions eventually helped fuel the revolt of the French citizens against the monarchy. French Revolution
1778–1861 Coronation 12 August 1829	Ranavalona I (Rabodoandrian-ampoinlmerina); Queen of Madagascar	Engaged in warfare with Europeans. Opposed Christian missionaries on her island. Implemented policies that resulted in subjects being forced into labor, military service, or both to pay for the policies.

Modern-Day Leaders (Since 1850)

Years	Rulers	What They Were Known For
1809–1865	Abraham Lincoln, 16th president of the United States of America	American Civil War—bloodiest war[21] on US soil, with 2.5 percent of the American population killed. Two-thirds of all deaths were from infectious disease. Increased federal government power, suspended habeas corpus[22] in Maryland. Took blue mass[23] pills for depression, malaria, and smallpox, while others suffered.
1835–1909	Leopold II of Belgium (Léopold Louis Phillippe Marie Victor) Dutch	Exploited the Congo population for their ivory and rubber. Used torture including: beatings, limb amputations, and other atrocities to enforce quotas, resulting in the deaths of approximately 10 million Congolese.
1832–1867	Maximilian I of Mexico; Austrian archduke	Essentially an Austrian invasion of Mexico, which United States could not address (Monroe[24] Doctrine) because of Civil War.
1868–1918	Nicholas II,[25] tsar of Russia (Nikolai Alexandrovich Romanov)	Two decades of economic collapse, military defeats, crop failures, antisemitic programs, execution of dissidents, and denial of reforms. Resulted in resignation and execution by Bolsheviks in July 1918.

Years	Rulers	What They Were Known For
1887–1945	Minister President Vidkun Abraham Lauritz Jonssøn Quisling; Norway	Collaborated with the Nazis forming the Nasjonal Samling. Placed on trial after WWII and found guilty of embezzling, murder, and high treason; executed as a result.[26] Supported the forced exile of Norwegian Jews.
1889–1945	Führer Adolf Hitler; Germany	Through Hitler's Schutzstaffel (SS), more than 11 million noncombatants were executed, including 6 million Jews. Used chemical gases on prisoners and conducted illegal research, including genetic studies on men, women, and children. Euthanasia program known as Tiergartenstraße 4[27] (Aktion T4).
1878–1953	Joseph Stalin	Sent millions into exile and executed at least eight hundred thousand. Endorsed rape. Sent assassins to kill enemies around the world. Famines under his rule resulted in policies that resulted in millions of deaths. Presented himself as "a man of the people."
1908–1973	Lyndon Baines Johnson, 36th president of the United States of America	Supported the war in Vietnam and the draft.[28] About 282,000 Americans soldiers died in the war. All together roughly 1,353,000 American soldiers, People's Army of Vietnam, Viet Cong, and civilians died.
1892–1975	Francisco Franco (Francisco Franco Bahamonde) head of Spain	Spanish Civil War. Spanish children were taught Franco was sent by divine providence to save the nation from anarchy and atheism. Hundreds of thousands of his opposition were imprisoned and executed by firing squads.
1919–1980	Mohammad Reza Pahlavi (shah of Iran)	Utilized secret police (SAVAK) to arrest opposition, including more than two thousand political prisoners. Banned opposition political parties and meetings.
1925–2005	2nd President Milton Obote (Apollo Milton Obote) of Uganda	Smuggled gold and ivory. Suspended Ugandan parliament, declared martial law, issued emergency decrees, sentenced opposition to life imprisonment without trials.
1925–2003	3rd President Idi Amin (Idi Amin Dada Oumee) of Uganda	Military officer who implemented a coup to become president of Uganda. Persecuted ethnic groups and executed up to a half million Ugandans before he fled to Saudi Arabia.

This is by no means an exhaustive list of humanity's self-appointed deities. Part of the purpose of this list is to make the reader aware of the history of atrocities committed by people over the centuries, around the world, involving all religions and countries, over time. You probably either don't think about many of them or haven't heard of them. In every instance, the atrocities were committed through the use of government and military forces. In every instance, the self-appointed leaders used the differences between people to galvanize brutality and control, until the people rose up and demanded accountability. We are all ultimately accountable and responsible for standing up against those who would harm the innocent.

> Rise and Rise again and again, like the Phoenix from the Ashes, until the Lambs become Lions and The Rule of Darkness is no more.[29]

The Military-Industrial Complex (MIC)

The United States has been no exception to the use of military power. Following the Second World War, former five-star General Dwight David Eisenhower was elected the thirty-fourth president of the United States of America. Prior to leaving office, President Eisenhower found himself concerned with the direction of the country. The military he had led during the war, as the supreme commander of the Allied Expeditionary Force into Europe, had, after recruiting Nazis in Operation Paperclip, changed to become a potential threat to liberty and the American way of life.

President Eisenhower's 17 January 1961 Farewell Warning of the MIC

"In the councils of Government we must guard against the acquisition of unwarranted influence ... by the Military / Industrial Complex. The potential for the disastrous rise of misplaced power exists and will persist."

"We must never let the weight of this combination endanger our Liberties or Democratic processes. We should take NOTHING for granted."

"We must be alert to the equal and opposite danger that public policy could, itself, become the captive of a Scientific / Technological Elite."

https://www.youtube.com/watch?v=OyBNmecVtdU

As we track the use of chemical, nuclear, and biologic weapons by the US military, we can see subsequent efforts to prevent other countries from using these weapons upon the United States.

While we criticize other countries for their barbaric use of such weapons, we need to recognize that only the United States has used nuclear weapons, and the first country in which we detonated a nuclear device was the United States,[30] in Alamogordo, New Mexico, prior to Hiroshima[31] and Nagasaki, and we have repeatedly demonstrated our willingness to continue doing so.[32]

No one can deny the use of chemical weapons by the United States or our continued involvement in biological weapons development, despite having signed treaties and passed federal legislation prohibiting such weapons of mass destruction. The US military-industrial complex is criminally responsible for the development and consequences of such bioweapons. In fact, the United States has been constantly running war-game scenarios to determine what the impact of such weapons would be.

War Games—Would You Like to Play a Game?

Long before the general public heard about Event 201,[33] the United States was actively playing out war-game scenarios. In fact, when JFK began an accelerated education program, just as other countries, including what the USSR had been doing, part of the training involved was the weekly exercise of being given what appeared to many to be a no-win scenario.[34]

The selection of individuals for the Kennedy Kids program was not an exact science, but it did follow certain information gleaned from the US efforts to infiltrate Nazi Germany. At that time the hubs of psychology in the United States were Harvard and Iowa. Programs had been developed to measure response to interrogation and the ability to adapt to conditions, to survive, and to get the specified job done. Candidates, including individuals who had escaped Nazi concentration camps, were selected and tested. They were given new identities, new clothing, and instructions on whom they could or could not talk to and what they could say. This developed out of Project 63 into the OSS,[35] which eventually became the CIA.

The original goal was to drop "agents" behind German lines, with money to bribe high-ranking German officials and military, in an effort to assassinate Adolf Hitler. The efforts failed, and most agents fled to Switzerland or elsewhere with the money. It turned out, the selection of such individuals was flawed. During the next several decades efforts to refine the selection process resulted in changes in testing and selection, but the end goal was the same. President Kennedy, I believe, was hoping he could use these efforts to the benefit of the country by trying to find young people who would place country above self.

Kennedy's famous statement "ask not what your country can do for you—ask what you can do for your country" was based on his thought about sacrifice that he had long held and had expressed in various ways in campaign speeches.[36]

Eventually the selection process began with selecting children as soon as they could be tested for aptitude, physical strength and agility, creativity, teamwork, independence, and fortitude. Tests were conducted multiple times during each year, before and after selection. Funding was made available to allow the development of these skills, depending upon areas of aptitude. Individuals were encouraged to take on intelligence and police roles, to complete doctoral training, to compete athletically—and win. In the end, I believe JFK knew there was a threat from the military-industrial complex, just as Eisenhower had warned. JFK was questioning the Vietnam War, the CIA, and other US agencies. For this, it is clear, he paid with his life.

> We shall pay any price, bear any burden, meet any hardship, support any friend, oppose any foe to assure the survival and success of liberty.

But Kennedy knew that if he could help develop such individuals, they might someday save the country, if they survived what lay ahead of them. For those of us who followed, the choice was clear. We would either follow and honor the man, or we would abandon our country and president.

Survival and tenacity for survival and winning was honed through the use of survival scenarios. In these scenarios, survival meant life; failure meant death. The first person to successfully render a successful result was the winner. Second was first loser, and losing was unacceptable. The game was simple. Here is where you are, and here is where you need to get to survive. These are your supplies; here are the people with you. To get from point A (where you are) to point B (where you will either be able to survive or be rescued), you will need to decide what supplies you can take with you and how you will use them to get you there. You also will only be able to take so many people with you, though they will be critical for you to get to point B, and they will be the only surviving humans. How will you get there? What will you take? You will not be able to take everything you want or need. Who will go with you? Who will live, and who will die? Begin![37]

On 18 October 2019, a group of individuals conducted a pandemic exercise (Event 201) at the Johns Hopkins Center for Health Security, in conjunction with the World Economic Forum, the Bill & Melinda Gates Foundation, and others. The scenario involved a global pandemic—how to respond, how to control health care, how to control media coverage and reporting of the pandemic, and so forth.

As you think through any problem, as I have recommended throughout this book, it's important to place the events and people in context: who knows whom, who works with whom, what is the endgame. To do that, let's look at a few comments that have been made by some of the major players, including the Rockefeller Foundation, World Economic Forum, Bill & Melinda Gates Foundation, and others.

Absolute Power and One World Order

1. David Rockefeller, founder of the Trilateral Commission, shared certain beliefs with Henry Kissinger, Klaus Schwab, and the World Economic Forum, evident from statements made in his book *Memoirs*, published in 2003.[38]
   A. "For more than a century, ideological extremists, at either end of the political spectrum, have seized upon well-publicized incidents, such as my encounter with Castro, to attack the Rockefeller family for the inordinate influence they claim we wield over American political and economic institutions. Some even believe we are part of a secret cabal, working against the best interests of the United States, characterizing my family and me as 'internationalists,' and of conspiring with others around the world to build a more integrated global political and economic structure—one world, if you will. **If that's the charge, I stand guilty, and I am proud of it.**" [emphasis added]
   B. "We are grateful to the *Washington Post*, the *New York Times*, *Time* Magazine and other great publications whose directors have attended our meetings and respected their promises of discretion for almost 40 years. . . . **It would have been impossible for us to develop our plan for the world if we had been subjected to the lights of publicity during those years**. But, the world is more sophisticated and prepared to march towards a world government. **The supernational sovereignty of an intellectual elite and world bankers is surely preferable to the national auto determination practiced in past centuries.**" [emphasis added]
2. Zbigniew Brzezinski, a notable member of the Council on Foreign Relations and cofounder of the Trilateral Commission, authored a book titled *Between Two Ages: America's Role in the Technetronic Era*, first published on 16 July 1970. Two popular quotes[39] from the book reveal shared dystopian ideologies with Trilateral Commission cofounder David Rockefeller, Henry Kissinger, Klaus Schwab's World Economic Forum, and others:

   The technotronic era involves the gradual appearance of a more controlled society. Such a **society would be dominated by an elite, unrestrained by traditional values**. Soon it will be possible to assert almost **continuous surveillance over every citizen** and maintain up-to-date complete files containing even the most personal information about the citizen. These files will be subject to instantaneous retrieval by the authorities. [emphasis added]

   In the technotronic society the trend would seem to be towards the aggregation of the individual support of millions of uncoordinated citizens, easily within the reach of magnetic and attractive personalities exploiting the latest communications techniques **to manipulate emotions and control reason**. [emphasis added]

3. James Warburg, son of Paul Warburg and director of the Council on Foreign Relations at its founding, stated the following to the Senate Foreign Relations Committee[40] on 17 February 1950 in Washington, DC:

    The past 15 years of my life have been devoted almost exclusively to studying the problem of world peace and, especially, the relation of the United States to these problems. These studies led me, 10 years ago, to the conclusion that the great question of our time is not whether or not one world can be achieved, but whether or not one world can be achieved by peaceful means. We shall have world government, whether or not we like it. **The question is only whether world government will be achieved by consent or by conquest.** [emphasis added]

4. Hillary Clinton, while secretary of state for the United States,[41] stated the following as part of her introductory speech at the Council on Foreign Relations in Washington, DC, on 15 July 2009.

    And I am delighted to be here in these new headquarters. I have been often to, I guess, the mothership in New York City, but it's good to have an outpost of the council right here down the street from the state department. **We get a lot of advice from the council, so this will mean I won't have as far to go to be told what we should be doing and how we should think about the future**. [emphasis added]

Let the Games Begin!

The recognition of Event 201 was soon followed by the March 2022 discovery of the November 2021 Nuclear Threat Initiative[42] (NTI) war game that played out the hypothetical release of a gain-of-function dual-use monkeypox virus that had been intentionally released for the war game.

The NTI war game followed the recognition of the term *dual-use* in 2015.[43] The recognition pointed out that the US government had previously recognized the problem of potential misuse of this technology in 2005, resulting in yet another agency, which specifically focuses on biosecurity and dual-use research. The website to this agency appears to work only intermittently. During the original presentations I gave on the dual-use topic the site worked, then it stopped working following those presentations and during the initial writing of this book, even though the United States is actively involved in the preparation of such gain-of-function biological weapons.[44] During the editing of this book, following a period of time when I had stopped talking about dual-use, the editors noticed the link was once again working. I cannot say for certain why the link stopped working and once again has appeared, but I do find it interesting and for the record, I have copies of the material from the website. Also for the record, the figure below is what I found when the link had stopped working.

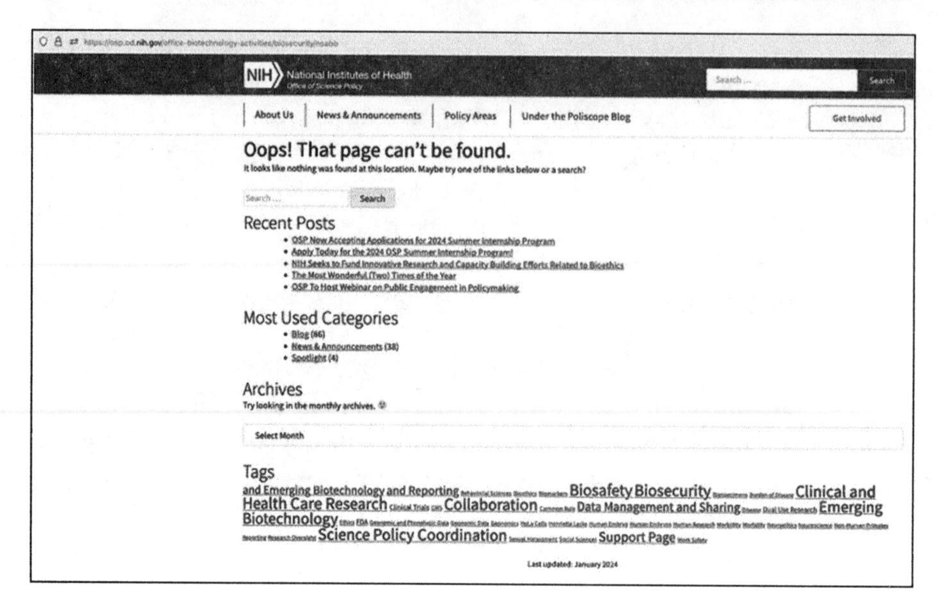

In the 2021 NTI tabletop exercise, gain-of-function bioweapons were officially front and center in bio–war game scenarios. Margaret A. Hamburg, MD, interim vice president, Global Biological Policy and Programs, Nuclear Threat Initiative; former twenty-first commissioner for the US Food and Drug Administration; and daughter of the president emeritus of the Carnegie Corporation of New York, proposed the following:

> [A] new mechanism would operate at the "seam" between existing mechanisms—including World Health Organization (WHO) outbreak investigation capabilities and the United Nations Secretary-General's Mechanism for investigating alleged deliberate bioweapons use—thereby strengthening UN system capabilities to investigate pandemic origins.

The report emphasized the need to develop a "no-regrets" approach; that is, an approach for which national governments adopted a pandemic response directed by a centralized World Health Organization (WHO) and United Nations (UN) control, without requiring scientific evidence before independently acting. The report found that the more financially independent the country—for example, the United States—the more likely it would resist this one world order control by the WHO and UN. Cochairs for the NTI included the UN, Janssen, Chinese CDC, Jeremy Farrar of the Welcome Trust, WHO, Merck, and others.

Both of these recent bioweapon war games, Event 201 and the 2021 NTI (gain of function monkeypox virus) exercises, emphasized replacing sovereign national control by elected officials within the country, with a one world order under UN and WHO control. That world order would be controlled, not by those elected within a country like the United States, but by self-appointed individuals ready to assume power and control over the country.

EXERCISE CO-CHAIRS

Dr. Ernest J. Moniz
Co-Chair and CEO
Nuclear Threat Initiative
Former U.S. Secretary of Energy

Ambassador Wolfgang Ischinger
Chairman
Munich Security Conference

PARTICIPANTS

Mr. Arnaud Bernaert
Head, Health Security Solutions
SICPA

Dr. Beth Cameron
Senior Director, Office of Global Health Security and Biodefense
U.S. National Security Council

Mr. Luc Debruyne
Strategic Advisor to the CEO
Coalition for Epidemic Preparedness

Dr. Ruxandra Draghia-Akli
Global Head
Johnson & Johnson Global Public Health R&D
Janssen Research & Development

Dr. Chris Elias
President, Global Development Division
Bill & Melinda Gates Foundation

Sir Jeremy Farrar
Director
Wellcome Trust

Dr. George Gao
Director-General, Chinese Center for Disease Control and Prevention (China CDC)
Vice President, the National Natural Science Foundation of China (NSFC)
Director and Professor, CAS Key Laboratory of Pathogenic Microbiology and Immunology, Institute of Microbiology, Chinese Academy of Sciences
Dean, Medical School, University of Chinese Academy of Sciences

Dr. Margaret (Peggy) A. Hamburg
Interim Vice President
Global Biological Policy and Programs, Nuclear Threat Initiative
Former Commissioner of the U.S. Food and Drug Administration

Ms. Angela Kane
Visiting Professor
Paris School of International Affairs (SciencesPo), and Tsinghua University

Dr. Emily Leproust
CEO and Co-Founder
Twist Biosciences

Dr. Elisabeth Leiss
Deputy Director of the Governance and Conflict Division
German Corporation for International Cooperation (GIZ)

Ms. Izumi Nakamitsu
Under-Secretary-General and High Representative for Disarmament Affairs
United Nations Office for Disarmament Affairs

Dr. John Nkengasong
Director
Africa Centres for Disease Control and Prevention

Sam Nunn
Founder and Co-Chair
Nuclear Threat Initiative
Former U.S. Senator

Dr. Michael Ryan
Executive Director
WHO Health Emergencies Programme

Dr. Joy St. John
Executive Director
CARPHA

Dr. Petra Wicklandt
Head of Corporate Affairs
Merck KGaA

These individuals are connected with the World Economic Forum (WEF), Trilateral Commission, and Rockefeller Foundation, and they have gone so far as to state on the record that they favor a one world order. A number of our elected leaders are as eager to hand over control of our country to these would-be dictators, as the self-appointed American Medical Association (AMA) was to hand over control of medicine to the lawyers and US federal and state governments, as you shall see in chapter 4.

Before we turn to the issue of the UN and WHO treaties, which do not authorize these individuals to take control of the United States, I want to make you aware of another biological weapon war game: the October 2017 SPARS[45] Pandemic 2025–2028: A Futuristic Scenario for Public Health Risk Communicators.

The 2017 SPARS bio–war game considered the following:

> [The] major socioeconomic, demographic, technological, and environmental trends likely to have emerged by that period . . . [and] . . . two dominant trends likely to influence regulatory and public responses to future public health emergencies were selected: one, varying degrees of access to information technology; and two, varying levels of fragmentation among populations along social, political, religious, ideological, and cultural lines. A scenario matrix was then constructed, illustrating four possible worlds shaped by these trends, with consideration given to both constant and unpredictable driving forces.

As the name clearly indicates, the scenario surrounded a coronavirus pandemic. The scenario focused on HHS, CDC, and FDA response, including control of media, treatments, and vaccines. The epidemic proceeded to a worldwide pandemic and WHO response.

The WHO-promoted responses of distancing, handwashing, and isolation failed, and concerns with drugs available in the Strategic National Stockpile (SNS) proved inadequate. The general public response deemed the FDA to be less than forthcoming, specifically with potential treatment options. Initial vaccine trials showed potential animal benefit but problems when given to people, including neurologic problems and death. The vaccine was heavily funded by NIH and HHS, which agreed with invoking the Public Readiness and Emergency Preparedness Act (PREP Act). All this should sound quite familiar to the reader, as the events of the actual COVID-19 pandemic followed this to a T. The only exception is that the pandemic in the scenario began in St. Paul, Minnesota, instead of Wuhan, China.

The study's report showed the public response to the vaccine was that it had been "rushed" and was inherently flawed. As a demonstration of the public's response not only to the vaccine but also to treatments, the report published the following headline examples of what should be expected using this rushed approach.

Fatality from the infectious SPARS virus, originally estimated at 4.7 percent was later recalculated to 1.1 percent, which was later revised to 0.6 percent. Efforts at encouraging people to become vaccinated proved to be less than successful due to resistance groups, following a CDC message:

> See your health care provider if you experience SPARS-like symptoms. . . . SPARS can kill you.

This led to a non-fear-based message:

> Seeing health care providers for SPARS-like symptoms can help you and your family members live long and happy lives.

The report emphasized problems with FDA, CDC, and HHS approaches, leading to a targeting of younger students with ads focused on encouraging them to become vaccinated or receive the government-approved treatment. Today, we would simply replace the drug in this war game, Kalocivir (fictitious for the purpose of this scenario), with Remdesivir or Paxlovid. The report even discussed what we later saw with COVID, including those who sought to receive vaccination first (head-of-the-line privileges) and "anti-vaccinators."

The 2017 SPARS bio–war game even had a power-grid failure, as we saw in 2021 in Texas.

IMPORTANT HEALTH ADVISORY!

Grant County Health District and Okanogan County Public Health will provide **COROVAX** for the general public from **8 AM - 7 PM** this **Saturday, July 18** at their local offices (see below).

GET VACCINATED AGAINST SPARS!

The report emphasized promoting vaccines and antivirals, while discouraging any other treatments proposed by nongovernment sources. It even went so far as to find a prominent individual who had opposed the government's approach to treatment and whose son became ill. She brought her son to the hospital, where he was given government-approved treatment and had a miraculous recovery that was "almost instantaneous." This was used as a method to encourage Americans to follow government advice. It proved successful.

The report discussed vaccine injury compensation, and the HHS decision to postpone evaluation of long-term vaccine injury side effects. This was followed by asking the population to wait and be patient as more data was collected and discussions about substance abuse and mental health problems in the country.

The funding partners for the Center for Health Security[46] that carried out this scenario are shown on the next page.

Chronologically, the 2017 SPARS scenario followed the 2005 US federal government official recognition of dual-use, gain-of-function biological agents and preceded the October 2019 Event 201 scenario, which was then followed by the November 2021 Nuclear Threat Initiative (NTI), which intentionally released a gain-of-function bioweapon, monkeypox (Mpox). The progression of these bioweapon war game scenarios showed the almost, if not exact, behavior seen with the SARS-CoV-2 and COVID pandemic response: a repeated set of behaviors that had previously been shown to fail but this time played out for real, allowing real-time results and data, from which the NTI war game was played out.

The NTI yielded the desired information needed for a one world order. Find a scenario, the circumstances under which the people of the United States and other financially stable and independent countries would cede control of their countries to the UN and WHO, using the treaties and International Health Regulations (IHR) that had already been signed. That would give the sovereign authority and power that countries like the United States have to the UN, WHO, and a one world order. The only thing that stands in the way is the United States Constitution and the American people.

The United States Constitution, Treaties, and Resolutions

The Founding Fathers knew that the United Kingdom ignored the treaties it signed, like most every other nation, and wanted to establish a country that honored its treaties. To do that, it was important that treaties couldn't just be signed off by the head of state; they required, at least indirectly, the approval of the people, through an elected body of representatives, thus making it impossible for the president to make a treaty with another country. The process wherein the United States of America enters into a treaty requires that the executive branch (the president) signs a treaty that must then be ratified by the legislative branch (Congress).

> Article II, Section 2, Clause 2:[47]
>
> He [the president] **shall have Power, by and with the Advice and Consent of the Senate, to make Treaties, provided two thirds of the Senators present concur**; and he shall nominate, and by and with the Advice and Consent of the Senate, shall appoint Ambassadors, other public Ministers and Consuls, Judges of the supreme Court, and all other Officers of the United States, whose Appointments are not herein otherwise provided for, and which shall be established by Law: but the Congress may by Law vest the Appointment of such inferior Officers, as they think proper, in the President alone, in the Courts of Law, or in the Heads of Departments. [emphasis added]

When Congress ratifies a treaty, it has the option of ratifying without limitation. However, if the Congress either wants to make a statement that does not alter the treaty—a declaration—or wants to place limits on a treaty—a reservation or understanding—Congress must do so at the time of ratification.[48] Accordingly, the SCOTUS has held this to be the law of the land.

> A declaration is not part of a treaty in the sense of modifying. . . . **The treaty is law. The Senate's declaration is not law**. . . . The **Senate's power under Article II extends only to the making of reservations** that require changes to a treaty before the Senate's consent will be efficacious. [emphasis added] *See INS v. Chadha,* 462 U.S. 919, 103 S.Ct. 2764, 77 L.Ed.2d 317 (1983). *Igartua-De La Rosa v. U.S.*, 417 F.3d 145, 190–91 (1st Cir. 2005)
>
> The courts must undertake their own examination of the terms and context of each provision in a treaty. . . . The Senate's declaration is not law. . . . The **Senate's power** under Article II **extends only** to the making of **reservations.** [emphasis added] *See INS v. Chadha,* 462 U.S. 919, . . . *Igartua-De La Rosa v. U.S.*, 417 F.3d 145, 190–91 (1st Cir. 2005)
>
> The United States ratified the Covenant on the **express understanding** that it was not self-executing and so did not itself create obligations enforceable in the federal courts. [emphasis added] See *supra,* at 2763. *Sosa v. Alvarez-Machain*, 542 U.S. 692, 735 (2004)
>
> [The] concluded . . . treaty [was] . . . self-executing . . . because **"the language of" the Spanish translation** (*brought to the Court's attention for the first time*) **indicated the parties' intent to ratify and confirm the landgrant "by force of the instrument itself."** [emphasis added] *Id.,* at 89. *Medellin v. Texas,* 552 U.S. 491, 514 (2008)
>
> There is something, too, which **shocks the conscience** in the idea that a treaty can be put forth as embodying the terms of an arrangement with a foreign power or an Indian tribe, **a material provision of which is unknown to one of the contracting parties**, and is **kept in the background to be used by the other** only when the exigencies of a particular case may demand it. . . . In short, . . . [that] **cannot be considered a part of the treaty.**" [emphasis added] *The Diamond Rings*, 183 U.S. 176, 183–84 (1901)

Neither Congress nor the president can unilaterally change a treaty once implemented.

> The Senate **has no right to ratify the treaty and introduce new terms.** . . . It may refuse its ratification, or make such ratification conditional upon the adoption of **amendments** to the treaty. [emphasis added] *The Diamond Rings*, 183 U.S. 176, 183 (1901)
>
> By the Constitution (art. 2, § 2) . . . the treaty must contain the whole contract between the parties, **and the power of the Senate is limited to a ratification of such terms as have already been agreed upon between the President, acting for the United States, and the commissioners of the other contracting power. The Senate has no right to ratify the treaty and introduce new terms into it**, which shall be obligatory upon the other power, although it may refuse its ratification, or make such ratification conditional upon the adoption of amendments to the treaty. [emphasis added] *The Diamond Rings*, 183 U.S. 176, 183–85 (1901)

In recent decades, presidents have frequently entered the United States into international agreements without the advice and consent of the Senate. These are called *executive agreements*. Though not brought before the Senate for approval, executive agreements are still binding on the parties under international law.[49]

Presumably, the legal basis for executive agreement authority resides in[50] Article II, § 2, Cl 3:

> The President shall have Power to fill up all Vacancies that may happen during the Recess of the Senate, by granting Commissions **which shall expire at the End of their next Session**. [emphasis added]

Under this US constitutional clause, it could be argued that executive agreements, in and of themselves, are potentially time limited.

However, independent of any time limitations that may or may not exist, many lower courts have held that "international [executive] agreements" are deemed non-self-executing:

> Determining whether a provision is self-executing or non-self-executing can be a complicated process. However, the Supreme Court has "deemed a treaty non-self-executing when the text manifest[s] an intent that the treaty . . . not be directly enforceable in US courts, or when the Senate conditioned its advice and consent on the understanding that the treaty was non-self-executing." Furthermore, while the Supreme Court has not opined directly on this topic, **"many [lower] courts and commentators agree that provisions in international agreements that would require the United States to exercise authority that the Constitution assigns to Congress exclusively must be deemed non-self-executing.**[51] [emphasis added]

The Congress of the United States has also very explicitly stated that the United States may withdraw from the WHO as stipulated.[52]

> For congressional-executive agreements, Congress may dictate how termination occurs in the statute authorizing or implementing the agreement. The legislation authorizing the United States to join the World Health Organization, for example, provides that the "United States reserves its right to withdraw from the Organization on a one-year notice."[53]

A review of the World Health Organization Resolution shows very clearly that it is not an executive agreement, as the president signed and Congress approved. Accordingly, neither the president nor Congress can unilaterally modify this resolution, which the International Health Regulations fall under, as we shall see as follows.

The World Health Organization Joint Resolution as signed by the thirty-third president of the United States, Harry S. Truman, on 14 June 1948 was approved by Congress on 16 June 1948:

> Sec. 5. In adopting this joint resolution, the Congress does so with the understanding that nothing in the Constitution of the World Health Organization in any manner commits the United States to enact any specific legislative program regarding any matters referred to in said Constitution.
>
> Signed by Speaker of the House of Representatives, Joseph W. Martin Jr. (Republican, Massachusetts) and President of the Senate pro tempore, Arthur H. Vandenberg (Republican, Michigan).

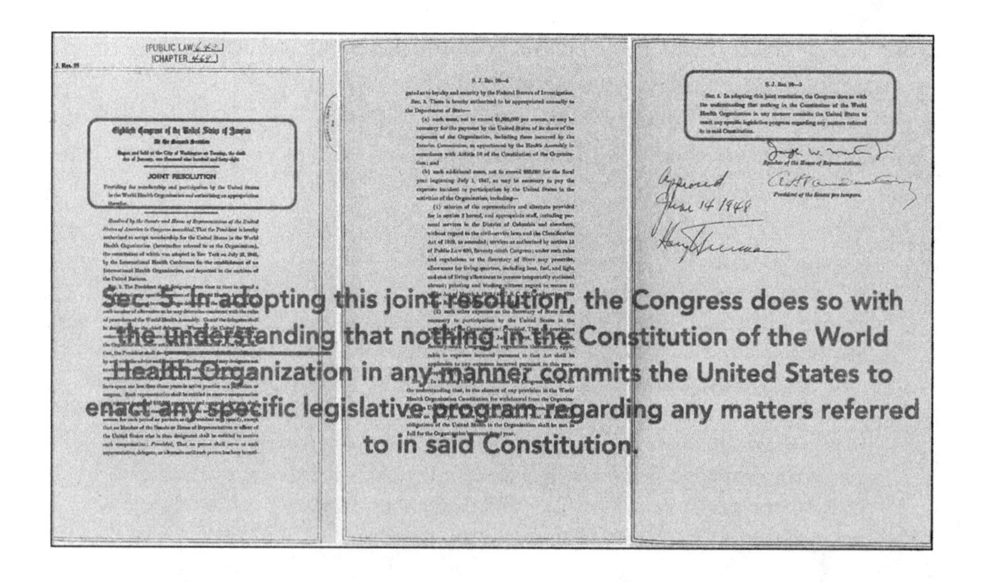
JOINT RESOLUTION

Sec. 5. In adopting this joint resolution, the Congress does so with the understanding that nothing in the Constitution of the World Health Organization in any manner commits the United States to enact any specific legislative program regarding any matters referred to in said Constitution.

Clearly, Congress did not intend for the WHO Resolution to impose any legal effect upon the United States, independent of anything in the WHO Constitution. Full-page copies of the WHO Resolution are provided in chapter 6, exhibit E.

On 2 February 2022[54] the forty-sixth president of the United States, Joseph Robinette Biden Jr., emphasized his plan to "reengage with the World Health Organization (WHO)".

> Last week, the United States once again demonstrated that commitment, by leading a successful decision at the WHO Executive Board meeting to strengthen the International Health Regulations (2005).

The IHR[55] (2005), appendix 2, beginning on page 60, displays the "Reservations and Other State Party Communications in Connection with the International Health Regulations." We can clearly see that the United States has placed certain reservations and understandings on the record:

> The Mission, by means of this note, informs the Acting Director-General of the World Health Organization that the Government of the United States of America accepts the IHRs, subject to the reservation and understandings referred to below.
>
> The Mission, by means of this note, and in accordance with Article 22 of the Constitution of the World Health Organization and Article 59(1) of the IHRs, submits the following **reservation** on behalf of the Government of the United States of America:
>
> The Government of the United States of America reserves the right to assume obligations under these Regulations in a manner **consistent with its fundamental principles of federalism**. With respect to obligations concerning the development, strengthening, and maintenance of the core capacity requirements set forth in Annex 1, these Regulations shall be implemented by the Federal Government or the state governments, as appropriate and in accordance with our Constitution, to the extent that the implementation of these obligations comes under the legal jurisdiction of the Federal Government. To the extent that such obligations come under the legal jurisdiction of the state governments, the Federal Government shall bring such obligations with a favorable recommendation to the notice of the appropriate state authorities.
>
> The Mission, by means of this note, also submits **three understandings** on behalf of the Government of the United States of America. The **first** understanding relates to the application of the IHRs to incidents involving natural, accidental or deliberate release of chemical, biological or radiological materials:
>
> In view of the definitions of "disease," "event," and "public health emergency of international concern" as set forth in Article 1 of these Regulations, the notification requirements of Articles 6 and 7, and the decision instrument and guidelines set forth in Annex 2, the United States understands that States Parties to these Regulations have assumed an obligation to notify to WHO potential public health emergencies of international concern, irrespective of

origin or source, whether they involve the natural, **accidental or deliberate** release of biological, chemical or radionuclear materials.

The **second** understanding relates to the application of Article 9 of the IHRs:

Article 9 of these Regulations obligates a State Party "as far as practicable" to notify the World Health Organization (WHO) of evidence received by that State of a public health risk occurring outside of its territory that may result in the international spread of disease. Among other notifications that could prove to be impractical under this article, **it is the United States' understanding that any notification that would undermine the ability of the U.S. Armed Forces to operate effectively in pursuit of U.S. national security interests would not be considered practical for purposes of this Article.**

The **third** understanding relates to the question of whether the IHRs create judicially enforceable private rights. Based on its delegation's participation in the negotiations of the IHRs, the Government of **the United States of America does not believe that the IHRs were intended to create judicially enforceable private rights:**

The United States understands that the provisions of the Regulations do not create judicially enforceable private rights.

[emphasis added to all paragraphs]

The fundamentals of federalism, the basis of the US Constitution, is defined as shown in the following graphic. Pursuant to the Tenth Amendment of the US Constitution, this "shield and sword" is designed to prevent the federal government from overextending its constitutional authority, an authority reserved for the states respectively, or to the people.

The reservations and understandings placed upon the WHO and IHR by the United States cannot unilaterally be changed by either the president or Congress. It doesn't matter which political party is in office. All political parties, courts, lawyers, and major parts of the medical system in favor of promoting control of US health care under the WHO and the IHR are doing so in direct opposition of the US Constitution and have demonstrated their unquenchable lust for power and control using the WHO and IHR to further their control over American health care.

Who Is in Control of the World Health Organization?

Given the efforts by too many people to surrender control of the United States to the WHO, it is important to simply ask, Who are the people behind the WHO and consequently the IHR?

In April 2020, the largest funder of the WHO was the United States, followed by the Bill & Melinda Gates Foundation,[56] which provided more money to the WHO than any other nation or even than the Global Alliance for Vaccines and Immunization (GAVI Alliance).[57]

Concerns over private funding of the WHO[58] cannot merely be focused on the Gates Foundation,[59] given the ever-increasing role that banks, philanthropic

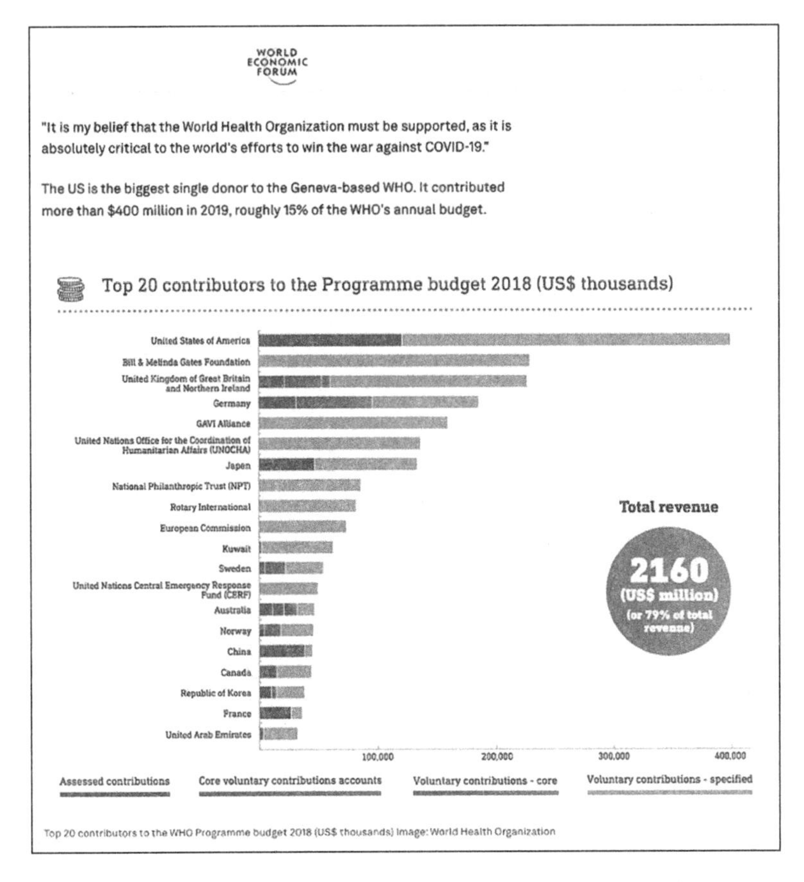

Top 20 contributors to the WHO Programme budget 2018 (US$ thousands) Image: World Health Organization

foundations, nongovernmental organizations, intergovernmental organizations, private sector, and "partnerships" like the World Economic Forum (WEF) are playing in the funding and, consequently, influence over the WHO.

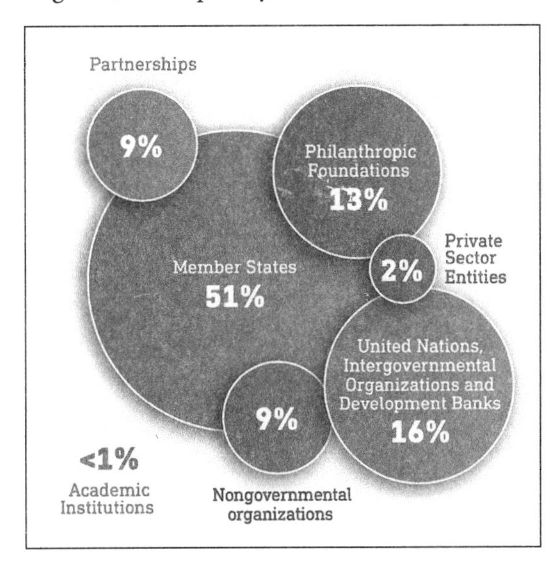

On 13 June 2019, the UN and WEF signed an agreement.[60]

> The United Nations–World Economic Forum
>
> Strategic Partnership Framework for the 2030 Agenda
>
> The United Nations and the World Economic Forum are committed to accelerate implementation of the 2030 Agenda for Sustainable Development—the world's plan for peace, prosperity, and a healthy planet.
>
> Recognising the ambition of the 2030 Agenda, the United Nations and the World Economic Forum seek to strengthen their partnership by focusing on jointly selected priorities and by pursuing a more strategic and coordinated collaboration, by leveraging their respective strengths and broadening their combined impact, building on existing and new collaborations by UN entities.
>
> The areas addressed in the agreement include (1) financing, (2) climate change, (3) health, (4) digital cooperation, (5) gender, and (6) education.
>
> Subject to their respective mandates, competencies, institutional settings and legal and operational frameworks, the United Nations and the World Economic Forum may inform and consult with each other, as appropriate, on additional issues of mutual interest in which cooperation may foster their respective and collective purposes.
>
> The partnership between the UN and the Forum is well placed to facilitate and encourage the multi-stakeholder engagement necessary to accelerate progress on the 2030 Agenda. The success of the partnership depends on strategic guidance, coordinated and coherent institutional arrangements for planning, collaboration, and a strong mechanism for knowledge development, learning, and accountability.
>
> The leadership of the UN and the Forum will meet annually to review the partnership. In parallel, the technical teams of the UN and the Forum will meet to seek to ensure effective implementation of commitments assumed under this Strategic Partnership Framework, including by enhancing coordination between the teams at the regional and country levels.

What if the People You Trust Are the People Causing the Problem?

It is clear that in signing the WHO and the IHR, the United States did not, and has not, constitutionally surrendered its sovereignty or citizen rights to the WHO comprised of people who were not elected to represent *We the People*, and the International Health Regulations. As such, nothing Congress nor the president can do gives the WHO or IHR authority over what happens in the United States, and it is up to the American people to remind our elected politicians and judges, who receive their salaries from our taxes, that they work for us and not the WHO.

Since the beginning of human history, men, women, and families have grouped together to take control over others. Sometimes enslaving people, sometimes torturing them, other times killing them. Their best argument is that they are doing this for the betterment of humanity. Purging the world of inferior people, using our differences to turn us against each other, all the while usurping more power and control over other people. It is their belief that they and they alone know what is best for everyone; just as the dictators and narcissists of old believed.

The SPARS, Event 201, and NTI Biological war games showed just how easy it is to manipulate people and drive the selected narrative—by deciding what information is correct and what information is misinformation. These war-game scenarios began with either a coronavirus or a gain-of-function biological agent, coupled with food and fuel shortages, lack of access to medicine, mandates, and the hope of a vaccine cure. Fear was used to turn people against each other, using differences in political parties, race, religion, region of the country, education, confidence in medicine and vaccines, and desire, to make us turn a blind eye toward problems in the hope that the scenario will somehow pass us by. Hopefully this sounds familiar, because this is exactly what has played out since December 2019—seven weeks after the DoD paid for the COVID report from Ukrainian Biolabs. More on the DoD and Ukrainian labs later in the book, including the final chapter on obstruction of justice.

On 5 July 2021 I began my four-and-a-half-hour presentation to a group of people in Dallas, Texas[61] with the following slide, which rings as true today as it did then.

When the people[62] who are funding gain-of-function biological-weapons development are the same people funding genetic vaccines, biological war-game scenarios, the WHO, gene therapy, and the military-industrial complex, then we have a problem! This is same problem Presidents Eisenhower and Kennedy warned us about.

It is time the American people, indeed people around the world, stand up against these criminals, whose lust for power threatens to return us to the atrocities we witnessed seventy-five years ago and multiple times since. The consequences of the actions of these people have resulted in more deaths and injuries than the United States military has experienced since 1776; more deaths worldwide than the Nazis inflicted on men, women, and children in concentration camps; more genetic manipulation and eugenics than Mengele's experiments—both before and in the concentration camps; more destruction and usurping of human rights than any dictator, emperor, or narcissistic leader in history.

Let us reflect on the words of a theologian born in Germany in 1892 who reminds us of just how easy it is to lose everything important to us, by failing to act:

> First they came for the Communists
> And I did not speak out
> Because I was not a Communist
> Then they came for the Socialists
> And I did not speak out
> Because I was not a Socialist
> Then they came for the trade unionists
> And I did not speak out
> Because I was not a trade unionist
> Then they came for the Jews
> And I did not speak out
> Because I was not a Jew
> Then they came for me
> And there was no one left
> To speak out for me.
>
> Martin Niemöller
> "First They Came"[63]

For us, like for our ancestors, the men and women who gave us the Constitution of the United States of America, the battle against these people will continue until we force them back into the primordial ooze from which they came, along with their unquenchable lust for power and control. That includes their usurping, of medicine, which we will look at in our next chapter.

CHAPTER 4

The Hijacking of Medicine

The four oldest professions on the planet are prostitution, theology, law, and medicine, and although the first probably gave rise to politicians and the second gave rise to organized religion, we shall concentrate on the latter two in this chapter. But first, my apologies to prostitutes, who appear to be working for a living, doing what they actually say they will do. So, perhaps the American people, working for a living, are being prostituted by the pimps elected to office.

Ancient Medicine (3000 BCE to 500 BCE)

Before there was organizational thought and training of individuals in the practice of medicine, there were people who, through experience, found cures for ailments and illness. This information was eventually put in writing and saved for people to learn from, rather than reinventing the wheel of medicine with every generation. Some of my most valuable textbooks include such texts and discussions of differences in people:

- *The Healing Art—Being a Book of Medical Recipes*, Reuben Andrews, MD, 1842
- *General Therapeutics and Materia Medica: Adapted for a Medical Text-Book*, Robley Dunglison, MD, Blanchard and Lea, 1853
- *History of Medicine from Its Origin to the Nineteenth Century*, P. V. Renouard, MD, Moore, Wilstach, Keys, 1856
- *On Wounds and Fractures*, Guy de Chauliac, ca. 1363 AD, Translated by W. A. Brennan, Royal University of Ireland, 1923
- *Preface to Eugenics, Revised Edition*, Frederick Osborn, Honorary Associate in Anthropology, American Museum of Natural History, Harper & Brothers Publishers, 1940

In the Torah or Christian Bible, very little is written of the need for medicine. Perhaps because, as discussed previously, illness and short lives were not an issue. There are even passages in these texts that admonish people for seeking medical attention, encouraging people to consider health problems a reflection of their distancing themselves from their God.

> I am Adonai, your healer.[1]

There are instances of kings being disapproved of for seeking the help of physicians, rather than religious leaders:

> In the thirty-ninth year of his reign Asa was diseased in his feet, and his disease became severe. Yet even in his disease he did not seek the Lord, but sought help from physicians.[2]

In fact, even today some people will tell you that "God does not make mistakes" and that any illness reflects a spiritual problem, for which one needs to repent.

I do not consider illness merely a reflection of some "sin" committed by someone. Not only do I consider this an excuse—blaming God for doing something to you[3]—but I also think it gives whatever your version of god is a bad name.

The term *stroke*—which is technically a cerebral vascular accident (CVA), meaning your brain has suffered a loss of blood flow resulting in damage—means God was so displeased with what you did that he struck you down. I think it's time mankind stopped trying to find scapegoats to take the blame for mankind's actions. God did not make the bioweapon SARS-CoV-2, people did.

With the passage of time, we have appreciated that our theological texts do look at ways to avoid or prevent disease, including certain health practices that should be avoided, as well as certain practices that should be implemented, including but not limited to the preparation of food and cleanliness.

Ancient cultures in the Mesopotamian region (e.g., Iran, Iraq, Kuwait, Syria, and Turkey) combined medicine, science, magic, and religion. There are records of Babylonian medical textbooks[4] focusing on the diagnosis and treatment of disease, dating back to the reign of King Adadaplaiddina (1069–1046 BCE). In the days of Egypt, Herodotus (ca. 484–425 BCE) believed[5] that a physician could only be an expert in one disease.[6]

> The practice of medicine is so specialized among them that each physician is a healer of one disease and no more.

Add in components of traditional Chinese medicine (TCM) from Taoist physicians and the *Huangdi Neijing*, written somewhere between the fifth and third century BCE covering herbal treatments and acupuncture, along with the understanding of energy chakras and herbal Ayurveda treatments from 600 BCE, and we have laid the foundations for modern medicine.

Modern Medicine

When you think of medicine's history, Greek physician Hippocrates of Kos (460 to 377 BCE) and the Hippocratic Oath often comes to mind. Taken by physicians upon completion of their medical training, this oath is frequently misquoted.

The original Hippocratic Oath is the seminal document addressing the ethics of medical practice. It predates all other oaths in existence and all governments. There are three versions of the Hippocratic Oath,[7] showing changes that have occurred with time. There is also the Declaration of Geneva Physicians' Oath—developed and adopted by the World Medical Association at Geneva in 1948—intended to address the ethical breaches occurring during Nazi Germany. The oath my medical class voted to take was the Geneva Oath.

The differences between the oaths are as follows:

1. The classic oath involves the physician, patient, and gods.
2. The revised oath involves the physician and patient; relieving the gods[8] of a few responsibilities.
3. The modern oath is an effort to put the classic oath into modern English.
4. The Declaration of Geneva is similar to the revised oath.

The Modern Oath and Declaration of Geneva Oath are shown here for you to compare. It is important for everyone to see that the oath I took is to my patients and those who trained me—not to a government, government agency, court, insurance company, BigPharma, or something else. I have kept this oath and will take it with me to my grave.

The Hippocratic Oath
(Modern Version)

I SWEAR in the presence of the Almighty and before my family, my teachers and my peers that according to my ability and judgment I will keep this Oath and Stipulation.

TO RECKON all who have taught me this art equally dear to me as my parents and in the same spirit and dedication to impart a knowledge of the art of medicine to others. I will continue with diligence to keep abreast of advances in medicine. I will treat without exception all who seek my ministrations, so long as the treatment of others is not compromised thereby, and I will seek the counsel of particularly skilled physicians where indicated for the benefit of my patient.

I WILL FOLLOW that method of treatment which according to my ability and judgment, I consider for the benefit of my patient and abstain from whatever is harmful or mischievous. I will neither prescribe nor administer a lethal dose of medicine to any patient even if asked nor counsel any such thing nor perform the utmost respect for every human life from fertilization to natural death and reject abortion that deliberately takes a unique human life.

WITH PURITY, HOLINESS AND BENEFICENCE I will pass my life and practice my art. Except for the prudent correction of an imminent danger, I will neither treat any patient nor carry out any research on any human being without the valid informed consent of the subject or the appropriate legal protector thereof, understanding that research must have as its purpose the furtherance of the health of that individual. Into whatever patient setting I enter, I will go for the benefit of the sick and will abstain from every voluntary act of mischief or corruption and further from the seduction of any patient.

WHATEVER IN CONNECTION with my professional practice or not in connection with it I may see or hear in the lives of my patients which ought not be spoken abroad, I will not divulge, reckoning that all such should be kept secret.

WHILE I CONTINUE to keep this Oath unviolated may it be granted to me to enjoy life and the practice of the art and science of medicine with the blessing of the Almighty and respected by my peers and society, but should I trespass and violate this Oath, may the reverse by my lot.

Physician's Oath

From the Declaration of Geneva

At the time of being admitted as a member of the medical profession

I solemnly pledge myself to consecrate my life to the service of humanity.

I will give to my teachers the respect and gratitude which is their due;

I will practice my profession with conscience and dignity;

The health of my patient will be my first consideration;

I will respect the secrets which are confided in me;

I will maintain, by all the means in my power, the honor and the noble traditions of the medical profession;

I will not permit considerations of religion, nationality, race, party politics or social standing to intervene between my duty and my patient;

I will maintain the utmost respect for human life; even under threat, I will not use my medical knowledge contrary to the laws of humanity.

I make these promises solemnly, freely and upon my honor.

Oath sworn on May sixteenth
Nineteen hundred and eighty-six
The University of Iowa
College of Medicine

Medicine Is Not a Business—It's a Profession, a Calling

Speaking as a physicist, physician (internist, cardiologist, nuclear cardiologist), and attorney, I can say that there is no greater opportunity to serve humanity than to take care of someone dying and keep them from dying; that is what drove me to cardiology to begin with—the ability to make a difference. My work has focused on trying to figure out why people are dying and what can be done to not only keep them alive but also improve their health, vitality, and quality of life. It has also largely centered around academic medical centers and the care of our military men and women, including my involvement in evaluating and reporting[9] on the care, or lack thereof, delivered to our military and veteran troops.

Although the eugenicists we are dealing with in this book may believe they almost have the power of physical immortality, I can assure you that when people are dying, they are not looking for a geneticist, eugenicist, elected official, government agency, the FDA, HHS, CDC, WHO, or any other three-letter agency to keep them alive. They are looking to their physician, the person who took the oath that transcends all other oaths in human history.

How Medicine Was Hijacked

One of the fundamental lessons I tried to teach my children was for them to make the world a better place, to raise the standards of the world for everyone. My children understand that some people focus on making the world a better place for themselves—making themselves look and feel better. This allows them to provide the illusion of a better world at the cost of dragging other people down. That's not really a better world, even if those people believe they are moving up in the world. Alternatively, people can genuinely make the world better for themselves by working hard at self-improvement and, in so doing, helping others rise with them into a better world.

You can either be the C student stealing the A student's project, so that person becomes a C student and you look relatively better, or you can be the A student who helps others understand and raise their C to a B or an A. The choice, and how you approach the world, is completely up to you. Medicine faces the same set of dilemmas and choices.

Some people who want to control medicine and physicians, for their own benefit, not for the benefit of medicine, doctors, or patients. We have seen too much of this, and it accounts for some, but not all, of what has been happening for many decades. As you can see, this has accelerated during the last several years.

We are going to take a look at some of the players involved in the hijacking of physicians and medicine, and in chapters 6–8, we are going to address what you and I need to do to hold these people accountable and to restore medicine in the United States of America and the world.

Terminology

Before I discuss the constitutional limitations of the US federal government—which I am certain politicians will not agree with, but then again, Hitler didn't see any limitations to his power or that of the Third Reich—I want to touch on two topics. First is the difference between a democracy and a republic, of which the United States is the latter, and second is the fundamental difference between a declaration and a constitution.

1. The Fundamental Difference Between a Democracy and a Republic

Democracies, of which Greece was the last, are forms of government in which the people directly participate in the act of governing. Republics are forms of government in which people elect others to act on their behalf. Republics reduce the workload of the citizen but are created at the expense of being able to trust those they elect to protect their rights. In both types of government, the supreme power resides in the people. Many of you have undoubtedly recited the Pledge of Allegiance, which includes the words "and to the republic, for which it stands."

Outside Independence Hall on 17 September 1787, a woman asked Benjamin Franklin, "Well, doctor, what have we got, a republic or a monarchy?"

He responded, "A republic, if you can keep it."[10]

2. The Fundamental Difference Between a Declaration and a Constitution

The term *declaration* means exactly what it says; you are declaring something. The Declaration of Independence was essentially plagiarized by Thomas Jefferson from two different sources. The first source, representing two thirds of Declaration of Independence, came from the Flemish Normans' *Act of Abjuration* (26 July 1581, Plakkaat van Verlatinge). The remaining third came from complaints raised by the people of Britain in the mid-1600s, declaring certain grievances with the then Kings Charles I and Charles II[11] which would ultimately become the English Bill of Rights.

The declaration did not establish a country, laws, or a governing body. When it was made 4 July 1776, the colonies were at war with England. Not until Monday 17 September 1787 did the Constitution of the United States of America get signed.[12] As with treaty law and federal law, the Constitution and its amendments must be both signed and ratified. Prior to the ratification of the Constitution, the signing of it was carried out by those in attendance at Independence Hall in September 1787. Ratification[13] by the representatives of the people (Republic) from nine of the thirteen states was completed 21 June 1788, when the Constitution became the supreme law of the United States of America.

The Constitution and its amendments replaced the Articles of Confederation, which had failed to unify the people behind the military and its leader, General George Washington. The Constitution stripped the citizens of power to enable Washington to run the military. The Founding Fathers knew the country could never be the one they envisioned, in which the people governed, not the government or military, without the amendments.

Appreciating what other countries had done, the Founding Fathers amended the US Constitution. Beginning on 24 August 1789 and finishing 15 December 1791, they approved ten of seventeen proposed amendments. These ten are known as the Bill of Rights,[14] and they restored rights back to the people from whom they had been taken.

The United States Constitution[15] established three branches of the government (legislative, executive, judiciary),[16] in addition to Article IV (relationships between each state and other states, in addition to the relationship between states and the federal government); Article V (the amendment process); Article VI (supremacy clause), establishing the Constitution, treaties and federal law as supreme over other laws; and Article VII (ratification), defining the process for establishing a new form of government.

Within Article I (legislative branch) of the US Constitution, the Founding Fathers detailed what powers explicitly were being granted to the legislative (US Congress) branch of the government. Congress was established as the branch of the federal government responsible for writing bills that could be passed into law, following the signing by the elected executive (Article II) branch of the government. Alternatively, if the executive (president) refuses to sign, the bill (statute) could become law with a two-thirds majority vote[17] by Congress. The judiciary (Article III), known as the Supreme Court of the United States[18] (SCOTUS), is the final arbitrator of the constitutionality of any legislation (law) passed by Congress—whether singed by the president or not.

This power of the SCOTUS was determined by *Marbury v. Madison*, 5. US (1 Cranch) 137 (1803). The importance of *Marbury v Madison* cannot be overstated. Simply put, according to Chief Justice Marshall[19] and the SCOTUS, if what is expressly stated as a power in the US Constitution disagrees with a statute passed by Congress, it is the Constitution that ultimately wins—Congress loses.

With that, we will now turn our attention to what the Constitution expressly states the powers of Congress and the president to be. Anything that disagrees with this, according to *Marbury v. Madison*, stare decisis,[20] is unconstitutional and, thus, invalid or perhaps even criminal.

WHAT ARE THE EXPRESSED POWERS OF CONGRESS AND THE PRESIDENT?

I. Article I, § 8, clauses 1–17: The enumerated (expressed) powers of Congress

> **Clause 1:** The Congress shall have Power to lay and collect Taxes, Duties, Imposts, and Excises, to pay the Debts and provide for the common Defence and general Welfare of the United States; but all Duties, Imposts and Excises shall be uniform throughout the United States;
> **Clause 2:** To borrow Money on the credit of the United States;

Clause 3: To regulate Commerce with foreign Nations, and among the several States, and with the Indian Tribes;
Clause 4: To establish a uniform Rule of Naturalization, and uniform Laws on the subject of Bankruptcies throughout the United States;
Clause 5: To coin Money, regulate the Value thereof, and of foreign Coin, and fix the Standard of Weights and Measures;
Clause 6: To provide for the Punishment of counterfeiting the Securities and current Coin of the United States;
Clause 7: To establish Post Offices and Post Roads;
Clause 8: To promote the Progress of Science and useful Arts, by securing for limited Times to Authors and Inventors the exclusive Right to their respective Writings and Discoveries;
Clause 9: To constitute Tribunals inferior to the Supreme Court;
Clause 10: To define and punish Piracies and Felonies committed on the high Seas, and Offences against the Law of Nations;
Clause 11: To declare War, grant Letters of Marque and Reprisal, and make Rules concerning Captures on Land and Water;
Clause 12: To raise and support Armies, but no Appropriation of Money to that Use shall be for a longer Term than two Years;
Clause 13: To provide and maintain a Navy;
Clause 14: To make Rules for the Government and Regulation of the land and naval Forces;
Clause 15: To provide for calling forth the Militia to execute the Laws of the Union, suppress Insurrections and repel Invasions;
Clause 16: To provide for organizing, arming, and disciplining the Militia, and for governing such Part of them as may be employed in the Service of the United States, reserving to the States respectively, the Appointment of the Officers, and the Authority of training the Militia according to the discipline prescribed by Congress;
Clause 17: To exercise exclusive Legislation in all Cases whatsoever, over such District (not exceeding ten Miles square) as may, by Cession of particular States, and the Acceptance of Congress, become the Seat of the Government of the United States, and to exercise like Authority over all Places purchased by the Consent of the Legislature of the State in which the Same shall be, for the Erection of Forts, Magazines, Arsenals, dock-Yards, and other needful Buildings;

The Implied Power of Congress

The final clause of Article I, § 8, clause 18 is known as the Necessary and Proper Clause is the source of the implied powers of Congress.

Clause 18: To make all Laws which shall be necessary and proper for carrying into Execution the foregoing Powers, and all other Powers vested by this Constitution in the Government of the United States, or in any Department or Officer thereof.

* * *

The Tenth Amendment to the US Constitution—Ratified 15 December 1791
Powers Not Specified: The Tenth Amendment

The powers not delegated to the United States by the Constitution, nor prohibited by it to the states, are reserved to the states respectively, or to the people.

All powers not granted to the US Congress by Article I, § 8 are left to the states. Worried that these limitations to the powers of the federal government were not clearly enough stated in the original Constitution, the first Congress adopted the Tenth Amendment, which clearly states that all powers not granted to the federal government are reserved to the states or the people.

Congress's Commerce Clause Powers

In passing many laws, Congress draws its authority from the commerce clause of Article I, § 8, granting Congress the power to regulate business activities "among the states."

Over the years, Congress has relied on the commerce clause to pass environmental, gun-control, and consumer-protection laws because many aspects of business require materials and products to cross state lines.

However, the scope of the laws passed under the commerce clause is not unlimited. Concerned about the rights of the states, the US Supreme Court in recent years has issued rulings limiting the power of Congress to pass legislation under the commerce clause or other powers specifically contained in Article I, § 8. For example, the Supreme Court has overturned the federal Gun-Free School Zones Act of 1990 and laws intended to protect abused women, on the grounds that such localized police matters should be regulated by the states.

Perhaps the most important powers reserved to Congress by Article I, § 8 are those to create taxes, tariffs, and other sources of funds needed to maintain the operations and programs of the federal government and to authorize the expenditure of those funds. In addition to the taxation powers in Article I, the Sixteenth Amendment authorizes Congress to establish and provide for the collection of a national income tax. The power to direct the expenditure of federal funds, known as the "power of the purse," is essential to the system of checks and balances by giving the legislative branch great authority over the executive branch, which must ask Congress for all of its funding and approval of the president's annual federal budget.

* * *

II. Article II, § 2 and 3: The enumerated powers of the president

Section 2: The President shall be Commander in Chief of the Army and Navy of the United States, and of the Militia of the several States, when called into the actual Service of the United States; he may require the Opinion, in writing, of the principal Officer in each of the executive Departments, upon any Subject relating to the Duties of their respective Offices, and he shall have Power to grant Reprieves and Pardons for Offences against the United States, except in Cases of Impeachment.

He shall have Power, by and with the Advice and Consent of the Senate, to make Treaties, provided two thirds of the Senators present concur; and he shall nominate, and by and with the Advice and Consent of the Senate, shall appoint

> Ambassadors, other public Ministers and Consuls, Judges of the supreme Court, and all other Officers of the United States, whose Appointments are not herein otherwise provided for, and which shall be established by Law; but the Congress may by Law vest the Appointment of such inferior Officers, as they think proper, in the President alone, in the Courts of Law, or in the Heads of Departments.
>
> The President shall have Power to fill up all Vacancies that may happen during the Recess of the Senate, by granting Commissions which shall expire at the End of their next Session.
>
> **Section 3:** He shall from time to time give to the Congress Information of the State of the Union, and recommend to their Consideration such Measures as he shall judge necessary and expedient; he may, on extraordinary Occasions, convene both Houses, or either of them, and in Case of Disagreement between them, with Respect to the Time of Adjournment, he may adjourn them to such Time as he shall think proper; he shall receive Ambassadors and other public Ministers; he shall take Care that the Laws be faithfully executed, and shall Commission all the Officers of the United States.
>
> Nowhere in the expressed powers of either the legislative branch (Congress) or the executive branch (president) is there the mention of giving the federal government power over health care.

When the expressed powers of congress and the president are coupled with the Tenth Amendment[21] to the United States Constitution,

> The powers not delegated to the United States by the Constitution, nor prohibited by it to the states, are reserved to the states respectively, or to the people.
>
> The Tenth Amendment helps to define the concept of federalism, the relationship between federal and state governments. As federal activity has increased, so too has the problem of reconciling state and national interests as they apply to the federal powers to tax, to police, and to regulate, such as wage and hour laws, disclosure of personal information in recordkeeping systems, and laws related to strip-mining.

and combined with the SCOTUS ruling established in *Cruzan v. Director, Missouri Department of Health*, where the SCOTUS specifically held that competent US citizens cannot be forced to undergo a medical treatment, one can only conclude that forced vaccination of a competent person, in light of these limitations in presidential and congressional powers, is a violation of the 14th amendment to the US constitution.

Cruzan v. Director, Missouri Department of Health, 497 U.S. 261 (1990)[22]

Facts of the Case

In 1983 Nancy Beth Cruzan was involved in an automobile accident that left her in a "persistent vegetative state." She was sustained for several weeks by artificial feedings through an implanted gastronomy tube. When Cruzan's parents attempted to terminate the life-support system, state hospital officials refused to do so without court approval. The Missouri Supreme Court ruled in favor of the state's policy over Cruzan's right to refuse treatment.

The Question Posed

Did the due process clause of the Fourteenth Amendment permit Cruzan's parents to refuse life-sustaining treatment on their daughter's behalf?

Conclusion

In a 5-to-4 decision, the Court held that while individuals enjoyed the right to refuse medical treatment under the due process clause, incompetent persons were not able to exercise such rights. Absent "clear and convincing" evidence that Cruzan desired treatment to be withdrawn, the Court found the State of Missouri's actions designed to preserve human life to be constitutional. Because there was no guarantee family members would always act in the best interests of incompetent patients and because erroneous decisions to withdraw treatment were irreversible, the Court upheld the state's heightened evidentiary requirements.

There is absolutely no valid expressed US constitutional authority for any branch of the federal government, be it Congress or the president, to mandate health care[23] in the United States of America; in particular, there is no such authority to force or coerce a competent US citizen to undergo any specific treatment as held by the SCOTUS.

A RETURN TO THE EUGENIC GREEKS

For almost eighty years, American medicine was practiced with little government oversight, with practitioners following their Hippocratic Oath. Physicians saw patients using methods of diagnosis and treatment that had been handed down for thousands of years. The number one cause of death was and remained infectious disease until World War II. Prior to the introduction of germ theory[24] in 1865 by Louis Pasteur and later by Robert Koch, the proposal of handwashing with chlorinated solutions in obstetrical clinicals for disinfection by Dr. Ignaz Semmelweis[25] in 1847 and the discovery of penicillin by Sir Alexander Fleming in 1928 allowing for the treatment of bacteria, little could be done except for the isolation of the diseased person from the healthy. Following the plague of Athens, the Greek historian Thucydides (circa 460–400 BCE) wrote about diseases spreading from person to person:

> There was the awful spectacle of men dying like sheep, through having caught the infection in nursing each other. This caused the greatest mortality. On the one hand, if they were afraid to visit each other, they perished from neglect; indeed many houses were emptied of their inmates for want of a nurse; on the other, if they ventured to do so, death was the consequence.[26]

Medical and scientific advancements[27] are frequently unwelcome at first, viewed as disruptive, despised by some and met with uncompromising resistance by others, only to later be welcomed as if they had never been rejected in the first place. While not fully appreciated at the time, the Hungarian physician Ignaz Semmelweis was concerned over infant mortality in hospitals. He tried to change the practice of medicine by encouraging handwashing of hospital personnel between each patient seen and examined. Although Semmelweis was rejected by the medical community of the day, costing him his medical practice and eventually, many believe, leading to an early death, once handwashing was practiced, stopping the transmission of infections, infant mortality in hospitals dropped from 18 percent to 2.2 percent during the first year.[28]

The early to mid-1800s saw changes in many fields of science and medicine, some for the better and some for the worse. The commonly taught theory of the Bohr model, with electrons (particles) orbiting the nucleus of an atom, is not exactly correct. Electrons are actually more energy than particle. For those of you interested in learning more about this, a good place to begin is the double-slit experiment that shows electrons behave both like matter and energy (waves). For a simple explanation of this, I refer you to Study.com,[29] where you can read and watch more about electrons.

The mid-1800s saw positive change, with an improved understanding of bacterial, fungal, and parasitic infections. Beginning with the work of Anton van Leeuwenhoek, who pioneered early microscopes, scientists were finally able to understand the cause of various infectious diseases, although it would not be until the beginning of the twentieth century when German physician Paul Ehrlich and Scottish physician Alexander Fleming would discover arsphenamine and penicillin, respectively. Given the information you have learned about the Ukrainian biolabs, you might be interested in knowing that the term *antibiotic* was first used by the Ukrainian American inventor and biologist Selman Waksman,[30] who is reported to have discovered more than twenty antibiotics.

Additional advancements during the mid-1800s in the field of infectious diseases included the frequently misunderstood work of British physician Gideon Mantell, French microbiologist Louis Pasteur, German bacteriologist Robert Koch, and British surgeon Joseph Lister. Their work helped pioneer our understanding of various bacteria and treatments, with Koch's postulates frequently being used to misrepresent germ theory.[31]

Koch's postulates were developed around the time of the Bohr model for the atom, and like the Bohr model, it is partially correct in that the Koch postulates are useful for bacteria but do not address other infectious diseases. The diagnosis of viruses, including SARS-CoV-2, is a two-step process. Unlike bacteria, which are large enough to be seen with a simple light microscope, viruses are too small to be seen without the use of an electron microscope. Although light microscopes can make bacteria, cells, and so forth visible without harming the object, electron microscopes actually kill the tissue (cells, viruses, bacteria, and so forth), making it impossible to do anything else with what you are looking at.

To identify a virus from a sample taken from the lungs or anywhere else in the body requires you to first clean up the sample to remove the garbage—what you are not interested in—from the viral sample. This is the same process used when you open a puzzle box, shake out all the pieces, and throw away the dust and debris in the box.

The viral sample is then sent for two different tests: The first is for Sanger Sequencing.[32] It allows the genetic material, which defines every living thing, including viruses to be replicated, producing fragments of the DNA of different lengths that can then be pieced together like the pieces of a jigsaw puzzle, to give you the final picture of the viral DNA or RNA. The other testing uses the electron microscope. Here we can see what the virus, or something else, looks like, but in the process, the virus is killed.[33] This change in the appearance of HIV allowed my friend and colleague, Professor Luc Montagnier, and me to conclude that the protease inhibitors used for HIV[34] made the AIDS[35] infection worse, due to denuding of the gp120 part of HIV, resulting in changes in the appearance of the HIV virus under the electron microscope. Professor Montagnier and I speculated, and have ultimately been proven correct, that the use of Paxlovid[36] and other 3CL protease inhibitors would make SARS-CoV-2 worse, with rebound infections[37] frequently reported.[38]

In addition to the infectious-disease advancements of the mid-1800s, other scientists, as discussed in chapter 2, were trying to advance their understanding of differences in species. This work, while significant in opening up the discussion of genetic differences and eventually understanding DNA, returned the British and American societies to the eugenic approach of the Greeks. Differences were seen as weaknesses that should not be condoned, and more important, if the opportunity presented itself, such people should be eliminated, as the Greeks did to a quarter of their babies, who were deemed inferior.

The true power of the eugenics movement in Britain and the United States—which you might think is gone but has merely changed names—is seen in the records of the likes of the Rockefellers and their foundation. The materials in appendix E provide just a sampling of the efforts the Rockefeller family went through to fund eugenics. The Rockefellers' efforts extended throughout the United States, England, and Germany, giving rise to Adolf Hitler's obsession with eradicating those of Jewish faith from Germany. Ignorant of the fact that Judaism is a religion and not genetic, Hitler was encouraged by American eugenicists and others discussed in chapter 2.

By the end of the Second World War, the United States and Russia frantically sought out the German scientists and intelligence agents involved in the quest for genetic purity, bringing these scientists, doctors and government agents to the United States and Russia, including Operation Paperclip,[39] where these

Nazis could continue their work in secret. But the quest for genetic purity by families like the Rockefellers, Carnegies, and others would not rely solely on post–World War II scientists. These families, determined to control the direction of medical science and, with it, genetic engineering, set out to become heavily involved, financially and politically, in the education and training of physicians in the United States. They would begin by determining what should be taught, by whom, and which medical programs should be eliminated, for the good of the country and mankind.

SEIZING MEDICAL EDUCATION IN AMERICA

The American Medical Association (AMA)[40]

We are so used to hearing the acronym AMA and associating that with "doctor" (allopathic[41] medicine), that we seldom think about how the AMA began. Who were the founders, and under what authority did they begin the AMA? What makes someone a physician? Early in my career I belonged to the AMA. I was even vice president and then president of Caduceus, the student medical association connected with the AMA, while I was a medical student. I decided not to continue with my membership after one of our professors pointed out that the AMA owned stock in cigarette companies. The very organization telling you to quit smoking was profiting from your smoking—a win-win scenario for the AMA, not so much for you.

To fully appreciate history and the motives of people, it is important to understand the context[42] of events occurring at the time these people were alive and taking action. As with the eugenics movement, in the mid-1800s the US military was beginning to grasp the use of chemical and biological weapons,[43] as was Napoleon, all of which began the process of people trying to rein in the use of these weapons by the militaries of the world, as shown in the following timeline.

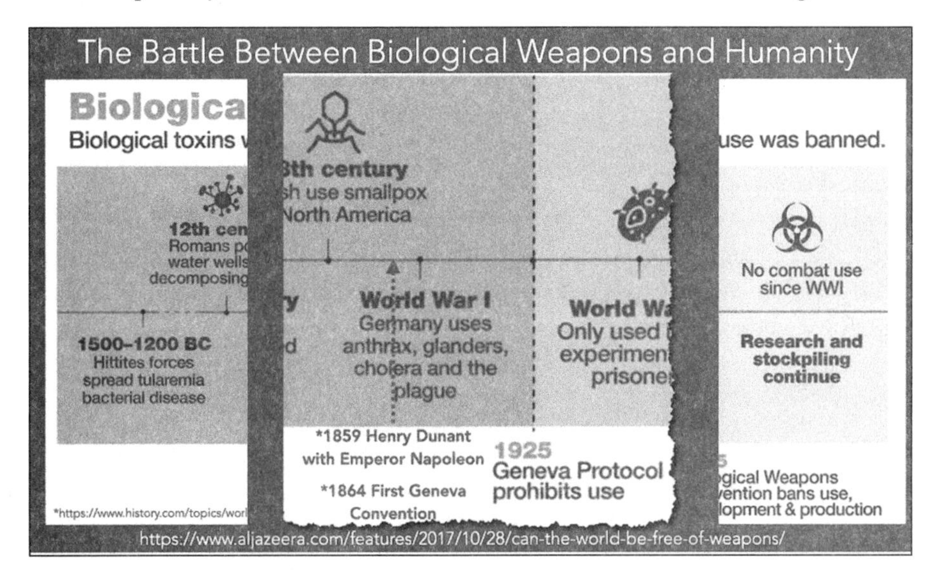

During this period, people like Dr. Nathan Smith Davis Sr. helped establish the AMA. He not only helped establish the AMA but also became the first AMA president. He was also the first editor in chief of the *Journal of the American Medical Association* (JAMA). Dr. Davis, like me, was both an attorney and medical doctor (MD and LLD, or legum doctor). The degrees one receives from either a medical or legal college (part of a university) are what determines if one is a doctor, lawyer, and so forth,[44] not the medical or bar associations.

The AMA established the physician code of ethics discussed in chapter 6 of this book. Also, the AMA developed the current procedural terminology (CPT) codes[45] (not the federal government), which are used to bill insurances for a medical procedure done on a patient. So, for the federal or state governments to complain about physician billing using the CPT codes without the AMA stating there was a billing issue and especially when the AMA states there wasn't a mistake, is a demonstration of the federal government's abuse of power. But I digress.

To try to establish a more organized approach to health care and consistency in medical education, the first organizational meeting of the AMA in 1846 dealt with asking the states to adopt measures to record information about births, deaths, and marriages. The AMA also began an education program, trying to make Americans aware of the potential dangers of certain patient medications.[46] However, these patient medications were not the medications of pharmaceutical companies as we know them today but rather the use of nonprescription drugs handed down from Greece, Rome, and Medieval Arab and Islamic treatments. Because of this effort, the 1848 legislation known as the Drug Importation Act[47] was passed, requiring the federal government to inspect drugs coming into the United States for adulteration.[48] This was the beginning of the FDA, which we will discuss below under Federal Administrative Agencies.

At the second meeting of the AMA held in 1849, Thomas Wood suggested that a committee be formed to address quack medicine and that this information, as determined by the committee, should be published and made available to the general public. The consequence of this action was the passage of the 1906 Pure Food and Drug Act.[49] By 1901 the AMA had established the Committee on Medical Education and the Committee on Medical Legislation, increasing the political effect of the AMA.[50]

Discussions surrounding the development of these early medical licensing laws give a better understanding of the motives of those involved in this legislative effort. Economic concerns and wanting to control how physicians must practice medicine drove the initiative:

> The aims of orthodox medicine and its most effective and tireless spokesman, the American Medical Association, were threefold:
>
> (1) the establishment of medical licensing laws in the various states to restrict entry into the profession and thus secure a more stable economic climate for physicians than that which existed under uninhibited competition;

(2) the destruction of the proprietary medical school and its replacement with fewer, non-profit institutions of learning, providing extensive and thorough training in medicine with a longer required period of study to a smaller and more select student body;

(3) the elimination of heterodox medical sects as unwelcome and competitive forces within the profession.

Page 75

In 1867 at the AMA meeting in Cincinnati, the AMA endorsed a resolution urging

> the members of the profession in the different States to use all their influence in securing such immediate and positive legislation as will require all persons, whether graduates or not, desiring to practice medicine, to be examined by a State Board of Medical Examiners, in order to become licensed for that purpose, [and further recommended that] said board be selected from members of the State Medical Society, who are not at the same time members of college faculties.[51]

This was the beginning of a campaign to solicit the aid of the respective state legislatures to obtain the goal of limiting the number of medical doctors in the United States by establishing medical boards controlling entry into the practice of medicine.

In 1873 Dr. Jerome Cochran used the Medical Association in Alabama—fractured by the Civil War—to obtain ultimate administrative control over Alabama's public health, molding it into a politically effective group. Explaining his purpose for doing this, Cochran wrote:

> It is well that we should understand that the primary and principal object of the Association is not the cultivation of the science and art of medicine. Truly, that is not a matter to be neglected, and we hope to accomplish much in this line. But it is not this that we have chiefly at heart. We will appreciate most adequately the real character of the Association if we regard it as a medical legislature, having for its highest function the governmental direction of the medical profession of the State, while its other functions, important as they are, in themselves, are, in comparison with this, of quite subordinate rank.[52]

The following was summarized in a report written by historian Ronald Hamowy:

> During the period 1875 to 1900, the groundwork had been laid. The legislatures and the courts had accepted the principle that medical practice laws constituted a legitimate and salutary extension of the police powers of the states. Medical examining boards, in all instances composed of physicians who had taken active roles in securing their creation, existed in almost all the states and territories; and public health boards, also staffed by the more outspoken representatives of organized medicine, could be relied upon to add pressure for stricter requirements for licensure.

> The direction future legislation would have to take if the supply of new physicians were to be significantly diminished was, by the end of the period, apparent. If the state examining boards were to require for licensure graduation only from those schools whose requirements for the issuance of a degree were particularly rigorous, whose instructional staff and facilities were only of the highest calibre, and whose standards of admission were unusually high, than the other medical schools, whose diplomas would go unrecognized, would be forced to close their doors. This was to prove the weapon with which the medical profession eventually succeeded in drastically reducing the number of physicians entering practice.
>
> Pages 103–4

Within two decades, the AMA had consolidated power over the medical profession and weaponized the legal and political system to control those elements promoting treatments not accepted by those in the AMA. Although beneficial at the time, and everyone would argue for the best practice of medicine, it provided the very tool that allowed the legal and political systems to take control of medicine away from the doctors and place it in the hands of the lawyers and politicians, thus laying the foundation for what would come next. Power makes strange, syphilitic bedfellows.

The Flexner Report and the Influence of the Rockefeller and Carnegie Foundations[53]

In 1904 the AMA Council on Medical Education (CME), whose objective was to reconstruct the educational system for medicine in the United States, reached out to the Carnegie Foundation[54] for the Advancement of Teaching. The then president of the Carnegie Foundation, Henry Pritchett, an advocate of changing medical education, selected Abraham Flexner to investigate and write a report on the American medical system of education. Flexner, it should be noted, was neither a physician nor a scientist. In fact, he had a bachelor of arts degree in classics and founded a college-preparatory school in Louisville, Kentucky.[55]

Abraham Flexner helped fund the medical education of one of his brothers, Simon, who went on to train as a pathologist, bacteriologist, and researcher at the Rockefeller Institute for Medical Research from 1901 to 1935.

In 1908, at the request of the AMA-CME, the Carnegie Foundation authorized the funding of the study authored by Abraham Flexner. Flexner issued the five[56] following recommendations in his report:

1. Reduce the number of medical schools (from 155 to 31) and the number of poorly trained physicians;
2. Increase the prerequisites to enter medical training to include a minimum of High School diploma and two years of collegiate scientific studies;

3. Train physicians to practice in a scientific manner and engage medical faculty in research increasing the length of medical education to 4 years;
4. Give medical schools control of clinical instruction in hospitals; and
5. Strengthen state regulation of medical licensure.

It is important to consider Flexner's personal perspective on African American, or as he worded it "Negro," doctors, taken in the context of eugenics positions expressed and funded by the Rockefeller and Carnegie families and foundations, which not only funded the Flexner *Report on American Medical Education* but also later funded Flexner's reports on European medical education:[57]

> The practice of the Negro doctor will be limited to his own race, which in its turn will be cared for better by good Negro physicians than by poor white ones. But the physical well-being of the Negro is not only of moment to the Negro himself. Ten million of them live in close contact with sixty million whites. Not only does the Negro himself suffer from hookworm and tuberculosis; he communicates them to his white neighbors, precisely as the ignorant and unfortunate white contaminates him. Self-protection not less than humanity offers weighty counsel in this matter; *self-interest seconds philanthropy*. The Negro must be educated not only for his sake, but for ours. He is, as far as the human eye can see, a permanent factor in the nation.[58]

And so, with the AMA, the AMA-CME, and the Flexner Report, with support and funding by the Carnegie and Rockefeller Foundations, the course and direction of the training of American physicians had changed. It had done so, not at the request of medicine, but at the request of those seeking to control the future of medicine. These people believed in eugenics and helped lay the foundation of the Third Reich and a eugenics program that would eradicate those Hitler felt were inferior to the Nazi German race. Ignoring the fact that Hitler was from Austria, a country he quickly annexed at the beginning of his reign, thus, making Austria part of Germany and in Hitler's mind making him German.

Having examined how the reins of medicine were handed over to those with the least knowledge of science and medicine, we will now look at the establishment of US federal administrative agencies designed to further control what physicians can and cannot do and say.

HANDCUFFING THE AMERICAN PHYSICIAN.

Under Article II, § 2 (executive branch), the president of the United States has the power to establish executive departments. Currently, President Biden's cabinet, includes fifteen heads of executive departments,[59] including those of departments of State, Treasury, Defense, Justice, Interior, Agriculture, Commerce, Labor, Housing and Urban Development (HUD), Transportation, Energy, Education, Veterans Affairs, Homeland Security, Management and Budget, and Health and

Human Services (HHS). Also on the president's cabinet is his chief of staff, the vice president, ambassador to the United Nations, and US trade representative, and the heads of the Environmental Protection Agency, National Intelligence, Office of Management and Budget, Council of Economic Advisers, Office of Science and Technology Policy, and Small Business Administration.

These executive departments secretaries and directors, and the agencies under their control (the FDA falls under HHS), are appointed by the president with the consent and approval of the Senate, with Congress passing any legislation (statutes and laws) controlling what these executive departments can or cannot do—article I, § 8, clause 18 (implied power).

As we have already seen, the AMA—along with support from the Rockefeller and Carnegie Foundations—provided the nidus for Congress to begin passing laws that the first seventeen clauses of Article I § I did not grant them power over. It is also clear from what we've covered that the AMA asked for the states (Tenth Amendment) to become involved in healthcare, which is where any efforts to impact medicine and health care would be relegated to under the US Constitution. However, just because it isn't expressly stated as a power of the federal government does not mean it should be a power of a state government.

> Government is not a solution to our problem; government is the problem.
>
> President Ronald Reagan

Health and Human Services (HHS)

On 3 April 1939 the US Congress passed the first major planned reorganization of the executive branch of the government since 1787.[60] The 1939 Reorganization Act[61] created the president's executive office. Several agencies were formed under the Reorganization Act of 1939, including the Department of Health, Education, and Welfare (HEW) on 11 April 1953. HEW included six major programs:

i. The Public Health Service
ii. The Office of Education
iii. The Food and Drug Administration
iv. The Social Security Administration
v. The Office of Vocational Rehabilitation, and
vi. St. Elizabeth's Hospital.[62]

HEW was renamed the Department of Health and Human Services (HHS) on 17 October 1979.

The most recent HHS "historical highlight"[63] listed on the U.S. Department of Health and Human Services website, "putting in place comprehensive US health insurance reforms," was the Affordable Care Act (ACA) of 2010.[64]

During the same time HHS was preparing for the ACA, the FDA, CDC, DoD, and other federally funded agencies were providing grants to EcoHealth Alliance and others to carry out gain-of-function biological-weapons research.[65]

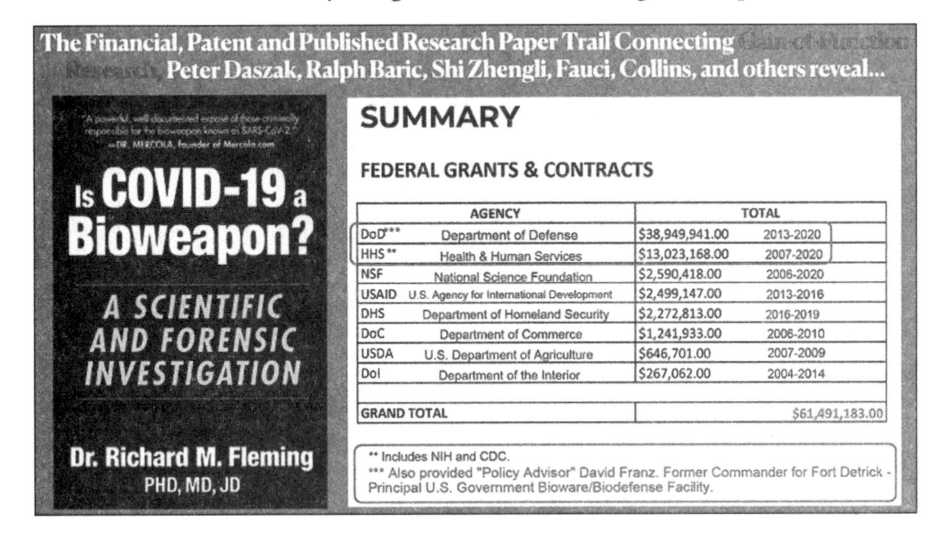

SUMMARY

FEDERAL GRANTS & CONTRACTS

AGENCY		TOTAL	
DoD***	Department of Defense	$38,949,941.00	2013-2020
HHS**	Health & Human Services	$13,023,168.00	2007-2020
NSF	National Science Foundation	$2,590,418.00	2006-2020
USAID	U.S. Agency for International Development	$2,499,147.00	2013-2016
DHS	Department of Homeland Security	$2,272,813.00	2016-2019
DoC	Department of Commerce	$1,241,933.00	2006-2010
USDA	U.S. Department of Agriculture	$646,701.00	2007-2009
DoI	Department of the Interior	$267,062.00	2004-2014
GRAND TOTAL			$61,491,183.00

** Includes NIH and CDC.
*** Also provided "Policy Advisor" David Franz. Former Commander for Fort Detrick - Principal U.S. Government Bioware/Biodefense Facility.

While these agencies were attacking me[66] behind the scenes, they were publicly courting me, asking me to hold a public town hall meeting on the ACA.[67] My first communication from the Obama-Biden (pre-inauguration)[68] transition team came from John D. Podesta, cochair for the transition team and ACA transition project, as noted in the email screen shot on p. 116.

After the town hall meeting, we took the meeting notes and assembled and submitted the response of the public back to Mr. Podesta and the Obama-Biden administration, including the public's concerns with access to the physician of their choice, abortion, insurance, drug costs, research, and so forth. None of these concerns were included in the final ACA of 2010. More on this in chapter 8, where we discuss what needs to be done to address problems with medicine, health care, medical and health-care education, research, research publication, the practice of medicine, and the FDA, CDC, and HHS.

The Food and Drug Administration (FDA)[69]

The Food and Drug Administration (FDA) is an executive-branch agency that reports to Health and Human Services (HHS). However, the FDA began decades before HHS on, 30 June 1906, following efforts by the AMA to stop what the AMA termed quack medicine, or medicine the members of the AMA disagreed with. Formation of the FDA began at the same time the Carnegie and Rockefeller Foundations were supporting Abraham Flexner and the Flexner Report previously discussed.

Within the FDA is the Center for Biologics Evaluation and Research (CBER), which, along with HHS and the FDA, wrote the two reports on genetic

From: "John D. Podesta, Obama-Biden Transition Project" <info@change.gov>
Date: December 15, 2008 2:17:54 PM PST
To: "Richard M. Fleming" [redacted]
Subject: Your turn to lead
Reply-To: info@change.gov

Obama-Biden Transition Team

Dear Richard M.,

Over the coming weeks, thousands of Americans will be leading Health Care Community Discussions -- small local gatherings in which Americans are sharing thoughts and ideas about reforming health care. President-elect Obama and Health and Human Services Secretary-designate Tom Daschle are counting on Americans from every walk of life to help identify what's broken and provide ideas for how to fix it.

You can help shape that reform by leading your own Health Care Community Discussion anytime between now and December 31st.

Secretary-designate Daschle recorded a short message about these important discussions. Watch the video and sign up today to lead a discussion in your community:

Watch the video and sign up to lead a discussion

Secretary-designate Daschle is committed to reforming health care from the ground up, which is why he won't just be reading the results of these discussions -- he'll be attending a few himself.

When you sign up to lead a discussion, we'll provide everything you need to make your conversation as productive as possible, including a Moderator's Guide with helpful tips. All you have to do is reach out to friends, family, and members of your community and ask them to attend -- and, when it's over, tell us how it went. The Transition's Health Policy Team will gather the results of these discussions to guide its recommendations for the Obama-Biden administration.

No transition has tried something like this before, and your participation is essential to our success.

Thank you,

John

John D. Podesta
Co-Chair
The Obama-Biden Transition Project

vaccine shedding discussed throughout this book. Shedding, you will recall, is something played down by these agencies, yet they issued reports to the pharmaceutical industry about genetic vaccines, gene therapy, and shedding in 2015 and again in 2020.

The responsibility of the FDA includes informed consent,[70] and research with children.[71] These considerations are noted on FDA websites, including *The Belmont Report: Ethical Principles and Guidelines for the Protection of Human Subjects of Research*, created on 18 April 1979 as a result of the 1974 National Research Act (Pub. L. 93–348).[72] The extent of the FDA's reach into health care is evident from the organizational chart shown in the QR code slides.[73]

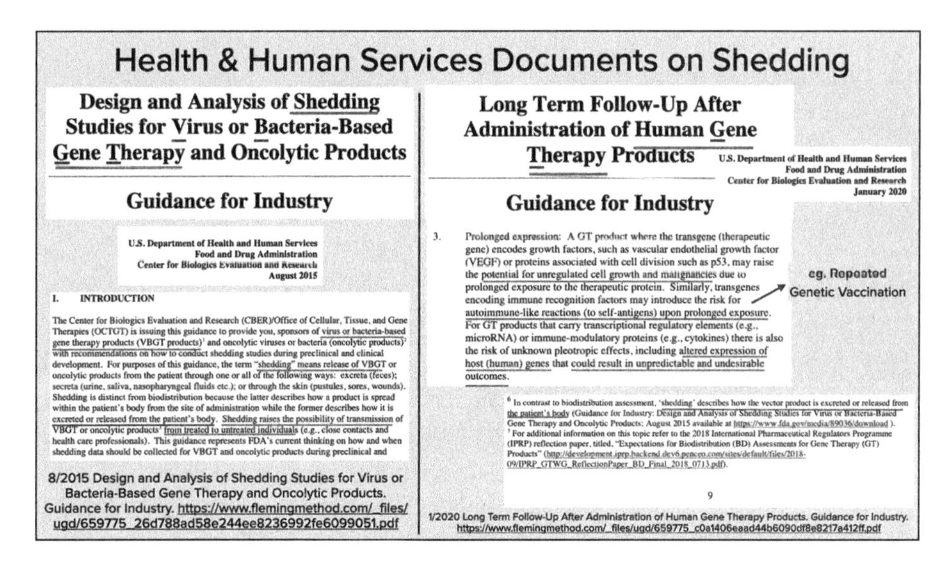

Health & Human Services Documents on Shedding

Design and Analysis of Shedding Studies for Virus or Bacteria-Based Gene Therapy and Oncolytic Products

Guidance for Industry

U.S. Department of Health and Human Services
Food and Drug Administration
Center for Biologics Evaluation and Research
August 2015

I. INTRODUCTION

The Center for Biologics Evaluation and Research (CBER)/Office of Cellular, Tissue, and Gene Therapies (OCTGT) is issuing this guidance to provide you, sponsors of virus or bacteria-based gene therapy products (VBGT products)[1] and oncolytic viruses or bacteria (oncolytic products)[2] with recommendations on how to conduct shedding studies during preclinical and clinical development. For purposes of this guidance, the term "shedding" means release of VBGT or oncolytic products from the patient through one or all of the following ways: excreta (feces); secreta (urine, saliva, nasopharyngeal fluids etc.); or through the skin (pustules, sores, wounds). Shedding is distinct from biodistribution because the latter describes how a product is spread within the patient's body from the site of administration while the former describes how it is excreted or released from the patient's body. Shedding raises the possibility of transmission of VBGT or oncolytic products[3] from treated to untreated individuals (e.g., close contacts and health care professionals). This guidance represents FDA's current thinking on how and when shedding data should be collected for VBGT and oncolytic products during preclinical and

8/2015 Design and Analysis of Shedding Studies for Virus or Bacteria-Based Gene Therapy and Oncolytic Products. Guidance for Industry. https://www.flemingmethod.com/_files/ugd/659775_26d788ad58e244ee8236992fe6099051.pdf

Long Term Follow-Up After Administration of Human Gene Therapy Products

U.S. Department of Health and Human Services
Food and Drug Administration
Center for Biologics Evaluation and Research
January 2020

Guidance for Industry

3. Prolonged expression: A GT product where the transgene (therapeutic gene) encodes growth factors, such as vascular endothelial growth factor (VEGF) or proteins associated with cell division such as p53, may raise the potential for unregulated cell growth and malignancies due to prolonged exposure to the therapeutic protein. Similarly, transgenes encoding immune recognition factors may introduce the risk for autoimmune-like reactions (to self-antigens) upon prolonged exposure. For GT products that carry transcriptional regulatory elements (e.g., microRNA) or immune-modulatory proteins (e.g., cytokines) there is also the risk of unknown pleotropic effects, including altered expression of host (human) genes that could result in unpredictable and undesirable outcomes.

[6] In contrast to biodistribution assessment, 'shedding' describes how the vector product is excreted or released from the patient's body (Guidance for Industry: Design and Analysis of Shedding Studies for Virus or Bacteria-Based Gene Therapy and Oncolytic Products; August 2015 available at https://www.fda.gov/media/89036/download).
[7] For additional information on this topic refer to the 2018 International Pharmaceutical Regulators Programme (IPRP) reflection paper, titled, "Expectations for Biodistribution (BD) Assessments for Gene Therapy (GT) Products" (http://development.iprp.backend.dev6.penceo.com/sites/default/files/2018-09/IPRP_GTWG_ReflectionPaper_BD_Final_2018_0713.pdf).

9

1/2020 Long Term Follow-Up After Administration of Human Gene Therapy Products. Guidance for Industry. https://www.flemingmethod.com/_files/ugd/659775_c0a1406eaad44b6090df8e8217a412ff.pdf

Although the responsibility of the FDA is purportedly that of overseeing the safety of drugs, devices, and biologics used by or given to the American citizen, the concern about a conflict of interest regarding the movement of employees between the FDA and Big Pharma is hard to miss, as discussed in "Trust Issues Deepen as Yet Another FDA Commissioner Joins the Pharmaceutical Industry."[74]

Commissioner name	rs as FDA Commissio	naceutical venture ar	Year
Arthur Hayes	1981-1983	E. M. Pharmaceuticals, president	1986
Frank Young	1984-1989	Braeburn Pharmaceuticals, executive vice president	2013
David Kessler	1990-1997	None	1997
Jane Henney	1999-2001	AmerisourceBergen Corp, member of board	2002
Mark McClellan	2002-2004	Johnson and Johnson, member of board	2006
Lester Crawford	2005-2005	Bexion Pharmaceuticals, member of board	2011
Andrew von Eschenbach	2006-2009	Bausch Health, member of board	2018
Margaret Hamburg	2009-2015	Alnylam Pharmaceuticals, director	2018
Robert Califf	2016-2017	Verily, advisor	2017
Scott Gottlieb	2017-2019	Pfizer, member of board	2019

Given the FDA claims of responsibility for the safety and efficacy of biologics, including the three genetic vaccines for SARS-CoV-2/COVID, it is interesting to note the following:

- The data presented in the three EUAs to the FDA, when analyzed, failed to show a statistical reduction in COVID or death cases.
- A significant number of adverse events were reported in the VAERS database, and pressure selection appears to have been placed on the SARS-CoV-2 viruses by the genetic vaccines.[75]

- The FDA has had to pull a number of drugs off the market due to safety concerns in recent years.
- A relationship between FDA employees and the pharmaceutical industry (Big Pharma) exists.
- When all of this is coupled with the gain-of-function research money spent by HHS, to which the FDA belongs, is it any wonder why Americans have trust issues with the FDA?

The Centers for Disease Control (CDC)—Part of the Public Health Service[76]

The Centers for Disease Control (CDC) began on 1 July 1946 as the Communicable Disease Center, evolving out of the Second World War Malaria Control in War Areas program, in which the Rockefeller Foundation had a vested financial interest. The Rockefeller Foundation has helped fund gain-of-function and CRISPR research as noted in appendix B and eugenics programs as shown in appendix E. In 1951[77] the CDC had an Epidemic Intelligence Service (EIS) in Korea, where the CDC field tested a program addressing biological-warfare concerns. It also has an interest in birth defects, including intellectual deficiencies and chronic inflammatory diseases.

In my discussions with the DoD in 2002–2003 regarding a cancer research fund I had applied to the department for, I gave the representative data I had collected, and I resolved some questions the DoD had not been able to answer. They included my ability to measure changes in tissue metabolism and blood flow (FMTVDM),[78] changes in homocysteine levels, and changes in thymic (immunity) activity following InflammoThrombotic Response(s) (ITR) to infectious diseases, including the Neu5Ac sialic acid receptor and HIV gp120 attachment, cancers, and other chronic diseases. As a result of those conversations, the CDC director's office asked me to present my work[79] at the 2005 "Vascular Endothelium: Translating Discoveries into Public Health Practice" conference in Crete, Greece. Two of the research materials presented are shown here and by QR code.

Centers For Disease Control And Prevention (CDC)

8th International Conference

VASCULAR ENDOTHELIUM:
TRANSLATING DISCOVERIES INTO PUBLIC HEALTH PRACTICE

June 25 – July 2, 2005 - Knossos Royal Village, Crete, Greece

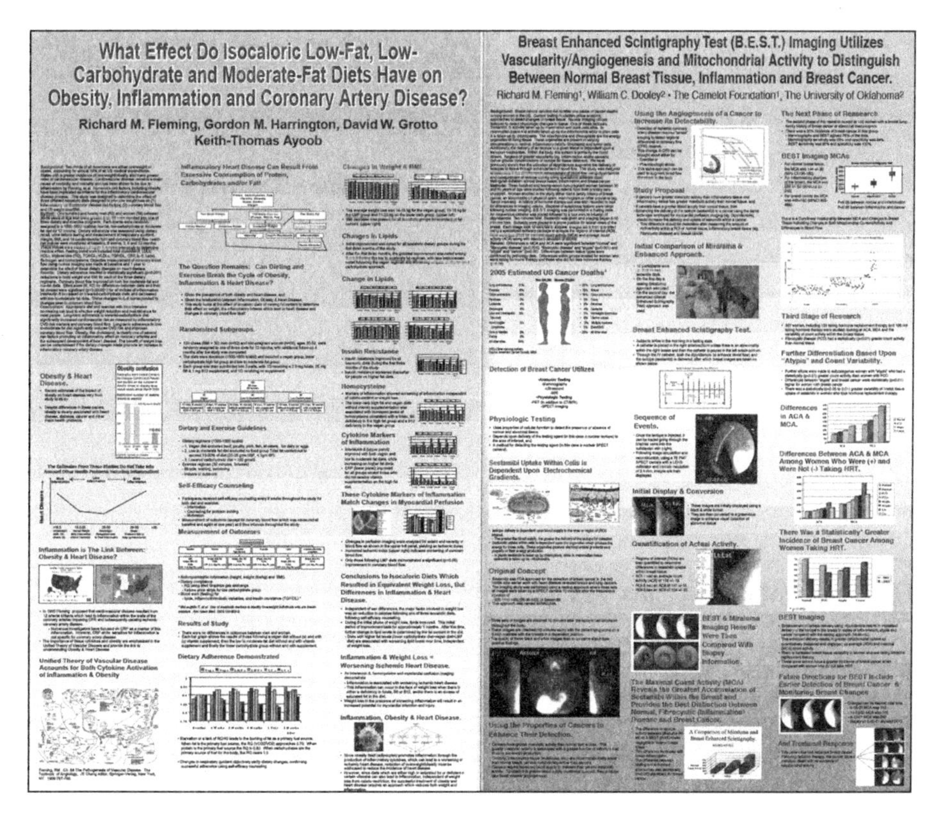

It would be an understatement to say that the CDC had an interest in my work and that Dr. Seffrin of the American Cancer Society (ACS), who also presented there, had an interest in learning more. Unfortunately, efforts by Big Pharma and the government to suppress me and my work accelerated,[80] and my efforts to later locate Dr. Seffrin proved impossible once he left the ACS.

Clearly, both the DoD, which later tried to recruit me to Fort Detrick—see emails in chapter 2 and elsewhere in the book—and the CDC had an intense interest in my work, including the ITR Theory, which was becoming accepted as the model explaining heart disease and other diseases, and the homocysteine, thymus, Neu5Ac-gp120 information. That information, as we now know, is critical to understanding the gain-of-function viruses known as SARS-CoV-2 and the resulting ITR disease known as COVID-19.

Multiple Additional Factors

In addition to the multiple federal and state overreach of power just examined, many other controls have been implemented to prevent physicians from practicing medicine in the way they see most appropriate for their patients. These constraints have been applied so physicians will do what the government, and those who handed medicine over to the lawyers and government, want physicians to do.

These encumbrances include, but are not limited to, the entry of the government into the insurance industry. These insurance plans essentially began when the Centers for Medicare and Medicaid Services (CMS) law[81] was signed by President Lyndon B. Johnson on 30 July 1965. With each passing year, CMS continued to pay less and demand more from physicians and health-care providers. The dental community was smarter and refused to be controlled, deciding it would rather see fewer patients than be controlled by the government. In the end, the dentists were right. In conjunction with the AMA, CMS set out to cut deeper and deeper into physician income until physicians left private practice and became paid employees of hospitals, major corporations, and the government. As a result, physicians were becoming paid employees, and medicine was being turned into a business that could be run by people who are not physicians.

In response to the Patriot Act and the massive amount of money set aside for the federal government to use, the Office of the Inspector General (OIG) was given some of the money to go after physicians[82] in an effort to scare them into further submission to the federal government.

Given the power of the OIG, targeted physicians—particularly those pointing out that Big Pharma was less interested in patient care than profits—didn't stand a chance, even when physicians were using the correct CMS billing codes for procedures as defined by both CMS[83] the AMA. As it turned out, that applied when the AMA tells physicians they underbilled[84] and could have billed for other procedures.

As the stranglehold on physicians increased, the ability of physicians to control their own practice of medicine decreased. As physicians became paid employees, patients became the property of hospitals, corporations, and the government.

Failure of physicians to do what the hospitals, corporations, and government wanted them to do resulted in physicians being censored and reported to state medical boards and the National Practitioner Databank[85] of 1986, controlled by none other than HHS—a further draconian effort to leverage control over physicians with the threat of giving them a black mark on an otherwise pristine record of patient care, resulting in their exclusion from insurances, hospitals, and so forth.

The true leveraging power of the federal and state governments was tested for the first time during the COVID pandemic, when physicians were accused of misrepresenting facts and information around the diagnosis, treatment, care, and vaccination of patients. Lacking any more information than we had when HIV and AIDS began killing people in 1982–1983, younger physicians, indoctrinated by these above regulations and controls, proved relatively easy to control by federal and state agencies. Other physicians, however, were either old enough to either remember how we approached HIV/AIDS treatment, or heard from those of us who had lived through the beginnings of HIV about how we responded to a never-before-seen virus that was killing people.

Independent of whether the infectious disease, in this case a virus, is naturally occurring, or as all evidence in this instance demonstrates, a man-made biological viral weapon funded by the US government, a scientist's and a physician's job is to figure out what is happening, why, and what can be done about it. This requires the free-flowing exchange of ideas and outside-the-box thinking and efforts—or a better understanding of the box.[86] As with the anthrax scare of 18 September 2001 when packages laced with that harmful bacterium were mailed to several politicians, the treatments that worked were those tried by physicians thinking outside the usual box of treatment but inside the box of knowledge.[87] Those physicians realized that *Bacillus anthracis* was a type of bacterium with an endospore.[88] Ciprofloxacin, clindamycin, and doxycycline are a good drug combination for treating endospore infections, even though it wasn't among the accepted treatments at the time. As it turned out, patients who received this combination of drugs lived. Physicians who didn't question what they were being told to use for treatments and followed orders ended up with dead patients.[89]

If you had wondered, *What happened to physicians and the practice of medicine in 2020 and throughout the COVID pandemic?* now you know. Many were castrated, or strangulated, or handcuffed. The practice of medicine by physicians has been usurped by the lawyers, the government, and the military-industrial complex.

PERFORMING CPR ON MEDICINE: WE'RE INJURED BUT NOT DEAD!

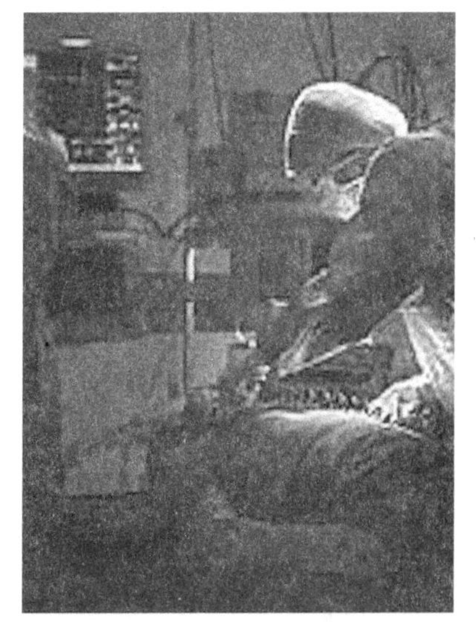

Medicine is one of the oldest professions on earth, and the oath taken by physicians is revered by most. Dedication to the oath and what it means, while not fully appreciated by everyone, is key to what it means to be a physician. Our medical training and oath deserves respect and recognition independent of government agencies. The practice of medicine is not for sale and it is not the result of a federal or state government granting permission for someone to practice medicine. The diploma received after satisfactorily obtaining entry into a medical college, graduating through the didactic and clinical training, and completing any post graduate training is what makes one a doctor; not the state or federal government.

The second thing the dean of my medical college told my class at the very beginning was that 90 percent of what we were to learn would someday be proven wrong. To the researchers, he asked us to research what is necessary to help find what is wrong, correct it, and advance medical science. Like all sciences, medicine is evolving. It is not static; if it were, it would be dead. Scientific growth frequently comes as a result of controversy and disagreement. Every theory I have proposed, every patent I have been issued, came after many years—frequently decades—of work, research, and criticism.

The evolution and advancement of science, including medicine, frequently means thinking outside the box. But if the medical system is so restrained that no one can do so, then medicine will die.

When the American Medical Association began, it was nothing more than an organization used by a handful of people to control what was and wasn't accepted. In the mid-1800s, the AMA began calling for a "purity" of thought and training at the same time eugenicists were calling for a "purity" of genes and people. The AMA turned to the lawyers and politicians to help them control those who thought and taught different ideas and approaches to medicine. Happy to oblige, the lawyers and politicians began, arbitrarily and capriciously, to pass laws.

As the legal and political system eventually took more control of medicine, federal and state governments began to control drugs; vaccines; insurance; the "right to health care," using the power to levy taxes[90] as the ultimate congressional tool; and physicians themselves, with physicians frequently defending Big Pharma and AMA billing codes, under the guise of protecting patients from being overbilled, while exposing patients to excess radiation and inferior diagnostic testing.[91]

The purpose of purity was completely thrown out the window when the SCOTUS decided that vaccine manufacturers could not be sued for producing a harmful product, stating vaccines are inherently harmful.[92]

> Plaintiff's design defect claims [were] expressly preempted by the Vaccine Act.
>
> *Bruesewitz v. Wyeth LLC*, 562 US 223 (2011)
>
> Justice Antonin Scalia, 22 February 2011

The SCOTUS affirmed US congressional laws that state vaccine manufacturers are not liable for vaccine-induced injury or death if they are "accompanied by proper directions and warnings."

> § 300aa-22. Standards of Responsibility[93]
>
> General Rule
>
> Except as provided in subsections (b), (c), and (e) of this section State law shall apply to a civil action brought for damages for a vaccine-related injury or death.

The ruling by the SCOTUS guaranteed a stable vaccine supply by limiting vaccine manufacturers' liability for vaccine harm.

When the FDA was called upon to evaluate the SARS-CoV-2 and COVID genetic (mRNA and DNA) vaccines, the FDA either failed to statistically analyze the data or decided not to recognize the statistical analysis[94] that demonstrated the three EUA FDA-approved genetic vaccines (Pfizer, Moderna, Janssen) statistically failed to reduce COVID cases or deaths.

There is also the issue of "Standards of Responsibility," according to Justice Scalia and the SCOTUS, because the vaccine package inserts and materials provided to doctors, pharmacists, hospitals, the military, and patients, providing "proper directions and warnings" didn't exist. They were "intentionally blank."

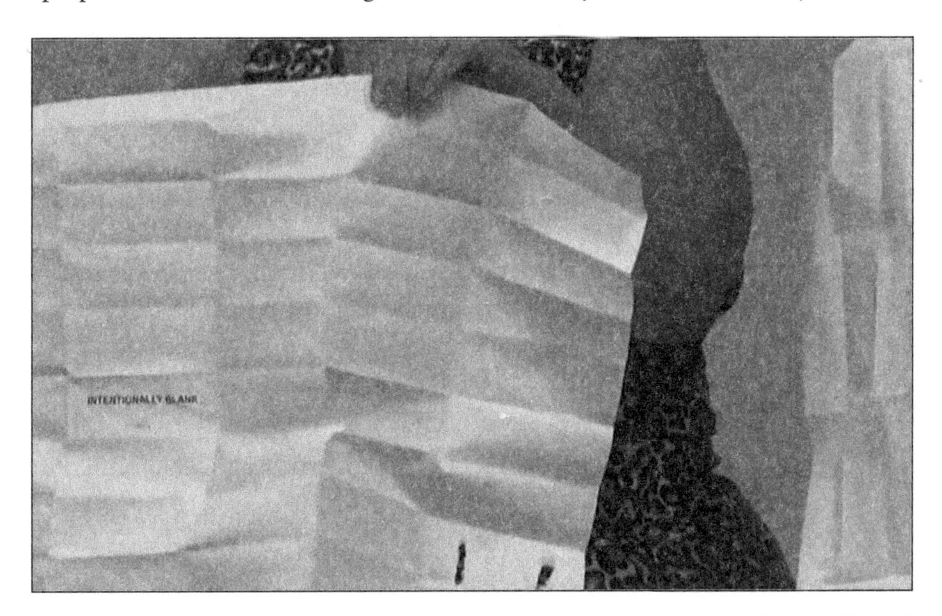

Questions regarding conflicts of interest or criminal intent of these federal agencies, including those responsible for funding the SARS-CoV-2 gain-of-function bioweapon research, in addition to those who were responsible for reviewing the safety and efficacy of the genetic vaccines[95] and drugs (e.g., remdesivir), loom over us, begging for answers and accountability. Questions arise about the genetic vaccines and drugs approved for use by the American people, ultimately, resulting in encouragement and use of these drugs and vaccines worldwide. This is just part of the questions and criminal case we will look at in chapter 6.

The federal agencies HHS, FDA, CDC, and such only exist because a group of physicians in the mid-1800s decided to usurp the power of medicine and then transfer that power to federal and state governments that would give the AMA, Rockefeller Foundation, and Carnegie Foundation ultimate nonpolitical control over the teachings and practice of medicine, effectively eliminating those physicians who disagreed with the AMA.

Not every physician receives the same level of training, and not every physician like myself is a scientifically trained physician. My medical faculty told us when we graduated to leave the University of Iowa and see how others practiced medicine. We were encouraged to incorporate the bits and pieces we could learn from others into what we had been taught. Most of my approach for practicing medicine follows the scientific framework I learned at Iowa, but I have added to that armamentarium, after seeing how others practice medicine. Additions applied only after I investigated why these other approaches worked and not blindly accepting them just because some other doctor said to.

The best question to ask your doctor, when looking for a physician to treat you, is, if this were you, or your significant other or close family member, which doctor would you go to and why? We know who the good and bad doctors are; as we have seen them in practice—the outcomes of their diagnosis and treatment of patients.

When I was a second-year internal medicine resident, I admitted a patient with heart disease and high blood pressure into the hospital. The blood-pressure medication he was taking was archaic. It had to be taken several times a day, and it had three different drugs in it. I thought I could make his life easier, so I prescribed a new drug that could be taken once a day. On the first morning when we made rounds on the patient, my attending cardiologist—noting how poorly the patient's blood pressure was controlled—asked my reason for changing the blood-pressure medications and then said, "Get it under control." I increased the dose. The next day saw the same story, with the same reply from the attending cardiologist. I increased the dose again. On the third day, there was some slight improvement, but the blood pressure was still higher than it had been when the patient was admitted three days earlier. "Get it under control."

On the morning of the fourth day, the attending cardiologist smiled at me and said, "Good job! You got the blood pressure under control. What did you do?"

I looked at him, smiled, and said, "I put him back on his original medication." This was a good cardiologist, and, yes, I remember his name and the lessons he and the patient taught me.

He smiled and said, "You're going to make a good doctor!"

This story points out the need for learning and for differences in opinions or ways of diagnosing and treating patients. No one way works for everyone. I was taught that anyone who uses *always* or *never* was probably wrong, because very seldom is something always or never.

When HIV hit the scene in the early 1980s, I was a medical student. We didn't have answers, although we had a lot of questions. The medical community worked together, discussing, debating, arguing, disagreeing, and finally coming together to treat our patients. I argue that what we saw then was much worse[96] than what we saw with SARS-CoV-2, except that with SARS-CoV-2 and COVID, we quit working together.

As of 2018 approximately seven hundred thousand Americans had died from HIV and AIDS since the virus was first reported in 1982–1983. Compare this with the almost 1.2 million Americans who have died from SARS-CoV-2 and COVID since January 2020. Something changed dramatically in the diagnosis and care of patients in America. Instead of asking each other what we thought or what we saw happening to our patients, we stopped collaborating and took orders from the federal and state governments and agencies issuing directives. Like the Nazi doctors in concentration camps, many physicians took no responsibility for how we diagnosed or treated patients. Doctors in Nazi Germany who didn't follow orders had a shortened career, and undoubtedly a shortened life.

In 2020 many physicians and hospitals (run in many but not all instances by bureaucrats who don't actually practice medicine) surrendered almost all of their control to the government. As a result, almost twice as many are dead in four years from COVID than from AIDS in almost four decades. There are more dead from COVID around the world than Hitler killed in his concentration camps.

Some physicians spoke out, asking questions, trying different treatment approaches. Many did so under threat by federal and state agencies. To be clear, I do not agree with everything proposed by some of these physicians, but it doesn't matter. In the end, the question you should be asking your doctor is, if this were you, or your significant other or close family member, which doctor would you go to and why?

Asking what the federal and state governments say a doctor should or must do and then only doing that has resulted in more American deaths than we have seen from AIDS, more American deaths than the US military has lost in battle since 1776, more maimed Americans from the genetic vaccines than the US military has had injured solders since 1776. Something is rotten in the United States of America. It's the smell of rotting American bodies, injured bodies, and governmental lust for power.

The question isn't whether the federal and state governments have the constitutional authority to run medicine. They do not. The question is, how much longer will the American people tolerate the abuse of authority by federal and state agencies in an area of their lives for which the US Constitution does not grant such invasion, into the day-to-day activities of the American people (masking, gatherings, religious practices, and such), into the practice of medicine, and into the lives of those trained to practice medicine, whose first and only obligation is to their patient—not the government, insurance companies, Big Pharma, hospitals or anything else?

CHAPTER 5

The Genetic Vaccines

One of the easiest ways to confuse people is to give them too much information. A friend of mine, another physicist, was instrumental in the development of the heads-up display for pilots, which provides pilots with real-time visual information on a variety of critical issues. Although you could argue that the more information the pilot had, the better job he or she could do, you would be wrong. It turns out that if you provide too much information, the combat jet fighter gets overwhelmed, and he or she soon becomes unable to function. Most people can only concentrate on three to five topics or ideas at a time. If you give them more, they will become confused.[1]

Excessive information has proven to be very effective at distracting people who are not actively involved in scientific research. As shown in the following figure, we have been discussing two separate pathways of concern. The first has been the actual criminal actions, which can and must be addressed in the state and federal courts if we are to recover our country and sanity. This is the path of biological weapons, which includes the dual use of gain-of-function. The second pathway is that of eugenics, which includes the dual use of gene therapy and the opportunity to accelerate a eugenics research program.

The Parallel Pathways of Eugenics and Biowarfare

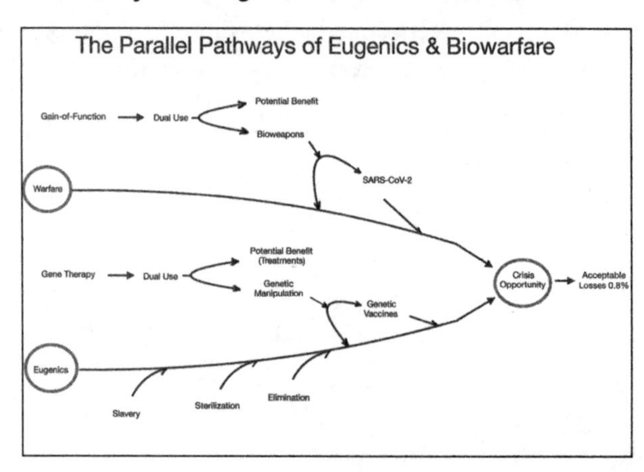

Most Americans and people around the world responded to their fear of dying or living with a body damaged by a bioweapon (SARS-CoV-2 and COVID). The general public was not actively involved in HIV and AIDS. Even though HIV and AIDS were covered in the news, most people saw this disease as something only gay people, intravenous drug abusers, and people with hemophilia had to worry about. Americans, for the most part, were not actively involved in these activities, so most people didn't feel like the illness could harm them—unlike those of us on the front lines in the hospitals, taking care of patients, exposing ourselves to an unknown deadly virus. When I was a senior internal medicine honors student, I presented a lecture explaining that HIV, based upon the data we had at the time, could and would be transmitted through heterosexual sex. I was scoffed at, ridiculed, and told I didn't understand the science. In some ways nothing has really changed. Explaining the obvious to people has always been a challenge.

> All truths are easy to understand once they are discovered; the point is to discover them.
>
> Galileo Galilei

In 2020 the story was different. People learned that SARS-CoV-2 could be spread relatively easily from person to person. Lockdowns (gently called "shelter at home"), masking (even in your car while driving by yourself), and handwashing increased anxiety and the willingness of Americans, and people elsewhere, to do whatever their government told them to do.

During his 1947 Nuremberg Trial Göring Said The Following.

... it is the leaders of the country who determine the policy and it is always a simple matter to drag the people along, whether it is a democracy or a fascist dictatorship or a Parliament or a Communist dictatorship.

...voice or no voice, the people can always be brought to the bidding of the leaders. That is easy. *All you have to do is tell them they are being attacked and denounce the pacifists for lack of patriotism and exposing the country to danger.* It works the same way in any country.

Everyone was being told there was no treatment, and doctors were instructed by their government and hospital owners to send people home. You only have to watch someone die once before you truly understand that dead is dead. People don't really appreciate this until they see it in person.

So, with the news covering body-bag counts the same way the media covered US soldier body-bag counts in Vietnam, people were more than willing to do what the government told them to do.

If there was no treatment; what was the answer? The government, including HHS and the FDA, had the solution: Emergency Use Authorization vaccines. EUAs could only be issued if there were no available treatments, and the general public was being told there were no treatments or alternatives.

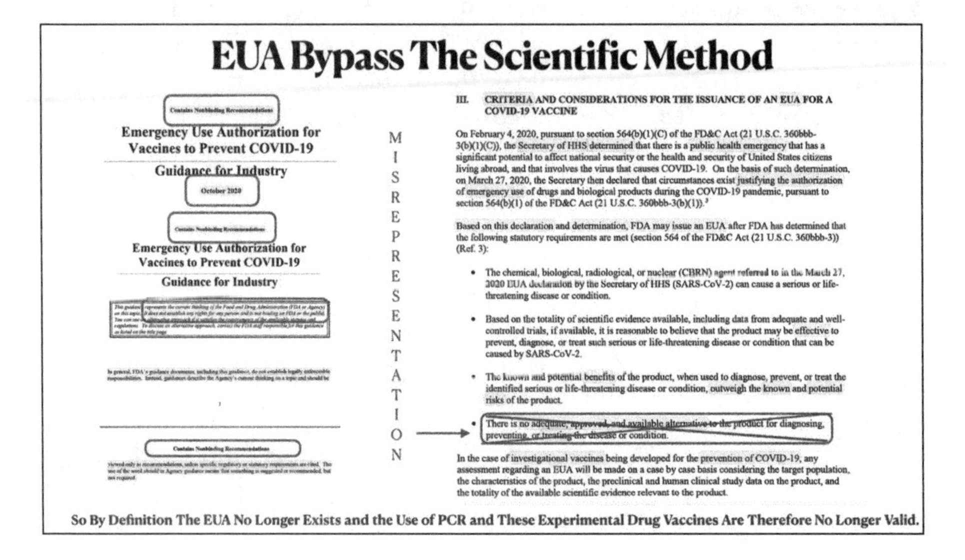

People were being divided, as fear and panic was used to weaponize and polarize those worried about dying against those experiencing doubt; resistance was unacceptable to the federal, state, and municipal governments. As history shows us (chapter 2), the tidal wave of fear and compliance swept over much of the world, including the United States, at a record blitzkrieg rate.

When President Trump announced a state of national emergency[2] on 13 May 2020, fears increased. The president called for and announced the answer: Operation Warp Speed.[3] Trump therefore made it possible to bypass the scientific and safety measures that had been put into place to protect people. How much Trump knew or didn't know, whether he knew what he was doing or was being played, is something only time will tell.

Many Americans were relieved and willing to accept that a new type of vaccine, a genetic vaccine, would allow the development of a SARS-CoV-2 and COVID vaccine in less than a year, even if it meant bypassing established standards and testing requirements laid out in treaty law, congressional legislation, and the established medical and scientific ethics and methods.

The promotion of genetic vaccines occurred in the absence of any scientific support for them, which had failed for decades to provide the desired outcome.

In fact, the published scientific work of Karikó and Weissman, who received the 2023 Nobel Prize[4] in Physiology or Medicine for their work on pseudouridine, showed that using pseudouridine for mRNA genetic vaccines increased the stability of the mRNA, making it a "promising tool" for gene replacement, but nonetheless resulted in a diminished interferon immunity response, compared with uridine mRNAs.

Nonetheless, many took on the behavior Göring talked about, attacking others for not getting vaccinated, declaring that they put the vaccinated at risk and calling for the unvaccinated to be refused medical treatment. Despite pressures, not everyone embraced the genetic vaccines.[5]

Strict Product Liability (SPL)

In the last chapter, we briefly looked at *Bruesewitz v. Wyeth LLC*, 562 US 223 (2011), the SCOTUS decision that upheld the National Childhood Vaccine Injury Act (NCVIA) of 1986, 42 U.S.C. §§ 300aa-1 to 300aa-34, for which Justice Antonin Scalia wrote the majority opinion absolving the vaccine manufacturer Wyeth from legal responsibility for any harm resulting from their vaccine.

In chapter 2, we looked at important issues surrounding the Pfizer, Moderna, and Janssen genetic vaccines, including the discoloration caused to human blood (prion effect) and the InflammoThrombotic Response (ITR) effect that our group investigated and published. It took more than a year to get this research published[6] because, as one reviewer noted,

> It may lead to undermine the Readers' trust in the pharmaceutical products and vaccination in general.

We also talked about the lack of EUA data—including the absence of any immunology (interferon, antibody production levels, and so forth) in the EUAs themselves—showing any benefit in reducing COVID cases or death, as well as the pressure selection produced by the genetic vaccines to promote changes in the virus (variants).

We looked at the documents issued by HHS, FDA, and CBER, revealing that these agencies recognized the use of these genetic vaccines using either pseudouridine mRNA or DNA, as gene therapy and that they knew these gene therapy biologics caused shedding. Those very agencies have been controlling the practice of medicine in the United States.

The premise of the SCOTUS decision in *Bruesewitz v. Wyeth LLC*, to uphold the law protecting vaccine manufacturers from liability for harm caused to people injected with their drugs, was based upon the presumption that the vaccines were "accompanied by proper directions and warnings." However, as we saw in the last chapter, these three EUA FDA-approved genetic vaccines from Pfizer, Moderna, and Janssen did not have directions or warnings included in the materials provided to health-care providers or patients.

Furthermore, we need to look at a critical problem associated with these genetic vaccines, an area of drug manufacturing that is not shielded by Congress or the SCOTUS decision on vaccines, namely, strict product liability[7] (SPL).

Let's Play Lawyer

For the remainder of this chapter, let's play attorney.[8] Let me show you how I would address a SPL case addressing these genetic vaccines if I were the attorney bringing that against these three genetic-vaccine manufacturers.

There are two fundamental types of legal cases. The first is criminal, which pertains to the defendant being suspected of having broken the law. Only prosecuting attorneys can bring criminal cases because they represent a harm done to society. In a criminal-law case, a prosecutor is essentially trying to send the defendant to jail. The second is civil, which relates to a person (plaintiff) claiming to have been wronged by another party (either a person or business, never the government). SPL falls into an area of the law known as tort law. *Torts* are a civil claim that the plaintiff has been harmed by the defendant as a result of something that person has done, excluding certain areas of the law, like contracts, real estate (property), family law, and others. In a civil-law case, a plaintiff is usually trying to get money.

To begin addressing a SPL civil lawsuit, we need to look at what SPL is. Two sets of rules need to be addressed in any case filed with a court; one is the procedural rules that deal with whether the court (judge) is happy with how you are moving ahead. Procedural rules include the Federal Rules of Civil Procedure (FRCP).

The second set of rules are substantive. Do you have the parts, of which there are eight required to prove, in this instance a SPL case? So, let's look at the substantive parts of strict Product liability. As we review these eight elements, we will analyze the case based upon the evidence[9] we have, including the following:

- Harmful priogenic and ITR were present when each of the three genetic vaccines were added directly to human blood under a microscope.
- Drugs such as normal saline (NS), atropine (cardiac medication), and medetomidine (Domitor; anaesthetic) do not cause the priogenic and ITR effect seen with these three genetic vaccines when added to people's blood under the microscope.
- Each of the genetic vaccines tested had debris detectable by microscopy that should not be present in a biological agent being injected into people.

EVIDENCE[10]

I. Evidence of harmful priogenic and ITR when each of the three genetic vaccines were added directly to human blood under a microscope.

Every time the Pfizer, Moderna, or Janssen vaccine was added directly to human blood, within fifteen to twenty-five seconds the blood changed from red (hemoglobin-oxygen) to gray (loss of hemoglobin ability to hold onto oxygen) and began to form clots (ITR) visible under the microscope.

II. Evidence that drugs such as normal saline (NS), atropine (cardiac medication), and medetomidine (Domitor; anaesthetic) do not cause the priogenic and ITR effect seen with these three genetic vaccines are added to patients blood under the microscope.

In every instance, the addition of normal saline, atropine, and medetomidine to blood had no adverse effect. The blood remained red (no prion effect with change of hemoglobin) and did not clot (no InflammoThrombotic Response, or ITR).

III. Evidence that each of the genetic vaccines tested had debris detectable by microscopy that should not be present in a biological agent being injected into people.

The microscopic evaluations of the three genetic vaccines as shown in the QR slide series revealed crystals (A, B), fibers (C, D), and lipid (E) debris crystals. Image F shows air droplets, which have been erroneously interpreted by some as nanotechnology. The air droplets coalesce when the microscope slide is tapped.

Frequently confused as microcircuitry, yet clearly not something that should be in these genetic vaccines for direct injection into a person, are crystals of sodium chloride (NaCl, a salt).

This debris can precipitate an InflammoThrombotic Response (ITR), leading to the formation of inflammation and blood clots that can cause direct harm to various parts of the body, including heart attacks, strokes, and such.

With this evidence in hand, let us look at the required elements of SPL civil litigation step-by-step.

1. Strict Product Liability (SPL)—It's the Product on Trial, Not the Defendant (Manufacturer)

When a manufacturer places a product on the market, knowing it is to be used without inspection for defect and a defect is present that then results in personal injury, SPL follows.[11]

A. The eight substantive elements, or what is required to bring a SPL case to court and win
   1. Proper plaintiff (user or bystander injured).
      Did you become injured as a result of being injected with one or more of these genetic vaccines? If so, you are a proper plaintiff.
   2. Proper defendant (anyone involved in business, production, or marketing chain).
      Did the defendant make, sell, market, or supply the genetic vaccine(s) you were injected with? If so, you have a proper defendant.
   3. Proper context. Did the product (e.g., genetic vaccine) cause personal or property damage beyond the product itself?
      If the product is defective but did not cause actual harm or damage, then it's a breach of warranty (contract-law) issue.
   4. Defect in product (e.g., genetic vaccine). This applies to manufacture, design, and warning and is the key element of SPL.
      i. Manufacturing defect.
         1. The product is different from the intended design, and it left the manufacturer that way.
            i. Majority[12] view (second restatement): ordinary consumer expectation.
            ii. Minority View (foreign vs. natural): for example, glass in chicken enchilada.

      Independent of whether you approach this from the majority view (ordinary consumer expectation) or from the minority view (foreign objects in a drug biologic product), I think a jury would agree that foreign material that is not part of the genetic vaccine, or any vaccine, shouldn't be in the vials of these vaccines.

The microscopic evidence shows the genetic vaccines had material within them clearly not listed in the patent materials, or within the list of materials claimed to be in the genetic vaccines by the manufacturers. Vaccines were delivered in this condition, and as evidenced in the published research and video materials, the seals were intact. There was no evidence of intervention by another party that could provide a defense (alteration of product) for the manufacturer, seller, marketer, or distributor.

ii. Design defect.
  1. There is a design defect when the final product came out as manufacturer intended, affecting the entire production line.
    i. Majority View (second restatement): the ordinary consumer expectation test. Is the product more dangerous than the ordinary consumer using the product would expect?
    ii. Minority View: the risk-utility balancing test (RUBT). Do the risks outweigh the benefited use?
      1. Does likelihood of harm exist?
      2. What is the gravity of harm?
      3. What is the feasibility of alternative designs?
      4. Under this minority view, prescription drugs are exempt from design defects, although jurisdictions are split over this.

Here the argument would simply be that if the genetic vaccines were intended to have crystals, fibers, and other material, the manufacturer should have listed this in the materials included in the genetic vaccines. However, under the majority view, an ordinary consumer of vaccines would not expect to find such debris in them. Such debris would cause inflammation and blood clotting (ITR), which would be harmful to the human body. As such, not only would this not be expected by the ordinary consumer, but also the likelihood of harm is increased, and the type of harm, including heart attacks, strokes, and so forth, would be considered severe, with the risk outweighing potential benefit. Under the minority RUBT approach, not every court agrees that prescription drugs should be considered.

iii. Warning defect (informed consent is a warning).
  i Lacks Warning.
    1. Majority View: the defendant manufacturer test. The manufacturer knew or should have known there was a significant enough risk to merit a warning.
      a. Likelihood of risk.
      b. Gravity of risk.
    2. No need to warn of "obvious" risk.

3. The learned intermediary doctrine.
    a. When the manufacturer of a prescription drug provides adequate warnings to physicians and physicians do not warn the patient, the manufacturer cannot be held liable under strict product liability because the physician is determined to be the "proximate cause" superseding the drug company.

ii. Inadequate warning.
    1. The warning must apprise a "reasonable" person of significant "risks."

Since informed consent is a warning provided to consumers of medical care and products, package inserts left intentionally blank cannot provide a warning to the consumer. Since a package insert is included with the genetic vaccines, the manufacturer certainly knew there was a significant-enough risk to alert consumers to, as this is part of the function of package inserts. Other functions of package inserts include informing physicians and other health-care providers, plus those administering and selling the vaccines (e.g., pharmacists), what the risks and benefits are to a patient, in addition to dosing, pharmacokinetics, pharmacodynamics, and the like. Since this information was not provided to those who would be considered "learned intermediaries," the manufacturer cannot use the learned intermediary doctrine to remove itself from liability. A reasonable person would want to know the risks and benefits and would certainly want to know about extraneous debris in the vaccines.

5. Cause-in-fact (a.k.a. actual cause of harm).
    1. Single defendant.
        i. But for the defect (manufacture, design, warning), the plaintiff (patient in this case) would not have been harmed.
    2. Multiple defendants.
        i. Substantial factor test Was it the action of the defendant that is substantially responsible for the damage?
    3. Multiple companies.
        i. Market-share liability: When multiple companies are involved in making the product (in this instance genetic vaccines) and the plaintiff does not know which manufacturer is responsible for the defective product, or part of the product, market-share liability allows the plaintiff to prove prima facie (on sight or first impression) causation against the manufacturers, based upon their share of the product on the market.

Your response here is determined by two factors. First, are you filing the suit for harm caused by one vaccine manufacturer (single defendant), or are multiple vaccine manufacturers causing the harm (multiple defendants)?

In the first instance, you need to be able to demonstrate that you were not harmed (specific physical injury or harm you are claiming due to vaccination) prior to receiving the vaccine and that the harm or injury occurred following vaccination.

With multiple defendants, you are trying to show that the injuries that occurred to the defendants were the result of the defendants' defective product. You are trying to show that the defendant manufacturer's conduct was substantially responsible for your injuries, either through something they did (commission), for example, making a vaccine with contaminants, or something they didn't do (omission), for example, a failure to have safety protocols or review of product to make certain the product (vaccine) was not defective.

The final factor, market-share liability, includes considerations in which more than one manufacturer is involved in the making of the product, covering supplies, equipment, various manufacturers, and locations. If it cannot be proven where the tainted vaccines came from, liability and responsibility with payments for damages (remedies), are determined by how much profit (share of the market), each company has. Since the companies work together to make the product, they are each liable to the extent they profit from their vaccines.

6. Proximate cause (a.k.a. legal cause of harm).
   1. Nothing is in between the plaintiff and defendant to account for the harm caused by the defect.

   The issue here is whether the manufacturer can make a case that it wasn't the manufacturer responsible for the debris or contaminants being in the vaccine that resulted in your injury. Someone or something else was. In this instance, the demonstration of chain-of-custody (hence the sixty-hours of video recordings made when we removed the iced vaccine vials from their containers, presented the vials to patients for inspection, verifying sealed vials with lot numbers and identifiers, sterilely drawing up the vaccines, and so forth) demonstrates no source of contamination came between the vaccine manufacturer and microscopic examination.

   You should be able to demonstrate such a chain-of-custody in which no source of contamination is between you and the manufacturer.
7. Damages (you were hurt).
   1. Property and personal injuries.
      i. If the product produced property damage only, then this is a contract-law issue.
      ii. For SPL to apply, there must be a personal injury.

You need to demonstrate a physical injury or harm. In the materials included with the 10Letters.org campaign to bring criminal charges against those involved in the development of bioweapons found at this QR code, we have military medical records of servicemen and -women, comparing their health and physical endurance testing before and after vaccination.

These records show deterioration in both physical health and physical endurance skills. The medical records, given after their written authorized release, provide evidence of proximate and cause-and-effect personal harm to these brave men and women serving our country.

8. SPL defenses (why the defendant shouldn't be found responsible)
   a. Assumption of risk.
      i. Second restatement and traditional common law.
         1. As with Strict Liability, this applies to "abnormally dangerous" activities.
            i. Plaintiffs cannot assume the risk if they didn't understand the danger and risk.
         2. The plaintiff
            i. Knew the risk, understood the risk, and voluntarily engaged in the activity.
   b. Comparative fault (objective).
      i. Majority view: most jurisdictions allow a sharing of fault between plaintiff and defendant, and recovery is based upon percentage of fault by each party.
   c. Misuse of product.
      ii. Plaintiff uses product in a manner not foreseeable by defendant manufacturer.
   d. Alteration of product.
      iii. When a third party alters the product in an unforeseeable way, the product will not be seen as defective in design.

The first seven elements demonstrate harm caused by a defective manufactured product, in this case, the genetic vaccine(s). The last element addresses potential defenses the manufacturer will try to raise in an effort to say that the harm you experienced was not their fault but yours.

The first approach is that you assumed the risk when you decided to take the injection. This is something that can apply if you, for example, are a scuba diver, like me. Although I am trained as an advanced and nitrox diver, I have signed waivers to not hold the dive shop and others responsible for my diving. You can bet that, unless something happened that an advanced diver would not be aware of, they would not be responsible for any problem I might encounter.

However, you probably did not get a degree in medicine or virology, and you probably were not involved in the development of these genetic vaccines. You therefore probably did not consider receiving a vaccine as an "abnormally dangerous activity." You have probably received several in your lifetime. In this instance, however, unless you received a different package insert—one not left "intentionally blank"—it is unconscionable that you could truly understand the risk and voluntarily engage in the taking of the vaccine. So, assumption of the risk is off the board.

Comparative fault presumes that you are at least partially responsible for the harm caused by the vaccination, which was tainted. If you opened the vial and put material into the vial before the vaccine was drawn up and injected into you, then you may be responsible for all or part of the injury. If that's the case, well, "you can't fix stupid."

The third defense is great if someone used a lawnmower to open a can of tuna and got cut. Again, you can't fix stupid, and we've all seen toothpaste and preparation-H tube warnings telling you not to do certain things with toothpaste and preparation-H. These warnings (outlined in Design defect—Warning defect) are probably not there because someone in an administrative meeting thought, We should add a warning about not putting preparation-H around your eyes to reduce wrinkles. No, they are there because stupid people did stupid things. If you do stupid things and misuse the product, that's a defense for the manufacturer.

The fourth and final SPL defense is alteration of product. If anyone other than the seller or manufacturer alters a product, the manufacturer or seller are not responsible for the injury resulting from the alteration. If vials were open (not sealed, or seal removed), or someone else put the contaminants into the vials, you clearly cannot hold the manufacturer responsible. As far as I know, the only people who can be held responsible for the actions of someone else are parents. Yet, school districts, lawyers, politicians, and special-interest groups want to make certain that parents don't know what they are doing to your children, while parents are still ultimately responsible for the bad outcomes—not the school districts, lawyers, politicians, or special-interest groups.

DO YOU HAVE A CASE?

Can you prove SPL from a genetic vaccine you were injected with?

Additional information on SPL can be found at:[13] https://rumble.com/v3su9e2-sv40-vs.-strict-product-liability.html. While civil litigation may help some people feel better about what has happened, it will not alter the course of history or the direction we are heading. It will not alter the dual-use gain-of-function research, or dual-use gene-therapy programs currently underway.

No amount of money will ever bring someone back from the dead or heal the maimed. No civil suit will ever restore these lives or the republic. Only indicting, prosecuting, and convicting these criminals will send a clear message that can reverberate through history. This time, if we do it right, there will be no saving the Nazi (American, Chinese, Ukrainian, and others) scientists, physicians, politicians, or intelligence agents. This time we need to put them, and the people promoting and paying for this, away for good (chapter 6), address the harm, and restore the republic (chapters 7 and 8).

CHAPTER 6

Putting It All Together—Indict, Prosecute, and Convict

Control over others lasts only as long as *We the People* allow it. Once *We the People* decide enough is enough, those in power, those who have abused *the People*, can, should, and will be brought to trial to pay for their crimes.

Prosecutors' Obligation to *We the People*

Every prosecuting attorney or agent in the United States, whether a federal or state attorney general, a district attorney, a Texas Ranger, or any other legally appointed person with the power to convene a grand jury, is obliged to present this material to a grand jury for the purpose of indicting the following individuals for the crimes stated below.

If you are one of *We the People*, you have the authority and the right to demand that these individuals with this legal authority convene a grand jury. You can do so in person, by telephone, by email, through www.10Letters.org, or by any other US and state constitutionally approved method, including the six sovereign states of Kansas, New Mexico, North Dakota, Nebraska, Nevada, and Oklahoma, which may convene a grand jury by consent of *the People*.[1] The following figure includes your state statutes (laws) that give you the legal authority to indict these criminals.

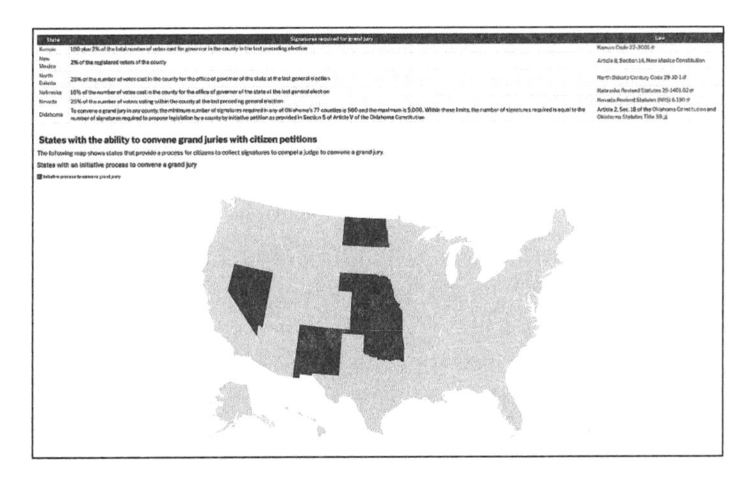

State	Signatures required for grand jury	Law
Kansas	100 plus 2% of the total number of votes cast for governor in the county in the last preceding election	Kansas Code 22-3001
New Mexico	2% of the registered voters of the county	Article II, Section 14, New Mexico Constitution
North Dakota	25% of the number of votes cast in the county for the office of governor of the state at the last general election	North Dakota Century Code 29-10.1
Nebraska	10% of the number of votes cast in the county for the office of governor of the state at the last general election	Nebraska Revised Statutes 29-1401.02
Nevada	25% of the number of voters voting within the county at the last preceding general election	Nevada Revised Statutes (NRS) 6.130
Oklahoma	To convene a grand jury in any county, the minimum number of signatures required in any of Oklahoma's 77 counties is 500 and the maximum is 5,000. Within these limits, the number of signatures required is equal to the number of signatures required to propose legislation by a county by initiative petition as provided in Section 5 of Article V of the Oklahoma Constitution	Article 2, Sec. 18 of the Oklahoma Constitution and Oklahoma Statutes Title 38

DEFENDANTS

As this book and its predecessor, *Is COVID-19 a Bioweapon? A Scientific and Forensic Investigation*, have demonstrated, and as will be laid out further in this criminal-complaint chapter, every prosecuting attorney in the United States is ethically bound to examine the evidence I've shown and submit it to a grand jury to fulfill his or her oath of office regarding the crimes committed by the following individuals, jointly and severally,[2] as well as individually and as part of a conspiratorial[3] crime:

1. Lloyd Austin, secretary, Department of Defense
2. Xavier Becerra, secretary of Health and Human Services
3. David Franz, former commander, Fort Detrick
4. Alejandro Mayorkas, secretary of the Department of Homeland Security
5. Chris Hassell, chairman of the HHS P3CO Review Committee
6. Rochelle P. Walensky, director of the Centers for Disease Control and Prevention
7. Janet Woodcock, commissioner of the US Food and Drug Administration
8. F. Fleming Crim, chief operating officer of the National Science Foundation
9. Francis Sellers Collins, MD, PhD, former director, National Institute of Health; former director of the National Human Genome Research Institute
10. Anthony Stephen Fauci, MD., former director, National Institute of Allergy and Infectious Diseases, second chief medical advisor to the president of the United States
11. Peter Daszak, PhD, president of EcoHealth Alliance
12. Ralph S Baric, PhD, University of North Carolina Chapel Hill
13. Shi Zhengli, PhD, Wuhan Institute of Virology
14. William Henry Gates III, Bill & Melinda Gates Foundation
15. Any and all other individuals found to be criminally culpable upon investigation of these crimes, including but not limited to presidents, present and past, of the United States of America.

CRIMES AND APPLICABLE COURTS

These individuals should be held responsible for federal criminal acts and ethical violations of the following:

1. The 1947 Nuremberg Code, including but not limited to
    a. Crimes of conspiracy, crimes against peace, war crimes, and crimes against humanity for
        i. The development, testing, and release of a biological weapon

ii. Interference with the treatment of patients and the practice of medicine
iii. Failure to follow adequate drug testing prior to use on people
iv. Failure to obtain informed consent
v. Experimentation upon the frail, elderly, those with physical or mental disabilities or diseases, prisoners, defendants in court, and children and pregnant women
vi. An attack on, with resulting loss of, personal liberties and protected rights
vii. Assault, threat, and coercion

2. The 1964 Declaration of Helsinki
3. The 1975 Biological Weapons Convention Treaty
4. The 1976 International Covenant on Civil and Political Rights Treaty
5. The 1979 *Belmont Report*
6. The 1989 Biological Weapons Anti-Terrorism Act
7. The 2012 federal law prohibitions with respect to biological weapons, 18 U.S.C. § 175
8. The 1965 Federal Perjury Statute 18 U.S.C. § 1001
   a. 18 U.S.C. § 2331 International and Domestic Terrorism
9. The 2018 Federal Perjury Statute 18 U.S.C. § 1621
10. The American Medical Association Code of Medical Ethics

In addition to criminal violations of state laws, using, for example, the State of Nevada:

1. Murder (NRS 200.030)
2. Attempted murder (NRS 200.030)
3. Manslaughter (NRS 200.050)
4. Reckless homicide (NRS 200.030)
5. Reckless endangerment (NRS 202.595)
6. Assault (NRS 200.471)
7. Battery (NRS 200.481)
8. False imprisonment (NRS 200.460)
9. Perjury (NRS 199.120)

These individuals, many of whom work for *We the People* in and through their official capacities within the United States government, appointed to positions of responsibility within our government, have repeatedly demonstrated certain behaviors,[4] including unethical experimentation on US citizens and military personnel. As my books and the following information demonstrate, these individuals have committed felonies for which they should be indicted by grand juries composed of the citizens of the United States of America. These individuals should then receive their day in court, where all evidence will be presented to the jury, not just what one party wants to present, as a jury cannot determine guilt or innocence if evidence is hidden[5] from them, as has been done previously in US courts.

On behalf of *People*, I present applicable jurisdictional authority for this case to be heard in federal and state courts, including but not limited to the following:

I. Evidence of habitual unethical behavior by the military-industrial complex and its agents acting under the guise of representing the *People of the United States of America*.

Unethical Human Experimentation in the U.S.

Numerous experiments which were performed on human test subjects in the United States are considered unethical, because they were illegally performed or they were performed without the knowledge, consent, or informed consent of the test subjects. Such tests were performed throughout American history, but most of them were performed during the 20th century. The experiments included the exposure of humans to many chemical and biological weapons (including infections with deadly or debilitating diseases), human radiation experiments, injections of toxic and radioactive chemicals, surgical experiments, interrogation and torture experiments, tests which involved mind-altering substances, and a wide variety of other experiments. Many of these tests were performed on children,[1] the sick, and mentally disabled individuals, often under the guise of "medical treatment". In many of the studies, a large portion of the subjects were poor, racial minorities, or prisoners.

Many of these experiments violated US law. Some others were sponsored by government agencies or rogue elements thereof, including the Centers for Disease Control, the United States military, and the Central Intelligence Agency, or they were sponsored by private corporations which were involved in military activities.[2][3][4] The human research programs were usually highly secretive and performed without the knowledge or authorization of Congress, and in many cases information about them was not released until many years after the studies had been performed.

The ethical, professional, and legal implications of this in the United States medical and scientific community were quite significant, and led to many institutions and policies that attempted to ensure that future human subject research in the United States would be ethical and legal. Public outrage in the late 20th century over the discovery of government experiments on human subjects led to numerous congressional investigations and hearings, including the Church Committee and Rockefeller Commission, both of 1975, and the 1994 Advisory Committee on Human Radiation Experiments, among others.

In 1987 the United States Supreme Court ruled in *United States v. Stanley*, 483 U.S. 669, that a U.S. serviceman who was given LSD without his consent, as part of military experiments, could not sue the U.S. Army for damages. Stanley was later awarded over $400,000 in 1996, two years after Congress passed a private claims bill in reaction to the case.[187] Dissenting the original verdict in *U.S. v. Stanley*, Justice Sandra Day O'Connor stated:

> No judicially crafted rule should insulate from liability the involuntary and unknowing human experimentation alleged to have occurred in this case. Indeed, as Justice Brennan observes, the United States played an instrumental role in the criminal prosecution of Nazi scientists who experimented with human subjects during the Second World War, and the standards that the Nuremberg Military Tribunals developed to judge the behavior of the defendants stated that the 'voluntary consent of the human subject is absolutely essential ... to satisfy moral, ethical, and legal concepts.' If this principle is violated, the very least that society can do is to see that the victims are compensated, as best they can be, by the perpetrators.

https://en.wikipedia.org/wiki/Unethical_human_experimentation_in_the_United_States

II. The 1975 Biological Weapons Convention (BWC) Treaty

Following more than a century of active ramped-up efforts by governments to use biological weapons against other countries—and as consistently demonstrated here, including experimentation upon US citizens and military men and women—the United States (following defeat in Vietnam) encouraged the development of the Biological Weapons Convention (BWC) Treaty to prevent the development, storage, use, or experimentation of biological agents.

> Whoever knowingly develops, produces, stockpiles, transfers, acquires, retains, or possesses any biological agent, toxin, or delivery system for use as a weapon or knowingly assists a foreign state or any organization to do so, or attempts, threatens, or conspires to do the same, shall be fined under this title or imprisoned for life or any term of years, or both.

As shown in the following image, *Nature* published two articles in July and December of 2001, revealing that efforts to institute verification procedures affirming compliance with the Biological Weapons Convention (BWC) Treaty were blocked by the United States. The articles reveal that the US believed there were violations of the BWC by Russia and that the US also believed inspections of US facilities could "threaten trade secrets," noting the Federation of American Scientists didn't "like the idea that someone could inspect our national labs." It is important to note that article XII of the BWC requires the exchange of information to prevent countries like the U.S., Russia, China, or the Ukraine from surprising other countries with potential scientific or technological developments relevant to biological weapons development.

US rejects bioweapon inspections

Jonathan Knight, San Francisco

A plan that would give teeth to the Biological Weapons Convention seems to be doomed. US negotiators will tell this week's talks on the convention in Geneva that they still strongly object to the inclusion of a verification procedure in the treaty.

In all, 140 countries have ratified the convention since it was hammered out almost 30 years ago. But the treaty contains no provision for verification, a loophole that allowed the Soviet Union to operate dozens of germ-warfare facilities in the 1970s and 80s.

Attempts to develop a verification plan began in 1995. It was hoped that the latest draft, released in March, would address the concerns of many participants, including the United States. But as *Nature* went to press, US representatives were set to announce that they would not sign it in its current form.

The United States has said in the past that the convention's inspections could threaten trade secrets in its biotechnology industry. A more sensitive issue may be the US biological defence programme, says Amy Rossi of the Federation of American Scientists in Washington. "We don't like the idea that someone could inspect our national labs."

China and Iran both made objections to the draft text in May, mainly over export restrictions. But as the current series of talks got under way on 23 July, both countries said they hoped the protocol would be finalized this year.

They and others may be waiting for the United States to kill the protocol to save themselves the embarrassment, says bioweapons expert Milton Leitenberg of the University of Maryland in College Park. "The Americans and the Chinese are playing this awful game of who's going to get the blame."

Nature | VOL 412 | 26 July 2001 | www.nature.com

Nature | VOL 414 | 13 December 2001 | www.nature.com

Bioweapons treaty in disarray as US blocks plans for verification

Shapes

In addition to any concerns about potential scientific or technological developments relevant to biological weapons development, as shown in the next figure, a report filed 22 November 2021 by the US to the BWC, revealed that beginning in 1997, the US DoD entered six former Soviet Union countries where instead of destroying "dangerous pathogens", the US "secured" them, in violation of the BWC Treaty.

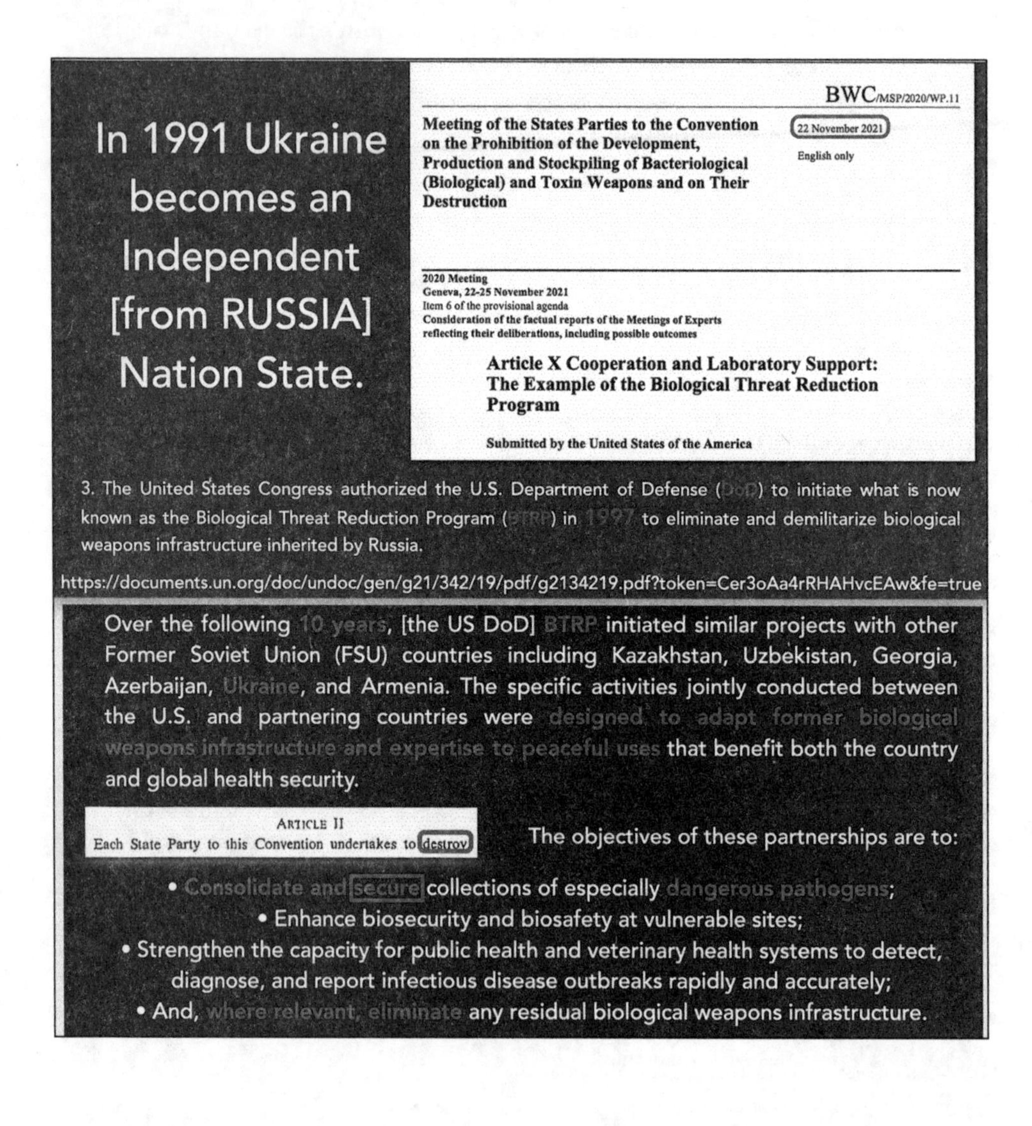

In 2020, after the beginning of the COVID pandemic, following the US DoD-funded request for COVID information from Ukrainian biological laboratories in late 2019, the United States once again agreed to and confirmed specific terms banning biological weapons.

This treaty and all treaties are given the full force of constitutional law as the supreme law of the land, contingent upon the provisions of declarations, reservations, and understandings,[6] as adjudicated by the United States Supreme Court. The US Congress imposed no reservations or understandings upon the BWC Treaty, thereby giving it full legal enforcement within the United States in and through the courts.

Article VI Clause 2 of the U.S. Constitution.
Treaties are the Supreme Law of the Land.

Article VI Article VI Annotated

All Debts contracted and Engagements entered into, before the Adoption of this Constitution, shall be as valid against the United States under this Constitution, as under the Confederation.

This Constitution, and the Laws of the United States which shall be made in Pursuance thereof; and all Treaties made, or which shall be made, under the Authority of the United States, shall be the supreme Law of the Land; and the Judges in every State shall be bound thereby, any Thing in the Constitution or Laws of any State to the Contrary notwithstanding.

The Senators and Representatives before mentioned, and the Members of the several State Legislatures, and all executive and judicial Officers, both of the United States and of the several States, shall be bound by Oath or Affirmation, to support this Constitution; but no religious Test shall ever be required as a Qualification to any Office or public Trust under the United States.

https://constitution.congress.gov/constitution/article-6/#article-6-clause-3

III. The 2012 Federal Law Prohibitions with Respect to Biological Weapons 18 U.S.C. § 175 and the 1989 Biological Weapons Anti-Terrorism Act.

As further evidence of the commitment of the United States Congress that there be no question as to the illegality of the development and use of biological weapons, Congress passed Federal Law 18, United States Code § 175, including all of the specific components (a.k.a. the Biological Weapons Anti-Terrorism Act of 1989), as shown on p. 146.

18 U.S. Code § 175 - Prohibitions with respect to biological weapons

U.S. Code Notes

prev | next

(a) In General.—
Whoever knowingly develops, produces, stockpiles, transfers, acquires, retains, or possesses any biological agent, toxin, or delivery system for use as a weapon, or knowingly assists a foreign state or any organization to do so, or attempts, threatens, or conspires to do the same, shall be fined under this title or imprisoned for life or any term of years, or both. There is extraterritorial Federal jurisdiction over an offense under this section committed by or against a national of the United States.

(b) Additional Offense.—
Whoever knowingly possesses any biological agent, toxin, or delivery system of a type or in a quantity that, under the circumstances, is not reasonably justified by a prophylactic, protective, bona fide research, or other peaceful purpose, shall be fined under this title, imprisoned not more than 10 years, or both. In this subsection, the terms "biological agent" and "toxin" do not encompass any biological agent or toxin that is in its naturally occurring environment if the biological agent or toxin has not been cultivated, collected, or otherwise extracted from its natural source.

(c) Definition.—
For purposes of this section, the term "for use as a weapon" includes the development, production, transfer, acquisition, retention, or possession of any biological agent, toxin, or delivery system for other than prophylactic, protective, bona fide research, or other peaceful purposes.

(Added Pub. L. 101–298, § 3(a), May 22, 1990, 104 Stat. 201; amended Pub. L. 104–132, title V, § 511(b)(1), Apr. 24, 1996, 110 Stat. 1284; Pub. L. 107–56, title VIII, § 817(1), Oct. 26, 2001, 115 Stat. 385; Pub. L. 107–188, title II, § 231(c)(1), June 12, 2002, 116 Stat. 661.)

Included within this federal statute is a requirement that the Department of Justice (DOJ) become actively involved in emergency situations involving biological weapons. On 25 March 2020, President Trump declared a state of emergency.[7]

18 USC 175a: Requests for military assistance to enforce prohibition in certain emergencies
Text contains those laws in effect on November 11, 2022

From Title 18-CRIMES AND CRIMINAL PROCEDURE
PART I-CRIMES
CHAPTER 10-BIOLOGICAL WEAPONS
Jump To:
Source Credit
Miscellaneous
References In Text

EVENT 2021:
What If The People You Trust Are The People Causing The Problem?

§175a. Requests for military assistance to enforce prohibition in certain emergencies

The Attorney General may request the Secretary of Defense to provide assistance under section 382 of title 10 [1] in support of Department of Justice activities relating to the enforcement of section 175 of this title in an emergency situation involving a biological weapon of mass destruction. The authority to make such a request may be exercised by another official of the Department of Justice in accordance with section 382(f)(2) of title 10.[1]

(Added Pub. L. 104–201, div. A, title XIV, §1416(c)(1)(A), Sept. 23, 1996, 110 Stat. 2723 .)

Editorial Notes

References in Text

Section 382 of title 10, referred to in text, was renumbered section 282 of title 10, Armed Forces, by Pub. L. 114–328, div. A, title XII, §1241(a)(2), Dec. 23, 2016, 130 Stat. 2497 .

[1] *See References in Text note below.*

https://uscode.house.gov/view.xhtml?req=granuleid:USC-prelim-title18-section175a&num=0&edition=prelim

Other countries, including China, were quick to point out that the United States has repeatedly violated the BWC Treaty.[8]

Meetings Coverage and Press Releases

Home | Secretary-General | General Assembly | Security Council | Economic and Social Council | International Court of Justice

MEETINGS COVERAGE
GENERAL ASSEMBLY » FIRST COMMITTEE

SEVENTY-SIXTH SESSION, 5TH MEETING (AM)

GA/DIS/3666
7 OCTOBER 2021

Use of Chemical, Biological Weapons Unacceptable in Any Context, Delegates Stress, as First Committee Continues General Debate

Condemning the use of chemical and biological weapons as unacceptable under any context or circumstances, delegates urged all States to abide by critical existing international instruments for their regulation, as the First Committee (Disarmament and International Security) continued its general debate today.

Emphasizing the importance of the Convention on the Prohibition of the Development, Production, Stockpiling and Use of Chemical Weapons and of Their Destruction (Chemical Weapons Convention), speakers also condemned use of those agents as an appalling violation of international law.

Turkey's representative said the Government of Syria used chemical weapons in that country, which constitutes a crime against humanity. Syria should cooperate fully with the Investigation and Identification Team and the Organisation for the Prohibition of Chemical Weapons (OPCW) regarding its chemical weapons programme and stockpiles, he stressed.

Canada's representative said his country is working with more than 20 States on strengthening biosafety and biosecurity and on enhancing surveillance and diagnostic capabilities. Also citing Syria as well as the Russian Federation's use of the nerve agent Novichok, she urged those countries to cooperate with OPCW and to comply with their obligations. Also referring to the poisoning of Alexei Navalny, Italy's delegate urged the Russian Federation to clarify its responsibility in the incident.

Syria's representative also condemned the use of chemical weapons, noting that his country has signed the Chemical Weapons Convention and fulfilled its obligations. However, some countries continue to politicize that issue and to create anti-Syria mechanisms within OPCW, he said. Recalling evidence that terrorist groups have used chemical weapons and blamed Damascus, he said that such information has not been considered.

Delegates noted that the Convention on the Prohibition of the Development and Stockpiling of Bacteriological (Biological) and Toxin Weapons and on Their Destruction (Biological Weapons Convention) has established norms in that domain, which must be implemented.

China's representative, however, pointed out that the United States unilaterally withdrew from the protocol to the Biological Weapons Convention in 2001 and operates more than 200 biological labs outside its territory. They function in an opaque manner, posing serious risks for the Russian Federation and China, he said. The United States and its allies also carry out activities in biological labs within their own territories, with the United States providing no information, he added, urging those States to operate in an open manner and to resume negotiations on a verifiable legally binding mechanism.

Call to UK Parliament & ICC

Due to our violation of US constitutional treaty commitments and federal statutory law, I reached out to our allies in the United Kingdom in an effort to encourage the UK Parliament to hold the United States accountable and encourage the US Congress to take action as shown in the Press Release on the next page.

In addition to requesting support from our allies, Professors Luc Montagnier, Kevin McCairn, and I provided affidavits in support of the International Criminal Court (ICC) case, filed by attorneys Melinda Mayne and Kaira S. McCallum (143/21); submitted by Slovakian attorneys Peter Weis, Marica

Press Release Monday, March 13, 2023

PARLIAMENT SHOULD HOLD USA ACCOUNTABLE

Andrew Bridgen, MP & Richard M Fleming, PhD, MD, JD

Gain-of-Function

Research investigation has shown that the United States is heavily invested, using money from NIAID and the DoD, in the development of viral Bioweapons in violation of The International (BWC) Treaty. Evidence indicates SARS-CoV-2 has resulted from that research Treaty violation.

Casualties From SARS-CoV-2

The world death count from the Viral Bioweapon is over 6.7 Million, including more than 203,000 deaths in the U.K.*.

COVID Gene Vaccine Casualties

Based upon the information as of 3 months ago (October 2022) there have been more than 2,400 deaths in the UK following the use of the genetic vaccine products, which are copies of the US Biological Viral Weapon. The Yellow Card report also shows more than 1.6 Million Vaccine Adverse Events following the injection of these Genetic Vaccines*.

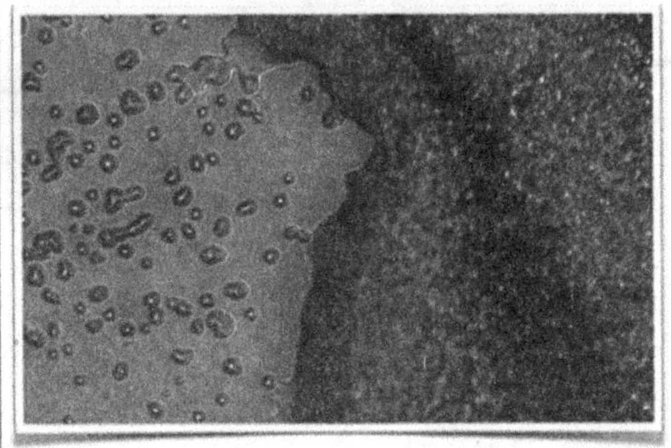

We encourage the US Congress to Investigate US Funding of COVID!

British citizens paying a price for the US Biological Viral Weapon and Genetic Vaccine Program funded by the US NIAID & DoD. Either the US will hold its Criminals Accountable or WE should!

We call upon the Houses of Parliament to demand accountability on the part of the US Government for their violation of the Biological Weapons Convention Treaty resulting in the COVID pandemic and unprecedented use of experimental genetic vaccines that turn red blood cells gray and cause blood to clot upon contact as shown in the image above; thus causing heart damage including prion disease (amyloidosis) and myocarditis, strokes, cancer, miscarriages and DEATH!

DrRichardMFleming https://10letters.org

* https://ukfreedomproject.org/covid-19-vaccines-yellow-card-analysis/

Pirosikova, and Erik Schmidt (133/21); submitted by French attorneys Patrice Lepiller and Raphael Cohen (271/21); and submitted by Czech Republic attorney Tomas Nielsen (326/21); as submitted to the ICC on 12 August 2021, to hold those responsible for developing and releasing these biological weapons criminally accountable in international court. These cases and affidavits have been supported by more than 73,000 people from around the world who have signed the petition for the case to be heard before the ICC, in addition to the letter of support for the case and the submitted affidavits. The letter was provided by Nazi concentration camp survivors Moshe Brown, Hillel Handler, and Vera Sharav which was submitted 20 September 2021.[9] To date, we still await action by the ICC.

IV. The 1979 *Belmont Report*

When people are experimented on without receiving "defined informed consent" (IC), including but not limited to an experimental mass genetic vaccination program, that is a violation as stipulated in the 1979 *Belmont Report* issued by the US Department of Health and Human Services (HHS).[10] The *Belmont Report* is the required understanding and agreement necessary to obtain US-sanctioned authorization to conduct human experimental research in the United States. Specifically of note is that informed consent requires information, comprehension, and voluntariness.

V. The 1976 International Covenant on Civil and Political Rights Treaty

In addition to the *Belmont Report*, the International Covenant on Civil and Political Rights (ICCPR) Treaty[11] covers informed consent and the right not to be involuntarily experimented on. US citizens' rights have been violated, as have the rights of people around the world whose governments signed and ratified the ICCPR Treaty. The ICCPR Treaty specifically states, in Article 7, "In particular, no one shall be subjected without his free consent to medical or scientific experimentation."

VI. The 1964 Declaration of Helsinki

As with the *Belmont Report*, satisfactory training and certification by all US researchers involved in human experimentation is required, including acknowledgment of the rules and regulations involved in human research experimentation, stipulating what is and is not accepted.

Several additional violations are noted as established in the 1964 World Medical Association Declaration of Helsinki Ethical Principles for Human Medical Research. These International Ethical Principles are included for all research conducted on human subjects, including the pharmaceutical industry, government agencies, and medical providers.

VII. The American Medical Association Code of Medical Ethics

These illegal and ethical violations committed by the defendants violate not only the BWC, federal and state statutes, the ICCPR Treaty, and the Declaration of

Helsinki but also the American Medical Association Code of Medical Ethics, which covers all individuals involved in the practice of medicine or who are members of the American Medical Association (AMA). During one of my first live presentations in Dallas, Texas, on 5 June 2021,[12] I demonstrated that the package inserts for the three EUA-authorized genetic vaccines (Pfizer, Moderna, and Janssen) violated informed consent (IC) by failing to provide any information on the risks, benefits, or mechanisms of action. They provided none of the components required to obtain IC, specifically information, comprehension, and voluntariness.[13]

As demonstrated during the live, televised[14] Event 2021 presentation, several participants, including a pharmacist, were asked to open the package inserts of the Pfizer, Moderna, and Janssen[15] genetic vaccines on stage. In every instance, the package inserts revealed no information that would be required by a physician, pharmacist, or any other health-care provider to give a patient so that the person can make informed consent regarding the risks and benefits of receiving a vaccination injection. One such participant was a pharmacist who resigned from her job because she could not ethically give these genetic vaccines to people when the package inserts were blank with no informed consent information to give to patients. Here is a picture of another woman who showed the group in attendance and viewers of *The Highwire Program* the package insert that was intentionally blank.

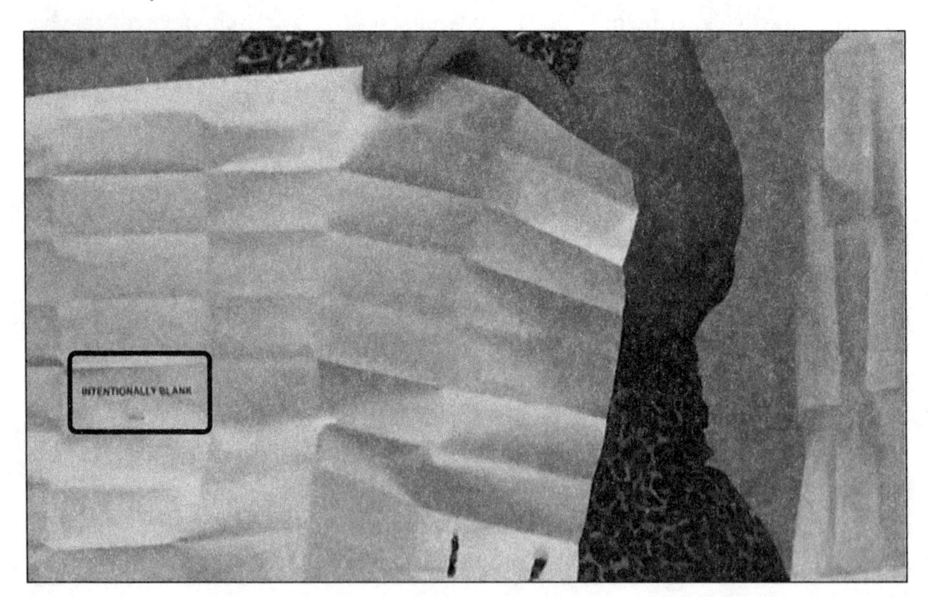

VIII. The 1947 Nuremberg Code

The 1947 Nuremberg code includes but is not limited to crimes of conspiracy, crimes against peace, war crimes, and crimes against humanity for

i. The development, testing, and release of a biological weapon.
ii. Interference with the treatment of patients and the practice of medicine.
iii. Failure to follow adequate drug testing prior to use on people.
iv. Failure to obtain informed consent.
v. Experimentation on the frail, elderly, those with physical or mental disabilities or diseases, prisoners, defendants in court, and children and pregnant women.
vi. Loss of personal liberties and protected rights.
vii. Assault, threat, and coercion.

In 1947 a series of war-crimes trials took place to hold criminally accountable those who had been involved in atrocities during the Second World War, including high-ranking government officials, lawyers, doctors, and nurses. These same violations listed were perpetrated against the American people and people around the world. The number of dead Americans (roughly 1.2 million) total more than the number of US military deaths in war since 1776. The number of wounded and maimed Americans (about 1.5 million) are more than the number of US military injuries since 1776. The number of dead people around the world (approaching 7 million) is greater than the number of people massacred by Hitler in his concentration camps.

SCAN ME

The following QR code slide shows the medical experimentation limitations set out in the 1947 Nuremberg Code.

For anyone who believes that the Nuremberg Code is not applicable in United States courts, I provide you with this ruling in 4:21-cv-01774 (*Bridges v. Houston Methodist Hospital*); in which US Federal Judge Lynn N. Hughes[16] held that the Nuremberg Code applies to "a government."

IX. The 1965 Federal Perjury Statute 18 USC § 1001[17]

a. Except as otherwise provided in this section, whoever, in any matter within the jurisdiction of the executive, legislative, or judicial branch of the Government of the United States, knowingly and willfully
   1. falsifies, conceals, or covers up by any trick, scheme, or device a material fact;
   2. makes any materially false, fictitious, or fraudulent statement or representation; or
   3. makes or uses any false writing or document knowing the same to contain any materially false, fictitious, or fraudulent statement or entry;

shall be fined under this title, imprisoned not more than 5 years or, if the offense involves international or domestic terrorism (as defined in "18 U.S.C. § 2331), imprisoned not more than 8 years, or both. If the matter relates to an offense under chapter 109A, 109B, 110, or 117, or section 1591, then the term of imprisonment imposed under this section shall be not more than 8 years.

b. Subsection (a) does not apply to a party to a judicial proceeding, or that party's counsel, for statements, representations, writings or documents submitted by such party or counsel to a judge or magistrate in that proceeding.
c. With respect to any matter within the jurisdiction of the legislative branch, subsection (a) shall apply only to
   1. administrative matters, including a claim for payment, a matter related to the procurement of property or services, personnel or employment practices, or support services, or a document required by law, rule, or regulation to be submitted to the Congress or any office or officer within the legislative branch; or
   2. any investigation or review, conducted pursuant to the authority of any committee, subcommittee, commission or office of the Congress, consistent with applicable rules of the House or Senate.

   (June 25, 1948, ch. 645, 62 Stat. 749; Pub. L. 103–322, title XXXIII, § 330016(1)(L), Sept. 13, 1994, 108 Stat. 2147; Pub. L. 104–292, § 2, Oct. 11, 1996, 110 Stat. 3459; Pub. L. 108–458, title VI, § 6703(a), Dec. 17, 2004, 118 Stat. 3766; Pub. L. 109–248, title I, § 141(c), July 27, 2006, 120 Stat. 603.)

Pursuant to 18 USC § 1001 Reference to 18 U.S.C. § 2331 Definition of International and Domestic Terrorism[18]

As used in this chapter,

1. the term "international terrorism" means activities that
   A. involve violent acts or acts dangerous to human life that are a violation of the criminal laws of the United States or of any State, or that would be a criminal violation if committed within the jurisdiction of the United States or of any State;
   B. appear to be intended
      i. to intimidate or coerce a civilian population;
      ii. to influence the policy of a government by intimidation or coercion; or
      iii. to affect the conduct of a government by mass destruction, assassination, or kidnapping; and
   C. occur primarily outside the territorial jurisdiction of the United States, or transcend national boundaries in terms of the means by

which they are accomplished, the persons they appear intended to intimidate or coerce, or the locale in which their perpetrators operate or seek asylum;

2. the term "national of the United States" has the meaning given such term in section 101(a)(22) of the Immigration and Nationality Act;
3. the term "person" means any individual or entity capable of holding a legal or beneficial interest in property;
4. the term "act of war" means any act occurring in the course of
   A. declared war;
   B. armed conflict, whether or not war has been declared, between two or more nations; or
   C. armed conflict between military forces of any origin;
5. the term "domestic terrorism" means activities that
   A. involve acts dangerous to human life that are a violation of the criminal laws of the United States or of any State;
   B. appear to be intended
      i. to intimidate or coerce a civilian population;
      ii. to influence the policy of a government by intimidation or coercion; or
      iii. to affect the conduct of a government by mass destruction, assassination, or kidnapping; and
   C. occur primarily within the territorial jurisdiction of the United States; and
6. the term "military force" does not include any person that
   A. has been designated as a
      i. foreign terrorist organization by the Secretary of State under section 219 of the Immigration and Nationality Act (8 U.S.C. 1189); or
      ii. specially designated global terrorist (as such term is defined in section 594.310 of title 31, Code of Federal Regulations) by the Secretary of State or the Secretary of the Treasury; or
   B. has been determined by the court to not be a "military force."

(Added Pub. L. 102–572, title X, § 1003(a)(3), Oct. 29, 1992, 106 Stat. 4521; amended Pub. L. 107–56, title VIII, § 802(a), Oct. 26, 2001, 115 Stat. 376; Pub. L. 115–253, § 2(a), Oct. 3, 2018, 132 Stat. 3183.)

Pursuant to 18 USC § 1001 reference to 18 U.S.C. § 2331, any individual who commits perjury and in doing so has also been involved in an act of terrorism is subject to criminal prosecution under both federal statutes.

X. The 2018 Federal Perjury Statute 18 USC § 1621.

Under Chapter 79[19] of this federal law, the US Congress made it clear what it considered lying under oath (perjury) to mean and what the penalties for doing so are.

> §1621. Perjury Generally Under Title 18 of Crimes and Criminal Procedure.
> Whoever-
> (1) having taken an oath before a competent tribunal, officer, or person, in any case in which a law of the United States authorizes an oath to be administered, that he will testify, declare, depose, or certify truly, or that any written testimony, declaration, deposition, or certificate by him subscribed, is true, willfully and contrary to such oath states or subscribes any material matter which he does not believe to be true; or
> (2) in any declaration, certificate, verification, or statement under penalty of perjury as permitted under section 1746 of title 28, United States Code, willfully subscribes as true any material matter which he does not believe to be true is guilty of perjury and shall, except as otherwise expressly provided by law, be fined under this title or imprisoned not more than five years, or both. This section is applicable whether the statement or subscription is made within or without the United States.
> (June 25, 1948, ch. 645, 62 Stat. 773; Pub. L. 88–619, §1, Oct. 3, 1964, 78 Stat. 995; Pub. L. 94–550, §2, Oct. 18, 1976, 90 Stat. 2534; Pub. L. 103–322, title XXXIII, §330016(1)(I), Sept. 13, 1994, 108 Stat. 2147.)

Applicable State Statutes

In addition to the foregoing federal crimes detailed subsequently and throughout this book and prior books, defendants should also be charged with having committed the following crimes, in and through their actions involving the development and subsequent release of biological agents and consequential activities and actions, including but not limited to the development and release of such agent(s). They should be charged with the following state crimes as defined by the independent states pursuant to their constitutions and laws (using Nevada as one such state, according to its statutes):

1. Murder (NRS 200.030)
2. Attempted murder (NRS 200.030)
3. Manslaughter (NRS 200.050)
4. Reckless homicide (NRS 200.030)
5. Reckless endangerment (NRS 202.595)
6. Assault (NRS 200.471)
7. Battery (NRS 200.481)

8. False imprisonment (NRS 200.460)
9. Perjury (NRS 199.120)

Jurisdiction[20]

For any court to hear a case, including a criminal case that needs to be brought by a prosecuting attorney through a grand jury indictment, two types of jurisdictions must be established.

1. Subject Matter Jurisdiction
   a. The particular court has the authority to hear this type of case—for example, family courts hear matters of family disputes, while criminal courts hear cases involving violations of the law.
   b. Here the subject matter, including violations of the law on *both* the national and state level, has been shown here.[21]
2. Personal Jurisdiction[22]
   a. The person accused of the crime must be a person the court is permitted to bring before the court.

Accordingly, the following material is being presented for grand jury indictment, prosecution, and conviction of, including but not limited to, the previously mentioned individuals for the listed crimes as supported by the following complaint.

Indictment Introduction

1. The SARS-Cov-2 viruses were engineered with gain-of-function (GoF) technology. These viruses biologically yield an InflammoThrombotic Response (ITR) and prion diseases, resulting from the spike protein.

2. The consequences of this spike protein are independent of whether the spike protein load is the result of person-to-person transmission or encoding of the spike protein within the genetic vaccines.[23]

3. The alteration of naturally occurring viral pathogens to make a chimeric virus (SARS-CoV-2) capable of infecting humans is a criminal act in violation of the previously noted treaties and statutes. Once infected, these viruses are then able to transmit by respiratory and gastrointestinal pathways. Untreated, these viruses (COVID-19) can kill people by producing inflammation and blood clotting (InflammoThrombotic Response; ITR) and priogenic diseases.

4. These untreated ITR and priogenic diseases are responsible for more than 568,000 excess American deaths, in addition to the 1,161,602 who died from the same priogenic and ITR caused by the virus, genetic vaccines,

or both and were pronounced dead due to COVID, with some 6,593,929 infected Americans.[24]

5. The evidence will demonstrate that SARS-CoV-2 meets the definition of a gain-of-function bioweapon, and the development of this bioweapon is a violation of US treaty and criminal law, making the defendants criminally responsible for the deaths of these Americans.

6. The evidence will show that the defendants are responsible for the funding and development of this bioweapon,[25] that they have interfered with the treatment of individuals infected with this bioweapon, and that they have promulgated the use of experimental genetic vaccine biologics, and as a result, these viruses and genetic vaccines have exacerbated underlying ITR diseases in individuals.[26]

7. The evidence will show that the Emergency Use Authorization (EUA) documents,[27] when statistically analysed to determine if these genetic vaccines reduced either (a) the number of COVID cases, or (b) the number of deaths, demonstrated no statistical reduction for the Pfizer or Moderna vaccines, while the Janssen vaccine demonstrated a reduction ($p<0.05$) at two-weeks postvaccination. This difference in Janssen vaccinated versus unvaccinated individuals no longer existed four weeks after vaccination.

8. The evidence will show that this interference in treatment of patients by physicians contributed to the injury and death of patients that could have been prevented by treating the ITR effects. This interference, not only promoted an environment of fear and manipulation but also was associated with federal and state governments overreaching their constitutional authority.

9. The evidence will also show that these actions fit the definition of crimes against humanity.

Applicable Laws Governing Gain-of-Function, Informed Consent, Perjury, and Terrorism

10. Since its founding in 1847, the American Medical Association (AMA) has played a crucial role in the development of medicine in the United States. Opinion 7.1.2[28] of the American Medical Association Code of Medical Ethics reads as follows:

    > Informed consent is an essential safeguard in research. The obligation to obtain informed consent arises out of respect for persons and a desire to respect the autonomy of the individual deciding whether to volunteer to participate in biomedical or health research. For these reasons, no person may be used as a subject in research against his or her will.

Physicians must ensure that the participant (or legally authorized representative) has given voluntary, informed consent before enrolling a prospective participant in a research protocol. With certain exceptions, to be valid, informed consent requires that the individual have the capacity to provide consent and have sufficient understanding of the subject matter involved to form a decision. The individual's consent must also be voluntary.

A valid consent process includes the following:

a. Ascertaining that the individual has decision-making capacity.
b. Reviewing the process and any materials to ensure that it is understandable to the study population.
c. Disclosing these elements:
    i. The nature of the experimental drug(s), device(s), or procedure(s) to be used in the research;
    ii. Any conflicts of interest relating to the research, in keeping with ethics guidance;
    iii. Any known risks or foreseeable hazards, including pain or discomfort that the participant might experience;
    iv. The likelihood of therapeutic or other direct benefit for the participant;
    v. Alternative courses of action open to the participant, including choosing standard or no treatment instead of participating in the study;
    vi. The nature of the research plan and implications for the participant;
    vii. The differences between the physician's responsibilities as a researcher and as the patient's treating physician.
d. Answering questions the prospective participant has.
e. Refraining from persuading the individual to enroll.
f. Not encouraging unrealistic expectations.
g. Documenting the individual's voluntary consent to participate.

Participation in research by minors or other individuals who lack decision-making capacity is permissible in limited circumstances when:

h. Consent is given by the individual's legally authorized representative, under circumstances in which informed and prudent adults would reasonably be expected to volunteer themselves or their children in research.
i. The participant gives his or her assent to participation, where possible. Physicians should respect the refusal of an individual who lacks decision-making capacity.
j. There is potential for the individual to benefit from the study. In certain situations, with special safeguards in keeping with ethics guidance, the obligation to obtain informed consent may be waived in research on emergency interventions.

AMA Principles of Medical Ethics: I, III, V

11. The 1947 Nuremberg Code[29] reads as follows:

1. The voluntary consent of the human subject is absolutely essential. This means that the person involved should have legal capacity to give consent; should be so situated as to be able to exercise free power of choice, **without** the intervention of any element of force, fraud, deceit, duress, **over-reaching**, or other ulterior form of constraint or **coercion**; and should have sufficient knowledge and comprehension of the elements of the subject matter involved, as to enable him to make an understanding and enlightened decision. This latter element requires that, before the acceptance of an affirmative decision

by the experimental subject, there should be made known to him the nature, duration, and purpose of the experiment; the method and means by which it is to be conducted; all inconveniences and hazards reasonably to be expected; and the effects upon his health or person, which may possibly come from his participation in the experiment. The duty and responsibility for ascertaining the quality of the consent rests upon each individual who initiates, health or person, which may possibly come from his participation in the experiment. The duty and responsibility for ascertaining the quality of the consent rests upon each individual who initiates, directs or engages in the experiment. It is a personal duty and responsibility which may not be delegated to another with impunity.

2. The experiment should be such as to yield fruitful results for the good of society, **unprocurable by other methods** or means of study, and not random and unnecessary in nature.
3. The experiment should be so designed and **based on the results of animal experimentation** and a knowledge of the natural history of the disease or other problem under study, that the anticipated results will justify the performance of the experiment.
4. The experiment should be so conducted as to **avoid all unnecessary physical and mental suffering and injury**.
5. No experiment should be conducted where there is an a priori reason to believe that death or disabling injury will occur; except, perhaps, in those experiments where the experimental physicians also serve as subjects.
6. The degree of risk to be taken should never exceed that determined by the humanitarian importance of the problem to be solved by the experiment.
7. Proper preparations should be made and adequate facilities provided to protect the experimental subject against even remote possibilities of injury, disability, or death.
8. The experiment should be conducted only by **scientifically qualified persons**. The highest degree of skill and care should be required through all stages of the experiment of those who conduct or engage in the experiment.
9. During the course of the experiment, the human subject should be at liberty to bring the experiment to an end, if he has reached the physical or mental state where continuation of the experiment seemed to him to be impossible.
10. During the course of the experiment, the scientist in charge must be prepared to terminate the experiment at any stage, if he has probable cause to believe, in the exercise of the good faith, superior skill and careful judgement required of him, that a continuation of the experiment is likely to result in injury, disability, or death to the experimental subject.

[emphasis added]

"Trials of War Criminals before the Nuremberg Military Tribunals under Control Council Law No. 10," Vol. 2, pp. 181–82, Washington, D.C.: U.S. Government Printing Office, 1949.

12. The Biological Weapons Convention (BWC) Treaty was signed by President Gerald R. Ford on 10 April 1972 and was ratified by Congress on 26 March 1975.[30]

In brief, the fifteen (XV) articles address the following.

Article I: Prohibition of Biological Weapons
Article II: Destruction of all Biological Weapons
Article III: Prohibition of Transfer of Biological Weapons
Article IV: Implementation within (in this instance) the United States
Article V: International Discussion and Cooperation
Article VI: Action to Hold Countries Accountable
Article VII: Assistance to Countries Exposed to Biological Weapons
Article VIII: Prior Agreements Against Use of Gases Remain in Effect
Article IX: Continued Agreement to Prevent Chemical Weapons
Article X: International Exchange of Data and Knowledge
Article XI: Any Country May Propose Additions to the Treaty
Article XII: Future Meetings Including Detailing of Any New Scientific and Technical Developments
Article XIII: Nations May Withdraw at Any Time
Article XIV: All Nations May Participate in This Treaty
Article XV: The Treaty Will Be Deposited in English, Russian, French, Spanish, and Chinese

In Summary, the BWC Treaty, as noted on the US Department of State[31] website (dated 4 August 2022) focuses on the following:

> The BWC is critical to international efforts to address the threat posed by biological and toxin weapons—whether in the hands of governments or non-state actors. To remain effective, it must be strengthened to address the biological threats we face in the 21st Century. An important objective shared by the United States and other BWC States Parties is universal adherence to the Convention, and—as of August 2022—only thirteen countries have not yet joined it.
>
> The core obligations of Parties under the Convention are:
>
> A. Never to develop, produce, stockpile, or otherwise acquire or retain: 1) biological agents or toxins of types and in quantities that have no justification for peaceful uses; and 2) weapons, equipment, or means of delivery designed to use such agents or toxins for hostile purposes (Article I).
>
> B. To destroy or divert to peaceful purposes all agents, toxins, weapons, equipment, and means of delivery specified in Article I in their possession, or under their jurisdiction or control (Article II).
>
> C. Not to transfer or in any way to assist, encourage, or induce any entity to manufacture or otherwise acquire any of the agents, toxins, weapons, equipment or means of delivery specified in Article I (Article III).
>
> D. To take any necessary measures to prohibit and prevent the development, production, stockpiling, acquisition, or retention of any of the agents, toxins, weapons, equipment, and means of delivery specified in Article I under its jurisdiction or control (Article IV).

In furtherance of ratifying the BWC Treaty, and in accord with US federal law, applicable to the states, the Congress of the United States passed legislation in accord with honouring the BWC Treaty:

A. The 1989 Biological Weapons Anti-Terrorism Act (BWATA).[32] Passed Senate 21 November 1989. Passed House 8 May 1990. Signed by President George H. W. Bush 22 May 1990.

Within the provisions of the BWATS, Congress stipulated the following:

Whoever knowingly develops, produces, stockpiles, transfers, acquires, retains, or possesses any biological agent, toxin, or delivery system for use as a weapon, or knowingly assists a foreign state or any organization to do so, shall be fined under this title or imprisoned for life or any term of years, or both.

"Any micro-organism, virus, infectious substance, or biological product that may be engineered as a result of biotechnology, or any naturally occurring or bioengineered component of any such microorganism, virus, infectious substance, or biological product, capable of causing death, disease, or other biological malfunction in a human, an animal, a plant, or another living organism; deterioration of food, water, equipment, supplies, or material of any kind or deleterious alteration of the environment."

Within this law, the terms toxin, delivery system, and vector are specifically defined and directly applicable for both the gain-of-function (GoF) viruses and genetic vaccines that encode for the GoF SARS-CoV-2[33] spike protein.

- Toxin: "whatever its origin or method of production—any poisonous substance produced by a living organism; or any poisonous isomer, homolog, or derivative of such a substance."
- Delivery system: "any apparatus, equipment, device, or means of delivery specifically designed to deliver or disseminate a biological agent, toxin, or vector."
- Vector: "a living organism capable of carrying a biological agent or toxin to a host."

B. The 2012 Prohibitions with respect to Biological Weapons 18 USC § 17[34] with subdivisions. This law went into effect 3 January 2012. It specifically states the following:

a. In General

Whoever knowingly develops, produces, stockpiles, transfers, acquires, retains, or possesses any biological agent, toxin, or delivery system for use as a weapon, or knowingly assists a foreign state or any organization to do so, or attempts, threatens, or conspires to do the same, shall be fined under this title or imprisoned for life or any term of years, or both. There is extraterritorial Federal jurisdiction over an offense under this section committed by or against a national of the United States.

b. Additional Offense

Whoever knowingly possesses any biological agent, toxin, or delivery system of a type or in a quantity that, under the circumstances, is not reasonably justified by a prophylactic, protective, bona fide research, or other peaceful purpose, shall be fined under this title, imprisoned not more than 10 years, or both. In this subsection, the terms "biological agent" and "toxin" do not encompass any biological agent or toxin that is in its naturally occurring environment, if the biological agent or toxin has not been cultivated, collected, or otherwise extracted from its natural source.

c. Definition

For purposes of this section, the term "for use as a weapon" includes the development, production, transfer, acquisition, retention, or possession of any biological agent, toxin, or delivery system for other than prophylactic, protective, bona fide research, or other peaceful purposes.

13. The International Covenant on Civil and Political Rights (ICCPR) Treaty was signed by President James Earl Carter Jr. on 5 October 1977, ratified by the US Congress on 8 June 1992, and effective 8 September 1992.[35] The ICCPR Treaty has full force in US jurisdiction, including enforcement against crimes committed by American citizens.[36]

    While several of the articles are important, Article 7[37] specifically states,

    > No one shall be subjected to torture or to cruel, inhuman or degrading treatment or punishment. In particular, no one shall be subjected without his free consent to medical or scientific experimentation.

14. The 1979 Belmont Report: Establishing the Ethical Principles and Guidelines for the Protection of Human Subjects of Research is available.[38]

    In summary, the report reads as follows:

    > On July 12, 1974, the National Research Act (Pub. L. 93–348) was signed into law, thereby creating the National Commission for the Protection of Human Subjects of Biomedical and Behavioral Research. One of the charges to the commission was to identify the basic ethical principles that should underlie the conduct of biomedical and behavioral research involving human subjects and to develop guidelines that should be followed to assure that such research is conducted in accordance with those principles. In carrying that out, the commission was directed to consider
    >
    > **i.** the boundaries between biomedical and behavioral research and the accepted and routine practice of medicine,
    > **ii.** the role of assessment of risk-benefit criteria in the determination of the appropriateness of research involving human subjects,
    > **iii.** appropriate guidelines for the selection of human subjects for participation in such research, and
    > **iv.** the nature and definition of informed consent in various research settings.
    >
    > *The Belmont Report* attempts to summarize the basic ethical principles identified by the commission in the course of its deliberations. It is the outgrowth of an intensive four-day period of discussions held in February 1976 at the Smithsonian Institution's Belmont Conference Center, supplemented by the monthly deliberations of the commission held over a period of nearly four years. It is a statement of basic ethical principles and guidelines that should assist in resolving the ethical problems surrounding the conduct of research with human subjects. By publishing the report in the *Federal Register* and providing reprints upon request, the secretary intends that it may be made readily available to scientists, members of institutional review boards, and federal employees. The two-volume appendix, containing the lengthy reports of experts and specialists who assisted the commission in fulfilling this part of its charge, is available as DHEW Publication No. (OS) 78–0013 and No. (OS) 78–0014, for sale by the Superintendent of Documents, U.S. Government Printing Office, Washington, D.C. 20402.
    >
    > Unlike most other reports of the commission, *The Belmont Report* does not make specific recommendations for administrative action by the secretary of health,

education, and welfare. Rather, the commission recommended that *The Belmont Report* be adopted in its entirety, as a statement of the department's policy. The department requests public comment on this recommendation.

15. The 1964 Declaration of Helsinki.[39]

This declaration was put to writing in 1974 by the World Health Organization because of no established ethical guidelines for the practice of medicine or research.

The declaration is morally[40] binding and is considered the cornerstone document on human research ethics. It includes the following articles:

A. Articles 2, 3, 10: Investigator's Duty Is to the Patient
B. Article 8: Respect for the Individual
C. Article 11: Responsibility for Thorough Scientific Knowledge of Research
D. Articles 16, 17: Careful Assessment of Risks and Benefits
D. Articles 20, 21, 22: Informed Consent
E. Article 27: Conflict of Interest

16. 18 U.S.C. § 1001 Reference to 18 U.S.C. § 2331

Under these federal codes detailed supra, violation of ths law is perjury and domestic-terrorism statute which is subject to criminal prosecution.

17. 18 U.S.C. § 1621 Perjury

Under Chapter 79[41] of this federal law, the US Congress clarified what it considered lying under oath (perjury) to mean and what the penalties for doing so are.

§1621. Perjury Generally

Whoever

(1) having taken an oath before a competent tribunal, officer, or person, in any case in which a law of the United States authorizes an oath to be administered, that he will testify, declare, depose, or certify truly, or that any written testimony, declaration, deposition, or certificate by him subscribed, is true, willfully and contrary to such oath states or subscribes any material matter which he does not believe to be true; or

(2) in any declaration, certificate, verification, or statement under penalty of perjury as permitted under section 1746 of title 28, United States Code, willfully subscribes as true any material matter which he does not believe to be true; is guilty of perjury and shall, except as otherwise expressly provided by law, be fined under this title or imprisoned not more than five years, or both. This section is applicable whether the statement or subscription is made within or without the United States.

(June 25, 1948, chap. 645, 62 Stat. 773; Pub. L. 88–619, §1, Oct. 3, 1964, 78 Stat. 995; Pub. L. 94–550, §2, Oct. 18, 1976, 90 Stat. 2534; and Pub. L. 103–322, title 33, §330016(1)(I), Sept. 13, 1994, 108 Stat. 2147.)

Factual Background Evidence: Gain-of-Function[42]

18. Gain-of-function (GoF) research has been funded for many years by the United States Department of Health and Human Services (HHS), as confirmed by former director of the National Institutes of Health (NIH) Francis S. Collins, MD, PhD in a statement issued on 19 May 2021, as follows.[43] Coupled with other evidence, this not only confirms funding of coronavirus research but also, under oath and in contrast to the data showing grants involved in GoF research, demonstrates perjury.

Statement on misinformation about NIH support of specific gain-of-function research:

> Based on outbreaks of coronaviruses caused by animal to human transmissions such as in Asia in 2003 that caused Severe Acute Respiratory Syndrome (SARS), and in Saudi Arabia in 2012 that caused Middle East Respiratory Syndrome (MERS), NIH and the National Institute of Allergy and Infectious Diseases (NIAID) have for many years supported grants to learn more about viruses lurking in bats and other mammals that have the potential to spill over to humans and cause widespread disease. However, neither NIH nor NIAID have ever approved any grant that would have supported **"gain-of-function" research on coronaviruses that would have increased their transmissibility or lethality for humans**. [emphasis added]

19. On 21 May 2000 Dr. Ralph Baric at the University of North Carolina at Chapel Hill published research showing he had successfully cloned SARS-CoV-Urbani using recombinant (chimeric) DNA to produce "an infectious, replication defective, coronavirus."

    This gain-of-function research was funded by the NIH via grant numbers AI23946, GM63228, and AI26603.[44] SARS-CoV-Urbani, later indexed on GenBank as SARS-CoV-Urbani 2003 and 2016[45] is a SARS-CoV-2 PCR match known as CoV-RsSHC014.

20. Following the 11 September 2001 attacks, the Patriot Act—which included an amendment to the biological-weapons legislation via Section 817—was passed and signed by the Bush administration on 26 October 2001. The amendment reopened the global bioweapons arms race, resulting in increased funding of biological weapons research by US government agencies; more than thirty years after President Richard M. Nixon's historic "Statement on Chemical and Biological Defense Policies and Programs of 1969,"[46] which included the following:

> Biological weapons have massive, unpredictable and potentially uncontrollable consequences. They may produce global epidemics and impair the health of future generations. I have therefore decided that:
> - The United States shall renounce the use of lethal biological agents and weapons, and all other methods of biological warfare.
> - The United States will confine its biological research to defensive measures such as immunization and safety measures.
> - The Department of Defense has been asked to make recommendations as to the disposal of existing stocks of bacteriological weapons.

In the spirit of these decisions, the United States associates itself with the principles and objectives of the United Kingdom Draft Convention ("We will seek, however, to clarify specific provisions of the draft to assure that necessary safeguards are included"), which would ban the use of biological methods of warfare.

The following text in Section 817, Expansion of the Biological Weapons Statute, was the catalyst for biological weapons research programs, including gain-of-function research being funded by various US government agencies since 2001:

> Whoever knowingly violates this section shall be fined as provided in this title, imprisoned not more than 10 years, or both, but the prohibition contained in this section shall not apply with respect to any duly authorized United States governmental activity.

21. In 2002 Dr. Shi Zhengli of the Wuhan Institute of Virology published her work showing increased infectivity and transmissibility of coronaviruses by inserting HIV pseudovirus segments into SARS-CoV-1.[47]

22. A patent was issued to Dr. Ralph S. Baric and Boyd Yount in 2003 for their invention of "an infectious, replication defective, coronavirus" titled *Directional assembly of large viral genomes and chromosomes* (United States Patent No. US006593111B2).[48]

    This patented method can be used to reconstruct and manipulate genetic sequences and chromosomes, in vitro, for introduction into a living host. The patented method makes possible selected mutagenesis and genetic manipulation of nucleoside gene sequences, in vitro, prior to assembly into a full-length genetic (RNA, DNA) molecule, after which the molecule can be introduced into a living host. The abstract specifically states this is for the purpose of manipulating genes of viruses, bacteria, higher plants, and animals (includes people).

23. In 2003 Dr. Ralph Baric received additional NIH funding and began working on synthetically altering *Coronaviridae* to increase their pathogenicity.[49]

24. Between 2004 and 2020 Peter Daszak at EcoHealth Alliance received more than $61 million in funding from the Department of Defense (DoD), Health and Human Services (HHS), the National Science Foundation (NSF), the US Agency for International Development (USAID), the Department of Homeland Security (DHS), the Department of Commerce (DOC), the US Department of Agriculture (USDA), and the Department of the Interior (DOI)[50] as listed:

Federal Grants and Contracts Awarded to EcoHealth Alliance

Agency	**Total**
Department of Defense (DoD)	$38,949,941.00
Health and Human Services (HHS)	$13,023,168.00
National Science Foundation (NSF)	$2,590,418.00
US Agency for International Development (USAID)	$2,499,147.00
Department of Homeland Security (DHS)	$2,272,813.00
Department of Commerce (DOC)	$1,241,933.00

US Department of Agriculture (USDA)	$646,701.00
Department of Interior (DOI)	$267,062.00
Grand Total	$61,491,183.00

25. David Franz, a former commander at Fort Detrick,[51] was assigned as science and policy advisor for EcoHealth Alliance, providing a direct link to the United States military.[52]

26. US government funding of gain-of-function research continues to this day, at facilities including but not limited to University of Texas, University of Washington, University of Iowa, University of Wisconsin, University of North Carolina, University of Maryland, University of Tennessee, University of Georgia and Boston University. Johns Hopkins University Bloomberg Distinguished Professor of Biomedical Engineering, Computer Science, and Biostatistics Steven Salzberg authored an article for *Forbes* magazine on 24 October 2022 titled "Gain-of-Function Experiments at Boston University Create a Deadly New Covid-19 Virus. Who Thought This Was a Good Idea?"

    This article highlighted a GoF study called "Role of Spike in the Pathogenic and Antigenic Behavior of SARS-COV-2 BA.1 Omicron" released on the preprint server bioRxiv on 14 October 2022.[53]

27. By 2011, *both* the University of Wisconsin and the Erasmus Medical Center in Rotterdam, the Netherlands, mutated the lethal H5N1 avian influenza (a.k.a. bird flu) virus, increasing its transmissibility between ferrets, which closely mimic the human response to flu.

    Both gain-of-function research studies triggered considerable debate within the scientific community and were submitted for review to the US National Science Advisory Board for Biosecurity (NSABB).[54]

28. In 2013, with funding from the NIH, Dr. Zhengli and the Wuhan Virology Team isolated three bat viruses with HKU4 spike proteins unable to infect humans.[55]

29. In 2015 Dr. Zhengli and others admitted to having "reengineered [the] HKU4 spike aiming to build its capacity to mediate viral entry into human cells. . . . To this end, we introduced two single mutations. . . . Mutations in these motifs in coronavirus spikes have demonstrated dramatic effects on viral entry into human cells."[56] The NIH grant numbers include R01AI089728, R21AI109094, R01AI089728, and R01AI110700.[57]

30. In 2014 the Middle East Respiratory Virus (MERS) outbreak occurs in Saudi Arabia and a year later (2015) in South Korea. Animal research

demonstrated that successful treatment of this MERS corona virus, must begin immediately upon symptomatic infection, and required a multidrug treatment approach.[58]

31. From 2014 to 2018, gain-of-function research was placed on hold by the Obama administration following pressure by the scientific community. Scientific concern included (1) the accidental exposure of a CDC lab worker to anthrax, (2) the NIH discovery of a fifty-year-old vial of smallpox left unattended on a laboratory countertop, and (3) the shipment of the deadly H5N1 avian influenza virus from that lab to another laboratory expecting a benign strain.[59]

32. During this "pause" in GoF research, Dr. Ralph Baric and others received an international patent (WO2015/143335 A1) on 24 September 2015, entitled Methods and Compositions for Chimeric Coronavirus Spike Proteins," funded by NIH Grant No. U54AI057157.

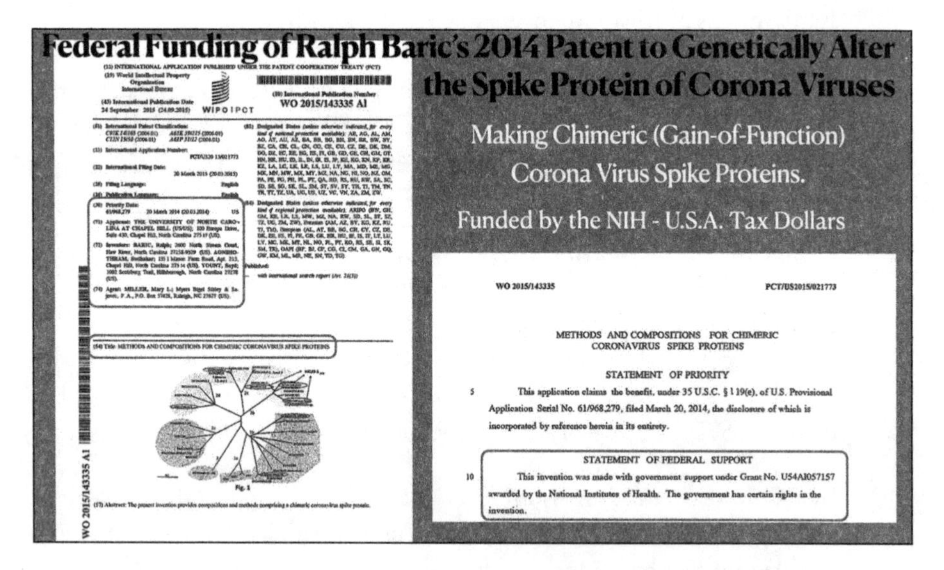

This international patent clearly shows that the purpose is to alter coronavirus spike proteins using gain-of-function chimeric methods. It is clearly funded by the National Institutes of Health (NIH), which means NIH was funding GoF research and development. Any statement to the contrary is contradicted by this patent.

33. During the moratorium, from which EcoHealth was exempt, at least five research grants were issued to Peter Daszak at EcoHealth Alliance by the National Institute of Allergy and Infectious Diseases (NIAID),[60] as shown:

Federal Grants and Contracts Awarded to EcoHealth Alliance During GoF Pause

Year	Project Number	Project Name	Amount
2014	1R01AI110964–01	Understanding the Risk of Bat Coronavirus Emergence	$666,442
2015	5R01AI110964–02	Understanding the Risk of Bat Coronavirus Emergence	$630,445
2016	5R01AI110964–03	Understanding the Risk of Bat Coronavirus Emergence	$611,090
2017	5R01AI110964–04	Understanding the Risk of Bat Coronavirus Emergence	$597,112
2018	5R01AI110964–05	Understanding the Risk of Bat Coronavirus Emergence	$581,646
5 years	5 grants	Understanding the Risk of Bat Coronavirus Emergence	$3,086,735

34. Despite GoF research concerns raised by the scientific community, GoF research continued, including the awarding of NIH grants to modify bat-based *Coronaviridae* (coronaviruses). The grants included funding that went to the Wuhan Institute of Virology (WIV) Laboratory, which Dr. Anthony Fauci and NIAID failed to flag.[61]

35. On 14 November 2018 Dr. Shi Zhengli presented her research, titled *Studies on Bat Coronavirus and its cross-species infection* at Shanghai Jiao Tong University. By 2019, Dr. Zhengli's research presentation was deleted from the university website. [62]

36. On 16 September 2019 Ralph S. Baric published a preprint confirming that he and teams of researchers at the University of North Carolina at Chapel Hill, the University of Texas in Galveston, Columbia University, and the National Institute of Allergy and Infectious Diseases had combined the spike protein of SARS-CoV-RsSHC014 (Urbani) with the backbone of another WIV lab virus (RsWIV1-CoV) to produce a chimeric (GoF) virus capable of infecting people, involving trypsin-like proteases (PRRA),[63] with the patent rights[64] for inserting the PRRA (furin cleavage site) owned by the NIH. Furthermore, in the supplementary text[65] of the research publication, Baric and others noted that besides combining the spike protein of one coronavirus with the backbone of another, they had changed the genetic code sequence in five specific sites, including the envelope protein critical for virus-cell fusion, genetically altering (GoF) the coronavirus. That produced increased human infectivity.

This published research, shown as follows, not only connects Daszak at EcoHealth with Baric at North Carolina but also connects to Zhengli at the WIV, all of whom received this GoF funding from the NIAID, USAID-PREDICT, NIH, and the National Natural Science Foundation of China. It is clearly defined by the authors in the paper as gain-of-function research, making any statements by any person from these agencies, institutions, or governments that GoF research has not been funded an intentional and knowing lie—that is, perjury.

ACKNOWLEDGMENTS

Research in this manuscript was supported by grants from the National Institute of Allergy & Infectious Disease and the National Institute of Aging of the US National Institutes of Health (NIH) under awards U19AI109761 (R.S.B.), U19AI107810 (R.S.B.), AI085524 (W.A.M.), F32AI102561 (V.D.M.) and K99AG049092 (V.D.M.), and by the National Natural Science Foundation of China awards 81290341 (Z.-L.S.) and 31470260 (X.-Y.G.), and by USAID-EPT-PREDICT funding from EcoHealth Alliance (Z.-L.S.). Human airway epithelial cultures were supported by the National Institute of Diabetes and Digestive and Kidney Disease of the NIH under award NIH DK065988 (S.H.R.). We also thank M.T. Ferris (Dept. of Genetics, University of North Carolina) for the reviewing of statistical approaches and C.T. Tseng (Dept. of Microbiology and Immunology, University of Texas Medical Branch) for providing Calu-3 cells. Experiments with the full-length and chimeric SHC014 recombinant viruses were initiated and performed before the GOF research funding pause and have since been reviewed and approved for continued study by the NIH. The content is solely the responsibility of the authors and does not necessarily represent the official views of the NIH.

AUTHOR CONTRIBUTIONS

V.D.M. designed, coordinated and performed experiments, completed analysis and wrote the manuscript. B.L.Y. designed the infectious clone and recovered chimeric viruses; S.A. completed neutralization assays; L.E.G. helped perform mouse experiments; T.S. and J.A.P. completed mouse experiments and plaque assays; X.-Y.G. performed pseudotyping experiments; K.D. generated structural figures and predictions; E.F.D. generated phylogenetic analysis; R.L.G. completed RNA analysis; S.H.R. provided primary HAE cultures; A.L. and W.A.M. provided critical monoclonal antibody reagents; and Z.-L.S. provided SHC014 spike sequences and plasmids. R.S.B. designed experiments and wrote manuscript.

COMPETING FINANCIAL INTERESTS

The authors declare no competing financial interests.

37. On 12 November 2019, the US Department of Defense paid $369,511 and $50,000 (total of $419,511) to laboratories in Ukraine via Labyrinth Global Health, described as "SME Manuscript Documentation and COVID 19 Research" and "Task Order 1" (subaward ID: 19–6192, parent award unique key: CONT_IDV_HDTRA108D0007_9700). Labyrinth Global Health is a subcontractor to Black & Veatch Special Projects Corporation, which is listed as the registered parent award holder. The award profile contract summary lists the grant as not compliant with the regulatory Clinger-Cohen Act.[66]

    Google, Hunter Biden's Rosemont Seneca Technology Partners, and Nathan Wolfe's[67] Metabiota heavily invested in the work at the Black & Vetch Ukrainian Biolabs.[68] According to Russian politician Vyacheslav Volodin, "U.S. President Joe Biden himself is involved in the creation of biolaboratories in Ukraine."

    Since 2005, the United States has partnered with the Ukrainian Biolabs via the Pentagon's Defense Threat Reduction Program, which became part of the Defense Threat Reduction Agency (DTRA) shortly after the 1991 USSR dissolution.

38. In April 2020[69] President Trump placed restrictions on funding for gain-of-function research to Peter Daszak at EcoHealth Alliance. Conversely, on 17 June 2020 EcoHealth Alliance received $1,546,744 for project number 1U01AI151797–01, "Understanding Risk of Zoonotic Virus Emergence in EID Hotspots of Southeast Asia."[70]

39. In a 19 May 2020 interview, Peter Daszak of EcoHealth admits to having funded Dr. Ralph Baric and Dr. Shi Zhengli, both of whom admit to performing genetic research on *Coronaviridae*—altering the viral spike proteins.

    You can manipulate them [coronaviruses] in the lab pretty easily. . . . Spike protein drives a lot of what happens with the coronavirus. Zoonotic risk. So, you can get the sequence, you can build the protein . . . and we work with Ralph Baric at [the University of North Carolina] to do this . . . and insert the backbone of another virus and do some work in the lab.[71]

    Peter Daszak, PhD

40. The presence of a furin cleavage site in the SARS-CoV-2 genome is prima facie evidence of laboratory intervention with manipulation of the coronavirus, per Markus Hoffman, Hannah Kleine-Weber, and Stefan Pöhlmann in a study published on 21 May 2020, titled "A Multibasic Cleavage Site in the Spike Protein of SARS-CoV-2 Is Essential for Infection of Human Lung Cells."[72] The study includes the following statements:

    The pandemic coronavirus SARS-CoV-2 threatens public health worldwide. The viral spike protein mediates SARS-CoV-2 entry into host cells and harbors a S1/S2 cleavage site containing multiple arginine residues (multibasic) not found in closely related animal coronaviruses. However, the role of this multibasic cleavage site in

SARS-CoV-2 infection is unknown. Here, we report that the cellular protease furin cleaves the spike protein at the S1/S2 site and that cleavage is essential for S-protein-mediated cell-cell fusion and entry into human lung cells. Moreover, optimizing the S1/S2 site increased cell-cell, but not virus-cell, fusion, suggesting that the corresponding viral variants might exhibit increased cell-cell spread and potentially altered virulence. Our results suggest that acquisition of a S1/S2 multibasic cleavage site was essential for SARS-CoV-2 infection of humans and identify furin as a potential target for therapeutic intervention.

The Furin Cleavage site is responsible for cleavage of four (4) specific amino acids (Proline-Arginine-Arginine-Alanine; identified by four letters of the alphabet as P-R-R-A or PRRA) thereby splitting apart the two parts of the spike protein identified as S1 and S2. This cleavage is the critical step for entry of the viruses into human cells. The patent [US 7,223,390 B2] for the "Insertion of Furin Protease Cleavage Sites in Membrane Proteins and Uses Thereof" was issued to the US Government on 29 May 2007.

The Furin Cleavage site which is critical for the entry of the SARS-CoV-2 virus into the cells of humans, resulting in disease and death, is not found in any other coronavirus at the critical S1/S2 location. The U.S. Government funds Gain-of-Function research and owns the patent for this Furin Protease Cleavage Site, also associated with the HIV glycoprotein (HIV gp120) 120 (sialic acid raft receptor) and cancer progression.

Further understanding of the "Significance of Spike Protein and Furin Cleavage Site" by Richard M. Fleming PhD, MD, JD, is attached and marked as Exhibit A.

41. During a recorded interview on 11 November 2020, Dr. Ralph Baric responded to questions from an Italian investigative researcher with details of collaborative gain-of-function research with Dr. Shi Zhengli[73] of the Wuhan Institute of Virology, involving genetically altering *Coronaviridae*. The following statements were made by Dr. Ralph Baric and scientific journalist and author Matt Ridley:

Dr. Ralph Baric: In sequence databases, there were sequences for a large number of bat coronaviruses that were SARS-like reported out of China. And in that massive pool, you can imagine there are strains that might be able to use human cells just fine, and so the question in the scientific community is going, is there going to be, if a new strain emerges, is it going to be able to, you know, is it preprogrammed to cause an outbreak of disease, or does it have to go through the sort of sequential steps of mutations. So, the important parts, the important take-home messages from that paper, were that there are viruses that exist in bat species preprogrammed to jump between species, replicate just fine in humans. We had no access to the viruses in China; all we had was access to the sequence, and so you can chemically synthesize the sequence of the virus in a laboratory and make the virus sequence and then recover the virus.

Matt Ridley: There were two teams in the world who were very good at making chimeric viruses. They were both working on SARS-like coronaviruses, one in North Carolina under Ralph Baric and one in Wuhan under Shi Zhengli, and they both developed techniques for combining two different parts of different viruses into one virus, the backbone of one virus and the spike protein from another virus. These experiments were done to understand how virulent, how dangerous coronaviruses were, particularly SARS-like coronaviruses, in order to be able to be ready to combat a pandemic. They warned the world in a publication in 2015 jointly between Professor Baric's team [and] Shi Zhengli's team, that they were capable of making more dangerous pathogens and that this was a risky line of research.

Interviewer to Dr. Ralph Baric: Did you give the virus a boost, did you strengthen it?

Dr. Ralph Baric: The only gain-of-function that occurred in that virus is that we changed its antigenicity and what that data tells you is that any vaccine or antibody that you'd made against the original virus from 2003 wasn't going to protect the public against any new—this new virus—if it should emerge in the future.

Interviewer to Dr. Ralph Baric: If we looked at the genome of your chimera, would we realize it was made in a laboratory?

Dr. Ralph Baric: Anything that we build in the laboratory has a, has what I call "signature mutation." It's like a little, it's where you sign your name almost, it says, you put in these mutations and it says, "This, this is built from material in the Baric laboratory."

Interviewer to Dr. Ralph Baric: But if you don't want to leave this signature, you can artificially construct a virus indistinguishable from a natural one, right?

Dr. Ralph Baric: It is correct, you can do it without leaving a signature, yes, using any one of three or four different approaches for coronaviruses that were developed by different researchers, you can leave no trace that it was made in a laboratory.

Interviewer to Dr. Ralph Baric: I imagine you know why I insist on asking.

Dr. Ralph Baric: If you think that people can engineer a virus genome and that it be infectious and leave no trace, they have to start from a template somewhere. You have to know, you have to have a sequence to begin with, you can't just put a sequence together to make a virus.

Interviewer to Dr. Ralph Baric: Couldn't this model to construct a chimera from be the virus found in 2013: its cousin, RaTG13?

Dr. Ralph Baric: So, there's twelve hundred mutations in RaTG13 and the . . . sequence of the virus is not complete; generating twelve hundred mutations in subculture passage is not as easy as it may sound.

Interviewer to Dr. Ralph Baric: So, do you rule out SARS-CoV-2's being a chimera made in a laboratory?

Dr. Ralph Baric: Not with the viruses that have been sequenced and reported to date.

Interviewer to Dr. Ralph Baric: Can there be viruses we know nothing about in laboratories?

Dr. Ralph Baric: That's certainly possible. If you are asking about intent or whether the virus existed beforehand, it would only be within the records of the Institute of Virology in Wuhan.

Interviewer to Dr. Ralph Baric: Are the databases public?

Dr. Ralph Baric: When they publish the papers, they download the sequences that they have associated with the paper. Do I know that they put every single sequence in that? How would I know? You end up with millions of sequences so—

Commentary: Looking in web archives we discovered that Professor Shi made a vast database available to the scientific community, a databank specialized in bat and rodent viruses containing data relating to over twenty thousand samples gathered over the years in various parts of China. It reported very detailed information: the sampling place's GPS coordinates, the type of virus found, whether the virus was sequenced or isolated—that is, grown in cell cultures. Database access required a password to consult the as-yet-unpublished virus-related data, with the sole duty of not divulging the information until the publication date. As of June 2020, the whole page was removed from the web but based on this portal, which monitors China's science-related databanks, the data had been inaccessible even since September 12, 2019.

Matt Ridley: The rest of the world, including the World Health Organization, needs to ask the Chinese authorities very politely, to be much more transparent about what experiments were done at the Wuhan Institute of Virology. If this virus was not created there, or cultured there, then they have nothing to hide and they should be able to help us clear up this mystery.

Dr. Baric's statement "If you are asking about intent or whether the virus existed beforehand, it would only be within the records of the Institute of Virology in Wuhan" was correct; however, the Wuhan Institute of Virology databank was inaccessible since 12 September 2019, and the entire page subsequently removed from June 2020. As Dr. Baric mentioned earlier in the interview, "We had no access to the viruses in China; all we had was access to the sequence, and so you can chemically synthesize the sequence of the virus in a laboratory and make the virus sequence and then recover the virus."

42. Using SnapGene[74] to run genetic sequences and assign primers for polymerase chain reaction (PCR), nucleotide base sequences of the SARS-CoV-2 virus with the nucleoside base sequences of several viruses recorded in the GenDataBank were analyzed, again using the SnapGene program. This revealed multiple primer matches for three Gain-of-Function (GoF) viruses. (See Exhibit A for further details.)

Date	**Virus Name**	**Authors**	**GenBank Ref.**
11/22/2013	SARS-CoV-RsSHC014 (Urbani)	Peter Daszak, Shi Zhengli, and others	KC881005
09/26/2014	SARS-CoV-MA15	Ralph Baric and others	JF292920
11/22/2013	SARS-CoV-Rs3367	Peter Daszak, Shi Zhengli, and others	KC881006

All three gain-of-function viruses were funded with federal money via NIH and NIAID and other federal agencies. The SARS-CoV-2 virus has been presented and is referred to by the scientific community, government agencies, and media as a single virus, but the nucleotide base sequences and PCR primers of the SARS-CoV-2 virus reveal matches with three US-government-funded GoF viruses. Refer to "Significance of Spike Protein and Furin Cleavage Site" document by Richard M. Fleming PhD, MD, JD, attached as Exhibit A.

43. On 19 July and again on 22 July 2021, representatives on behalf of Fort Detrick reached out to the author in an effort to recruit him to Fort Detrick to work as a physicist on viruses, funded by NIAID. All communications

ceased when Dr. Fleming's work exposed the funding of SARS-CoV-2 (COVID-19) a few months later.

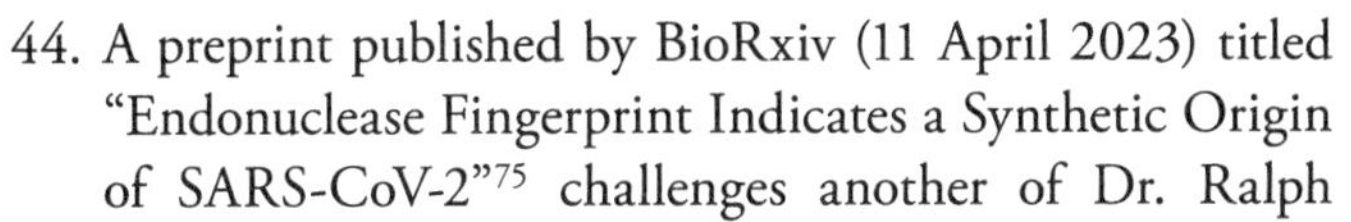

44. A preprint published by BioRxiv (11 April 2023) titled "Endonuclease Fingerprint Indicates a Synthetic Origin of SARS-CoV-2"[75] challenges another of Dr. Ralph Baric's statements during the 11 November 2020 interview. When asked if a laboratory created virus could be created without leaving a signature trace, with the virus "indistinguishable from a natural one," Dr. Baric responded, as previously noted:

> It is correct, you can do it without leaving a signature, yes, using any one of three or four different approaches for coronaviruses that were developed by different researchers, you can leave no trace that it was made in a laboratory.

The preprint by Valentin Bruttel, Alex Washburne, and Antonius VanDongen revealed that laboratory created viruses do in fact carry a signature trace. The preprint contains the following statements:

> To prevent future pandemics, it is important that we understand whether SARS-CoV-2 spilled over directly from animals to people, or indirectly in a laboratory accident. The genome of SARS-COV-2 contains a peculiar pattern of unique restriction endonuclease recognition sites allowing efficient dis- and re-assembly of the viral genome characteristic of synthetic viruses. Here, we report the likelihood of observing such a pattern in coronaviruses with no history of bioengineering. We find that SARS-CoV-2 is an anomaly, more likely a product of synthetic genome assembly than natural evolution. The restriction map of SARS-CoV-2 is consistent with many previously reported synthetic coronavirus genomes, meets all the criteria required for an efficient reverse genetic system, differs from closest relatives by a significantly higher rate of synonymous mutations in these synthetic-looking recognitions sites, and has a synthetic fingerprint unlikely to have evolved from its close relatives. We report a high likelihood that SARS-CoV-2 may have originated as an infectious clone assembled in vitro.
>
> Lay Summary: To construct synthetic variants of natural coronaviruses in the lab, researchers often use a method called in vitro genome assembly. This method utilizes special enzymes called restriction enzymes to generate DNA building blocks that then can be "stitched" together in the correct order of the viral genome. To make a virus in the lab, researchers usually engineer the viral genome to add and remove stitching sites, called restriction sites. The ways researchers modify these sites can serve as fingerprints of in vitro genome assembly.
>
> We found that SARS-CoV has the restriction site fingerprint that is typical for synthetic viruses. The synthetic fingerprint of SARS-CoV-2 is anomalous in wild coronaviruses, and common in lab-assembled viruses. The type of mutations (synonymous or silent mutations) that differentiate the restriction sites in SARS-CoV-2 are characteristic of engineering, and the concentration of these silent mutations in the restriction sites is extremely unlikely to have arisen by random evolution. Both the restriction site fingerprint and the pattern of mutations generating them are extremely unlikely in wild coronaviruses and nearly universal in synthetic viruses. Our findings strongly suggest a synthetic origin of SARS-CoV2.

Discussion: Gain-of-Function

The SARS-CoV-2 viruses bear the fingerprints of laboratory manipulation as shown by the aforementioned preprint (point 44). The conclusion of the preprint corroborates the initial view of many medical professionals, including the five highly respected and influential scientists Kristian G. Andersen, Andrew Rambaut, W. Ian Lipkin, Edward C. Holmes, and Robert F. Garry (authors of "The Proximal Origin of SARS-CoV-2") and Richard M. Fleming, as shown in his research on nucleotide base sequences analysis. That research positively identifies the people responsible for the gain-of-function research that resulted in the creation of SARS-CoV-2 by matching the SARS-CoV-2 PCR primers with PCR primers of three GoF viruses funded by United States government agencies, including the NIH, DoD, and NIAID. As per the GenDataBank portal, the virus known as SARS-CoV-2 is, in fact, the consequence of three GoF viruses[76] created by none other than Peter Daszak (EcoHealth Alliance), Shi Zhengli (Wuhan Institute of Virology), and Dr. Ralph Baric (University of North Carolina at Chapel Hill).

The funding by US agencies is not limited to that carried out by the previously mentioned researchers and institutions. It permeates many of our top universities.

The evidence is very clear: the defendants have engaged in the development of biological weapons, of which SARS-CoV-2 is but one. The development or modification of either a naturally occurring or completely manufactured biologic, or anything in between, is a criminal act. The development of such biological weapons is also, by definition as noted in the referenced statutes, an action of terrorism and as such is a criminal act, which has been carried out by multiple individuals working together. Such intentional and knowing cooperation in violation of the law is by definition a criminal conspiracy, with all those involved criminally culpable.

These criminal acts have been carried out over decades. The release of such a weapon, beyond the violations resulting from the development of such, either intentionally or accidentally, is an act of war, which other countries have called upon the international community to address. The United States has the opportunity to address these crimes and criminals before the international community demands accountability or reacts to this act of war.

Factual Background Evidence: SARS-CoV-2 Pandemic of 2019 (COVID-19)

45. On 12 November 2019 the DoD paid $369,511 and $50,000 (total of $419,511) to laboratories in Ukraine via Labyrinth Global Health, described as "SME Manuscript Documentation and COVID 19 Research" and "Task Order 1" (sub-award ID: 19–6192, parent award unique key: CONT_IDV_HDTRA108D0007_9700).

46. On 18 December 2019, Global MIT (Massachusetts Institute of Technology) published an article by Anne Trafton, titled "Storing Medical Information Below the Skin's Surface,"[77] which included the following statements:

> Every year, a lack of vaccination leads to about 1.5 million preventable deaths, primarily in developing nations. One factor that makes vaccination campaigns in those nations more difficult is that there is little infrastructure for storing medical records, so there's often no easy way to determine who needs a particular vaccine.
>
> MIT researchers have now developed a novel way to record a patient's vaccination history: storing the data in a pattern of dye, invisible to the naked eye, that is delivered under the skin at the same time as the vaccine.
>
> "In areas where paper vaccination cards are often lost or do not exist at all, and electronic databases are unheard of, this technology could enable the rapid and anonymous detection of patient vaccination history to ensure that every child is vaccinated," says Kevin McHugh, a former MIT postdoc who is now an assistant professor of bioengineering at Rice University.
>
> The researchers showed that their new dye, which consists of nanocrystals called quantum dots, can remain for at least five years under the skin, where it emits near-infrared light that can be detected by a specially equipped smartphone.
>
> McHugh and former visiting scientist Lihong Jing are the lead authors of the study, which appears today [18 December 2019] in Science Translational Medicine. Ana Jaklenec, a research scientist at MIT's Koch Institute for Integrative Cancer Research, and Robert Langer, the David H. Koch Institute Professor at MIT, are the senior authors of the paper.

The research was funded by the Bill & Melinda Gates Foundation and the Koch Institute Support Grant from the National Cancer Institute. Other authors of the paper include Sean Severt, Mache Cruz, Morteza Sarmadi, Hapuarachchige Surangi Jayawardena, Collin Perkinson, Fridrik Larusson, Sviatlana Rose, Stephanie Tomasic, Tyler Graf, Stephany Tzeng, James Sugarman, Daniel Vlasic, Matthew Peters, Nels Peterson, Lowell Wood, Wen Tang, Jihyeon Yeom, Joe Collins, Philip Welkhoff, Ari Karchin, Megan Tse, Mingyuan Gao, and Moungi Bawendi.

47. The US government announced on 21 January 2020[78] that a coronavirus, later identified as SARS-CoV-2, first appeared in China in December 2019, and spread to the United States in January 2020.

48. On 22 January 2020 Christian Drosten MD, PhD, and his team at the Charité–Berlin University of Medicine reported on "the establishment and validation of a diagnostic workflow for 2019-nCov screening and specific confirmation." The report titled "Detection of 2019 Novel Coronavirus (2019-nCoV) by Real Time RT-PCR"[79] confirmed the polymerase chain reaction (PCR) test invented by 1993 Nobel Prize winners Kary B. Mullis, PhD., and Michael Smith, PhD, as the global standard for SARS-CoV-2 detection at 45 (heat and cooling) cycles (Ct, or cycle threshold).

The Mullis patent (USPTO Patent No. 4,683,202)[80] issued 28 July 1987 defined the correct number of PCR cycles to be used for his process for amplifying nucleic acid sequences (later known as polymerase chain reaction, or PCR) was twenty. Twenty cycles produced 1,048,555 copies of the original nucleic acid sequence (adenine, cytosine, guanine, thymine, uracil), reaching 100 percent efficiency. Subsequent cycles produce what is known in physics as an increase in noise-to-signal ratio, producing false reports.

Use of the PCR is critical to finding SARS-CoV-2 genetic sequences, demonstrating the presence of the virus. The diagnosis of any disease, including COVID, requires a combination of diagnostic skills taught to physicians that include laboratory testing, physical examination, and patient history.[81]

49. The National Science Foundation (NSF) funded a research study in 2007[82] that advised that, in the event of a pandemic, the first action to be taken is the immediate cessation of international travel to provide containment of the pathogen. In January 2020 Dr. Fauci advised President Trump against cessation of international travel. Trump initiated[83] travel restrictions 31 January 2020.

50. On 31 January 2020, US Health and Human Services Secretary Alex Azar declared a public health emergency,[84] per the Public Health Service Act, section 319, which allows HHS to declare emergencies in the case of significant infectious diseases.

51. The International Committee on Taxonomy of Viruses (ICTV) announced "Severe Acute Respiratory Syndrome Coronavirus 2 (SARS-CoV-2)" as the officially recognized name of the new virus on 11 February 2020. The World Health Organization (WHO) announced "COVID-19" as the name of the new ITR disease[85] on 11 February 2020.

52. On 19 February 2020 *The Lancet* published a statement authored by twenty-seven medical professionals titled "Statement in Support of the Scientists, Public Health Professionals, and Medical Professionals of China Combatting COVID-19." Peter Daszak, Christian Drosten, and Jeremy Farrar[86] are listed among the authors. The statement included the following:

> We are public health scientists who have closely followed the emergence of 2019 novel coronavirus disease (COVID-19) and are deeply concerned about its impact on global health and well-being. We have watched as the scientists, public health professionals, and medical professionals of China, in particular, have worked diligently and effectively to rapidly identify the pathogen behind this outbreak, put in place significant measures to reduce its impact, and share their results transparently with the global health community. This effort has been remarkable. We sign this

statement in solidarity with all scientists and health professionals in China who continue to save lives and protect global health during the challenge of the COVID-19 outbreak. We are all in this together, with our Chinese counterparts in the forefront, against this new viral threat.

The rapid, open, and transparent sharing of data on this outbreak is now being threatened by rumours and misinformation around its origins. We stand together to strongly condemn conspiracy theories suggesting that COVID-19 does not have a natural origin. Scientists from multiple countries have published and analysed genomes of the causative agent, severe acute respiratory syndrome coronavirus 2 (SARS-CoV-2), and they overwhelmingly conclude that this coronavirus originated in wildlife . . . as have so many other emerging pathogens. . . . This is further supported by a letter from the presidents of the US National Academies of Science, Engineering, and Medicine and by the scientific communities they represent. Conspiracy theories do nothing but create fear, rumours, and prejudice that jeopardise our global collaboration in the fight against this virus. We support the call from the Director-General of WHO to promote scientific evidence and unity over misinformation and conjecture. . . . We want you, the science and health professionals of China, to know that we stand with you in your fight against this virus.

We invite others to join us in supporting the scientists, public health professionals, and medical professionals of Wuhan and across China. Stand with our colleagues on the frontline.

53. On 8 March 2020, in an interview with *60 Minutes*, Dr. Anthony Fauci[87] stated that the general population need not wear masks in public places during the pandemic:

> When it comes to preventing coronavirus, public health officials have been clear: Healthy people do not need to wear a face mask to protect themselves from COVID-19. There's no reason to be walking around with a mask. While masks may block some droplets, they do not provide the level of protection people think they do. Wearing a mask may also have unintended consequences: People who wear masks tend to touch their face more often to adjust them, which can spread germs from their hands. But there is another risk to healthy people buying disposable masks as a precaution. The price of face masks is surging, and Prestige Ameritech, the nation's largest surgical mask manufacturer, is now struggling to keep up with the increased demand. It could lead to a shortage of masks for the people who really need it.

54. On 13 March 2020 President Trump issued[88] two national-emergency declarations. The first was based on the Stafford Act, which led to the creation of the National Response Framework (NRF) so state efforts could be coordinated. The Stafford Act was used as the authority for appropriation of up to $7 billion for long-term, low-interest loans for small businesses financially impacted by pandemic measures.

55. President Trump invoked[89] the National Emergencies Act,[90] section 201 and 301, on 13 March 2020. This in turn allowed Secretary Azar to modify[91] certain Medicare and Medicaid rules.

56. On 16 March 2020, the World Health Organization (WHO) advised[92] all members of the United Nations to employ certain methods of controlling

the virus, including wearing face masks, using a social distancing of six feet, washing hands, and staying home (shelter in place) as much as possible.

57. On 17 March 2020, an article published by *Nature Medicine* titled "The Proximal Origin of SARS-CoV-2" was authored by five highly respected and influential scientists: Kristian G. Andersen, Andrew Rambaut, W. Ian Lipkin, Edward C. Holmes, and Robert F. Garry. The article was a theoretical[93] comparative analysis of genomic data that discussed scenarios by which the SARS-CoV-2 genome could have arisen (lab origin vs. natural origin). The authors stated that "SARS-CoV-2 is not the product of purposeful manipulation." In the conclusion, they reiterated, "We do not believe that any type of laboratory-based scenario is plausible."

 This paper became the foundation upon which the official narrative regarding the possible origin of SARS-CoV-2 was established and promoted by government agencies, the media, and the scientific community. Any information to the contrary was regarded as COVID-19 misinformation by Big Tech companies such as YouTube, Facebook, Twitter, and Instagram and online fact-checkers.

 It was later revealed by emails obtained under the Freedom of Information Act (FOIA) that the authors of the article were initially convinced that the SARS-CoV-2 virus was, in fact, the result of laboratory manipulation or gain-of-function research. The correspondence contradicts the natural occurring origin of SARS-CoV-2 and speaks directly to the appearance of man-made GoF manipulation.

 Evidence suggests that the cover-up of US involvement in the development of a GoF virus that violates the BWC Treaty, 18 U.S.C. § 175, and the other federal and state statutes previously detailed goes all the way to the White House (as shown in the following Droegemeier letter of 3 February 2020), predating the announcements of natural origins and demonstrating concerns over the evidence of HIV gp120 insertion in the SARS-CoV-2 spike protein.

58. On 19 March 2020 President Trump named Federal Emergency Management Agency (FEMA), part of Homeland Security Department, as the lead[94] agency in the COVID-19 emergency-response efforts. Under the Stafford Act, he pledged billions of dollars in aid[95] to the states, to be federally coordinated through FEMA.

59. On 27 March 2020, the Trusted News Initiative (TNI) announced "plans to tackle harmful Coronavirus disinformation"[96] via an article published by the BBC.[97] The global partners within the TNI include the BBC, Facebook, Google, YouTube, Twitter, Microsoft, Agence France-Presse

Thanks, Kristian. Talk soon on the call.

From: Kristian G. Andersen (b)(6)>
Sent: Friday, January 31, 2020 10:32 PM
To: Fauci, Anthony (NIH/NIAID) [E] (b)(6)
Cc: Jeremy Farrar (b)(6)>
Subject: Re: FW: Science: Mining coronavirus genomes for clues to the outbreak's origins

Hi Tony,

Thanks for sharing. Yes, I saw this earlier today and both Eddie and myself are actually quoted in it. It's a great article, but the problem is that our phylogenetic analyses aren't able to answer whether the sequences are unusual at individual residues, except if they are completely off. On a phylogenetic tree the virus looks totally normal and the close clustering with bats suggest that bats serve as the reservoir. The unusual features of the virus make up a really small part of the genome (<0.1%) so one has to look really closely at all the sequences to see that some of the features (potentially) look engineered.

We have a good team lined up to look very critically at this, so we should know much more at the end of the weekend. I should mention that after discussions earlier today, Eddie, Bob, Mike, and myself all find the genome inconsistent with expectations from evolutionary theory. But we have to look at this much more closely and there are still further analyses to be done, so those opinions could still change.

Best,
Kristian

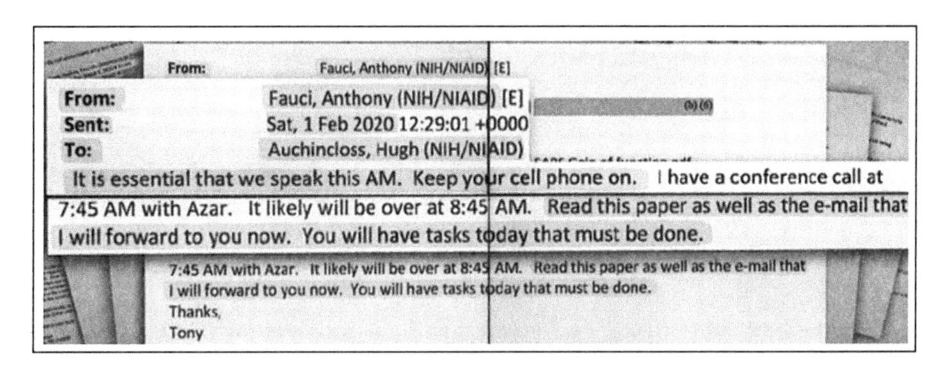

From: Fauci, Anthony (NIH/NIAID) [E] (b)(6)
Sent: Sat, 1 Feb 2020 12:29:01 +0000
To: Auchincloss, Hugh (NIH/NIAID)

It is essential that we speak this AM. Keep your cell phone on. I have a conference call at 7:45 AM with Azar. It likely will be over at 8:45 AM. Read this paper as well as the e-mail that I will forward to you now. You will have tasks today that must be done.

Thanks,
Tony

(AFP), Reuters, European Broadcasting Union (EBU), *Financial Times*, *Wall Street Journal*, *The Hindu*, CBC/Radio Canada, First Draft News, and Reuters Institute for the Study of Journalism. The decision was made by TNI partners, during the 2019 Trusted News Summit, to "work collectively, where appropriate, to agree collaborative actions on various initiatives." Tony Hall, director general of the BBC, stated, "Disinformation and so-called fake news is a threat to us all. At its worst, it can present a serious threat to democracy and even to people's lives."

An antitrust lawsuit was filed against TNI on 10 January 2023 by a coalition of outspoken critics of mainstream media narratives, led by Robert F. Kennedy Jr., who had openly criticized the Biden administration's COVID-19 vaccination policies.[98] The lawsuit claims that "TNI members have been working with Big Tech to censor what they

EXECUTIVE OFFICE OF THE PRESIDENT
OFFICE OF SCIENCE AND TECHNOLOGY POLICY
WASHINGTON, D.C. 20502

February 3, 2020

Dr. Marcia McNutt
President, National Academy of Sciences
2101 Constitution Ave, N.W.
Washington, D.C., U.S. 20418

SUBJECT: Rapid Response Assessment of 2019-nCoV Data Needs

In support of the Office of Science and Technology's (OSTP) National Science and Technology Committee (NSTC) rapid research response work for the 2019-nCoV response, and the Administration's efforts to characterize and provide evidence-based assessments for outbreak response efforts, I am writing to ask the National Academies of Sciences, Engineering, and Medicine (NASEM) to rapidly examine information and identify data requirements that would help determine the origins of 2019-nCoV, specifically from an evolutionary/structural biology standpoint. I also ask NASEM to consider whether this should include more temporally and geographically diverse clinical isolates, sequences, etc.

Although a widely-disputed paper, "Uncanny similarity of unique inserts in the 2019-nCoV spike protein to HIV-1 gp120 and Hag," posted on the pre-print server bioRxiv last week has been withdrawn, the response to that manuscript highlights the need to determine information and data requirements as quickly as possible to better perform and validate such analyses of origin. These questions are important not only for this current situation, but to inform future outbreak preparation and better understand animal/human and environmental transmission aspects of coronaviruses. As part of a broader deliberative process, this review will aid preparedness for future events by establishing a process that quickly assembles subject matter experts for evaluating other potentially threatening organisms.

OSTP requests NASEM convene a meeting of experts, particularly world class geneticists, coronavirus experts, and evolutionary biologists, to assess what data, information, and samples are needed to address the unknowns, in order to understand the evolutionary origins of 2019-nCoV and more effectively respond to both the outbreak and any resulting misinformation. I request a letter statement from the National Academies be prepared and provided in response to this solicitation. A more in-depth examination of the issues will be established as a follow up as needed.

Sincerely,

Kelvin K. Droegemeier
Director

condemned as 'misinformation,' such as reports that COVID-19 may have originated in a laboratory in the Chinese city of Wuhan, that the COVID-19 vaccines do not prevent infection, and that vaccinated people may still transmit COVID-19 to others." Time, the emergence of recent evidence, and more detailed information has proven the alleged 'misinformation' to have been true.[99]

60. By mid-to-late March 2020, governors, mayors, and others were ordering[100] closures of schools, businesses, places of worship, and public gatherings and ordering the quarantining of some people, placing restrictions on travel, closing borders, and so forth.

61. In April 2020 Google and Apple announced a joint effort to support contact tracing: "Apple and Google will be launching a comprehensive solution that includes application programming interfaces (APIs) and operating system–level technology to assist in enabling contact tracing."[101]

62. The CDC released official guidance in April 2020 (expanded in February 2023), titled "Guidance for Certifying Deaths[102] Due to Coronavirus Disease 2019 (COVID-19)."[103] The document stated,

> In cases where a definite diagnosis of COVID-19 cannot be made, but it is suspected or likely (e.g., the circumstances are compelling within a reasonable degree of certainty), it is acceptable to report COVID-19 on a death certificate as "probable" or "presumed." In these instances, certifiers should use their best clinical judgement in determining if a COVID-19 infection was likely. However, please note that testing for COVID-19 should be conducted whenever possible. . . . Intermediate causes are those conditions that typically have multiple possible underlying etiologies and thus, a UCOD (underlying cause of death) must be specified on a line below in Part I. For example, pneumonia is an intermediate cause of death since it can be caused by a variety of infectious agents or by inhaling a liquid or chemical. Pneumonia is important to report in a cause-of-death statement but, generally, it is not the UCOD. The cause of pneumonia, such as COVID-19, needs to be stated on the lowest line used in Part I. Additionally, the reported UCOD should be specific enough to be useful for public health and research purposes. For example, a "viral infection" can be a UCOD, but it is not specific. A more specific UCOD in this instance could be "COVID-19." An accurate count of the number of deaths due to COVID-19 infection, which depends in part on proper death certification, is critical to ongoing public health surveillance and response. When a death is due to COVID-19, it is likely the UCOD and thus, it should be reported on the lowest line used in Part I of the death certificate. Ideally, testing for COVID-19 should be conducted, but it is acceptable to report COVID-19 on a death certificate without this confirmation if the circumstances are compelling within a reasonable degree of certainty."

63. On 3 April 2020, CBS News updated its website with the following statement regarding face masks[104] and the 8 March 2020 interview with Dr. Anthony Fauci:

> Update: On Friday, April 3, President Trump announced that the CDC now recommends Americans wear a "basic cloth or fabric mask" in public.

This sudden change of direction regarding the wearing of masks drew a strong response from the scientific community, independent media outlets, and social media.

64. On 9 April 2020, Fox News asked Dr. Scott Jensen, a Minnesota family physician who is also a Republican state senator, to respond[105] to the

CDC's "Guidance for Certifying Deaths Due to Coronavirus Disease 2019 (COVID-19)." Dr. Jensen responded as follows:

> Well, in short, it's ridiculous. I spent some time earlier today just going through the CDC's manual on how to complete death certificates and the part that was specifically written for physicians, and in that manual, it talks about precision and specificity, and that's what we are trained with. The determination of cause of death is a big deal. It has impact on estate planning, it has impact on future generations, and the idea that we are going to allow people to massage and sort of game the numbers is a real issue because we are going to undermine the trust and right now, as we see politicians doing things that aren't necessarily motivated on fact and science. The public's gonna—Their trust in politicians is already wearing thin. . . . Well, let's just take influenza. If I have a patient who died a month ago, had fever, cough, and died after three days and maybe had been an elderly, fragile individual. And there happened to be an influenza epidemic around our community, I wouldn't put influenza in the death certificate; I have never been encouraged to do so. I would put probably respiratory arrest would be the top line, and the underlying cause of this disease would be pneumonia, and in the contributing factors, I might well put emphysema, congestive heart failure. But I would never put influenza down as the underlying cause of death. That is what we are being asked to do here.
>
> Laura Ingraham: Dr. Fauci was asked about the COVID death count today. Here's what he said in part:
> Question to Dr. Fauci: What do you say to those folks who are making the claim without really any evidence that these deaths are being padded, that the number of COVID-19 deaths are being padded?
> Dr. Fauci: You will always have conspiracy theories when you have a very challenging public health crisis. They are nothing but distractions.
> Laura Ingraham: Conspiracy theories, doctor. So, you are engaging in conspiracy theories. What do you say to Dr. Fauci tonight?
> Dr. Jensen: Well, I would remind him that any time health care intersects with dollars, it gets awkward. Right now, Medicare is determined that if you have a COVID-19 admission to the hospital, you'll get paid $13,000. If that COVID-19 patient goes on a ventilator, you'll get $39,000—three times as much. Nobody can tell me after thirty-five years in the world of medicine that sometimes those kinds of things impact on what we do. Some physicians really have a bend towards public health, and they will put down influenza or whatever because that's their preference. I try to stay very specific, very precise; if I know I've got pneumonia, that's what's going on the death certificate. I'm not going to add stuff just because it's convenient.

65. Dr. Rick Bright, head of the Biomedical Advanced Research and Development Authority at the HHS, was demoted or reassigned to a less prestigious[106] role on 20 April 2020.

Widely known as a top vaccine official, Dr. Bright filed a complaint with the Office of Special Counsel (OSC) claiming he was demoted for opposing the broad use of hydroxychloroquine, an antimalaria drug. The Rockefeller Foundation announced the appointment of Dr. Rick Bright[107] as senior vice president of Pandemic Prevention and Response on 8 March 2021.

66. On 23 April 2020 an article was published on the Foundation for Economic Education (FEE) website, titled "YouTube to Ban Content That Contradicts WHO on COVID-19, Despite the UN Agency's Catastrophic Track Record of Misinformation" by Dan Sanchez, director of content at the Foundation for Economic Education (FEE) and editor in chief of FEE.org.[108] YouTube CEO Susan Wojcicki is a respected speaker and agenda contributor at World Economic Forum (WEF) events.

67. On 1 May 2020 Representative Bobby Rush introduced HR 6666, the TRACE Act[109]—the COVID-19 Testing, Reaching, and Contacting Everyone Act. It would award grants for testing and tracing. The bill, which proposed $100 billion[110] for testing and tracing Americans, was reintroduced on 15 September 2022. It has not yet been voted on as of 30 December 2023.

68. On 13 May 2020 the Wisconsin Supreme Court struck down part of Governor Tony Evers's stay-at-home order because the governor and state Department of Health Services enacted it without any oversight from the legislature, which violated[111] the state constitution. Supreme Court justices declared the statewide mask mandate invalid and ruled that Governor Tony Evers did not have the authority to issue multiple emergency declarations amid the COVID-19 pandemic.

69. On 15 May 2020 President Donald Trump announced Operation Warp Speed (OWS), and on that date Moncef Slaoui was appointed chief advisor and General Gustave Perna chief operating officer of the initiative. OWS would also handle the distribution of the vaccine. The president announced that he would use the military[112] to deliver the vaccines. Under a putative emergency, or martial law, President Trump believed he had the authority to potentially declare vaccination mandatory.

70. On 3 June 2020 an article titled "Now Is the Time for a 'Great Reset'"[113] appeared on the World Economic Forum (WEF) website. The article states the following:

> COVID-19 lockdowns may be gradually easing, but anxiety about the world's social and economic prospects is only intensifying. There is good reason to worry: a sharp economic downturn has already begun, and we could be facing the worst depression since the 1930s. But, while this outcome is likely, it is not unavoidable.
>
> To achieve a better outcome, the world must act jointly and swiftly to revamp all aspects of our societies and economies, from education to social contracts and working conditions. Every country, from the United States to China, must participate, and every industry, from oil and gas to tech, must be transformed. In short, we need a "Great Reset" of capitalism.

71. On 19 June 2020, an article was published on the Brookings Institution[114] website, titled "Rebuilding Toward the Great Reset: Crisis, COVID-19, and the Sustainable Development Goals" by Zia Khan and John W.

McArthur. The article promotes the pandemic as an opportunity[115] for the great reset: "The world needs to make the most of the moment at hand." Specifically noted with this article was a form of action:

> Reset systems for the long term: where the objective is to establish, wherever possible, a new equilibrium among political, economic, social, and environmental systems toward common goals. Ultimately, the only limit within this category is our collective imagination. As we emerge from a moment of great crisis, we can imagine a "great reset."

72. The World Economic Forum (WEF) published an article on 6 July 2020, titled "The U.S. Employment-Population Ratio Drops to a Historic Low," outlining the impact[116] of the pandemic on employment and job losses in the United States of America:

    According to the U.S. Bureau of Labor Statistics' latest jobs report, roughly 17.75 million Americans were unemployed in June 2020, resulting in an unemployment rate of 11.1 percent on a seasonally adjusted basis. The number of people that are currently not employed is much higher than 17.75 million, however, as a look at another indicator, the employment-population ratio, reveals.

73. WEF Executive Chairman Klaus Schwab and Thierry Malleret released a book titled *COVID-19: The Great Reset*[117] on 9 July 2020. Founded in 1971 by Klaus Schwab as the European Management Forum, the organization changed its name to the World Economic Forum in 1987 and has since become one of the most powerful organizations in the world and what some consider the premier channel of communication for financial, corporate, and political elites.

74. At the July 2020 WEF conferences in Davos, Switzerland, attendees and speakers included world leaders in medicine, media, business, politics, governments, and other key global industry sectors.[118] Among them were the following:

Dr. Anthony Fauci (former director of NIAID)	Jim Smith (president and CEO, Reuters)
Francis S. Collins (former director, NIH)	George Stephanopoulos (ABC News)
Albert Bourla (Pfizer CEO)	Anderson Cooper (CNN)
Alex Gorsky (Johnson & Johnson chairman)	Ian Bremmer (Eurasia Group)
Stéphane Bancel (Moderna CEO)	Gretchen Whitmer (governor, Michigan)
Marc Dunoyer (AstraZeneca CEO)	Joseph Biden (president of the USA)
Susan Wojcicki (YouTube CEO)	George Soros (Open Society Foundations)
Jeremy Farrar (WHO chief scientist)	Rupert Murdoch (international media)
Meghan O'Sullivan (Trilateral Commission)	Benjamin Netanyahu (prime minister of Israel)

Andrew Ross Sorkin (CNBC, *New York Times*)	Donald Trump (former president of the USA)
Bret Stephens (*New York Times*)	Larry Fink (BlackRock CEO)
Leana Wen (CNN medical analyst)	Sanjay Gupta (CNN chief medical correspondent)
Priya Basu (head, COVID-19 Taskforce Secretariat, World Bank Group)	

A. Klaus Schwab is also founder of Global Leaders for Tomorrow (1992), renamed in 2004 to Young Global Leaders. An extensive list of the five-year World Economic Forum elite program[119] includes individuals actively involved in world health decision making, in addition to policy making. Among these graduates are the following:

William Henry Gates (Microsoft)	Adam Kinzinger (U.S. House of Representatives)
Larry Page (cofounder of Google)	Samantha Power (administrator, USAID)
Jeff Bezos (Amazon)	Huma Abedin (Hillary Clinton aide)
Mark Zuckerberg (Meta)	Nikki Haley (former US ambassador to UN)
Gavin Newsom (governor, California)	Elise Stefanik (congresswoman, New York)
Tom Cotton (senator, Arkansas)	Colin Allred (congressman, Texas)
Richard L. Scott (senator, Florida)	Kate Gallego (mayor, Phoenix, AZ)
Jared Polis (governor, Colorado)	Ivanka Trump (Trump Organization)
Steven Fulop (mayor, Jersey City)	Chelsea Clinton (vice chair, Clinton Foundation)
Vivek Murthy (21st US surgeon general)	Justin Trudeau (prime minister, Canada)
Aja Brown (former mayor, Compton, CA)	Tony Blair (former Prime Minister, UK)
Evan Bayh (former senator, Indiana)	Angela Merkel (former chancellor, Germany)
Luke Ravenstahl (former mayor, Pittsburgh)	Emmanuel Macron (president of France)
Matt Blunt (former governor, Missouri)	Jacinda Ardern (former prime minister, New Zealand)
Bobby Jindal (former governor, Louisiana)	Viktor Orban (prime minister, Hungary)
Jeffrey Zients (White House coronavirus response coordinator)	William R. Steiger (USAID director of Global Affairs at the US Department of HHS)
Nathan Wolfe (Metabiota, DARPA, EcoHealth Alliance)	

B. David Rockefeller, founder of the Trilateral Commission, shared certain beliefs with Henry Kissinger, Klaus Schwab, and the World Economic Forum, evident from statements made in his book *Memoirs,* published on 15 October 2002:[120]

For more than a century, ideological extremists, at either end of the political spectrum, have seized upon well-publicized incidents, such as my encounter with Castro, to attack the Rockefeller family for the inordinate influence they claim we wield over American political and economic institutions. Some even believe we are part of a secret cabal, working against the best interests of the United States, characterizing my family and me as 'internationalists,' and of conspiring with others around the world to build a more integrated global political and economic structure—one world, if you will. **If that's the charge, I stand guilty, and I am proud of it.**

We are grateful to the *Washington Post*, the *New York Times*, *Time* Magazine and other great publications whose directors have attended our meetings and respected their promises of discretion for almost 40 years. . . . It would have been impossible for us to develop our plan for the world if we had been subjected to the lights of publicity during those years. But, the world is more sophisticated and prepared to march towards a world government. **The supernational sovereignty of an intellectual elite and world bankers is surely preferable to the national autodetermination practiced in past centuries.** [emphasis added]

C. The importance of the influence of the WEF cannot be overemphasized. Their control of world governments applies to a variety of issues, including but not limited to GoF viruses, genetic vaccines, and mandates exercised by leaders in the free world. During a 20 September 2017 discussion at Harvard's John F. Kennedy School of Government, Klaus Schwab openly stated,[121]

Yes, actually, there's this notion to integrate young leaders as part of the World Economic Forum since many years. And I have to say, when I mention now, names like Mrs. Merkel, even Vladimir Putin, and so on, they all have been Young Global Leaders of the World Economic Forum. But what we are very proud of now, the young generation like Prime Minister Trudeau, President of Argentina, and so on. So, we penetrate the cabinets. So, yesterday, I was at a reception for Prime Minister Trudeau, and I know that half of his cabinet, or even more than half of his cabinet, are actually Young Global Leaders of the World Economic Forum.

Interviewer: That's true in Argentina as well?

Klaus Schwab: It's true in Argentina, and its true in France now, I mean, with the president who is a Young Global Leader. But what is important for me is those Young Global Leaders have an opportunity to come here [Harvard]. And, we have established a course now since several years. And I think it has, this cooperation, has a tremendous impact because being here for a week really creates a strong community. And we, in addition to the Young Global Leaders, we have now the Global Shapers in 450 cities around the world. . . . And what is astonishing is to see how those young people really have a different mindset, and I have great, I mean, admiration because when I have a group of global shapers in the room and ask them, "Are you thinking in global terms or in national terms?" **the majority would say in global terms.** [emphasis added]

D. An article titled "Welcome to 2030: I Own Nothing, Have No Privacy and Life Has Never Been Better" written by WEF contributor Ida Auken and published by *Forbes* magazine on 10 November 2016, provides valuable insight into the goals and ideas promoted by Klaus Schwab and the World Economic Forum via its international network of Young Global Leaders and Global Shapers. In the future outlined in the article, there are no products to be owned, only services that are rented and delivered by drones. In this system, humans are completely dependent on WEF-controlled corporations for every basic need, with absolute exclusion of autonomy, freedom, and privacy. The article has since been removed from the *Forbes* website but can be viewed on the Medium.com website.[122]

75. On 17 July 2020 Microbe TV published the video interview "TWiV641: COVID-19 with Dr. Anthony Fauci" that, according to the website, covered "SARS-CoV-2 transmission, testing, immunity, pathogenesis, vaccines, and preparedness." At the 4:23 mark, Dr. Fauci states,

> What is now evolving into a bit of a standard, that if you get cycle threshold of 35 or more, that the chances of it being replication competent are miniscule. . . . We have patients, and it's very frustrating for the patients as well as for the physicians, somebody comes in and they repeat their PCR and it's like 37 cycle threshold, but you never, you almost can never, culture virus from a 37 threshold cycle.[123] So, I think if somebody does come in with 37, 38, even 36, you got to say, "You know, it's just, it's just dead nucleotides, period."

76. On 22 September 2020 President Donald J. Trump made the following statements during his address to the 75th Session of the United Nations General Assembly:[124]

> It is my profound honor to address the United Nations General Assembly. Seventy-five years after the end of World War II and the founding of the United Nations, we are once again engaged in a great global struggle. We have waged a fierce battle against the invisible enemy—the China virus—which has claimed countless lives in 188 countries.
>
> In the United States, we launched the most aggressive mobilization since the Second World War. We rapidly produced a record supply of ventilators, creating a surplus that allowed us to share them with friends and partners all around the globe. We pioneered life-saving treatments, reducing our fatality rate 85 percent since April.
>
> Thanks to our efforts, three vaccines are in the final stage of clinical trials. We are mass-producing them in advance so they can be delivered immediately upon arrival. We will distribute a vaccine, we will defeat the virus, we will end the pandemic, and we will enter a new era of unprecedented prosperity, cooperation, and peace.
>
> As we pursue this bright future, we must hold accountable the nation which unleashed this plague onto the world: China. In the earliest days of the virus, China locked down travel domestically while allowing flights to leave China and infect the world. China condemned my travel ban on their country, even as they cancelled domestic flights and locked citizens in their homes.
>
> The Chinese government and the World Health Organization—which is virtually controlled by China—falsely declared that there was no evidence of

human-to-human transmission. Later, they falsely said people without symptoms would not spread the disease. The United Nations must hold China accountable for their actions.

77. The value of the Mullis PCR test and the importance of following the patent instructions were further emphasized in a 28 September 2020 peer reviewed study funded by the French[125] government. Published in Oxford Academic's *Clinical Infectious Diseases*, "Correlation Between 3790 Quantitative Polymerase Chain Reaction–Positives Samples and Positive Cell Cultures, Including 1941 Severe Acute Respiratory Syndrome Coronavirus 2 Isolates" states the following:

> It can be observed that at Ct = 25, up to 70% of patients remain positive in culture and that at Ct = 30 this value drops to 20%. At Ct = 35, the value we used to report a positive result for PCR, <3% of cultures are positive.

The study revealed that a person receiving a positive PCR test result at a cycle threshold (Ct) of 35 or higher has less than 3 percent possibility of being actively infected. Concurrently, the probability of a false positive in this case is 97 percent or higher. The study revealed that the *reliability of PCR testing is greatly affected by the threshold of amplification cycles* (Ct). At a Ct of 25, the accuracy of the test result is up to 70 percent. At a cycle threshold of 30, the accuracy of the test result drops to 20 percent. At a cycle threshold of 35, the accuracy of the test result drops to 3 percent. The cycle threshold cutoffs in the United States vary between 35 to 40 Ct.

78. On 29 September 2020, Canadian Prime Minister Justin Trudeau appeared as part of a United Nations videoconference in which he discussed Canada's planned contribution to helping the global fight against the COVID-19 pandemic. Statements made publicly during the conference by Trudeau[126] linking the "build back better" slogan of President Joe Biden's administration with ideas reminiscent of Klaus Schwab's World Economic Forum received international attention via social media and Canadian media networks, including the *Toronto Sun*.

> Canada believes that a strong coordinated response across the world and across sectors is essential. This pandemic has provided an opportunity for a reset; this is our chance to accelerate our pre-pandemic efforts to reimagine economic systems that actually address global challenges like extreme poverty, inequality, and climate change. . . . Building back better means getting support to the most vulnerable while maintaining our momentum on reaching the 2030 Agenda for Sustainable Development and the SDG's [sustainable development goals].

79. On 2 October 2020 Michigan's Supreme Court ruled[127] that Democratic Governor Gretchen Whitmer did not have the authority to extend a state of emergency past 30 April 2020. Pursuant to the separation of powers doctrine, before the governor of a state may declare a lockdown, that power must be granted to the executive branch by the state legislature.

Many governors circumvented that rule, as in the case of the Michigan governor, who attempted to unilaterally extend her emergency powers.

A. Justice Markman, joined by Justices Zahra and Clement, concluded,

> The Governor lacked the authority to declare a "state of emergency" or a "state of disaster" under the EMA after April 30, 2020, on the basis of the COVID-19 pandemic and that the EPGA violated the Michigan Constitution because it delegated to the executive branch the legislative powers of state government and allowed the executive branch to exercise those powers indefinitely. First, under the EMA, the Governor only possessed the authority or obligation to declare a state of emergency or state of disaster once and then had to terminate that declaration when the Legislature did not authorize an extension; the Governor possessed no authority to redeclare the same state of emergency or state of disaster and thereby avoid the Legislature's limitation on her authority.

B. Later, on 11 January 2023, Fox News aired the segment "Michigan Gov. Whitmer headed to Europe, WEF meeting in Davos." The subheading states, "The MI Governor Is Traveling Abroad as her Profile Rises on the National Stage."

80. On 4 October 2020, the Great Barrington Declaration[128] was authored and signed by the following medical professionals:

A. Dr. Martin Kulldorff, professor of medicine at Harvard University, a biostatistician and epidemiologist with expertise in detecting and monitoring infectious disease outbreaks and vaccine safety evaluations.

B. Dr. Sunetra Gupta, professor at Oxford University, an epidemiologist with expertise in immunology, vaccine development, and mathematical modelling of infectious diseases.

C. Dr. Jay Bhattacharya, professor at Stanford University Medical School, a physician, epidemiologist, health economist, and public health policy expert focusing on infectious diseases and vulnerable populations.

The declaration, initially cosigned by forty-three medical professionals, continues to attract signatures by medical professionals and the general public. As of 18 November 2023, the declaration had been signed by 16,039 medical and public health scientists, 47,456 medical practitioners, and 872,942 concerned citizens. The declaration stated that the signees had "grave concerns about the damaging physical and mental health impacts of the prevailing COVID-19 policies, and [they] recommend an approach we call Focused Protection."

81. On 22 October 2020 a US Food and Drug Administration (FDA) online meeting to "discuss the general matter of the development, authorization, and/or licensure of vaccines indicated to prevent COVID-19" was streamed live via the FDA YouTube channel.

During a slideshow presentation by Steven A. Anderson, PhD, MPP, the director of Office of Biostatistics and Pharmacovigilance, at precisely 2:33:40 of the video recording, a slide was momentarily presented showing adverse events associated with the genetic COVID-19 vaccines. Though the slide was not discussed in detail, an immediate response by those in attendance arose. The slide, noted as follows [during Dr. Anderson's slide presentation], included the current "Working List of Possible Adverse Event Outcomes." Despite the extensive list of known possible side effects, Pfizer was granted an EUA on 11 December 2020, seven weeks after the meeting.[129]

FDA Safety Surveillance of COVID-19 Vaccines: Draft Working List of Possible Adverse Event Outcomes (Subject to Change)

Guillain-Barré syndrome	Deaths
Acute disseminated encephalomyelitis	Pregnancy and birth complications
Transverse myelitis	Other acute demyelinating diseases
Encephalitis / myelitis / encephalomyelitis / meningoencephalitis / meningitis / encephalopathy	Nonanaphylactic allergic reactions Thrombocytopenia Disseminated intravascular coagulation
Convulsions / seizures	Venous thromboembolism
Stroke	Arthritis and arthralgia / joint pain
Narcolepsy and cataplexy	Kawasaki disease
Anaphylaxis Acute myocardial infarction	Multisystem inflammatory syndrome in children
Myocarditis / pericarditis	Vaccine-enhanced disease
Autoimmune disease	

Another slide, from that same FDA meeting by CDC Immunization Safety Officer Tom Shimabukuro, noted at 2:06:29 of the video recording,[130] showed a similar list of adverse events from the genetic vaccines. Details follow.

Preliminary list of VAERS AEs of special interest

COVID-19 disease	Seizures / convulsions
Death	Stroke
Vaccination during pregnancy and adverse pregnancy outcomes	Narcolepsy / cataplexy Autoimmune disease
Guillain-Barré syndrome (GBS)	Anaphylaxis

Other clinically serious neurologic AEs (group AE) shown below -	Non-anaphylactic allergic reactions
Acute disseminated encephalomyelitis (ADEM)	Acute myocardial infarction
Transverse myelitis (TM)	Myocarditis / pericarditis
Multiple sclerosis (MS)	Thrombocytopenia
Optic neuritis (ON) Chronic inflammatory demyelinating	Disseminated intravascular coagulation (DIC)
Polyneuropathy (CIDP) Encephalitis Myelitis	Venous thromboembolism (VTE) Arthritis and arthralgia (not osteoarthritis or traumatic arthritis)
Encephalomyelitis	Kawasaki disease
Meningoencephalitis Meningitis	Multisystem Inflammatory Syndrome (MIS-C, MIS-A)
Encephalopathy	
Ataxia	

A. In December 2020, the CDC launched its v-safe (vaccine safe) program. The CDC website noted the following:

> V-safe is a safety monitoring system that lets you share with CDC how you, or your dependent, feel after getting a COVID-19 vaccine. After you enroll, v-safe will send you personalized and confidential health check-ins via text messages and web surveys to ask how you feel, including if you experience any side effects after vaccination. Completing health check-ins and sharing how you feel, even if you don't experience side effects after vaccination, helps CDC's vaccine safety monitoring efforts. Your personal information in v-safe is protected so it is safe and private.

B. A short video[131] was posted on the CDC YouTube account promoting the v-safe "after vaccination health checker" on 29 March 2021. Users are required to click or tick checkboxes from a list of questions regarding symptoms experienced after vaccination and a list of health impacts. The symptom list consists of reactions that the CDC referred to as reactions commonly experienced by people post vaccination or "normal reactions," including fever, pain, redness, swelling, itching, chills, headache, joint pains, muscle or body aches, fatigue or tiredness, nausea, vomiting, diarrhea, abdominal pain, rash outside the immediate area around the injection site, any other health symptoms users want to report.

The health impact list of questions to be ticked by users included the following questions:

Did any of the symptoms or health conditions you reported today cause you to (select all that apply):

- Be unable to work.
- Be unable to do your normal daily activities.

- Get care from a doctor or other healthcare professional.
- None of the above.
- Were you pregnant at the time of your COVID-19 vaccination? (Yes/No/I don't know).

Please note that you cannot change your responses after you submit today's health check-in.

The FDA and the CDC were fully aware of potential side effects of the COVID-19 genetic vaccines by 22 October 2020, as confirmed by the lists of adverse reactions presented by Steven A. Anderson and Tom Shimabukuro. The v-safe program was launched in December 2020, more than a month after the 22 October 2020 online meeting, yet it does not include material or references to any of the side effects presented at the FDA meeting.

In fact, only questions regarding mild symptoms have been listed on the v-safe system, with no mention of potentially life-threatening adverse events that have been seen with the COVID-19 genetic vaccines. This raises certain questions. Why did the FDA and CDC withhold information presented at their meeting, information that was clearly in their possession by 22 October 2020 of potential serious, even fatal, adverse reactions? Why was the phrase "safe and effective" repeatedly used by the media, government agencies, health authorities, and other institutions throughout the pandemic? Why were some states incentivizing COVID-19 genetic vaccinations, including but not limited to prizes and cash payments? Why would incentives be needed if the genetic vaccines were truly safe and effective? Why would you need to mandate these genetic vaccines if they were indeed so safe and effective?

82. On 16 November 2020 Fox News published the segment "Tucker Carlson: The Elites Want COVID-19 Lockdowns to Usher in a 'Great Reset' and That Should Terrify You." The subheading states, "The Most Intimate Details of Our Lives Are Being Controlled by Our Leadership Class."[132]

83. On the 19 November 2020 an article written by Sainath Suryanarayanan of US Right to Know was published in *Biotechnology, Health, News*, discussing the conflict of interests surrounding the efforts by Peter Daszak of EcoHealth to discourage any perception that SARS-CoV-2 was a gain-of-function virus built in a human laboratory.

84. On 26 November 2020 the American Institute for Economic Research (AIER) published "The Lasting Consequences[133] of Lockdowns" by Ethan Yang, Adjunct Research Fellow at AIER.

   There has been much discussion over the immediate effects of public health interventions such as business closures and restrictions on social activity in response to Covid-19. It is clear that lockdowns have led to a number of adverse consequences such as unprecedented economic retraction, psychological stress, suicides, and disruptions to all sorts of important social institutions. . . . It is abundantly clear that lockdown policies such as nonessential business closures and movement restrictions have ravaged the economy in the short term. It is also clear that in the near future the security of small businesses remains uncertain as Yelp reports that 60% of restaurants will never reopen. Such developments are certainly painful but perhaps one of the most important and least discussed issues is the potential economic crisis that may result years into the future. A crisis that will not just affect small businesses and vulnerable families but the entire country. . . . Closing down the country has forced the US government to implement trillions of dollars' worth of quantitative easing to prop up Wall Street and stimulus checks to prop up Main Street. If lockdowns continue such policies will need to continue. The result is an unprecedented level of government debt. . . . Although nobody is sure to what extent the government can continue printing money without severe consequence, and it is entirely possible that it could go for much longer, we are testing the boundaries of monetary policy.

85. On 3 December 2020 Florida Department of Health (FDOH), Division of Disease Control and Health Protection Bureau of Epidemiology sent a memorandum to laboratories,[134] titled "Mandatory Reporting of COVID-19 Laboratory Test Results: Reporting of Cycle Threshold Values." It states the following:

    Laboratories are subject to mandatory reporting to the Florida Department of Health (FDOH) under section 381.0031, Florida Statutes, and Florida Administrative Code, Chapter 64D-3.

    • All positive, negative and indeterminate COVID-19 laboratory results must be reported to FDOH via electronic laboratory reporting or by fax immediately. This includes all COVID-19 test types—polymerase chain reaction (PCR), other RNA, antigen and antibody results. For a list of county health departments and their reporting contact information, please visit www.FLhealth.gov/chdepicontact.

    • Cycle threshold (CT) values and their reference ranges, as applicable, must be reported by laboratories to FDOH via electronic laboratory reporting or by fax immediately. If your laboratory is not currently reporting CT values and their reference ranges, the lab should begin reporting this information to FDOH within seven days of the date of this memorandum.

86. By 12 December 2020 Pfizer had its SARS-CoV-2 (COVID) genetic vaccine approved by the FDA, and on 18 December 2020 Moderna's genetic vaccine received approval from the FDA. Full review of the EUA documents, including Pfizer,[135] Moderna,[136] and Janssen[137] (which received FDA EUA approval on 26 February 2021), revealed there was no statistically significant reduction in COVID cases or deaths among the vaccinated versus the unvaccinated (p=NS). Refer to "Statistical Analysis of the Emergency Use Authorization Documents" by Richard M. Fleming, PhD, MD, JD; see attached Exhibit B.

    There was no published comparison of possible health risks of the COVID vaccines that were produced in record time versus the risk of getting COVID-19. This disease has 99.95% recovery rate in patients below the age of 70.[138]

Absent a statistically significant benefit (efficacy) demonstrated by the material within the documents submitted to the FDA for EUA genetic vaccine clearance, questions of safety must be considered. A review of published materials raises several concerns that were not discussed during the FDA EUA clearance meetings, including these:

> A. According to a report on Vaccine Injury[139] Compensation Data from Health Resources & Services Administration (HRSA) published on 1 March 2023, over $4.9 billion ($4,984,801,879.65) was paid to petitioners between 1989 and 2023 because of "alleged vaccines having caused alleged injury." The report states that since 1988 over 25,961 petitions were filed with the National Vaccine Injury Compensation Program (VICP), 12,366 petitions were dismissed, and 9,664 petitions deemed to be compensable after adjudication.
>
> B. A study (2007 to 2010) funded by Harvard Pilgrim Healthcare, *Electronic Support for Public Health—Vaccine Adverse Event Reporting System (ESP: VAERS)*, concluded that less than 1 percent[140] of vaccine adverse events are reported to the FDA. The report states, "Low reporting rates preclude or delay the identification of 'problem' vaccines, potentially endangering the health of the public."

Further review of the VAERS and CDC data available on safety concerns and deaths reveals more adverse events (as of 31 December 2023, approximately 1.5 million) from these three genetic vaccines given FDA EUA clearance than there have been US military casualties since 1776, and more deaths (as of 31 December 2023, over 34,500) admittedly[141] associated with these three genetic vaccines than US military deaths since 1776 in any single war fought, except for the five bloodiest. The following two graphics show the adverse events, including deaths, following injection with these three genetic vaccines, followed by a graphic looking at FDA removal of drugs from the US market, due to drug-related deaths.

87. On 13 December 2020, Sky News host Rowan Dean presented the segment "'You Will Own Nothing, and You Will Be Happy': Warnings of 'Orwellian'[142] Great Reset." Here's the summary: A terrifying coalition of big business and Big Tech are so confident and brazen that they are promising the public "you will own nothing, and you will be happy" in an advertising campaign for a global reset, according to Sky News host Rowan Dean. "What they should have added is 'We the very rich will own everything and be even happier,'" he said.

    The Great Reset is a proposal set out by the World Economic Forum for a new globalized fiscal system that would allow the world to effectively tackle the so-called climate crisis. Dean said the plan intends to use the "tools of oppression" implemented during the pandemic, such as lockdowns and forced business closures, as well as other measures destroying

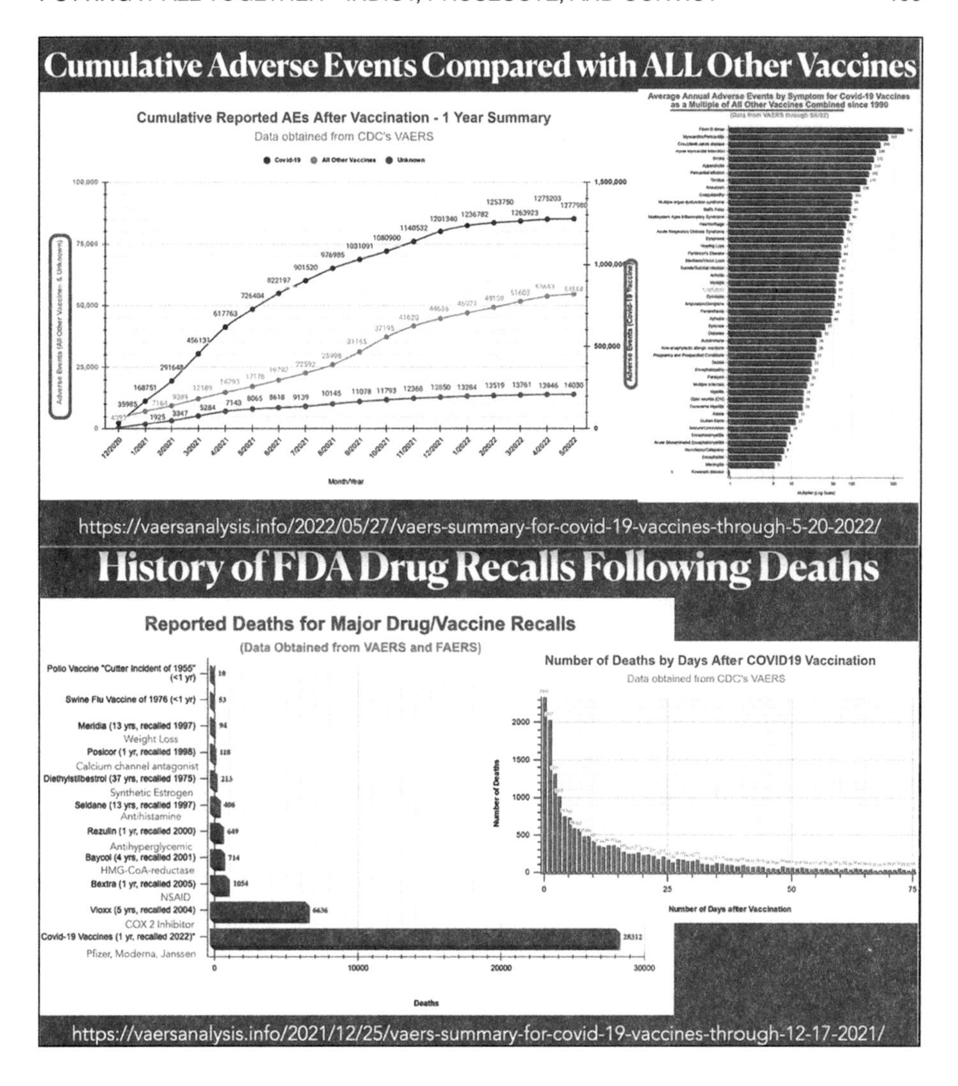

private property rights, to combat the coronavirus to achieve climate outcomes. "I've spoken before about the insidious phrase 'Build Back Better,' which sounds like common sense but is in fact just one of several slogans for the Great Reset, another being the Orwellian phrase the fourth industrial revolution." "This is as serious and as dangerous a threat to our prosperity and freedom as we have faced in decades. . . . This garbage is already deeply embedded into our state and federal governments."

88. On December 20, 2020, the CDC recommended[143] that the vaccine be administered to frontline workers and people over age seventy-five.

89. On 4 January 2021 the FDA released a statement from Stephen M. Hahn, MD (commissioner of food and drugs, December 2019 to January 2021) and Peter Marks, MD, PhD (director, Center for Biologics Evaluation

and Research, or CBER), "FDA Statement on Following the Authorized Dosing Schedules for COVID-19 Vaccines." It[144] includes the following:

> Two different mRNA vaccines have now shown remarkable effectiveness of about 95% in preventing COVID-19 disease in adults. As the first round of vaccine recipients become eligible to receive their second dose, we want to remind the public about the importance of receiving COVID-19 vaccines according to how they've been authorized by the FDA in order to safely receive the level of protection observed in the large, randomized trials supporting their effectiveness. . . . The available data continue to support the use of two specified doses of each authorized vaccine at specified intervals. For the Pfizer-BioNTech COVID-19 vaccine, the interval is 21 days between the first and second dose. And for the Moderna COVID-19 vaccine, the interval is 28 days between the first and second dose. . . . We have committed time and time again to make decisions based on data and science. Until vaccine manufacturers have data and science supporting a change, we continue to strongly recommend that health care providers follow the FDA-authorized dosing schedule for each COVID-19 vaccine.

However, statistical analysis of the EUA data, shown in Exhibit B, demonstrated no statistically significant difference between those who received either of the vaccines and those who received placebo (p=NS)[145].

90. On the 3 February 2021 Thomson Reuters Foundation[146] published an article by Nina Kheladze and Elena Stepanova, "The Thomson Reuters Foundation Partners with Sabin Vaccine Institute to Deliver 'Reporting on Immunisation and Vaccination' Journalism Training." The following is taken from the article:

> With news about coronavirus vaccines dominating headlines and misinformation spreading around the world, the Thomson Reuters Foundation—in partnership with the Sabin Vaccine Institute—launched an eight-day "Reporting on Immunisation and Vaccination" programme for journalists in Eastern Europe.
>
> The online programme covered a wide range of topics—from the role of vaccines in public health, to new and emerging issues around COVID-19 vaccines, to the responsibility of the media in educating the public about vaccine and immunisation related misinformation.
>
> Over the course of the training, the cohort of journalists were briefed by science and health experts and also received one-to-one mentoring from leading media trainers. . . .
>
> The programme also provided us with the unique opportunity to be connected with experts from the World Health Organization and the Sabin Vaccine Institute.

A. Following the completion of the training, participants continued to receive mentorship from the course's trainers as they developed their stories.

The Sabin Vaccine Institute has "fortified relationships with global organizations such as the World Health Organization; Gavi, the Vaccine Alliance; UNICEF; the Bill & Melinda Gates Foundation and Wellcome Trust, as well as with country leadership." On 26 June 2014, Pfizer announced the election of James C. Smith to its Board of Directors.[147] He is also president and CEO of Thomson Reuters and serves on its board of directors, in addition to the boards of the International Business Council of the World Economic Forum, the International Advisory Boards of British American Business, and the Atlantic Council.

In April 2020 Thomson Reuters Foundation launched its Coronavirus Crisis Reporting Hub initiative,[148] a training course to educate journalists in Eastern Europe and Africa on how to report on COVID-19 according to the standards set by Thomson Reuters Foundation. Partners of the COVID 19 Crisis Reporting Hub includes the World Economic Forum, BBC Media Action, National Endowment for Democracy, International Fund for Agricultural Development, and Fritt Ord Foundation. The Coronavirus Crisis Reporting Hub website contains the following statements:

> As the world faces an unprecedented set of challenges posed by the global COVID-19 pandemic, the Thomson Reuters Foundation is leveraging its unique combination of journalism and legal skills to strengthen the global response.
>
> With the sudden emergence of COVID-19, coverage of this new disease became a priority for newsrooms, often leaving journalists with no or limited health reporting experience struggling to provide accurate and reliable information on this complex issue to their audiences.
>
> In this context, the Thomson Reuters Foundation launched its Coronavirus Crisis Reporting Hub. The global initiative was first launched in April 2020 in English-speaking African nations before scaling around the world.
>
> Our 8-week training, delivered to cohorts of no more than 15 participants, equips journalists with the core skills and information they need to report on the pandemic and its impact on economies, health care systems and communities. It also gives cohorts access to key experts, as well as reporters who have been covering the crisis in other countries.
>
> The free programme includes live video masterclasses on reporting and production techniques, as well as the economic and social impact of the pandemic, the science behind it, the race to find a vaccine, countering misinformation, the safety of journalists online and offline, and ethical standards and legal dangers. It also includes briefings from journalists at the frontline of the crisis, and discussion of potential story ideas and how to approach them in the local context. Finally, a skills lab offers the latest digital and multimedia tools and techniques to enhance the production of stories.
>
> Our Coronavirus Reporting Hub also focusses on the wellbeing of participants by fostering peer-to-peer support, as well as upskilling journalists with digital and multimedia learning to help them future-proof their careers.

B. The Defender published an article on 11 August 2021 by Megan Redshaw, titled "Conflict of Interest: Reuters 'Fact Checks' COVID-Related Social Media Posts, but Fails to Disclose Ties to Pfizer, World Economic Forum."[149] The subheading stated, "Reuters is now in the business of 'fact-checking' Facebook and Twitter posts about COVID vaccines—despite having ties to Pfizer, World Economic Forum and Trusted News Initiative" and includes the following text:

> But here's a less-publicized fact some social media users—and consumers of online news—may not know: Reuters, owned by the $40 billion international multimedia company, Thomson Reuters Corporation, is also in the business of "fact checking"

social media posts. Reuters publishes its fact-checking commentary online in a format designed to resemble new stories, which turn up in online searches.

Last week, Reuters announced a new collaboration with Twitter to "more quickly provide credible information on the social networking site as part of an effort to fight the spread of misinformation." In February, Reuters announced a similar partnership with Facebook to "fact check" social media posts.

However, when announcing its fact-checking partnerships with Facebook and Twitter, Reuters made no mention of this fact: The news organization has ties to Pfizer, World Economic Forum (WEF) and Trusted News Initiative (TNI), an industry collaboration of major news and global tech organizations whose stated mission is to "combat spread of harmful vaccine disinformation."

Reuters also failed to provide any criteria for how information would be defined as "misinformation" and did not disclose the qualifications of the people responsible for determining fact versus false or misleading "misinformation."

C. Here is one example of a physician removed from LinkedIn for discussing Reuters' ties to TNI, Pfizer, and the WEF:

Dr. Robert Malone, a physician and inventor of mRNA vaccines and RNA drugs, said he thinks anyone reading Reuters "fact check" articles about COVID-related content should know about, as Malone says, the obvious conflicts of interest. . . .

Malone stated: "What we have here is this horizontal integration across pharma, big tech, big media, government and traditional media. It's not just the Trusted News Initiative. It goes beyond. The same thing is true with Merck and all the others. Pfizer is really playing quite aggressively here."

In addition to the conflicts of interest Malone identified in his Twitter and LinkedIn posts, Malone said it's the lack of transparency—whether on the part of Reuters, Facebook or TNI—around who defines "misinformation," based on what criteria, that is concerning to him.

Malone told The Defender, based on his research, most fact checkers don't have a background in science or health. Yet even without such qualifications, and without working off of a transparent definition of "misinformation," fact checkers are able to shut down online communication between scientists and physicians by flagging or deleting posts.

Worse yet, Malone said, if the fact-checkers label posts by a physician, like himself, as "disinformation," the fact-checker's claim potentially could be used as a justification for revoking a physician's license.

"How is this in the public's best interest?" Malone asked.

91. On 8 February 2021, nuclear cardiologist and physicist Dr. Richard M. Fleming published research[150] demonstrating several treatments that were successful in both the outpatient and inpatient setting, following the inflammation and vascular disease theory he originally presented at the American Heart Association meetings in 1994. The theory details the role multiple factors, including viruses, play in stimulating an InflammoThrombotic response (ITR) and disease, and explains why SARS-CoV-2 produces the ITR disease identified as coronavirus disease first identified in 2019 (COVID-19). The treatments were measured using FMTVDM, a patented[151] method for quantitatively measuring disease resulting from changes in tissue blood flow and metabolism associated

with infectious diseases and the ITR of SARS-CoV-2. Treatments were successful 99.83 percent of the time in people infected with SARS-CoV-2 and those critically ill with COVID-19.

Treatment of each patient with SARS-CoV-2 should focus on the stage of infection and InflammoThrombotic response to the virus with measurement of the extent and severity of the disease and response to treatment. Once patients required hospitalization, they responded favorably (99.83 percent) to treatments focusing on reducing the ITR resulting from the body's immune response to SARS-CoV-2. These ITR drugs proved most promising when initiated upon admission and when used in combination, reducing hospitalization time from 30 to 45 days to as little as 18 to 25 days with 0.17 percent mortality.

This research[152] has since been presented at the American Society of Nuclear Cardiology Conference and referenced in several other research publications that discuss efforts to diagnose and treat viral[153] infections, including SARS-CoV-2.[154]

92. During a congressional hearing on 18 March 2021, Senator Rand Paul questioned Dr. Anthony Fauci "over whether masks are necessary for people who have been vaccinated" or have been infected and recovered from the virus.

Senator Paul argued that no studies could prove masks were required[155] after vaccination or infection from COVID-19. A transcript of that Senate testimony by Dr. Anthony Fauci, titled "Masks Are Not Theatre" includes the following exchange:

Senator Paul: (02:42) You're telling everybody to wear a mask, whether they've had an infection or a vaccine. What I'm saying is they have immunity and everybody agrees they have immunity. What studies do you have that people that have had the vaccine or have had the infection are spreading the infection? If we're not spreading the infection, isn't it just theater?

Senator Paul: (05:25) You can't get it again. There's virtually 0% chance you're going to get it. And yet you're telling people that have had the vaccine who have immunity, you're defying everything we know about immunity by telling people to wear masks who have been vaccinated. Instead, you should be saying, there is no science to say we're going to have a problem from the large number of people being vaccinated. You want to get rid of vaccine hesitancy, tell them they can quit wearing their mask after they get the vaccine. You want people to get the vaccine? Give them a reward, instead of telling them that the nanny state's going to be there for three more years and you got to wear a mask forever. People don't want to hear there's no science behind it.

Dr. Fauci: (05:59) Well, let me just state for the record that masks are not theater. Masks are protective and we ask—

Senator Paul (06:07) If they have immunity, they are theater. If you already have immunity, you're wearing a mask to give comfort to others. You're not wearing a mask because of [inaudible 00:06:14].

93. On 11 May 2021, Dr. Anthony Fauci, CDC Director Rochelle Walensky, and other health officials testified[156] before the US Senate to answer questions regarding the ongoing COVID-19 prevention measures and guidance.

During the hearing, Senator Rand Paul accused the National Institutes of Health, of funding gain-of-function research at the Wuhan Institute of Virology and challenged Dr. Fauci about the possibility of the novel coronavirus originating in a laboratory. The transcript, titled "Dr. Fauci, CDC Director Testify Before Senate on COVID-19 Guidelines Transcript," reveals that contrary to the evidence, Dr. Fauci perjured himself, under oath, before the US Senate. The transcript includes the following statements:

Senator Rand Paul: (57:58) Dr. Fauci, we don't know whether the pandemic started in a lab in Wuhan or evolved naturally, but we should want to know. Three million people have died from this pandemic, and that should cause us to explore all possibilities. Instead, government authorities, self-interested in continuing gain-of-function research say there's nothing to see here. Gain-of-function research, as you know, is juicing up naturally occurring animal viruses to infect humans. To arrive at the truth, the US government should admit that the Wuhan Virology Institute was experimenting to enhance the coronavirus' ability to infect humans. Juicing up super-viruses is not new; scientists in the US have long known how to mutate animal viruses to infect humans. For years, Dr. Ralph Baric, a virologist in the US, has been collaborating with Dr. Shi Zhengli of the Wuhan Virology Institute, sharing his discoveries about how to create super-viruses. This gain-of-function research has been funded by the NIH. The collaboration between the US and the Wuhan Virology Institute continues. Doctors Baric and Shi worked together to insert bat virus spike protein into the backbone of the deadly SARS virus and then use this man-made super-virus to infect human airway cells. Think about that for a moment. The SARS virus had a 15% mortality. We're fighting a pandemic that has about a 1% mortality. Can you imagine, if a SARS virus that's been juiced up and had viral proteins added to it, to the spike protein, if that were released accidentally? Dr. Fauci, do you still support funding of the NIH funding of the lab in Wuhan?

Dr. Anthony Fauci: (59:49) Senator Paul, with all due respect, you are entirely and completely incorrect that the NIH has not ever and does not now fund gain-of-function research in the Wuhan Institute of Virology.

Senator Rand Paul: (01:00:06) Do they fund Dr. Baric?

Dr. Anthony Fauci: (01:00:09) We do not fund gain—

Senator Rand Paul: (01:00:11) Do you fund Dr. Baric's Gain-Of-Function research?

Dr. Anthony Fauci: (01:00:13) Dr. Baric does not do Gain-Of-Function research, and if he is, it's according to the guidelines and it is being conducted in North Carolina, not in China.

Senator Rand Paul: (01:00:24) You don't think inserting a bat virus spike protein that he got from the Wuhan Institute into the SARS virus is gain-of-function? You would be in the minority because at least 200 scientists have signed a statement from the Cambridge Working Group saying that it is gain of function.

Dr. Anthony Fauci: (01:00:39) Well, it is not. If you look at the grant and you look at the progress reports, it is not gain-of-function, despite the fact that people tweet that, they write about it—

Senator Rand Paul: (01:00:50) Do you support sending money to the Wuhan Virology Institute?

Dr. Anthony Fauci: (01:00:53) We do not send money now to the Wuhan Virology Institute.

Senator Rand Paul: (01:00:55) Do you support sending money? We did under your tutelage. We were sending it through EcoHealth. It was a sub-agency and a sub-grant. Do you support that the money from NIH that was going to the Wuhan Institute.

Dr. Anthony Fauci: (01:01:08) Let me explain to you why that was done. The SARS-CoV-1 originated in bats in China. It would have been irresponsible of us if we did not investigate the bat viruses and the serology to see who might have been infected in China.

Senator Rand Paul: (01:01:29) Or perhaps it would be irresponsible to send it to the Chinese government that we may not be able to trust with this knowledge and with these incredibly dangerous viruses. Government scientists like yourself who favor gain-of-function research maintain the disease arose naturally.

94. On 03 June 2021, *Vanity Fair* published[157] "The Lab-Leak Theory: Inside the Fight to Uncover COVID-19's Origins," by Katherine Eban, with the following subheading: "Throughout 2020, the Notion That the Novel Coronavirus Leaked from a Lab was Off-Limits. Those Who Dared to Push for Transparency Say Toxic Politics and Hidden Agendas Kept Us in the Dark." The report included information on the use of CRISPR technology uncovered by the US National Security Council (NSC):

As the NSC tracked these disparate clues, U.S. government virologists advising them flagged one study first submitted in April 2020. Eleven of its 23 co-authors worked for the Academy of Military Medical Sciences, the Chinese army's medical research institute. Using the gene-editing technology known as CRISPR, the researchers had engineered mice with humanized lungs, then studied their susceptibility to SARS-CoV-2. As the NSC officials worked backward from the date of publication to establish a timeline for the study, it became clear that the mice had been engineered sometime in the summer of 2019, before the pandemic even started. The NSC officials were left wondering: Had the Chinese military been running viruses through humanized mouse models, to see which might be infectious to humans?

Believing they had uncovered important evidence in favor of the lab-leak hypothesis, the NSC investigators began reaching out to other agencies. That's when the hammer came down. "We were dismissed," said Anthony Ruggiero, the NSC's senior director for counterproliferation and biodefense. "The response was very negative.

The Eban article also discussed money flowing from NIH to Peter Daszak of EcoHealth Alliance:

In October 2014, the Obama administration imposed a moratorium on new funding for gain-of-function research projects that could make influenza, MERS, or SARS viruses more virulent or transmissible. But a footnote to the statement announcing the moratorium carved out an exception for cases deemed "urgently necessary to protect the public health or national security."

In the first year of the Trump administration, the moratorium was lifted and replaced with a review system called the HHS P3CO Framework (for Potential Pandemic Pathogen Care and Oversight). It put the onus for ensuring the safety of any such research on the federal department or agency funding it. This left the review process shrouded in secrecy. "The names of reviewers are not released, and the details of the experiments to be considered are largely secret," said the Harvard epidemiologist Dr. Marc Lipsitch, whose advocacy against gain-of-function research helped prompt the moratorium. (An NIH spokesperson told Vanity Fair that "information about individual unfunded applications is not public to preserve confidentiality and protect sensitive information, preliminary data, and intellectual property.")

Inside the NIH, which funded such research, the P3CO framework was largely met with shrugs and eye rolls, said a long-time agency official: "If you ban gain-of-function research, you ban all of virology." He added, "Ever since the moratorium, everyone's gone wink-wink and just done gain-of-function research anyway."

British-born Peter Daszak, 55, is the president of EcoHealth Alliance, a New York City–based non-profit with the laudable goal of preventing the outbreak of emerging diseases by safeguarding ecosystems. In May 2014, five months before the moratorium on gain-of-function research was announced, EcoHealth secured a NIAID grant of roughly $3.7 million, which it allocated in part to various entities engaged in collecting bat samples, building models, and performing gain-of-function experiments to see which animal viruses were able to jump to humans. The grant was not halted under the moratorium or the P3CO framework.

By 2018, EcoHealth Alliance was pulling in up to $15 million a year in grant money from an array of federal agencies, including the Defense Department, the Department of Homeland Security, and the U.S. Agency for International Development, according to 990 tax exemption forms it filed with the New York State Attorney General's Charities Bureau. Shi Zhengli herself listed U.S. government grant support of more than $1.2 million on her curriculum vitae: $665,000 from the NIH between 2014 and 2019; and $559,500 over the same period from USAID. At least some of those funds were routed through EcoHealth Alliance.

EcoHealth Alliance's practice of divvying up large government grants into smaller sub-grants for individual labs and institutions gave it enormous sway within the field of virology. The sums at stake allow it to "purchase a lot of omertà" from the labs it supports, said Richard Ebright of Rutgers. (In response to detailed questions, an EcoHealth Alliance spokesperson said on behalf of the organization and Daszak, "We have no comment.")

As the pandemic raged, the collaboration between EcoHealth Alliance and the WIV wound up in the crosshairs of the Trump administration. At a White House COVID-19 press briefing on April 17, 2020, a reporter from the conspiratorial right-wing media outlet Newsmax asked Trump a factually inaccurate question about a $3.7 million NIH grant to a level-four lab in China. "Why would the U.S. give a grant like that to China?" the reporter asked.

Trump responded, "We will end that grant very quickly," adding, "Who was president then, I wonder."

A week later, an NIH official notified Daszak in writing that his grant had been terminated. The order had come from the White House, Dr. Anthony Fauci later testified before a congressional committee. The decision fueled a firestorm: 81 Nobel Laureates in science denounced the decision in an open letter to Trump health officials, and *60 Minutes* ran a segment focused on the Trump administration's short-sighted politicization of science.

Daszak appeared to be the victim of a political hit job, orchestrated to blame China, Dr. Fauci, and scientists in general for the pandemic, while distracting from the Trump administration's bungled response. "He's basically a wonderful, decent human being" and an "old-fashioned altruist," said the NIH official. "To see this happening to him, it really kills me."

In July, the NIH attempted to backtrack. It reinstated the grant but suspended its research activities until EcoHealth Alliance fulfilled seven conditions, some of which went beyond the non-profit's purview and seemed to stray into tinfoil-hat territory. They included: providing information on the "apparent disappearance" of a Wuhan Institute of Virology researcher, who was rumored on social media to be patient zero, and explaining diminished cell phone traffic and roadblocks around the WIV in October 2019.

95. On 04 June 2021, Informed Consent Action Network (ICAN) released a report[158] on their website titled, "ICAN OBTAINS OVER 3,000 PAGES OF TONY FAUCI'S EMAILS".

The article stated, "Last year, ICAN made FOIA requests to NIH for documents regarding COVID-19, including two requests for Anthony Fauci's emails. ICAN has received nearly 3,000 emails sent by Fauci from early February 2020 through May 2020. Read what Fauci was saying privately about masks, therapeutics, vaccines, ventilators, and many other COVID-19 topics."

The emails reveal an invitation from Richard Fontaine, executive director of the Trilateral Commission-North America, for Dr. Fauci to "participate in a conversation at the meeting."

A. 10 February 2020 at 11:30AM (Email Ref: ICAN_000103), From Richard Fontaine to Dr. Anthony Fauci, cc. Meghan O'Sullivan, Torrey Taussig. Subject: Invitation to Speak at the March 13–15 Trilateral Commission Meeting in Washington, D.C.

Dear Dr. Fauci, The Trilateral Commission will hold its invitation-only annual plenary meeting in Washington, D.C., on March 13–15, 2020. I would like to invite you to participate in a conversation at the meeting on responding to the coronavirus and global pandemics. As you may know, the Trilateral Commission was launched by David Rockefeller in 1973 to think through the shared challenges and leadership responsibilities of the three principal industrialized democratic areas of the world: Europe, North American, and Japan (now Asia). Today, the Commission believes its original mission of bringing democratic countries together to tackle international challenges should once again be a major priority for our country and our partners. This year's plenary meeting is a major, three-day gathering of our global membership during which we will explore the theme "Democracy and Capitalism at a Crossroads." Your participation would make a great difference if you are available. Attached to this email you will find an invitation from North American chair Meghan O'Sullivan. Please do let us know if you have any questions about the event or your potential role in it, and we hope that you are able to accept our invitation. Best wishes, Richard.

Dr. Anthony Fauci emailed Patricia Conrad, NIH/NIAID special assistant to the director, outside of office hours expressing his willingness to attend the Trilateral Commission meeting. He also instructed her to withhold details of the invitation from other members of the team.

B. 10 February 2020 at 19:19 (Email Ref: ICAN_000103), From Dr. Anthony Fauci to Patricia Conrad (NIH/NIAID). Subject: Dr. Anthony Fauci—Trilateral Plenary Meeting Speaking Invitation.pdf.

Let us discuss. We do not need to bring before the OD AM group. I would like to do this if possible. It is an invitation from Meghan O'Sullivan[159] who was one of the security crew from Bush 43.

C. 21 February 2020 at 4:52PM (Email Ref: ICAN_ 000274–000277), From Jennifer Routh (NIH/NIAID) to Patricia Conrad (NIH/NIAID), cc. NIAID COGCORE, NIAID Media Inquiries. Subject: interview request: Draft responses for Greek newspaper reporter Theodora Tsoli, for *The Tribune* (Βήμα; VIMA), a weekly Greek newspaper VIMA in late February 2020.

As noted in an email[160] approved by Fauci, in response to written questions submitted by Theodora Tsoli, Dr. Anthony Fauci expressed his opinion that the Chinese had taken "extreme measures to control the outbreak" and that "people outside of China do not need to wear a mask":

> The vast majority of people outside of China do not need to wear a mask. A mask is more appropriate for someone who is infected than for people trying to protect against infection.

D. On 02 April 2020 at 9:58AM (Email[161] Ref: ICAN_001778), in an email reply from Dr. Anthony Fauci to Emilio A. Emini (CEO of Bill & Melinda Gates Medical Research Institute), cc. Patricia Conrad (NIH/NIAID), John Mascola (NIH/Vaccine Research Center), Emily Erbelding (NIH/NIAID), Rick Bright Office of Special Counsel (OSC)/Assistant Secretary for Preparedness and Response (ASPR)/Biomedical Advanced Research and Development Authority (BARDA). Subject: Connection Request per Bill Gates.[162]

Emilio: Thanks for your note. As I had mentioned to Bill [Gates] yesterday evening, I am enthusiastic about moving towards a collaborative and hopefully synergistic approach to COVID-19 on the part of NIAID/NIH, BARDA and the BMGF [Bill & Melinda Gates Foundation]. I will ask John Mascola and Emily Erbelding to connect with you to stay a conversation. Perhaps they can organize a call with you including BARDA. I will try to engage as much as I can given my current circumstances. We look forward to working with you. Best regards, Tony

E. In an email dated 21 April 2020 at 4:10PM (ICAN_002472), From Victor J. Dzau to Francis S. Collins (former NIH Director), cc. Morgan Kanarek` (NIH/NIAID). Subject: For your attention.

> Dear Francis, . . . I am sure you are aware of a global coordinating effort to accelerate vaccines, diagnostics and therapeutics. I have been part of the conversation and planning along with Jeremy Farrar, Richard Hatchett, Seth Berkley, Chris Elias, Paul Stoffels etc. Recently Who, Gates Foundation and European Commission have been leading the planning. This has advanced rapidly and is in the final stages in development that will soon be announced. It has involved European Commission, Germany, Japan, UK, Norway, France, Saudi as well as Gate Foundation, WHO, World Bank, Wellcome Trust, GAVI, Global Fund, CEPI, GPMB and private sector industry. This initiative will begin with a Pledge conference for $8B as a starting point. This will be led by President von der Leyen and is co-chaired by the above country leaders. This will occur on May 4. In addition by the end of this week or early next week there will be an announcement on the global coordinating structure which will involve Gates, WHO etc. .

Bill Gates and the Bill & Melinda Gates Foundation, the WHO, and European Commission are mentioned as being at the forefront of the global pandemic[163] response in an email from Victor J. Dzau, president of the National Academy of Medicine to Francis S. Collins, former NIH director.

Dr. Francis S. Collins forwarded the email to Dr. Anthony Fauci and other team members with the following remarks approximately three hours later:

F. 21 April 2020 at 7:28PM (ICAN_002471), From Francis S. Collins to Dr. Anthony Fauci, cc. Cliff Lane (NIH/NIAID), Lawrence Tabak (NIH/OD), Maria Freire (FNIH), David Wholley (FNIH). Subject: FW: For your attention.

> Hi all, See note below from Victor Dzau about a global effort on COVID-19. I can't tell if this is more than a fund-raising effort. I know we have Gates reps on our ACTIV working groups—has any of this plan come up, David? Francis.

96. On 5 June 2021 Fox News's Laura Ingraham interviewed epidemiologist and health economist Jay Bhattacharya,[164] professor of medicine at Stanford University as part of the segment "Ingraham: Are Fauci's days numbered?" Dr. Bhattacharya made the following responses and statements:

> [In response to Dr. Fauci's conflicting advice on face masks:] I think he's been all over the place on masks. There are emails you can find in the treasure trove of emails that have been released where he acknowledged the virus has been aerosolized. Well,

> the cloth masks people have been recommending, they're not particularly effective against aerosolized viruses. I really don't understand his back and forth, and his answer made absolutely no sense. . . . Yes, you should change your mind when the science changes. What is that science that changed that convinced him that masks are the most effective way? I think his credibility is entirely shot. In the early days of the epidemic, he was quite a sensible person, he understood immunity, he understood the necessity of not panicking the population. Something happened in late February where he just flipped on a dime. It wasn't the science changing. Something else happened where he just changed.

97. On 10 June 2021 NBC News aired "Evidence Grows Stronger for Covid Vaccine Link to Heart Issue,[165] CDC Says," with a corresponding article by Erika Edwards. The article quotes Dr. Tom Shimabukuro, deputy director of the CDC's Immunization Safety Office (see point 81 above—the list of adverse effects, including myocarditis and pericarditis):

> The condition, called myocarditis, is usually mild, but a handful of patients remain hospitalized.
>
> A higher-than-usual number of cases of a type of heart inflammation has been reported following Covid-19 vaccination, especially among young men following their second dose of an mRNA vaccine, the Centers for Disease Control and Prevention said Thursday.
>
> Overall, 226 cases of myocarditis or pericarditis after vaccination in people younger than age 30 have been confirmed, Dr. Tom Shimabukuro, deputy director of the CDC's Immunization Safety Office, said during a presentation to a Food and Drug Administration advisory group. Further investigation is needed, however, to confirm whether the vaccination was the cause of the heart problem.
>
> Normally, fewer than 100 cases would be expected for this age group.
>
> Teenagers and people in their early 20s accounted for more than half of the myocarditis cases reported to the CDC's safety monitoring systems following Covid-19 vaccination, despite representing a fraction of people who have received the shots.
>
> "We clearly have an imbalance there," Shimabukuro said.
>
> The vast majority of the cases were sent home following a visit to a hospital as of the end of May. It's unclear how many patients were admitted to the hospital, or, for example, were discharged following a visit to the emergency room. Fifteen patients remain hospitalized, with three in intensive care units. Two of the patients in the ICU had other health problems.
>
> The CDC had information on the recovery of patients in 220 cases; in more than 80 percent of these cases, patients got better on their own.
>
> Following the presentation, Dr. Cody Meissner, chief of pediatric infectious diseases at the Tufts Children's Hospital in Boston, said, "It is hard to deny that there's some event that seems to be occurring in terms of myocarditis."

This myocardial disease, along with cancer, diabetes, high blood pressure, strokes, and a variety of other InflammoThrombotic Response diseases, is in fact, exactly as laid out by Dr. Richard M. Fleming in 1994 at the American Heart Association meetings and later reduced to writing in a cardiology textbook[166] published in 1999, when Dr. Fleming added infectious diseases to the etiological list of causes for ITR diseases. This is a crucial fact, given that infected and hospitalized patients were not treated according to the theory laid out, published, and repeatedly proven. Fleming's theory is now accepted scientific and medical fact and explains why, in this instance, a gain-of-function viral (SARS-CoV-2) infection resulted in an InflammoThrombotic disease (COVID), responsible for killing and maiming so many.

98. On 16 July 2021, the *New York Post*[167] published "Government Dictating What Social-Media Bans Is Tyrannical" by Rachel Bovard. The article states the following:

    There is a dystopian element to telling social media platforms to control "misinformation" when the very definition of that keeps changing. In the early months of the pandemic, Facebook began banning anti-lockdown protest content. Not because it violated any laws, but because such gatherings might run afoul of local guidance and public health recommendations. YouTube began censoring any content that disagreed with the error-prone World Health Organization, removing videos from emergency room doctors and podcasts from Stanford University neuroradiologists alike.

99. On 20 July 2021, Dr. Anthony Fauci and CDC Director Dr. Rochelle Walensky testified before the US Senate[168] to provide updates on the COVID-19 response. Senator Rand Paul accused Dr. Fauci of lying[169] before Congress about the origins of COVID-19 and NIH funding of gain-of-function research at the Wuhan Institute of Virology.

    The transcript "Fauci, Welensky COVID-19 Response Testimony Senate Hearing Transcript July 20" included the following statements:

    Senator Rand Paul: (50:05) Dr. Fauci, as you are aware, it is a crime to lie to Congress. Section 1.0.0.1 of the US Criminal Code creates a felony and a five-year penalty for lying to Congress. On your last trip to our committee on May 11th, you stated that the NIH has not ever and does not now fund gain-of-function research in the Wuhan Institute of Virology. And yet, gain-of-function research was done entirely in the Wuhan Institute by Dr. Shi and was funded by the NIH. I'd like to ask unanimous consent to insert into the record the Wuhan virology paper entitled "Discovery of a Rich Gene Pool of Bat SARS Related Coronaviruses." Please deliver a copy of the journal article to Dr. Fauci.

    In this paper, Dr. Shi credits the NIH and lists the actual number of the grant that she was given by the NIH. In this paper, she took two bat coronavirus genes, spike genes, and combined them with a SARS related backbone to create new viruses that are not found in nature. These lab-created viruses were then shown to replicate in humans. These experiments combined genetic information from different coronaviruses that infect animals, but not humans, to create novel artificial viruses able to infect human cells. Viruses that in nature only infect animals were manipulated in the Wuhan lab to gain the function of infecting humans. This research fits the definition of the research that the NIH said was subject to the pause in 2014 to 2017, a pause in funding on gain-of-function, but the NIH failed to recognize this, defines it away, and it never came under any scrutiny.

Dr. Richard Ebright, a molecular biologist from Rutgers, described this research in Wuhan as the Wuhan lab used NIH funding to construct novel chimeric SARS related to coronaviruses able to infect human cells and laboratory animals. This is high risk research that creates new potential pandemic pathogens, potential pandemic pathogens that exist only in the lab, not in nature. This research matches . . . these are Dr. Ebright's words. This research matches, indeed epitomizes the definition of gain-of-function research, done entirely in Wuhan, for which there was supposed to be a federal pause. Dr. Fauci, knowing that it is a crime to lie to Congress, do you wish to retract your statement of May 11th, where you claimed that the NIH never funded gain-of-function research in Wuhan.

Dr. Anthony Fauci: (53:05) Senator Paul, I have never lied before the Congress, and I do not retract that statement. This paper that you were referring to was judged by qualified staff up and down the chain as not being gain of function. What was . . . let me finish!

Senator Rand Paul: (53:25) You take an animal virus and you increase its transmissibility to humans. You're saying that's not gain-of-function?

Dr. Anthony Fauci: (53:30) Yeah, that is correct. And Senator Paul, you do not know what you are talking about, quite frankly. And I want to say that officially, you do not know what you are talking about, okay? You get one person . . . can I answer?

Senator Rand Paul: (53:46) This is your definition that you guys wrote. It says that scientific research that increases the transmissibility among animals is gain-of-function. They took animal viruses that only occur in animals, and they increased their transmissibility to humans. How you can say that is not gain a function—

Dr. Anthony Fauci: (54:06) It is not.

Senator Rand Paul: (54:07) It's a dance, and you're dancing around this because you're trying to obscure responsibility for 4 million people dying around the world from a pandemic.

Dr. Anthony Fauci: (54:17) I have to . . . well, now you're getting into something. If the point that you are making is that the grant that was funded as a subaward from Eco . . . Health to Wuhan created SARS- CoV-2, that's where you were getting. Let me finish. Wait a minute.

Senator Rand Paul: (54:35) We don't know that it didn't come from the lab, but all the evidence is pointing that it came from the lab, and there will be responsibility for those who funded the lab, including yourself.

100. On 29 July 2021 the White House released the statement "Fact Sheet: President Biden to Announce New Actions to Get More Americans Vaccinated[170] and Slow the Spread of the Delta Variant." It included the following statements:

Six months into the Biden Administration's vaccination effort, 164 million Americans are fully vaccinated, including 80 percent of seniors and more than 60 percent of adults. . . . We are now faced with a much more transmissible strain of this virus—the Delta variant. . . . We know how to stop it: get more people vaccinated. . . . Today's actions include:

Strengthening Safety Protocols for Federal Employees and Federal Contractors. Today, the President will announce that to help protect workers and their communities, every federal government employee and onsite contractor will be asked to attest to their vaccination status. Anyone who does not attest to being fully vaccinated will be required to wear a mask on the job no matter their geographic location, physically distance from all other employees and visitors, comply with a weekly or twice weekly screening testing requirement, and be subject to restrictions on official travel. . . . President Biden is directing his team to take steps to apply similar standards to all federal contractors. The Administration will encourage employers across the private sector to follow this strong model.

Protecting Those Who Serve Our Country. Today, the President will announce that he is directing the Department of Defense to look into how and when they will add COVID-19 vaccination to the list of required vaccinations for members of the military. . . . Earlier this week, like many health care employers across the country, the Department of Veterans Affairs took the common-sense and important step of requiring their health care providers and personnel to be fully vaccinated against COVID-19. . . . On Monday, over 50 leading health care societies and organizations called for all health care and long-term care employers to require their workers to receive the COVID-19 vaccine.

Expanding Paid Leave to Get Families and Kids Vaccinated. The President will announce that small- and medium-sized businesses will now be reimbursed for offering their employees paid leave to get their family members, including their kids, vaccinated. . . . The federal government is fully reimbursing any small- or medium-sized business that provides workers with paid time off to get vaccinated. . . . The President will also call on employers who have not offered paid time off to their employees for vaccination to do so.

Calling on State and Local Governments to Offer $100 to Get Vaccinated. Today, the President will call on states, territories, and local governments to do more to incentivize vaccination, including offering $100 to those who get vaccinated. . . .

Increasing Vaccinations Among Adolescents as Kids Go Back to School. Today, the President will, in an effort to get more kids 12 and older vaccinated, call on school districts nationwide to host at least one pop-up vaccination clinic over the coming weeks. . . . Almost 90% of educators and school staff are vaccinated.

101. On 20 October 2021 CNN published an article by Laura Ly and Holly Yan—"New York City vaccine mandate extends to all city workers and includes a new $500 bonus,[171] mayor says." The article included the following:

Previous efforts to encourage Covid-19 vaccinations didn't result in enough city employees getting shots, the mayor told CNN. As of Wednesday, 46,000 New York City employees were still unvaccinated, he said. . . . "We're going to work with your union to figure out what happens next," de Blasio told MSNBC, noting medical and religious accommodations remain in place. "But the bottom line is we're not going to pay people unless they're vaccinated." However, the mandate faces legal challenges from two key city unions. The Police Benevolent Association of the City of New York, which represents about 24,000 police officers, plans to take legal action against the mandate, president Patrick Lynch said. "From the beginning of the de Blasio administration's haphazard vaccine rollout, we have fought to make the vaccine available to every member who chooses it, while also protecting their right to make that personal medical decision in consultation with their own doctor," Lynch said. "Now that the city has moved to unilaterally impose a mandate, we will proceed with legal action to protect our members' rights." The union that represents New York City firefighters announced similar plans Wednesday. "Putting people out of work for making a personal health choice is something we can never accept," Uniformed Firefighters Association President Andrew Ansbro announced.

102. On 21 October 2021 the *New York Post* published "NIH admits US Funded Gain-of-Function in Wuhan[172]—Despite Fauci[173] Denials," written by Emily Crane. The *New York Post* article included the following:

> The National Institutes of Health has stunningly admitted to funding gain-of-function research on bat coronaviruses at China's Wuhan lab—despite Dr. Anthony Fauci repeatedly insisting to Congress that no such thing happened. In a letter to Rep. James Comer (R-Ky.) on Wednesday, a top NIH official blamed EcoHealth Alliance—the New York City–based non-profit that has funnelled US funds to the Wuhan lab—for not being transparent about the work it was doing. NIH's principal deputy director, Lawrence A. Tabak, wrote in the letter that EcoHealth's "limited experiment" tested whether "spike proteins from naturally occurring bat coronaviruses circulating in China were capable of binding to the human ACE2 receptor in a mouse model."

103. On 1 November 2021, CNN published the article "2,300 NYC Firefighters Call Out Sick[174] as Vaccine Mandate Begins, but Mayor Says Public Safety Not Disrupted."

104. On 9 September 2021 NBC News[175] published "Biden Announces Sweeping Vaccine Mandates Affecting Millions of Workers." The subheading states, "The Administration Said the New Mandates Could Affect Around 100 Million People, More than Two-Thirds of the US Workforce." The article includes the following statements:

> President Joe Biden on Thursday issued two executive orders mandating vaccines for federal workers and contractors and announced new requirements for large employers and health care providers that he said would affect around 100 million workers, more than two-thirds of the U.S. workforce. **"We've been patient, but our patience is wearing thin,"** Biden said, making a direct appeal to the 80 million people who he said were still unvaccinated. "Your refusal has cost all of us." Biden also announced that he asked the Department of Labor to issue an emergency rule requiring all employers with 100 or more employees to ensure their workforce is fully vaccinated or require any unvaccinated workers to produce a negative Covid test at least once a week. The requirement could carry a $14,000 fine per violation and would affect two-thirds of the country's workforce, a senior administration official said. Employees working in health care facilities that receive Medicare or Medicaid reimbursement will also be required to be vaccinated, Biden said, a move that will impact 7 million workers at 50,000 health care providers. As of July, 27 percent of the country's health care workers were unvaccinated, according to a study by the Covid States Project. [emphasis added]

105. On 4 November 2021, the White House issued "Fact Sheet: Biden Administration Details of Two Major Vaccination Policies." The statement detailed new OSHA and CMS rules for vaccination, which will affect two-thirds[176] of all workers in the United States of America The following is taken from the statement:

Today's announcements include:

New Vaccination Requirement for Employers With 100 or More Employees: OSHA is issuing a COVID-19 Vaccination and Testing Emergency Temporary Standard (ETS) to require employers with 100 or more employees (i.e., "covered employers") to:

Get Their Employees Vaccinated by January 4th and Require Unvaccinated Employees to Produce a Negative Test on at Least a Weekly Basis: All covered employers must ensure that their employees have received the necessary shots to be fully vaccinated—either two doses of Pfizer or Moderna, or one dose of Johnson & Johnson—by January 4th. . . .

Pay Employees for the Time it Takes to Get Vaccinated: All covered employers are required to provide paid-time for their employees to get vaccinated and, if needed, sick leave to recover from side effects experienced that keep them from working.

Ensure All Unvaccinated Employees are Masked: All covered employers must ensure that unvaccinated employees wear a face mask while in the workplace.

Other Requirements and Compliance Date: Employers are subject to requirements for reporting and recordkeeping. . . . While the testing requirement for unvaccinated workers will begin after January 4th, employers must be in compliance with all other requirements—such as providing paid-time for employees to get vaccinated and masking for unvaccinated workers—on December 5th. The Administration is calling on all employers to step up and make these changes as quickly as possible.

New Vaccination Requirements for Health Care Workers: CMS is requiring workers at health care facilities participating in Medicare or Medicaid to have received the necessary shots to be fully vaccinated—either two doses of Pfizer or Moderna, or one dose of Johnson & Johnson—by January 4th. The rule covers approximately 76,000 health care facilities and more than 17 million health care workers—the majority of health care workers in America—and will enhance patient safety in health care settings. . . .

Streamlining Implementation and Setting One Deadline Across Different Vaccination Requirements: The rules released today ensure employers know which requirements apply to which workplaces. . . . To make it easy for all employers to comply with the requirements, the deadline for the federal contractor vaccination requirement will be aligned with those for the CMS rule and the ETS. Employees falling under the ETS, CMS, or federal contractor rules will need to have their final vaccination dose—either their second dose of Pfizer or Moderna, or single dose of Johnson & Johnson—by January 4, 2022.

106. Also on 4 November 2021 the *New York Post*'s article "Rand Paul Calls on Fauci to Resign[177] over Gain-of-Function Research," written by Natalie O'Neill, came out. The article includes the following:

Sen. Rand Paul blasted Dr. Anthony Fauci at a Senate hearing Thursday over gain-of-function research in Wuhan, China—calling for his resignation and alleging he's "learned nothing from this pandemic."

Paul (R-Ky.) grilled the director of the National Institute of Allergy and Infectious Diseases, part of the National Institutes of Health, on why he repeatedly denied that the virus research was funded by the NIH prior to the COVID-19 pandemic.

"Gain-of-function could cause a pandemic even worse next time," Paul said. "[It] could endanger civilization as we know it."

The firebrand conservative claimed NIH-funded scientists "created viruses not found in nature" and that Fauci "misled" the American public by refusing to admit it.

"Your repeated denials have worn thin and the majority of Americans, frankly, don't believe you," Paul said. "Your persistent denials are not just a stain on your reputation but are a clear and present danger to the country and to the world."

"You appear to have learned nothing from this pandemic," Paul said. "I think it's time you resign." Regardless of whether the viruses created at the Wuhan Institute of Virology were linked to COVID-19, similar research funded by the NIH could spark another pandemic, Paul said. He also alleged that Fauci changed the definition of the term "gain-of-function" in order to deny it had happened.

"You've changed the definition on your website to cover your ass," Paul said.

107. On 6 November 2021 CNN Politics published "Federal Appeals Court Issues Stay[178] of Biden Administration's Vaccine Mandate for Private Companies." The following is taken from the article:

> A federal appeals court temporarily blocked the Biden administration's new vaccine rules that could apply to larger employers, certain health care workers and federal contractors. In the brief order, a three-judge panel on the Fifth Circuit Court of Appeals said that the petitioners in the case—Republican-led states and private businesses—"give cause to believe there are grave statutory and constitutional issues with the Mandate."

108. On 8 December 2021 the CNBC piece, "Senate Votes to Block[179] Biden Vaccine Mandate, Which Has Already Hit Roadblocks in Court," by Jacob Pramuk and Spencer Kimball, came out. The segment reported as follows:

> The Senate voted Wednesday to block President Joe Biden's vaccine mandate on private employers in the latest blow to his push to flex federal power to boost vaccinations in the U.S. The measure to block the mandate heads to the Democratic-held House. It faces a tougher path to passage in the House, and the Biden administration has threatened a veto if it reaches the president's desk. Because the mandate itself has a slim chance of becoming law, the measure to overturn it will have little practical effect.

109. RealClearPolitics (RCP), reported on a Senate Health Committee hearing held on 11 January 2022 with an article written by Tim Hains: "Rand Paul Grills Fauci: Does Government Pay You to Discredit Other Scientists?"[180]

Dr. Anthony Fauci "refused to answer for published emails where he and Dr. Francis Collins discuss ways to discredit the 'fringe' Great Barrington Declaration, which called for an end to federal lockdown policies in fall 2020." The article included a partial transcript of the exchange between Senator Dr. Rand Paul and Dr. Fauci:

Senator Rand Paul: Dr. Fauci, the idea that a government official like yourself would claim unilaterally to represent science and that any criticism of you would be considered a criticism of science itself is quite dangerous. Central planning, whether of the economy or of science, is risky because of the fall built of the planner. It wouldn't be so catastrophic if it were one physician in Peoria. The mistake would only affect those patients who chose that physician.

But when the planner is a government official like yourself who rules by mandate, the errors are compounded and become much more harmful. A planner who believes he is The Science leads to an arrogance that justifies in his mind using government resources to smear and to destroy the reputations of other scientists who disagree with him. In an email exchange with Dr. Collins, and I quote directly from the email, to create a quick and devastating published takedown of three prominent epidemiologists from Harvard, Oxford and Stanford. There are a lot of fringe epidemiologists there at Harvard, Oxford, and Stanford. You quote in the email that they are fringe. Immediately there is this take-down effort. A published takedown doesn't conjure up the image of a dispassionate scientist. Instead of engaging them on the merits, you and Dr. Collins sought to smear them as fringe and take them down. Not in journals, in lay press. This is not only antithetical to the scientific method, it is the epitome of cheap politics and it is reprehensible Dr. Fauci. Do you think it's appropriate to use your $420,000 salary to attack scientists that disagree with you?

Dr. Anthony Fauci: The email you are referring to was an email of Dr. Collins to me. If you look at the email.

Senator Rand Paul: That you responded to and hurried and said I can do it. We got something in *Wired* magazine.

Dr. Anthony Fauci: No, no, no, I think in usual fashion, senator, you are distorting everything about me.

Senator Rand Paul: Did you ever object to Dr. Collins's characterization of them as fringe? Did you write back and say no they are not fringe, they're esteemed scientists and it would be beneath me to do that. You responded that you would do it. You immediately got an article in *Wired* and sent it back and said look, I've got them and I nailed them in *Wired*, of all scientific publications.

Dr. Anthony Fauci: That's not what went on. There you go again. You do the same thing every hearing.

Senator Rand Paul: That was your response. It wasn't the only time. Your desire to take down people.

Dr. Anthony Fauci: You are incorrect as usual, senator. You are incorrect almost everything you say. No. No. No.

110. On 13 January 2022, The Intercept published "House Republicans Release Text of Redacted Fauci Emails[181] on COVID Origins" by Maia Hibbett and Ryan Grim. The following statements are included in the article:

On Tuesday, Republicans on the House Committee on Oversight and Reform released a letter that paints a damning picture of U.S. government officials wrestling with whether the novel coronavirus may have leaked out of a lab they were funding, acknowledging that it may have, and then keeping the discussion from spilling out into public view.

The letter, signed by James Comer, R-Ky., and Jim Jordan, R-Ohio, was followed by pages of notes on emails that were first obtained through the Freedom of Information Act by BuzzFeed News and the *Washington Post*, but were heavily redacted when published in June 2021. The redacted emails included the agenda for a February 1, 2020, telephone conference between National Institute of Allergy and Infectious Diseases director Anthony Fauci; his then-boss, former National Institutes of Health director Francis Collins; and several of the world's leading virologists. The communications contained extensive notes summarizing what was said during the call, but their substance was hidden at the time. Oversight Committee staff were able to view the full emails "in camera," which means they could physically look at them and take notes but couldn't take copies with them. The information released Tuesday for the first time reveals the content of notes taken on the February 1 call.

On that call, virologists Michael Farzan and Robert Garry told Fauci and Collins the virus might have leaked from the Wuhan lab. It might have been genetically engineered, the transcription of Garry's notes suggests, but this now seems unlikely. Another possibility, put forward by Farzan, was that it could have been evolved in the lab through a process known as serial passage. "The email is out-of-context," Garry wrote Wednesday in an email to The Intercept. "This was one email among many I was sharing with my colleagues."

The day before the call, Scripps Research infectious disease expert Kristian Andersen had warned Fauci that the virus may have been engineered in a lab, noting that he and several other high-profile scientists "all find the genome inconsistent with expectations from evolutionary theory." The scientists agreed to have a conference call the next day. "It was a very productive back-and-forth conversation where some on the call felt it could possibly be an engineered virus," Fauci told Alison Young, writing for *USA Today*, in June 2021.

Not long after the call, Andersen was the lead author on a paper in *Nature Medicine* titled "The Proximal Origin of SARS-CoV-2." The paper proposed "two scenarios that can plausibly explain the origin of SARS-CoV-2: (i) natural selection in an animal host before zoonotic transfer; and (ii) natural selection in humans following zoonotic transfer." For the scientists and pundits who sought to discount the emerging lab-leak hypothesis, it offered the authoritative proof they needed. The paper has since been accessed more than 5.6 million times, with over 2,000 citations. . . .

On February 2, Jeremy Farrar, an infectious disease expert and the director of Wellcome, sent around notes, including to Fauci and Collins, summarizing what some of the scientists had said on the call. Farzan, a Scripps professor who studied the spike protein on the 2003 SARS virus, "is bothered by the furin site and has a hard time explain that as an event outside the lab (though, there are possible ways in nature, but highly unlikely)," Farrar's note reads, referring to a spike protein feature that aids interaction with furin, a common enzyme in human lung cells. Farzan didn't think the site was the product of "directed engineering," but found that the changes would be "highly compatible with the idea of continued passage of the virus in tissue culture."

According to the transcribed notes, Garry, a professor at the Tulane University School of Medicine, said on the call that he had aligned the SARS-CoV-2 genome with that of RaTG13, a 96-percent similar virus isolated from bats at the Wuhan Institute of Virology that was long regarded as the new virus's closest known relative—though a closer one has since been identified. Garry found that the spike proteins of RaTG13 and SARS-CoV-2, which makes the latter so infectious, were nearly identical. The key distinction was in the ability of the new virus's spike protein to interact with furin, which Garry found too perfect to make natural sense. "I just can't figure out how this gets accomplished in nature," he said. . . .

As they discussed what to present to the public, the scientists determined that questions of potential lab origin might prove more trouble than they're worth. "Given the evidence presented and the discussions around it, I would conclude that a follow-up discussion on the possible origin of 2019-nCoV would be of much interest," wrote Ron Fouchier, a virologist at the Erasmus MC Center for Viroscience in the Netherlands, on February 2. Years earlier, Fouchier's gain-of-function research had brought the discipline under fire for a 2011 experiment in which he infected ferrets in adjacent cages with the avian influenza virus, allowing it to become airborne and infect mammals. "However, further debate about such accusations would unnecessarily distract top researchers from their active duties and do unnecessary harm to science in general and science in China in particular," Fouchier wrote.

Farzan, Fauci, and Fouchier did not immediately respond to The Intercept's requests for comment. Several of the scientists on the email chain ended up co-authoring the *Nature Medicine* paper with Andersen and Garry. In a February 4 email, which House Republicans presented as a response to a first copy of the draft, Fauci wrote: "?? Serial passage in ACE2-transgenic mice . . ."

"Neither Drs. Fauci or Collins edited our Proximal Origins paper in any way. The major feedback we got from the Feb 1 teleconference was: 1. Don't try to write a paper at all—it's unnecessary or 2. If you do write it don't mention a lab origin as that will just add fuel to the conspiracists," Garry wrote on Wednesday.

When the paper appeared in *Nature Medicine* on March 17, 2020, it noted near the end that in order for the novel coronavirus to have emerged in a lab via serial passage, scientists would have to conduct those experiments using a relative with very high genetic similarity, but there was no evidence that such experiments had been done. The authors added, "Subsequent generation of a polybasic cleavage site," which lets the virus process furin, "would have then required repeated passage in cell culture or animals with ACE2 receptors similar to those of humans, but such work has also not previously been described."

Though the paper was publicly embraced by the scientific community and the mainstream media, Collins worried that its impact wasn't sufficient. "Wondering if there is something NIH can do to help put down this very destructive conspiracy," Collins wrote on April 16, 2020, in reference to a Fox News segment on the lab-leak theory. "I hoped the *Nature Medicine* article on the genomic sequence of SARS-CoV-2 would settle this. But probably didn't get much visibility. Anything more we can do?"

"I would not do anything about this right now," Fauci replied. "It is a shiny object that will go away in times."

111. On 14 January 2022, ABC News reported, "Supreme Court Blocks[182] Biden Vaccine-or-Test Mandate for Large Businesses," with the subheading "But the Justices Did Allow a Mandate for Certain Health Care Workers." The article included the following:

The Supreme Court on Thursday issued a stay of the OSHA vaccine-or-test requirement on private businesses of 100 or more workers, dealing a setback to the Biden administration's effort to control the COVID pandemic. By a 6–3 vote, with the three liberal justices—Stephen Breyer, Sonia Sotomayor, and Elena Kagan—dissenting, the court reasoned that the agency exceeded its authority to regulate workplace safety. "Although COVID-19 is a risk that occurs in many workplaces, it is not an occupational hazard in most," the majority wrote. At the same time, the justices voted 5–4—with Chief Justice John Roberts and Justice Brett Kavanaugh joining the three liberals—to allow the Biden administration to require vaccination of health care workers at facilities that treat Medicare and Medicaid patients, **subject to religious and medical exemptions**. Justices Samuel Alito and Clarence Thomas dissented, joined by Justices Neil Gorsuch and Amy Coney Barrett. [emphasis added]

112. On 7 February 2022, The Defender published an article by Michael Nevradakis, PhD, titled “Pfizer, FDA Lose Bid to Further Delay Release of COVID Vaccine Safety Data”[183] with the subheading “A Federal Judge Last Week Rejected a Bid by the U.S. Food and Drug Administration and Pfizer to Delay the Court-Ordered Release of Nearly 400,000 Pages of Documents Pertaining to the Approval of Pfizer’s COVID Vaccine.” Here is an excerpt:

Federal judge Mark Pittman of the U.S. District Court for the Northern District of Texas, in an order issued Feb. 2, said the FDA must release redacted versions of the documents in question according to the following disclosure schedule:

- 10,000 pages apiece, due on or before March 1 and April 1, 2022.
- 80,000 pages apiece, to be produced on or before May 2, June 1 and July 1, 2022.
- 70,000 pages to be produced on or before Aug. 1, 2022.
- 55,000 pages per month, on or before the first business day of each month thereafter, until the release of the documents has been completed.

The order grants the FDA the ability to “bank” excess pages as part of this release schedule—meaning that if the agency exceeds its monthly quota in any given month it can apply those extra pages to a subsequent month.

Last week’s ruling is the most recent development in an ongoing court case that began with a Freedom of Information Act (FOIA) request filed in August 2021 by Public Health and Medical Professionals for Transparency (PHMPT), a group of doctors and public health professionals.

PHMPT, a group of more than 30 medical and public health professionals and scientists from institutions such as Harvard, Yale, and UCLA, in September 2021 filed a lawsuit against the FDA after the agency denied its original FOIA request.

In that request, PHMPT asked the FDA to release “all data and information for the Pfizer vaccine,” including safety and effectiveness data, adverse reaction reports, and a list of active and inactive ingredients.

The FDA argued it didn’t have enough staff to process the redaction and release of hundreds of thousands of pages of documents, claiming it could process only 500 pages per month.

This would have meant the cache of documents would not be fully released for approximately 75 years.

In his Jan. 6 order, Pittmann rejected the FDA’s claim and instead required the agency to release 12,000 pages of documents by Jan. 31 and an additional 55,000 pages per month thereafter.

Pfizer responded to the Jan. 6 order by filing a memorandum with the court on Jan. 21, requesting to intervene in the case for the “limited purpose of ensuring that information exempt from disclosure under FOIA is adequately protected as FDA complies with this Court’s order.”

Pfizer claimed to support the disclosure of the documents, but asked to intervene in the case to ensure that information legally exempt from disclosure will not be “disclosed inappropriately.”

As reported by The Defender, this request, if granted, would have also meant further delay for the release of the next tranche of documents, until May 1. Lawyers for PHMPT, in a brief submitted Jan. 25, asked Pittman to reject Pfizer's motion, prompting Pittman's Feb. 2 order.

> The first batch of documents produced in Nov. 2021, which totalled a mere five hundred pages, revealed there were more than 1,200 vaccine-related deaths within the first ninety days following the release of the Pfizer-BioNTech COVID vaccine.

113. On 16 February 2022, the CDC posted a video, "How Nose Swabs Detect New Covid-19 Strains," on its Twitter account, along with the following text:[184]

> Remember that #COVID19 nose swab test you took? What happened to the swab? If it was processed with a PCR test, there's a 10% chance that it ended up in a lab for genomic sequencing analysis. Learn more about the process and its importance: https://bit.ly/3sJOkoC @WIRED @CDC_AMD.

The CDC announcement drew strong reactions from the public regarding DNA privacy, with some alleging that "people's DNA is being taken without consent." An article by Nicolle K. Strand, JD, MBioethics, in the *American Medical Association Journal of Ethics* in March 2016, "Shedding Privacy Along with our Genetic Material: What Constitutes Adequate Legal Protection against Surreptitious Genetic Testing?"[185] took up the issue:

> Surreptitious genetic testing happens when a sample containing a person's genetic information is accessed without the knowledge or consent of that person and when that sample is tested without the knowledge or consent of that person. . . . No matter the intended or actual use, surreptitious genetic testing is ethically and legally problematic. In each of the examples described above, the potential for harm—whether in the form of unjust discrimination or another consequence—is generated by the genetic material having been stolen. So, surreptitious genetic testing is ethically and legally problematic not only because of potential harmful consequences of testing, but also because both sample acquisition and the acquisition of information generated by testing the sample threaten privacy. . . . States have taken a variety of approaches to protecting against surreptitious genetic testing. . . . More stringent laws would also be ethically acceptable; what follows is the minimum level of protection that would adequately protect privacy.
>
> Perpetrators. First, to achieve an adequate standard of protection, the law should protect against surreptitious genetic testing regardless of where, how, or by whom the sample was obtained. For example, instead of only prohibiting health care workers from conducting unauthorized analyses on samples obtained with informed consent, as some states do, the law should protect against unauthorized genetic analysis or testing regardless of how or by whom the sample was obtained. . . . In addition, adequate protections would emphasize that informed consent should be obtained not only for an initial sample collection, but also for any subsequent uses. A person might consent to donate a sample for de-identified research but might object to certain analyses or tests of that sample or disclosures of information learned from that sample. Prohibiting the collection, analysis, and retention of samples containing genetic material and the disclosure of information about that sample by any person or entity without the knowledge and informed consent of

> the person whose sample is accessed, tested, and learned about seems to adequately cover many potential scenarios of surreptitious genetic testing, and it underscores the importance of detailed informed consent procedures. For example, biological samples are often collected from patients in clinical settings, creating the potential for genetic analysis and a variety of subsequent uses of the data and information obtained from those samples.

During the pandemic, Americans and people around the world allowed the mass mucosal collection of genetic material or cells containing their human DNA, to comply with testing requirements for COVID-19. The technologies recommended by medical authorities to test for COVID-19 included reverse transcription polymerase chain reaction (RT-PCR), loop-mediated isothermal amplification (LAMP), lateral flow, and enzyme-linked immunosorbent assay (ELISA).[186] All four technologies require the collection of genetic material, along with the person's personal details for identification and communications purposes. This mass collection of genetic data would have been impossible without the COVID-19 pandemic and could be used without consent to advance future pharmaceutical interests, such as genetic vaccine programs with the capability to modify or alter human DNA.

114. Free Thinker Fitness posted a video clip on YouTube on 25 February 2022, "Francis Collins Talking About Klaus Schwab's 4th Industrial Revolution and Collecting Genomic Data," which included the following statements by Dr. Francis Collins:[187]

> It's perhaps . . . traditional these days to talk about the **Fourth Industrial Revolution. Those of us who go to the World Economic Forum and Davos are used to this kind of conversation,** but maybe it is a useful organizing principle to consider these four industrial revolutions that people are identifying. **The kind of data that will be collected on these individuals is quite broad**. Some people have thought of this, "Oh, it's just a genomics project." **Well, genomics is in there, but it's also going to be accompanied by a wide variety of other data types about what we [NIH] should be doing in this space . . . to help assist in what seems to be that Fourth Industrial Revolution coming to bear on biomedical research** in a way that we have only really, I think, begun to appreciate in early fashion in many of the things we are doing, and **I think we need to rather quickly escalate our involvement and our investment**. [emphasis added]

115. On 26 February 2022, links to information and details regarding Ukrainian bioweapons labs were deleted[188] from the US embassy website. However, a media clip was archivally preserved on the US Armed Forces Counterproliferation Center's CPC Outreach Journal, number 818. The article by Tina Redlup was published via BioPrepWatch.com on 17 June 2010 and includes the following statements:

> U.S. Sen. Dick Lugar applauded the opening of the Interim Central Reference Laboratory in Odessa, Ukraine, this week, announcing that it will be instrumental in researching dangerous pathogens used by bioterrorists. The level-3 bio-safety lab, which is the first built under the expanded authority of the Nunn-Lugar Cooperative

Threat Reduction program, will be used to study anthrax, tularemia and Q fever as well as other dangerous pathogens. "The continuing cooperation of Nunn-Lugar partners has improved safety for all people against weapons of mass destruction and potential terrorist use, in addition to advancements in the prevention of pandemics and public health consequences," Lugar said.

Lugar said plans for the facility began in 2005 when he and then Senator Barack Obama entered a partnership with Ukrainian officials. Lugar and Obama also helped coordinate efforts between the U.S. and Ukrainian researchers that year in an effort to study and help prevent avian flu.

116. By 4 March 2022, numerous media reports of US DoD involvement in biolabs located in the Ukraine raised awareness of Russian concerns with US involvement in biological laboratories. Over four months, these media reports confirmed involvement with Western pharmaceutical companies, a minimum of forty-six biolabs, and more than $224 million in US funding, beginning during the Obama Administration's call for a "pause" in gain-of-function funding and research development.

117. During a Senate Foreign Relations Committee hearing on 8 March 2022, Senator Marco Rubio questioned Undersecretary of State Victoria Nuland about whether Ukraine possesses chemical or biological weapons:[189]

Senator Marco Rubio: Does Ukraine have chemical or biological weapons?

Victoria Nuland: Ukraine has biological research facilities which, in fact, we're now quite concerned Russian troops, Russian forces may be seeking to gain control of, so we are working with the Ukrainians on how they can prevent any of those research materials from falling into the hands of Russian forces should they approach.

Senator Marco Rubio: I'm sure you are aware that the Russian propaganda groups are already putting out there all kinds of information about how they have uncovered a plot by the Ukrainians to release biological weapons in the country, and with NATO's coordination. If there is a biological or chemical weapon incident or attack inside Ukraine, is there any doubt in your mind that 100 percent it would be the Russians that would be behind it?

Victoria Nuland: There is no doubt in my mind, senator. And in fact, it is a classic Russian technique to blame the other guy for what they are planning to do themselves.

118. On 11 March 2022, journalist Sharyl Attkisson posted the article "List of Ukraine Biolab Documents[190] Reportedly Removed by US Embassy," which is largely reproduced here:

The following is from information being circulated by journalists and advocates online and is published as received for general information purposes.

List of Ukraine Biolabs Documents Removed by US Embassy

Up until recently, the existence and details of these bioweapons labs were public knowledge. The US embassy had previously disclosed the locations and details of these laboratories in a series of PDF files online. On February 26, 2022, the official embassy website shut down the links to all 15 bioweapon laboratories.[191] All the documents associated with these labs have been removed from the internet. If you click on any of the links, the PDF files are no longer available. Thankfully, these files have been archived and can still be accessed.

119. Ryan Morgan wrote the article "Pentagon Funding Biolabs in Ukraine[192]—'Real Concern' of Pathogen Releases if Russia Attacks," published by American Military News on 14 March 2022. The article included the following:

The Department of Defense released a fact sheet on Friday detailing a program it has funded from 2005 "through the present day" in Ukraine which gives money to biological laboratories.

The DoD fact-sheet detailed the program known as the Biological Threat Reduction Program (BTRP), which is a subordinate program of the Cooperative Threat Reduction (CTR) Program aimed at reducing the threat of pathogens.[193] The fact sheet further said Ukraine has taken steps to secure biological samples, to prevent the "real concern" that dangerous pathogens could be released amid Russia's ongoing invasion of the country.

The fact sheet stated BTRP has "invested approximately $200 million in Ukraine since 2005, supporting 46 Ukrainian laboratories, health facilities, and diagnostic sites." The fact-sheet also said CTR "began its biological work with Ukraine to reduce the risk posed by the former Soviet Union's illegal[194] biological weapons program, which left Soviet successor states with unsecured biological materials after the fall of the USSR."

The fact-sheet states since 2005, the DoD program has worked with the Government of Ukraine to reduce the threats posed by pathogens, including by disposing of biological weapons materials left behind by the Soviet Union, the predecessor of the modern Russian Federation.

"After Russia launched its unlawful invasion of Ukraine, the Ukrainian Ministry of Health responsibly ordered the safe and secure disposal of samples," the fact-sheet states. "These actions **limit the danger of an accidental release of pathogens** should Russia's military attack laboratories, a real concern since they have attacked Ukraine's nuclear power plants and research facilities." . . .

The DoD further warned, "Russia propagates disinformation aimed at BTRP's laboratory and capacity building efforts in former Soviet Union countries—falsely claiming that the U.S. Department of Defense support is used to develop biological weapons." [emphasis added]

120. True North, a Canadian digital media platform, posted "Trudeau Government Gave $3 million to WEF[195] and $1.6 billion to UN in 2021" by Cosmin Dzsurdzsa on 10 May 2022. Here is part of the article:

The Liberal government funnelled more than a billion-and-a-half taxpayer dollars into various United Nations bodies, and millions into the World Economic Forum (WEF) last year, public accounts data shows. According to the transfer payments section of the 2020–2021 *Public Accounts of Canada*, the WEF received $2,915,095 from Canadian taxpayers in the form of grants and contributions. Funding was provided by two departments—the Department of Environment and the Department of Foreign Affairs, Trade and Development. The largest of the transfer payments to WEF was a $1,141,851 contribution from the International Development Assistance for Multilateral Programming. WEF also received another $1 million grant under the same program. Other payments were cited as "contributions in support of conserving nature" and for the "establishment and management of conservation measures."

The Trudeau government also generously funded the UN to the tune of $1.576 billion in the form of financial support, contributions and grants. Funding came primarily from Global Affairs, although other departments including Immigration, Refugees and Citizenship also gave the UN money. Six UN-affiliated organizations received transfer payments worth more than $100 million each. The largest payment was given to the United Nations Children's Fund, totalling $543 million. Meanwhile UN peacekeeping operations saw contributions worth $235 million, while the UN High Commissioner for Refugees received $139 million. Other large recipients include the UN Population Fund and the United Nations Organizations. . . .

According to Federal Director of the Canadian Taxpayers Federation Franco Terrazzano, the Trudeau government needs to do a better job accounting for its funding of international organizations like the WEF, while Canada deals with a debt of over $1 trillion. "That's a lot of money, and we can't just keep sending a ton of tax dollars to international organizations because we've been doing it for years," Terrazzano told True North. "The feds are more than $1 trillion in debt, so it's on the government to make a clear case for every cent it sends to international organizations, and if it can't make the case then we need to see reductions."

The controversial WEF has received renewed attention in the Conservative leadership race after candidate and MP Pierre Poilievre committed to boycotting the organization.

121. On 31 May 2022, Government Executive published "Feds' Vaccine Mandate Enforcement Could Be Days Away, but Agencies Are Not Yet Prepping" by Senior Correspondent Eric Katz. The subheading states: "The Clock Is Ticking on a Federal Court[196] to Either Hear Another Appeal on Biden's Mandate or Allow the Administration to Resume Suspensions and Firings." Here is an excerpt:

    The Biden administration could soon be able to start suspending and firing the remaining federal employees who have yet to receive a COVID-19 vaccine, but agencies are not yet taking steps to prepare for that outcome. A federal appeals court in April struck down a nationwide injunction that had paused the enforcement of President Biden's mandate, but enforcement was stalled due to a standard buffer period after the judges' ruling. The mandate was set to go into effect Tuesday, but a petition from those challenging the mandate for a rehearing from the entire U.S. Court of Appeals for the Fifth Circuit has further delayed the Biden administration from taking action on employees out of compliance with the requirement. The Justice Department had asked the court to immediately allow it to resume enforcement, but the court has opted to let the secondary appeal play out.

122. Also on 31 May 2022, Kelly Gooch's article "Lawsuits Still Piling Up over Hospital Vaccine Mandates"[197] was published by *Becker's Hospital Review* and is excerpted here:

    As more hospitals and health systems mandated COVID-19 vaccination for their employees, lawsuits arose related to the policies. Houston Methodist was the first large, integrated health system in the U.S. to implement a mandate, in spring 2021.

In June of that year, a federal judge dismissed a lawsuit brought by more than 100 Houston Methodist employees, marking the first decision by a court regarding such a requirement at a health system.

The lawsuit, filed May 28, 2021, argued the mandate is illegal and forces workers to get an experimental vaccine to keep their jobs. But U.S. District Judge Lynn Hughes ruled June 12 that Houston Methodist did not violate state or federal law or public policy with its requirement. Nearly a year after that lawsuit was dismissed, others have been filed against health systems.

Most recently, workers in Indiana filed a lawsuit against Indianapolis-based Ascension St. Vincent and its parent company, St. Louis-based Ascension, alleging religious discrimination, the Indianapolis Star reported May 31. The lawsuit, filed May 27 in the U.S. District Court for the Southern District of Indiana, comes on behalf of workers who were suspended without pay for refusing the vaccine on religious grounds.

Ascension "established a coercive process calculated to force healthcare workers and staff to abandon their religious objections to the COVID-19 vaccination and receive the vaccination against their will," the lawsuit claims. Ascension, a health system with more than 140 hospitals, announced its mandate in late July, saying at the time that tens of thousands of Ascension workers had already been vaccinated. Employees could request an exemption for medical or religious reasons.

But the lawsuit filed May 27 said St. Vincent and Ascension "failed to individually and properly assess each application for religious exemption." A St. Vincent spokesperson declined to comment to the *Indianapolis Star* about pending litigation. The newspaper and Becker's also requested comment from Ascension. The workers are asking the court to open a class-action lawsuit under Title VII of the Civil Rights Act of 1964. The lawsuit seeks damages, including lost back wages due to unpaid suspension.

Meanwhile, Rochester, Minn.–based Mayo Clinic could face a slew of lawsuits from employees alleging they were wrongly fired for refusing COVID-19 vaccines, the *Post Bulletin* reported May 18. Gregory Erickson, a Minneapolis attorney representing two former Mayo employees who recently filed such lawsuits, told the *Post Bulletin* at that time that the recently filed cases were among more than 100 similar suits he is filing against Mayo. Mr. Erickson represents fired Mayo employees in Wisconsin, Florida and Arizona, but about 80 to 100 of the cases against Mayo will be for former Mayo employees who live in Rochester, he added. . . . In January, Mayo estimated it would fire about 1 percent of its 73,000-person workforce because of noncompliance with the health system's required COVID-19 vaccination program.

123. RealClearPolitics published part of the transcript taken during the Senate Health, Education, Labor and Pensions (HELP) Committee hearing[198] in an article on 16 June 2022 written by Ian Schwartz, titled "Rand Paul Grills Fauci: Have You Received Royalties from a Company That You Later Oversaw the Funding Of?"

Senator Dr. Rand Paul: Over the period of time from 2010 to 2016, 27,000 royalty payments were paid to 18,000 NIH employees. We know that not because you told us, but because we forced you to tell us through the Freedom of Information Act. Over $193 million was given to these 18,000 employees.

Can you tell me that you have not received a royalty from any entity that you ever oversaw the distribution of money in research grants?

Dr. Anthony Fauci: You know, well first of all, let's talk about royalties—

Senator Rand Paul: That's the question. No, that's the question. Have you ever received a royalty payment from a company that you later oversaw money going to that company?

Dr. Anthony Fauci: You know, I don't know it as a fact, but I doubt it.

Senator Rand Paul: Why don't you let us know? Why don't you reveal how much you've gotten and from what entities?

124. On 22 August 2022 the National Institutes of Health announced Dr. Fauci's formal resignation, effective December 2022, as director of the National Institute of Allergy and Infectious Diseases (NIAID).[199]

125. On 8 September 2022, United States District Judge Robert Pitman ordered[200] the CDC to release data collected via the v-safe program due to a FOIA lawsuit filed by Siri & Glimstad on behalf of Informed Consent Action Network (ICAN). The initial legal request for access to the data collected by the CDC was made in June 2021.

126. On 14 September 2022, RealClearPolitics published[201] an article with part transcript, "Sen. Rand Paul vs Fauci: When We're in Charge, You Will Have to Divulge Where You Get Your Royalties From."

Senator Rand Paul told Dr. Anthony Fauci that "Republicans will change the rules on royalties, investigate where he receives his royalties from and if there are conflicts of interest." The partial transcript from the Congressional hearing noted in the article includes the following:

Senator Rand Paul: Do any of the guidelines for vaccines from the governments include previous infection as something to base your decision-making on with vaccines? Do any of the guidelines involve previous infection? That's why you're ignoring previous infection because it does not involve any of the guidelines. Furthermore, we've been asking you, and you refuse to answer, whether anybody on the vaccine committees gets royalties from the pharmaceutical companies. I asked you last time, what was your response? We don't have to tell you.

Dr. Anthony Fauci: Right.

Senator Rand Paul: We have demanded them through Freedom of Information Act, and what have you said? We're not going to tell you. But I'll tell you this. When we get in charge, we're going to change the rules, and you will have to divulge where you get your royalties from, from what companies, and if anyone in the committee has a conflict of interest, we are going to learn about it, I promise you that.

Dr. Anthony Fauci: Mister Chair, can I respond to that, please? There are two aspects for what you said. You keep saying you approve, you do this, you do that. The committees that give the approval are FDA through their advisory committee. The committees that recommend are CDC through their advisory committee. You keep saying I am the one that's approving a vaccine based on certain data. So I don't really understand with all due respect, Senator—

Senator Rand Paul: You would not reveal which companies gave you royalties or what company gave the other scientists royalties. That's what you told the committee.

Dr. Anthony Fauci: Can I please answer that? . . . You keep asking committees, they're not my committees. There is the VRBPAC committee for the FDA. And the ACIP for the CDC. So I don't have any idea what goes on.

127. Regarding the 14 September 2022 hearing, Breitbart published an article on 16 September 2022, "Rand Paul Confronts Anthony Fauci: 'You're Not Paying Attention to the Science,'" written by Hannah Knudsen.[202]

The article covers Senator Dr. Rand Paul's questions to Dr. Anthony Fauci on his "ever evolving opinions" on the reality of natural immunity when applied to the coronavirus, "ultimately concluding that the White House chief medical advisor is 'not paying attention to the science.'" The article included the following:

> The Kentucky Republican played a past clip of Fauci openly stating that someone who was infected with the flu would not need a vaccine, as "the most potent vaccination is getting infected yourself."
>
> "This is an ongoing question, and you know, we've had ever evolving opinions from you Dr. Fauci," Paul said, noting that roughly 80 percent of children have contracted the coronavirus, but there are "no guidelines coming from you or anybody in the government to take into account their naturally acquired immunity."
>
> "You seemed quite certain of yourself in 2004 but in 2022 there's a lot less certainty," Paul said, questioning why Fauci was so willing to accept the reality of natural immunity years ago but has a seemingly difficult time with that reality now.
>
> Fauci denied that he has dismissed natural immunity and cited the approvals from U.S. federal health agencies, stating a "vaccination following infection gives an added extra boost." He also asserted that the clip Paul played was out of context, although Paul dismissed Fauci's excuse and made it clear that studies "don't report anything on hospitalization or death or transmission."
>
> "They only report that if you give them the jab, they'll make antibodies and you can give kids hundreds of jabs and they'll make antibodies every time but that does not prove efficacy," Paul said, adding that Fauci is "denying the very fundamental premise of immunology that previous infection does provide some sort of immunity."
>
> "It's not in any of your studies. Almost none of your studies from the CDC or from the government have the variable of whether or not you've been previously infected," the Republican said.
>
> At the end of the day, Paul continued, it is Fauci and those of his ilk who are feeding vaccine hesitancy with their dishonesty.
>
> "You decry—people decry—vaccine hesitancy. It's coming from the gobbledygook that you give us. You're not paying attention to the science," Paul said.
>
> "The very basic science is that previous infection provides a level of immunity. If you ignore that in your studies, if you don't present that in your committees, you're not being truthful or honest with us," he added.

128. On 3 October 2022, Siri & Glimstad posted a news release[203] on PR Newswire that included the following statements regarding the released v-safe data:

> Out of the approximate 10 million v-safe users, 782,913 individuals, or over 7.7% of v-safe users, had a health event requiring medical attention, emergency room intervention, and/or hospitalization. Another 25% of v-safe users had an event that required them to miss school or work and/or prevented normal activities.
>
> There were also 71 million symptoms reported in the pre-populated fields. This is an average of more than 7 symptoms reported per v-safe user. Reported symptoms

include, for example, over 4 million reports of joint pain. While around 2 million of these joint pain reports were mild, over 1.8 million were for moderate joint pain and over 400,000 were for severe joint pain. It is noted that v-safe includes data from less than 4 percent of individuals who received a Covid-19 vaccine in the United States.

There were also around 13,000 infants under 2 years of age registered in v-safe. Among these infants, over 33,000 symptoms were reported, with the most common symptoms being irritability, sleeplessness, pain, and loss of appetite.

The data also reflects a disproportionate amount of negative health impacts, including medical events, following the Moderna vaccine versus the Pfizer vaccine and shows a disproportionate number of negative events reported by women versus men.

Since the data is voluminous, ICAN has generated a v-safe dashboard to present it in a user-friendly format for the public. This v-safe dashboard can also generate the statistics noted above and is available at www.icandecide.org/v-safe-data/.

V-safe provides users with a limited number of fields to choose from when reporting health events as well as free-text fields. The data produced thus far is limited to the pre-populated fields within v-safe. Siri & Glimstad's attorneys leading these lawsuits, Aaron Siri and Elizabeth A. Brehm, will continue to litigate to obtain the data submitted by v-safe users in the free-text fields.

An infographic on the "Vaccines for Your Children" page on the CDC website includes the following statements:[204]

FDA and CDC closely monitor vaccine safety after the public begins using the vaccine. The purpose of monitoring is to watch for adverse events (possible side effects). Monitoring a vaccine after it is licensed helps ensure that possible risks associated with the vaccine are identified. . . .

The data recorded by the V-safe system reveals "782,913 individuals, or over 7.7% of v-safe users, had a health event requiring medical attention, emergency room intervention, and/or hospitalization. Another 25% of v-safe users had an event that required them to miss school or work and/or prevented normal activities."

Why was action not taken by the CDC or FDA to launch an immediate investigation, and why did it take legal lawsuits to force the CDC to release this data to the public?

129. On 8 November 2022 "Rand Paul Promises[205] to 'Subpoena Every Last Document of Dr. Fauci' in Victory Speech" appeared on Fox News. The article includes the following:

U.S. Sen. Rand Paul sailed to re-election Tuesday night, promising to subpoena "every last document of Dr. [Anthony] Fauci" and focusing on whether COVID-19 can be traced to lab research in China. "I promise you this: the COVID cover-up will end" the Kentucky senator told supporters. Paul won a third six-year term in Congress on Tuesday, scoring a victory that the Associated Press called relatively early Tuesday evening as election results rolled in.

The libertarian-leaning senator ran on a staunchly conservative ticket, promising voters to investigate the origins of the COVID-19 virus, among other key issues. "I will not only hold Dr. Fauci accountable, we will finally investigate why your tax dollars were sent to fund dangerous research in Wuhan [in China]." Paul and Fauci, the White House chief medical adviser and director of the National Institute of Allergy and Infectious Diseases, have repeatedly sparred over this issue at several Senate hearings.

130. On 30 November 2022 *Sarasota Herald-Tribune* published "Sarasota Memorial Hospital's COVID Protocols to Be Investigated[206] After Emotional Meeting." The article included the following:

> In the first meeting of newly elected board members critical of the hospital's standard medical approach to fighting COVID, public testimony pushed them to open an investigation into Sarasota Memorial Hospital's protocols during the height of the pandemic.
>
> Nearly 50 people signed up to speak during public comment: grieving spouses, furious parents, exhausted doctors and former patients varying from commendation or condemnation toward their in-patient experience. Heartbreak, grief, and frustration were all on full display Tuesday night inside SMH's auditorium.
>
> Many also spoke out of decorum, booing one doctor who said that COVID vaccines work with another heckler shouting "murder keeps the building full." . . .
>
> Tanya Parus, president of Sarasota County Moms for America, said she collected over 100 verbal testimonies alleging mistreatment from patients at SMH.
>
> Citing a stipulation in the CARES Act that created a 20% premium, or add-on, for COVID-19 Medicare patients at hospitals, Parus accused the hospital of choosing profit over people by inflating COVID diagnoses to receive higher pay-outs.
>
> "It is blatantly obvious that there is a more sinister stream at hand and that this hospital is one of the countless hospitals to become a victim at the hands of government overreach," she said.

131. Also on 30 November 2022, "Long Covid[207] May Be 'the Next Public Health Disaster'[208]—with a $3.7 Trillion Economic Impact Rivalling the Great Recession," written by Greg Iacurci, appeared on CNBC's website. The article included the following key points and statements:

> Long Covid is a chronic illness resulting from a Covid-19 infection. It goes by many names, including long-haul Covid, post-Covid or post-acute Covid syndrome.
>
> Not much is yet known about the illness. Its symptoms number in the hundreds and can be debilitating. They can also be challenging to diagnose—for doctors even willing to do so.
>
> Long Covid has affected as many as 23 million Americans. It may cost the U.S. economy $3.7 trillion, roughly that of the Great Recession, according to one estimate.
>
> All told, long Covid is a $3.7 trillion drag on the U.S. economy—about 17% of our nation's pre-pandemic economic output, said David Cutler, an economist at Harvard University. The aggregate cost rivals that of the Great Recession, Cutler wrote in a July report.
>
> Cutler revised the $3.7 trillion total upward by $1.1 trillion from an initial report in October 2020, due to the "greater prevalence of long Covid than we had guessed at the time." Even that revised estimate is conservative: It is based on the 80.5 million confirmed U.S. Covid cases at the time of the analysis, and doesn't account for future caseloads.

Higher medical spending accounts for $528 billion of the total. But lost earnings and reduced quality of life are other sinister trickle-down effects, which respectively cost Americans $997 billion and $2.2 trillion.

132. On 19 January 2023, *The Nation* published "Unredacted NIH E-mails Show Efforts to Rule Out a Lab Origin of Covid." The subheading states, "In Early 2020, Top Scientists Told Anthony Fauci They Were **Concerned** That SARS-CoV-2 Appeared Potentially '**Engineered.**' Here's a Look at What Happened Next" [emphasis added]. The article reports on recently released unredacted email communications between Dr. Anthony Fauci and eminent biologists and virologists at the end January 2020 regarding the possible origins[209] of SARS-CoV-2.

Three of the email participants went on to coauthor "The Proximal Origin of SARS-CoV-2," published on 17 March 2020, the paper upon which the official narrative was based regarding the possible origin of SARS-CoV-2. Release of the unredacted emails reveals a sharp contrast to private communications held[210] between the coauthors, who were initially convinced that that "SARS-CoV-2 may have emerged from a lab."

The unredacted emails showed the scientists were concerned with (1) US funding of the WIV, (2) the furin cleavage site, and (3) the gain-of-function serial passage of SARS-related coronaviruses through humanized (chimeric) mice:

Among other things, the NIH helped fund experiments at WIV that infected genetically engineered mice with "chimeric" hybrids of SARS-related bat coronaviruses in what some scientists have described as unacceptably risky research. As The Intercept has reported, these particular experiments could not have sparked the pandemic—the viruses described in the research are too different from SARS-CoV-2—but it does raise questions about what other kinds of experiments were going on in Wuhan and haven't been disclosed. **Key details of these US-funded experiments were made public only after *The Intercept* filed a FOIA lawsuit.** [emphasis added]

Farrar then summarized the perspectives of several other scientists, including Michael Farzan, of UF Scripps Institute. Farzan, Farrar wrote, was particularly puzzled by the presence in the virus's genome of a *furin cleavage site*, which is a feature that has not been found in other SARS-related coronaviruses. *The furin cleavage site* plays an important role in helping the virus infect human airway cells. Farzan was "bothered by the furin site and has a hard time explaining that as an event **outside the lab** (though, there are possible ways in nature, but **highly unlikely**)." On the question of whether the virus had a natural origin or came from some sort of accidental lab release, Farrar reported that Farzan was "70:30" or "60:40" in favor of an "accidental-release" explanation and that "Bob"—an apparent reference to Robert Garry—was also surprised by the presence of a *furin cleavage site* in this virus. Farrar quoted Bob saying: "I just can't figure out how this gets accomplished in nature. . . . [I]t's stunning." [emphasis added]

Holmes sent the summary to Farrar, who forwarded it to Fauci and Collins. It sparked a speculative discussion among the three men about the kind of laboratory work that could have inadvertently created the virus. Their speculations centered on

"serial passage" or "repeated tissue culture passage," a practice in which a virus is evolved in a lab by repeatedly passaging it through mice, other lab animals, or cell culture. In some cases, **this technique involves passing viruses through the bodies of mice that have been genetically altered to express certain human proteins**. The technique can also make it possible for scientists to "fairly rapidly **select for more pathogenic** variants [of a virus] in the laboratory," as Garry would note in a later e-mail. [emphasis added]

Farrar replied in an early-morning e-mail: "Being very careful in the morning wording. 'Engineered' probably not. Remains very real possibility of accidental lab passage in animals to give glycans." The scientists seem by this point to have made a sharp distinction between a scenario in which the virus was deliberately engineered in a lab and a scenario in which the **virus was generated during serial passage**[211] **experiments in a lab**. [emphasis added]

133. On 3 February 2023, the CDC updated its website[212] for COVID-19 vaccine information and guidance. Under "How mRNA COVID-19 Vaccines Work," the website page stated,

mRNA COVID-19 vaccines are given in the upper arm muscle or upper thigh, depending on the age of who is getting vaccinated.

After vaccination, the mRNA will enter the muscle cells. Once inside, they use the cells' machinery to produce a harmless piece of what is called the spike protein. The spike protein is found on the surface of the virus that causes COVID-19. After the protein piece is made, our cells break down the mRNA and remove it, leaving the body as waste.

A. Whereas the currently updated[213] CDC website states the COVID vaccines stay at the site of injection and "side effects after a COVID-19 vaccine are common, however severe allergic reactions after getting a COVID-19 vaccine are rare," the following information, along with the more than 2.5 million reported adverse events on the Vaccine Adverse Events Reporting System (VAERS),[214] indicates that moderate or severe "nonallergic" reactions are anything but rare.

i. **Injected COVID vaccines stay at the site of injection**—Animal research by Bahl and others in 2017 revealed CDC's statement to be demonstrably false. The lipid nanoparticle vaccines do not stay at the site of injection (upper arm muscle or upper thigh) and instead travel throughout the body, including the bloodstream, heart, liver, brain, bone marrow, and reproductive organs.[215]

Exhibits C and D provide scientific and medical materials presented at medical and public conferences, in addition to published research material addressing the InflammoThrombotic Response, sudden cardiac death, cancer, and miscarriages resulting from the spike protein of SARS-CoV-2 and the spike protein derived from genetic vaccination.

ii. **Sudden cardiac death**—In November 2008 Karikó and Weissman published their research on incorporating pseudouridine into mRNA. Prior research found the mRNA molecules unstable. The replacement of uridine with pseudouridine not only solved the instability problem but also provided a product with two unique properties. First, there was greater gene transfection following the use of the pseudouridine mRNAs. Second, the modified genetic vaccines had a diminished immunogenetic

Research Published by Moderna in 2017 Clearly Demonstrated that LNP Genetic Vaccines Do NOT Stay at the Injection Site

Table 2. Number and Percentage of Subjects Who Experienced a Solicited Reactogenicity Event after Receiving 100 μg H10N8 mRNA IM or Placebo

Parameter	100 μg IM H10N8 mRNA n (%)	Placebo n (%)
Total number of subjects	23 (100)	8 (100)
Any reactogenicity event	23 (100)	5 (62.5)
Mild	23 (100)	3 (37.5)
Moderate	12 (52.2)	1 (12.5)
Severe	3 (13.0)	1 (12.5)
Any local reactogenicity event	12 (91.3)	2 (25.0)
Mild	20 (87.0)	2 (25.0)
Moderate	9 (39.1)	0
Severe	2 (8.7)	0
Any systemic reactogenicity event	21 (91.3)	5 (62.5)
Mild	21 (91.3)	3 (37.5)
Moderate	11 (47.8)	1 (12.5)
Severe	1 (4.3)	1 (12.5)

Table 1. Biodistribution of H10 mRNA in Plasma and Tissue after IM Administration in Mice

Matrix	t_{max} (hr)	C_{max} (ng/mL) Mean	C_{max} (ng/mL) SE	$AUC_{0-264\,h}$ (ng.hr/mL) Mean	$AUC_{0-264\,h}$ (ng.hr/mL) SE	$t_{1/2}$ (h)
Bone marrow	2.0	3.35	1.87	NA		NC
Brain	8.0	0.429	0.0447	13.9	1.61	NR
Cecum	8.0	0.886	0.464	11.1	5.120	NC
Colon	8.0	1.11	0.501	13.5	5.51	NC
Distal lymph nodes	8.0	177.0	170.0	4,050	2,060	28.0
Heart	2.0	0.799	0.225	6.76	1.98	3.50
Ileum	2.0	3.54	2.60	22.6	10.8	5.42
Jejunum	2.0	0.330	0.120	5.24	0.931	8.24
Kidney	2.0	1.31	0.273	9.72	1.44	11.4
Liver	2.0	47.2	8.56	276	37.4	NC
Lung	2.0	1.82	0.555	12.7	2.92	16.0
Muscle (injection site)	2.0	5,680	2,870	95,100	20,000	18.8
Plasma	2.0	5.47	0.829	35.5	5.41	9.67
Proximal lymph nodes	8.0	2,120	1,970	38,600	22,000	25.4
Rectum	2.0	1.03	0.423	14.7	3.67	NR
Spleen	2.0	86.9	29.1	2,270	585	25.4
Stomach	2.0	0.626	0.121	11.6	1.32	12.7
Testes	8.0	2.37	1.03	36.6	11.8	NR

Male CD-1 mice received 300 μg/kg (6 μg) formulated H10 mRNA via IM immunization. Two replicates of bone marrow, lung, liver, heart, right kidney, inguinal- and popliteal-draining lymph nodes, axillary distal lymph nodes, spleen, brain, stomach, ileum, jejunum, cecum, colon, rectum, testes (bilateral), and injection site muscle were collected for bDNA analysis at 0, 2, 8, 24, 48, 72, 120, 168, and 264 hr after dosing (n = 3 mice/time point). NA, not applicable AUC with less than three quantifiable concentrations; NC, not calculated; NR, not reported because extrapolation exceeds 20% or R-squared is less than 0.80.

Bahl K, et al. Preclinical and Clinical Demonstration of Immunogenicity by mRNA Vaccines against H10N8 and H7N9 Influenza Viruses. Molecular Therapy 2017;25(6):1316-1327.

effect—that is, these modified mRNAs using pseudouridine had the opposite effect desired if you were looking to develop a vaccine product. All of which made pseudouridine mRNA "a promising tool for . . . gene replacement." Karikó and Weissman were awarded the 2023 Nobel Prize in Physiology or Medicine for their work.

In summary, the researchers specifically showed a reduction in interferon compared with uridine mRNA, which results in an increase in vascular endothelial growth factor A (VEGF-A). The consequence of increasing VEGF-A is the altering of slow potassium channels found on the cells in our heart, responsible for carrying out the heart's electrical activity. By interfering with this potassium current, the mean action potential (MAP) is altered, increasing susceptibility to unstable heart rhythms that result in ventricular tachycardia, fibrillation, polymorphic tachydysrhythmia, torsade de pointes, and sudden cardiac death. These potassium channels account for the congenital risk for sudden death, known as long QT syndrome (LQTS). In such individuals with this syndrome, any irritation to the heart (e.g., inflammation from an infection) or increase in heart rate allows less time for these impaired potassium channels to work. For example, increased exertion or physical exercise typical of sporting events can result in sudden cardiac death. Full details of the biological pathways involved in long COVID and sudden cardiac death are detailed in Exhibits C and D.[216]

iii. **Vaccine contamination**[217]

a. "Viral Contamination in Biologic Manufacture and Implications for Emerging Therapies," a study by Paul W. Barone and others published by *Nature Biotechnology* on 27 April 2020, reveals information regarding the contamination of vaccine products, including the following statements:[218]

In the twentieth century, several vaccine products were unintentionally contaminated with unwanted viruses during their production. This included the contamination of poliovirus vaccine with simian virus 40 (SV40),[219] for which the health impacts were not fully known for many decades. In the early 1980s, unknowingly contaminated therapeutic proteins from human plasma caused widespread transmission of viruses such as human immunodeficiency virus (HIV) to people with hemophilia who received these treatments.

As a result, public trust in the plasma industry's ability to safely make these therapies declined. To ensure that current plasma-derived, vaccine, and recombinant biotherapeutics are safe, complementary safety strategies to reduce the risk of virus contamination were developed and implemented. . . . The biotechnology industry has a long history of supplying safe and effective therapies to patients owing to the extensive controls in place to ensure product safety. Despite these controls, viral infection of cell culture is a real risk with severe consequences. Learning from these events has historically been a challenge; the work presented here represents a comprehensive collection and analysis of previously unpublished industry-wide viral contamination information. The CAACB[220] study has identified five viruses that have been shown to contaminate CHO[221] cell culture and four viruses that have contaminated cell culture of human or primate cells. Importantly, the viruses that have been shown to contaminate human or primate cell lines can also infect humans. The choice of which cell line to use for recombinant protein or vaccine production is a complicated decision, of which viral contamination risks are just one consideration. However, manufacturers that are using human or primate cells should be aware of the difference in the potential risk to patients from a viral contaminant in products produced in those cells compared with CHO cells. . . . This is never more true than when faced with a previously unknown emerging virus, such as SARS-CoV-2, where the capacity of the virus to infect production cell lines or be detected in existing assays is not initially known. . . .

Finally, lessons from the CAACB study, applied to emerging biotech products, lead us to conclude that the viral safety of some ATMPs rely almost exclusively on preventing contamination through the use of rigorous process controls. Further, the short time frame associated with the use of many ATMPs, relative to their manufacture, is a challenge for current viral testing paradigms and offers a clear opportunity for technological advancement.

b. The CDC statement "After the protein piece is made, our cells break down the mRNA and remove it, leaving the body as waste." is incorrect according to findings published in a March 2023 article by Castruita and others, titled "SARS-CoV-2 Spike mRNA Vaccine Sequences Circulate in Blood up to 28 Days After COVID-19 Vaccination." The paper included the following:[222]

In Denmark, vaccination against SARS-CoV-2 has been with the Pfizer-BioNTech (BTN162b2) or the Moderna (mRNA-1273) mRNA vaccines. Patients with chronic hepatitis C virus (HCV) infection in our clinic received mRNA vaccinations according to the Danish roll-out vaccination plan. To monitor HCV infection, RNA was extracted from patient plasma, and RNA sequencing was performed on the Illumina platform. In 10 of 108 HCV patient samples, full-length or traces of SARS-CoV-2 spike mRNA vaccine sequences were found in blood up to twenty-eight days after COVID-19 vaccination. Detection of mRNA vaccine sequences in blood after vaccination adds important knowledge regarding this technology and should lead to further research into the design of lipid-nanoparticles and the half-life of these and mRNA vaccines in humans.

C. On 10 April 2023, a preprint by Kevin McKernan and others, concerning the contamination of Moderna and Pfizer vaccines, titled "Sequencing of Bivalent Moderna and Pfizer mRNA Vaccines Reveals Nanogram to Microgram Quantities of Expression Vector dsDNA per Dose" was published by OSF Preprints. The study includes the following statements on the contamination of expired vials of Moderna and Pfizer vaccines:[223]

Several methods were deployed to assess the nucleic acid composition of four expired vials of the Moderna and Pfizer bivalent mRNA vaccines. Two vials from each vendor were evaluated with Illumina sequencing, qPCR, RT-qPCR, Qubit 3 fluorometry and Agilent Tape Station electrophoresis. Multiple assays support DNA contamination that exceeds the European Medicines Agency (EMA) 330ng/mg requirement and the FDAs 10ng/dose requirements. These data may impact the surveillance of vaccine mRNA in breast milk or plasma as RT-qPCR assays targeting the vaccine mRNA cannot discern DNA from RNA without RNase or DNase nuclease treatments. Likewise, studies evaluating the reverse transcriptase activity of LINE-1 and vaccine mRNA will need to account for the high levels of DNA contamination in the vaccines. The exact ratio of linear fragmented DNA versus intact circular plasmid DNA is still being investigated. Quantitative PCR assays used to track the DNA contamination are described.

iv. **Impaired quality of semen**—The conclusion of a study conducted by Yanfei He and others, titled "Effect of COVID-19 on Male Reproductive System—a Systematic Review," published on 27 May 2021, states as follows:[224]

The likelihood of SARS-CoV-2 in the semen of COVID-19 patients is very small, and semen should rarely be regarded as a carrier of SARS-CoV-2 genetic material. However, COVID-19 may cause testicular spermatogenic dysfunction via immune or inflammatory reactions. Long-term follow-up is needed for COVID-19 male patients and fetuses conceived during the father's infection period.

The only published research[225] demonstrating the immediate effect of the Pfizer, Moderna, or Janssen vaccines when added directly to human blood or sperm showed reduction in sperm motility when examined microscopically.

v. **Blood clotting and inflammation (InflammoThrombotic Response; ITR)**

a. A study published on 27 August 2021 by Lize M. Grobbelaar and others, titled "SARS-CoV-2 Spike Protein S1 Induces Fibrin(ogen) Resistant to Fibrinolysis: Implications for Microclot Formation in COVID-19," opposes the CDC's description of the SARS-CoV-2 spike protein: "harmless piece of what is called the spike protein."

The study concluded that the SARS-CoV-2 spike protein causes blood clotting and had the following statements:[226]

Severe acute respiratory syndrome coronavirus 2 (SARS-Cov-2)–induced infection, the cause of coronavirus disease 2019 (COVID-19), is characterized by unprecedented clinical pathologies. One of the most important pathologies, is hypercoagulation and microclots in the lungs of patients. Here we study the effect of isolated SARS-CoV-2 spike protein S1 subunit as potential inflammagen sui generis. Using scanning electron and fluorescence microscopy as well as mass spectrometry, we investigate the potential of this inflammagen to interact with platelets and fibrin(ogen) directly to cause blood hypercoagulation. Using platelet-poor plasma (PPP), we show that spike protein may interfere with blood flow. Mass spectrometry also showed that when spike protein S1 is added to healthy PPP, it results in structural changes to β and γ fibrin(ogen), complement 3, and prothrombin. These proteins were substantially resistant to trypsinization, in the presence of spike protein S1. Here we suggest that, in part, the presence of spike protein in circulation may contribute to the hypercoagulation in COVID-19 positive patients and may cause substantial impairment of fibrinolysis. Such lytic impairment may result in the persistent large microclots we have noted here and previously in plasma samples of COVID-19 patients. This observation may have important clinical relevance in the treatment of hypercoagulability in COVID-19 patients.

b. A study by Yi Zheng and others, published 29 October 2021, "SARS-CoV-2 Spike Protein Causes Blood Coagulation and Thrombosis by Competitive Binding to Heparan Sulfate," found that the SARS-CoV-2 virus spike protein "can directly induce blood coagulation in addition to inflammation." The paper states as follows:[227]

Herein, we found that the S protein can competitively inhibit the bindings of antithrombin and heparin cofactor II to heparin/HS, causing abnormal increase in thrombin activity. SARS-CoV-2 S protein at a similar concentration (~10 μg/mL) as the viral load in critically ill patients can cause directly blood coagulation and thrombosis in zebrafish model.

c. A study by R. M. Fleming and others, published 13 January 2023,[228] is the only research published to date demonstrating the immediate effect of the Pfizer, Moderna, or Janssen vaccines when added directly to human blood or sperm as examined microscopically, in real time. The results of that study are summarized thusly:

Conclusion: Administration of Pfizer, Moderna, and Janssen vaccines resulted in the immediate loss of red coloration present in erythrocytes. This loss of red coloration indicates that there is a disruption of the hemoglobin binding of oxygen. Since atmospheric oxygen is immediately available to re-saturate the hemoglobin molecules restoring the red color responsible for the function and name of erythrocytes; the results of this investigation suggest that there is an alteration in the hemoglobin molecule preventing the hemoglobin from binding with oxygen. This alteration of the hemoglobin molecule could be explained if the vaccines merge with the erythrocytes and release their genetic material (RNA or DNA) directly into the erythrocytes; having a prion altering effect upon the hemoglobin molecule. Some of the vaccine samples included extraneous materials including crystals, fibrous material and precipitated lipid nanoparticles.

The following microscopic photographs in the QR code slides show the change in color from red-oxygenated blood, to grey, unable to bind to oxygen to hemoglobin with blood clotting. These findings occurred within 15–30 seconds, in every sample of blood when the Pfizer, Moderna and/or Janssen genetic vaccines were added to the blood under the microscop. This confirmed an almost immediate ITR in addition to priogenic response.

As shown in the QR codes slides, microscopic analysis of the Pfizer, Moderna, and Janssen genetic vaccines revealed no graphene oxide, nanotechnology, or biological organisms. The samples did contain extraneous material (A,B) including crystals (H), fibrous material (C, D) and precipitated lipid nanoparticles (E) in addition to air bubbles (F) frequently confused with nanotechnology. G, not shown here, provided readers with a graphene oxide (GO) sample to demonstrate GO is not present in the vaccines.

vi. **Myocarditis (InflammoThrombotic response of the heart)**

a. A study published on 14 December 2021 by Martina Patone and others, "Risks of Myocarditis, Pericarditis, and Cardiac Arrhythmias Associated with COVID-19 Vaccination or SARS-CoV-2 Infection," includes the following statements:[229]

> Our findings are consistent with those from a case-control study of 884,828 persons receiving the BNT162b2 vaccine in Israel. That study observed an association with myocarditis in the 42 days following vaccination (risk ratio of 3.24), but no association with pericarditis or cardiac arrhythmia. . . . Risk of myocarditis was restricted to males under the age of 40 years and only observed following the second dose. Similarly, two studies from the United States have reported an incident rate ratio of 2.7 for myocarditis in the 10 days following the second dose of both mRNA vaccines and an estimated 6.3 and 10.1 extra cases per million doses in the 1- to 21-day period following the first and second dose of both mRNA vaccines, respectively, in those younger than 40 years. . . . In summary, this population-based study quantifies for the first time the risk of several rare cardiac adverse events associated with three COVID-19 vaccines as well as SARS-CoV-2 infection. Vaccination for SARS-CoV-2 in adults was associated with a small increase in the risk of myocarditis within a week of receiving the first dose of both adenovirus and mRNA vaccines, and after the second dose of both mRNA vaccines. By contrast, SARS-CoV-2 infection was associated with a substantial increase in the risk of hospitalization or death from myocarditis, pericarditis and cardiac arrhythmia.

b. On 17 January 2023 a study by Lael A. Yonker and others, "Circulating Spike Protein Detected in Post–COVID-19 mRNA Vaccine Myocarditis" was published online and then in the journal *Circulation.* A summary of the study, written by Salim Hayek for the American College of Cardiology, makes the following points:[230]

> This is a great example of a study with mostly negative findings which are, however, insightful. The investigators used a thorough approach in teasing out the various aspects that could underlie vaccine-induced myocarditis. In summary, the data show that adaptive and T-cell immunity responses were normal in recipients of mRNA vaccines, both with and without myocarditis. Patients who developed postvaccine myocarditis had persistently elevated free spike protein in circulation, which correlated with evidence of cardiac injury and inflammatory cytokines. The implications of this finding are unclear, since it is yet unknown how the spike protein evades cleavage or clearance, especially in the setting of a normal adaptive immune response, or whether in itself is pathogenic. Given myocarditis also occurs after other vaccines, it is likely that the presence of circulating spike is a biomarker rather than the causal agent. Indeed, presence of viral proteins has been associated with hyperinflammatory responses such as in severe COVID-19 or the notorious multisystem inflammatory syndrome in children (MIS-C). We are left with several hypotheses and more questions, but with a clear direction.

vii. **Prion- and Amyloid-associated diseases**

a. On 17 May 2022 the *Journal of the American Chemical Society* published "Amyloidogenesis of SARS-CoV-2 Spike Protein" by Sofie Nystrom and Per Hammarstrom, which made the following statements:[231]

SARS-CoV-2 infection is associated with a surprising number of morbidities. Uncanny similarities with amyloid-disease associated blood coagulation and fibrinolytic disturbances together with neurologic and cardiac problems led us to investigate the amyloidogenicity of the SARS-CoV-2 spike protein (S-protein). Amyloid fibril assays of peptide library mixtures and theoretical predictions identified seven amyloidogenic sequences within the S-protein. . . . The prospective of S-protein amyloidogenesis in COVID-19 disease associated pathogenesis can be important in understanding the disease and long COVID-19. . . . Amyloidosis from several culprit proteins manifests as systemic and localized disorders with many phenotypes overlapping with reported COVID-19 symptoms. It has been proposed that severe inflammatory disease including ARDS in combination with SARS-CoV-2 protein aggregation might induce systemic AA amyloidosis. Neurotropic colonization and cross-seeding of S-protein amyloid fibrils to induce aggregation of endogenous proteins [have] been discussed in the context of neurodegeneration. Notably, blood clotting associated with extracellular amyloidotic fibrillar aggregates in the bloodstream has been reported in COVID-19 patients. Hypercoagulation/impaired fibrinolysis was demonstrated in blood plasma from healthy donors experimentally spiked with S-protein. . . . Amyloidosis is associated with cerebral amyloid angiopathy, blood coagulation disruption, fibrinolytic disturbance, FXII Kallikrein/Kinin activation, and thromboinflammation, suggesting potential links between S-protein amyloidogenesis and COVID-19 phenotypes. We therefore hypothesized a potential molecular link between S-protein and amyloid formation. Inspired by previous hypotheses about human and viral protein amyloids and interactions between them, (11–13) in particular SARS-CoV spike proteins, (6,14,15) we asked the question: Is SARS-CoV-2 S-protein amyloidogenic?

We found that all common coronaviruses infecting humans contain amyloidogenic sequences. . . . Recent studies demonstrate that COVID-19 recovered patients have an increased risk of type II diabetes, an amyloid associated disease. . . . While our study is limited to in vitro findings of pure preparations of peptides and proteins, the results propose taking S-protein amyloidogenesis into account when studying COVID-19 and long COVID-19 symptoms.

Common diseases associated with amyloids include Parkinson disease, Alzheimer disease, dementia, type 2 diabetes, kidney disease, and rheumatoid arthritis.

b. On 13 January 2023, Dr. Richard M. Fleming and others published "Pfizer, Moderna and Janssen Vaccine InflammoThrombotic and Prion Type Effect on Erythrocytes When Added to Human Blood" in *Haematology International Journal*, as previously discussed.

c. A short summary of a study published on 5 April 2023 by Zhouyi Rong and others, "SARS-CoV-2 Spike Protein Accumulation in the Skull-Meninges-Brain Axis: Potential Implications for Long-Term Neurological Complications in Post-COVID-19" reveals the following:

The accumulation of SARS-CoV-2 spike protein in the skull-meninges-brain axis presents potential molecular mechanisms and therapeutic targets for neurological complications in long-COVID-19 patients.

The Medline Plus website, from NIH, describes prion disease as follows:[232]

> Prion disease represents a group of conditions that affect the nervous system in humans and animals. In people, these conditions impair brain function, causing changes in memory, personality, and behavior; a decline in intellectual function (dementia); and abnormal movements, particularly difficulty with coordinating movements (ataxia). The signs and symptoms of prion disease typically begin in adulthood and worsen with time, leading to death within a few months to several years.

viii. **Vaccine-induced side effects**

a. On 31 August 2022 "Serious Adverse Events of Special Interest Following mRNA COVID-19 Vaccination in Randomized Trials in Adults" by Joseph Fraiman and others was published in the journal *Vaccine*. Here is the study's abstract:[233]

> Introduction: In 2020, prior to COVID-19 vaccine rollout, the Brighton Collaboration created a priority list, endorsed by the World Health Organization, of potential adverse events relevant to COVID-19 vaccines. We adapted the Brighton Collaboration list to evaluate serious adverse events of special interest observed in mRNA COVID-19 vaccine trials.
>
> Methods: Secondary analysis of serious adverse events reported in the placebo-controlled, phase III randomized clinical trials of Pfizer and Moderna mRNA COVID-19 vaccines in adults (NCT04368728 and NCT04470427), focusing analysis on Brighton Collaboration adverse events of special interest.
>
> Results: Pfizer and Moderna mRNA COVID-19 vaccines were associated with an excess risk of serious adverse events of special interest of 10.1 and 15.1 per 10,000 vaccinated over placebo baselines of 17.6 and 42.2 (95 % CI -0.4 to 20.6 and -3.6 to 33.8), respectively. Combined, the mRNA vaccines were associated with an excess risk of serious adverse events of special interest of 12.5 per 10,000 vaccinated (95 % CI 2.1 to 22.9); risk ratio 1.43 (95 % CI 1.07 to 1.92). The Pfizer trial exhibited a 36 % higher risk of serious adverse events in the vaccine group; risk difference 18.0 per 10,000 vaccinated (95 % CI 1.2 to 34.9); risk ratio 1.36 (95 % CI 1.02 to 1.83). The Moderna trial exhibited a 6 % higher risk of serious adverse events in the vaccine group: risk difference 7.1 per 10,000 (95 % CI –23.2 to 37.4); risk ratio 1.06 (95 % CI 0.84 to 1.33). Combined, there was a 16 % higher risk of serious adverse events in mRNA vaccine recipients: risk difference 13.2 (95 % CI -3.2 to 29.6); risk ratio 1.16 (95 % CI 0.97 to 1.39).
>
> Discussion: The excess risk of serious adverse events found in our study points to the need for formal harm-benefit analyses, particularly those that are stratified according to risk of serious COVID-19 outcomes. These analyses will require public release of participant level datasets.

b. The concluding remarks of a study published on 28 October 2022 by Marco Cosentino and Franca Marino, "The Spike Hypothesis in Vaccine-Induced Adverse Effects: Questions and Answers" includes the following material:[234]

> Such considerations as a whole support the possibility that COVID-19 mRNA vaccines under some circumstances induce high and possibly toxic amounts of S protein in organs and tissues, in turn leaking into the circulation. In animal models, it is well established that lipid nanoparticle-carried mRNAs undergo systemic disposition and expression in organs such as liver, skeletal muscle, and lungs. It can be suggested that

at least part of the risk to develop adverse reactions following vaccination with mRNA products depends on the organs/tissues where S protein production occurs, as well as on the total amount produced and on the production time course. For example, it was established long ago that distinct tissues widely differ in the efficiency of protein synthesis, but no one so far assessed whether and to what extent this is relevant for the efficacy and safety of mRNA vaccines. Therefore, we further recommend careful characterization of COVID-19 vaccines with regards to systemic disposition, including organs and tissues where S protein production occurs and interindividual variability in local protein synthesis efficiency, to provide a rational basis for dose individualization and to identify subjects at risk for adverse reactions due to either vaccine-induced S protein production in vulnerable sites, excessive S protein production, or both.

134. On 5 March 2023, the majority staff of the Select Subcommittee on the Coronavirus Pandemic released an official memorandum regarding "New Evidence[235] Resulting from the Select Subcommittee's Investigation into the Origins of COVID-19—'The Proximal Origin of the SARS-CoV-2.'" The following statements were included in the memorandum:

On February 1, 2020, Dr. Anthony Fauci, Dr. Francis Collins, and at least eleven other scientists convened a conference call to discuss COVID-19. It was on this conference call that Drs. Fauci and Collins were first warned that COVID-19 may have leaked from a lab in Wuhan, China and, further, may have been intentionally genetically manipulated.

Only three days later, on February 4, 2020, four participants of the conference call authored a paper entitled "The Proximal Origin of SARS-CoV-2" (Proximal Origin) and sent a draft to Drs. Fauci and Collins. Prior to final publication in *Nature Medicine*, the paper was sent to Dr. Fauci for editing and approval.

On April 16, 2020, slightly more than two months after the original conference call, Dr. Collins emailed Dr. Fauci expressing dismay that Proximal Origin—which they saw prior to publication and were given the opportunity to edit—did not squash the lab leak hypothesis and asks if the NIH can do more to "put down" the lab leak hypothesis. The next day—after Dr. Collins explicitly asked for more public pressure—Dr. Fauci cited Proximal Origin from the White House podium when asked if COVID-19 leaked from a lab.

New evidence released by the Select Subcommittee today suggests that Dr. Fauci "prompted" the drafting of a publication that would "disprove" the lab leak theory, the authors of this paper skewed available evidence to achieve that goal, and Dr. Jeremy Farrar went uncredited despite significant involvement.

New Evidence: The Drafting and Publication of "The Proximal Origins of SARS-CoV-2"

I. "Prompted by . . . Tony Fauci"

The evidence available to the Select Subcommittee suggests that Dr. Anthony Fauci "prompted" Dr. Kristian Andersen, Professor, Scripps Research (Scripps), to write Proximal Origin and that the goal was to "disprove" any lab leak theory.

On August 18, 2021, Scripps responded to then–Committee on Oversight and Reform Ranking Member, James Comer, and then-Committee on the Judiciary Ranking Member, Jim Jordan's, July 29, 2021, letter to Dr. Andersen. In this letter, Scripps asserts that Dr. Andersen "objectively" investigated the origins and that Dr.

Anthony Fauci did not attempt to influence his work. Both statements do not appear to be supported by the available evidence.

The Goal of Proximal Origin Was to "Disprove" A Lab Theory

In Scripps' August 18 letter, on behalf of Dr. Andersen, it stated: "In January 2020, Dr. Andersen began investigating the origins of SARS-CoV-2. At every point, *Dr. Andersen has objectively weighed all of the evidence available to him*. . . . Dr. Andersen's view evolved consistent with the evidence at his disposal. . . . Scientists must make conclusions supported by the available evidence, even when it conflicts with earlier assessments."

According to previously released e-mails, this assertion is also demonstrably false. On February 8, 2020, Dr. Andersen stated: "Our main work over the last couple of weeks has been focused on *trying to disprove any type of lab theory*. . ."

This e-mail directly contradicts Scripps' earlier statement that Dr. Andersen "objectively" weighed all the evidence regarding the origins of COVID-19. Instead, it appears that Dr. Andersen was given direction and sought to formulate a paper, regardless of available evidence, that would disprove a lab leak.

Dr. Anthony Fauci "Prompted" the Drafting of "The Proximal Origin of SARS-CoV-2"

In Scripps' August 18 letter, on behalf of Dr. Andersen, it stated: "As for the conference call of February 1, *Dr. Fauci did not, in Dr. Andersen's view, attempt to influence Dr. Andersen* or any other member of the ad hoc working group of international subject matter experts with respect to any aspect of the discussion."

According to new evidence obtained by the Select Subcommittee, this assertion is demonstrably false. On February 12, 2020, Dr. Andersen wrote to *Nature* to request the publication of what would become Proximal Origin. In this e-mail, Dr. Andersen wrote: "There has been a lot of speculation, fear mongering, and conspiracies put forward in this space and we thought that bringing some clarity to this discussion might be of interest to Nature [sic].

Prompted by Jeremy Farrah [sic], Tony Fauci, and Francis Collins, Eddie Holmes, Andrew Rambaut, Bob Garry, Ian Lipkin, and myself have been working through much of the (primarily) genetic data to provide agnostic and scientifically informed hypothesis around the origins of the virus."

This e-mail directly contradicts Scripps' earlier statement that Dr. Fauci did not influence Dr. Andersen.

II. The False Narrative of the Pangolin Sequences

It remains unclear what science changed, or new evidence was discovered to change the minds of the authors of Proximal Origin between the February 1 conference call and the February 4 draft. In a July 14, 2021 interview with the *New York Times*, Dr. Andersen was asked about how his view changed from possible lab leak to definitely zoonotic, "[c]an you explain how the research changed your view?" He replied: "The features in SARS-CoV-2 that initially suggested possible engineering were identified in related coronaviruses, meaning that features that initially looked unusual to us weren't. . . . Yet more extensive analyses, significant additional data and thorough investigations to compare genetic diversity more broadly across coronaviruses led to the peer-reviewed study published in Nature Medicine [sic]. *For example, we looked at data from coronaviruses found in other species, such as bats and pangolins, which demonstrated that the features that first appeared unique to SARS-CoV-2 were in fact found in other, related viruses.*"

According to new evidence obtained by the Select Subcommittee, while Proximal Origin was going through peer review with *Nature Medicine* more than a year earlier, Dr. Andersen actually did not find the pangolin data compelling. The first referee asked: "There are two recent reports about coronaviruses in pangolins. The authors might want to comment on these."

Dr. Andersen replied: "We have included these references as well as several others that have investigated pangolin CoV. *In addition . . . we should point out that these additional pangolin CoV sequences do not further clarify the different scenarios discussed in our manuscript.* There is nothing in these reports that changes our statements regarding a potential role of pangolins."

The second referee asked: "The paper itself is interesting, but unnecessarily speculative. It's not clear why the authors do not refute a hypothetical lab origin in their coming publication on the ancestors of SARS-CoV-2 in bats and pangolins. . . . Once the authors publish their new pangolin sequences, a lab origin will be extremely unlikely. It is not clear why the authors rush with a speculative perspective if their central hypothesis can be supported by their own data. Please explain."

Dr. Andersen replied: "Our manuscript is written to explore the potential origin of SARSCoV-2. We do not believe it is speculative. . . . *Unfortunately, the newly available pangolin sequences do not elucidate the origin of SARS-CoV-2 or refute a lab origin.* Hence, the reviewer is incorrect on this point. . . . *here is no evidence on present data that the pangolin CoVs are directly related to the COVID-19 epidemic.*"

Privately, Dr. Andersen did not believe the pangolin data disproved a lab leak theory despite saying so publicly. It is still unclear what intervening event changed the minds of the authors of Proximal Origin in such a short period of time. Based on this new evidence, the pangolin data was not the compelling factor; to this day, the only known intervening event was the February 1 conference call with Dr. Fauci.

III. Uncredited Involvement of Dr. Jeremy Farrar

The evidence available to the Select Subcommittee suggests that Dr. Farrar, the former Director of the Wellcome Trust and current Chief Scientist at the World Health Organization, was more involved in the drafting and publication of Proximal Original than previously known.

Dr. Eddie Holmes Sought Permission from Dr. Farrar to Involve Dr. W. Ian Lipkin

Dr. Lipkin, Professor of Epidemiology, Columbia University, was not on the February 1 conference call and was not involved in the drafting of Proximal Origin in the early stages. However, on February 10, 2020, Dr. Holmes sent a draft of Proximal Origin to Dr. Lipkin for his review. Dr. Holmes stated: "Here's the document we wrote a few days ago. Things are moving so quickly that is hard [sic] to keep up. Comments welcome. I favour natural evolution myself, but the furin cleavage site is an issue. *I'll have a chat with Jeremy [Farrar] in a little while to see if can [sic] get you more directly involved.*"

Dr. Lipkin responded with his thoughts on the draft of Proximal Origin: "It's well-reasoned and provides a plausible argument against genetic engineering. *It does not eliminate the possibility of inadvertent release following adaptation through selection in culture at the institute in Wuhan. Given the scale of the bat CoV research pursued there and the site of emergence of the first human cases we have a nightmare of circumstantial evidence to assess.*"

Dr. Holmes agreed with Dr. Lipkin's assessment of the possibility of a lab leak and reiterated that he was asking Dr. Farrar about including Dr. Lipkin in the drafting process: "I agree. *Talking to Jeremy (Farrar) in a few minutes and I'll get back in touch after.* It is indeed striking that this virus is so closely related to SARS yet is behaving so differently. *Seems to have been pre adapted for human spread since the get go.* It's the epidemiology that I find most worrying."

Dr. Farrar Led the Drafting Process and Made At Least One Uncredited Direct Edit to Proximal Origin

Dr. Farrar is not credited as having any involvement in the drafting and publication of Proximal Origin. According to new evidence obtained by the Select Subcommittee, Dr. Farrar led the drafting process and in fact made direct edits to the substance of the publication. Right before publication, on February 17, 2020, Dr. Lipkin emails Dr. Farrar to thank him for leading the process of drafting Proximal Origin: "*Thanks for shepherding this paper.* Rumors of bioweaponeering are now circulating in China."

Dr. Farrar responds, confirming and saying that he will pressure *Nature* to publish: "Yes I know and in US—why so keen to get out ASAP. *I will push nature.*"

In addition to leading the drafting and publication process, Dr. Farrar made at least one direct edit to Proximal Origin. On February 17, 2020, the day Proximal Origin was first published publicly, Dr. Farrar made an edit to the draft: "*Sorry to micro-manage/micro edit*! But would you be willing to change one sentence?

> From
> *It is unlikely* that SARS-CoV-2 emerged through laboratory manipulation of an existing SARS-related coronavirus.
> To
> *It is improbable* that SARS-CoV-2 emerged through laboratory manipulation of an existing SARS-related coronavirus."

To which, Dr. Andersen responds: "Sure, attached."

This evidence suggests that Dr. Farrar was more involved in the drafting and publication of Proximal Origin than previously known and possibly should have been credited or acknowledged for this involvement.

135. Also on 5 March 2023 "New Emails Show Dr. Anthony Fauci Commissioned[236] Scientific Paper in Feb. 2020 to Disprove Wuhan Lab Leak Theory," written by Miranda Devine, appeared in the *New York Post*. The article includes the following statements:

New emails uncovered by House Republicans probing the COVID-19 pandemic reveal the deceptive nature of Dr. Anthony Fauci. They show he "prompted" or commissioned—and had final approval on—a scientific paper written specifically in February 2020 to disprove the theory that the virus leaked from a lab in Wuhan, China. Eight weeks later, Fauci stood at a White House press conference alongside President Donald Trump and cited that paper as evidence that the lab leak theory was implausible while pretending it had nothing to do with him and he did not know the authors.

"There was a study recently," he told reporters on April 17, 2020, when asked if the virus could have come from a Chinese lab, "where a group of highly qualified evolutionary virologists looked at the sequences . . . in bats as they evolve and the mutations that it took to get to the point where it is now is totally consistent with a jump of a species from an animal to a human.

"So, the paper will be available. I don't have the authors right now, but we can make it available to you."

That paper, titled "The Proximal Origin of SARS-CoV-2," was sent to Fauci for editing in draft form and again for final approval before it was published in *Nature Medicine* on Feb. 17, 2020. It was written four days after Fauci, and his NIH boss Dr. Francis Collins, held a call with the four authors to discuss reports that COVID-19 may have leaked from the Wuhan lab and "may have been intentionally genetically manipulated."

The House Oversight subcommittee published emails Sunday in which the paper's co-author Dr. Kristian Andersen admits Fauci "prompted" him to write the paper with the goal to "disprove" the lab leak theory.

On Feb. 12, 2020, Andersen submitted the paper to *Nature Medicine* with a cover email: "There has been a lot of speculation, fear-mongering, and conspiracies put forward in this space. [This paper was] Prompted by Jeremy Farrah [sic], Tony Fauci, and Francis Collins." Farrar, then head of British nonprofit the Wellcome Trust, which has historic ties to the pharmaceutical industry and the Gates Foundation, was rewarded with the plum role of chief scientist at the World Health Organization last December.

On the day the "Proximal Origin" paper was published, emails show Farrar pushing through a crucial change: "Sorry to micromanage/micro edit! But would you be willing to change one sentence?" Farrar's change was to replace the word "unlikely" with "improbable" in a statement about the lab leak origin, so it would read: "It is improbable that SARS-CoV-2 emerged through laboratory manipulation of an existing SARS-related coronavirus." Improbable means having a probability too low to inspire belief; unbelievable, even ridiculous.

That's what Fauci and friends wanted us to think of the lab leak theory that looked probable from the get-go, as one dissenting scientist said at the time, and looks more probable by the day. The question of why Fauci went to such an effort to obscure the origins of COVID-19 is a major focus of the GOP-led committee.

136. On 24 March 2023 Government Executive and senior correspondent Eric Katz reported that "The Federal Employee COVID Vaccine Mandate Remains Blocked,[237] After Appeals Court Ruling." The article subheading states, "Court Says the President Overstepped in Mandating "Private, Irreversible Medical Decisions."

137. On 28 March 2023 a press release was published on Wisconsin Senator Ron Johnson's website—"Sen. Johnson Offers the No WHO Pandemic Treaty Act as an amendment to AUMF Repeal":[238]

On Tuesday, U.S. Sen. Ron Johnson (R-Wis.) introduced the No WHO Pandemic Treaty Without Senate Approval Act as an amendment to the repeal of the 1991 and 2002 Authorizations for Use of Military Force (AUMF). The No WHO Pandemic Treaty Act would require Senate advice and consent of any convention, agreement, or other international instrument on pandemic prevention, preparedness, and response reached by the World Health Assembly (WHA).

"Today, Democrats defeated my amendment to require Senate ratification for any pandemic agreement with the World Health Organization. Now we know Democrats are willing to relinquish U.S. sovereignty to a global entity. How sad," the Senator tweeted.

Mr. President, last December, the World Health Assembly established an Intergovernmental Negotiating Body to draft a new convention on pandemic prevention and preparedness. In its fourth meeting last month, the negotiating body accepted a draft of this new convention that would give the World Health Organization broad new powers in managing future pandemics. If accepted, it would cement the World Health Organization at the center of a global system for managing future pandemics, and it would erode U.S. sovereignty. Let me just list a few of the examples of some of the provisions of this draft, and I'll call it a treaty. Currently, it would require a substantial new financial—U.S. financial commitment—to an international body without proportional voting power. It would require the U.S. to give the World Health Organization 20% of vaccines and other pandemic-related products produced for future pandemics. It includes a heavy emphasis on the transfer of intellectual property rights to the World Health Organization. It gives the World Health Organization a leading role in fighting misinformation and disinformation. And as the Twitter Files reveal, that leads to censorship and the suppression and abridging of freedom of speech. It also promotes a global one health approach to health care, including harmonizing regulation under WHO guidance. The WHO has not earned this power. Far from it. At a critical moment in late 2019 and early 2020, the WHO utterly failed to detect the emerging COVID-19 pandemic and delayed informing its member states. Instead, it was kowtowing to Beijing. Unfortunately, there are indications that the Biden Administration is considering joining this new convention by executive agreement and avoiding the Senate. We should not let this happen. An agreement of such magnitude needs to be submitted to the Senate for advice and consent. This is not a partisan issue. This is about reclaiming the Senate's prerogatives on international agreements. Mr. President, I call up my amendment number 11 and ask the amendment to be reported by number.

138. On 5 May 2023, Politico ran the article "CDC Head resigns,[239] Blindsiding Many Health Officials," written by Krista Mahr and Adam Cancryn, with the following subheading: "Rochelle Walensky Acknowledged the Agency Did Not Meet Expectations During the Pandemic and Launched a Reorganization." The article includes the following statements:

Rochelle Walensky, the director of the Centers for Disease Control and Prevention who guided President Joe Biden's response to the Covid-19 pandemic from his first day in office, is leaving her post, the White House announced Friday. Her announcement comes days before the Biden administration plans to end the public health emergency in place since early 2020, and at a time when Covid fears have receded and life mostly returned to a pre-pandemic normal. . . . In an internal email announcing her departure, Walensky wrote that she would step down on June 30. . . . She gave no specific reason for the decision to resign, writing that "at this pivotal moment for our nation and public health, having worked together to accomplish so much over the last two-plus years, it is with mixed emotions that I will step down." . . . She took over the agency from Robert Redfield at a deeply fraught moment. One year into the pandemic, the CDC had been beset by devastating testing failures, confusing communication and health guidance and disruptive political meddling by Trump-era political appointees, among other problems. Morale inside the agency's ranks was low, and Americans' distrust of its work was already growing. . . .

But during her first year on the job, she, too, quickly came under fire, particularly for her communication and messaging. The agency was not holding regular Covid briefings, and Walensky did not hold a solo news conference until about a year into her directorship. . . . She also faced backlash over the agency's Covid public health guidance, which was criticized by both those who felt the agency wasn't doing enough to protect the public and those who felt it was doing too much and overstepping its mandate. . . . Both the Covid and mpox[240] crises raised fundamental questions about the way the CDC functions within the federal government. . . . At the same time, a string of lawsuits has challenged the CDC's authority to impose its guidance on Americans. A year ago, a Trump-appointed federal district court judge in Florida ordered an injunction against the CDC's mandate that people wear masks on public transportation, saying the order was outside the agency's authority. The administration appealed the case, which is still in the courts, a move aimed less at preserving the mandate under the current pandemic circumstances than preserving the agency's powers to make public health rules in future health emergencies.

139. On 8 May 2023 *The Spectator* in Australia published an article by Queensland Senator Malcolm Roberts, "Children Targeted by WHO 'Standards for Sexuality Education in Europe.'"[241] The article raises further concerns about the behaviors and motives of those involved in the development and response to the biological viral weapon SARS-CoV-2 and provides an opportunity to understand the mindset and behaviors being promoted by the World Health Organization. Given efforts by the WHO and IHR to determine what actions will be taken, when and by whom, the following statements are included for a better understanding of legislation currently under consideration in the United States:

The World Health Organisation has orchestrated a "framework for policy makers, educational and health authorities and specialists" titled, Standards for Sexuality Education in Europe.

Its purpose is to standardise (in other words override) the diverse teaching practices of each sovereign nation within Europe and the wider international community with regards to sexual education.

Having all-but forced European nations to comply, the United Nations is seeking to expand a similar framework to all UN member states—including Australia. This framework is called International Guidance on Sexuality Education, produced as part of UN Education 2030 and counter-signed by UNICEF. The WHO are now actively promoting the framework. In mid-April 2023, the Commission on Population and Development failed to reach a consensus on advancing the strategy, providing a reprieve . . . for now.

"Nobody is happy with this result," said a spokesperson representing Senegal. They went on to point out that people come from different "horizons and realities" and that the commission must "respect all cultures." The problem with communist-style policy is that it demands a uniform approach with identical ideological outcomes irrespective of culture.

And what sort of "vision" does the WHO have in mind for the world's children?

Their preferred framework demands that sex education begin at birth and be guided by the State via the relentless work of educators instead of the current model of parent-led development with catch-up assistance from schools.

European countries have already begun integrating the WHO agenda into their curricula with Germany, for example, using the WHO document "widely" for "development and revision, advocacy work, and training educators."

Quite frankly, the Standards for Sexuality Education in Europe is a "rapey" document that reads like the mind of a child-fiddling psychopath given control of public health.

The UN document makes their intention very clear that:

"This framework aims to empower children and young people to develop respectful social and sexual relationships. These skills can help children and young people form respectful and healthy relationships with family members, peers, friends and romantic or sexual partners."

The Framework also teaches children what consent consists of, meaning they assume a child can content [consent] to sex.

The WHO lays out its reason for teaching children aged 0–6 the detail of biological reproduction—that is, children who are still young enough to believe in Santa and the Tooth Fairy. By age 6, the WHO wants the education industry—and presumably their teachers—to expose children to the concepts of intercourse, masturbation, and pornography. By age 9, they are expected to reach an "adult" knowledge of sex including teaching of masturbation and viewing of online pornography.

At age 12—remembering that we are still talking about young children—the WHO wishes the official European education course to explore political and emotional responses to sex, puberty, and gender.

Starting sex education at birth is an indication of the mindset of these people. 0- to 4-year-olds should be able to distinguish between consensual and non-consensual sexual interaction and develop a "positive attitude" to the different sexual lifestyles of adults.

These standards, if you can call them that, form part of an initiative launched by the WHO Regional Office for Europe in 2008 and were further developed by the Federal Centre for Health Education with the collaboration of 19 "experts" from Western European countries.

In their own words, it was created as part of a "new need" for sexual education "triggered by various developments during the past decades." These include "globalisation and migration of new population groups with different cultural and religious backgrounds, the rapid spread of new media, particularly the internet and mobile phone technology, the emergence and spread of HIV/AIDS, increasing concerns about the sexual abuse of children and adolescents and, not least, changing attitudes towards sexuality and changing sexual behaviour among young people."

It sounds as though bad parenting, incompatible cultural practices, and a lacklustre policing of child abuse is being used as an excuse to do away with fundamental child protection standards and the innocence of children that the West used to pride itself in.

The original argument for introducing basic levels of sex education into the school system centred around child safety. These courses were designed as a catch-up, particularly for young girls who had reached an age where it was possible for them to get pregnant, to ensure they understood reproductive essentials in order to protect themselves. The point was to avoid dangerous adolescent pregnancies and abuse—not to encourage sexual behaviour in minors.

Now it appears that adults seeking affirmation for their sexual choices are flooding the education system with age-inappropriate content that is being solidified through the edicts of unelected global bureaucracies such as the WHO.

In this case, the education framework points out that there is an increase in the spread of sexual diseases among children and a rise in teen pregnancies across Europe—but what the report does not explain is that this is largely being seen among migrant demographics after coming from cultures where the abuse and sexualisation of children is common compared to European standards.

There are countless articles detailed a doubling of child abuse in recent years, with some publications describing Europe as "a hub of child abuse material" and Save the Children reporting that child migrants are being 'systematically abused by police, people smugglers, and other adults." . . .

Above all, you might imagine that parents and the education system would seek to shelter children from the sexual world in their formative years to ensure the cycle of degeneracy was broken.

That is not what is being proposed by the WHO.

In reference to traditional (and highly successful) sex education in schools, the WHO says:

"Traditionally, sexuality education has focused on the potential risks of sexuality, such as unintended pregnancy and STI. This negative focus is often frightening for children and young people: moreover, it does not respond to their need for information and skills and, in all too many cases, it simply has no relevance to their lives."

Well, yes, children should be frightened of pregnancy—it could kill them. As for a "need for information and skills," children do not need "skills" in sexual practice. Indeed, the document appears to lament that most Western children have their first sexual encounter between 16–18.

The WHO adds:

"A holistic approach based on an understanding of sexuality as an area of human potential helps children and young people to develop essential skills to enable them to self-determine their sexuality and their relationships at various developmental stages. It supports them in becoming more empowered in order to live out their sexuality and their partnerships in a fulfilling and responsible manner."

Remember, we are speaking of children, not teenagers.

There are significant ethical problems with this document that jump out of the page. For example, during its complaint about traditional "age appropriate" sexual education in schools, the WHO insists that "it is more correct to use the term 'development-appropriate' because not all children develop at the same pace."

As the document goes on, it appears to misuse the sacred concept of fundamental human rights to claim that the United Nations Convention on the Rights of the Child "clearly states the right to information and the State's obligation to provide children with educational measures" which includes "sexual rights as human rights related to sexuality" and that "everybody [has the right] to access sexuality education." . . .

It lists human rights as the "guiding principle" of the WHO Reproductive Health Strategy in search of those lofty and terrifying "international development goals" that have caused so much horror in Western nations in other aspects of society including—but not limited to—the bid to lock people into 15-minute cities.

"Sexual health needs to be promoted as an essential strategy in reaching the Millennium Development Goals."

This is followed by the dubious claim that "the fear that sexuality education might lead to more or earlier sexuality activity by young people is not justified, as research results show."

Regardless of this "research," real-world results show a generation of increasingly sexualised children and moves to normalise paedophilia among activist communities under the guise of terms such as "minor-attracted persons."

Germany, one of the early adopters of the framework, has seen a dramatic rise in sexualised violence against minors, listing 17,704 children in 2022 as victims of sexual violence. One of the leading causes of this abuse? Young people sharing sexual images on social media—which is unsurprising given they are being sexually encouraged by the State from infancy. . . .

Meanwhile, the saturation of kindergartens and classrooms with LGBTQ+ and trans ideology has led to a rapid increase in children—who are too young to be thinking about sexual relationships—identifying as part of these movements or becoming confused about their gender to the extent that they become severely distressed. In both Europe and the states, this has created a lucrative medical industry in the chemical and surgical interference of children's bodies the results of which children will never recover from.

Children are impressionable. Opening up their world to adult sexual content is wholly inappropriate.

"When talking about the sexual behaviour of children and young people, it is very important to keep in mind that sexuality is different for children and adults," says the WHO. "Adults give sexual significance to behaviour on the basis of their adult experiences and sometimes find it very difficult to see things through children's eyes. Yet it is essential to adopt their perspective. . . . The development of sexual behaviour, feelings, and cognitions begins in the womb and continues throughout a person's lifetime. Precursors of later sexual perception, such as the ability to enjoy physical contact, are present from birth."

Which sounds awfully like the WHO believes a baby enjoying its parent holding its hand is linked to sexual feelings.

"Children have sexual feelings even in early infancy. Between the second and third year of their lives, they discover the physical differences between men and women."

"During this time, children start to discover their own bodies (early childhood masturbation, self-stimulation) and they may also try to examine the bodies of their friends (playing doctor), . . . From the age of three, they understand that adults are secretive about this subject. They test adults' limits, for instance by undressing without warning or by using sexually charged language."

The conclusions this report draws is not that society and its adults should protect children from the complex and confusing process of growing up into an adult—keeping them safe from not only themselves, but from other adults who might seek to abuse them.

"Sexuality education starts at birth," claims the WHO and, "sexuality education is firmly based on gender equality, self-determination, and the acceptance of diversity."

The implementation of this horror show comes via the Sexuality Education Matrix and includes questions such as, "Why should sexuality education start before the age of four?"

Within the Matrix, 0–4-year-olds will be taught about pregnancy and birth, the enjoyment of child masturbation, gender identity, and different types of "love."

4–6-year-olds will be encouraged to "consolidate their gender identity" and acceptable feelings of love and understand that "all feelings are okay, but not all actions taken as a result of these feelings."

And so, it goes on.

As the UN proudly declares, "Teachers must equip children to have sexual relationships."

Why? Why is it the role of the State to encourage the sexual behaviour of children? More to the point, why would anyone allow the United Nations or World Health Organisation to be involved in the protection of children when their organisations have been repeatedly involved in sexual abuse and child rape in the third world?

A recent report found that the WHO failed in its obligation to tackle "widespread sexual abuse during the Ebola response in the Congo."

It was alleged that WHO staff were aware of the serious allegations in May 2019, but nothing was done about it until October of 2020—keep in mind this is the organisation that wants to micromanage the sexual education of Western children.

The investigation found that at least 83 victims said they had been lured into sex work, with the investigation finding that people were promised jobs in exchange for sexual relationships during a time of extreme vulnerability. At least 29 pregnancies resulted from this abuse.

"How many times do I have to speak before (the doctors) at WHO responsible for the sexual abuse are punished? If WHO does not take radical measures, we will conclude that the organisation has been made rotten by rapists," said a Congolese woman, who worked at an Ebola clinic in north-eastern Congo, as reported by AP News.

It's not the first time the WHO or the UN has been caught up abusing the people it is charged with helping, with one the co-director of the AIDS-Free World saying, "The process itself is the opposite of justice. The UN is the only institution in the world that is allowed to investigate itself. The WHO's head handpicked experts to lead a commission to look into criminal allegations against the agency's personnel and senior officials."

Further, according to The New Humanitarian, the independent commission criticised the WHO "for a 'systematic tendency' to reject all reports of sexual exploitation and abuse unless they were made in writing."

Let us not forget that an Associated Press investigation from 2017 accused 100 United Nations peacekeepers of running a child sex ring in Haiti for a decade with over 2,000 complaints of sexual abuse made against UN peacekeepers.

Why would any nation allow an organisation accused of the institutionalised abuse of women and children in third-world nations to dictate sexual education for minors?

The United Nations and the World Health Organisation are the last places on Earth that we should be taking advice from regarding the health and prosperity of our children.

Pursuant to the World Health Organization Treaty (see Exhibit E) signed by President Harry S. Truman and ratified by Congress 6 January 1948, this treaty was ratified with expressly stipulated "understandings." The subsequent 2005 (also in Exhibit E) has one "reservation" and "three understandings" limiting its use in the United States of America. "Reservations" and "understandings" cannot be removed by the executive branch of the federal government. Such unilateral actions by the president of the United States demonstrate yet another attempt at intentional, knowing, and wilful actions that could conceal involvement of gain-of-function criminal research and weapon deployment, either accidental or intentional.

140. On 26 May 2023 the World Economic Forum published an article on its website authored by Reuters Geneva correspondent Emma Farge, titled "WHO Pandemic Treaty: What Is It and How Will It Save Lives in the Future?"[242] The article includes the following:

> Negotiations on new rules for dealing with pandemics are underway at the World Health Organization (WHO), with a target date of May 2024 for a legally binding agreement to be adopted by the U.N. health agency's 194 member countries.
>
> A new pact is a priority for WHO chief Tedros Adhanom Ghebreyesus who called it a "generational commitment that we will not go back to the old cycle of panic and neglect" at the U.N. agency's annual assembly. It seeks to shore up the world's defences against new pathogens following the COVID-19 pandemic that has killed nearly 7 million people. . . .
>
> However, the proposed pandemic treaty has come under fire on social media, mostly from right-wing critics warning it could lead to countries ceding authority to the WHO. The body strongly refutes this, stressing that governments are leading the negotiations and are free to reject the accord. . . .
>
> After five rounds of formal negotiations, the latest 208-page draft of the treaty still includes thousands of brackets, which mark areas of disagreement or undecided language, including over the definition of the word "pandemic." With so many member countries involved, securing agreement may be tricky.
>
> It is also not yet clear what happens if the measures are not followed. A co-chair of the talks said it would be preferable to have a peer-review process, rather than sanction non-compliant states.

141. On 5 September 2023, U.S. Right to Know published "Fauci Was Told of NIH Ties to Wuhan Lab's Novel Coronaviruses by January 2020," by Emily Kopp.[243] That article contains the following:

> In an email dated January 27, 2020, Fauci received talking points for a press availability that evening. An aide laid out in detail the research at the Wuhan lab funded by the National Institute of Allergy and Infectious Diseases—the institute Fauci steered for decades.
>
> Through the nonprofit EcoHealth Alliance, its president Peter Daszak and its contracted lab, the Wuhan Institute of Virology, NIAID had funded the discovery of 52 novel sarbecoviruses, or coronaviruses related to SARS—the species that SARS-CoV-2 belongs to.
>
> These included the closest known relative of the "nCoV," or novel coronavirus, quickly spreading around the globe.
>
> "EcoHealth group (Peter Daszak et al), has for years been among the biggest players in coronavirus work, also in collaboration with Ralph Baric, Ian Lipkin and others," wrote Fauci's chief of staff Greg Folkers.

The 27 January 2020 email from Greg Folkers serves as evidentiary confirmation that Dr. Anthony Fauci perjured himself several times before Congress by claiming he had no knowledge of gain-of-function research taking place at the Wuhan Institute of Virology. Dr. Fauci was briefed in detail about novel coronavirus gain-of-function research projects that were commissioned and funded by his organization, NIAID, via EcoHealth Alliance.

Discussion: SARS-CoV-2 Pandemic of 2019 (COVID-19)

The evidence shows that components of, and individuals within, the United States government have been actively funding and carrying out research on gain-of-function in violation of the 1975 Biological Weapons Convention (BWC) Treaty made pursuant to Article VI of the United States Constitution, 18 USC Section 175 Prohibitions with respect to biological weapons, the 1989 Biological Weapons Anti-Terrorism Act, the 1976 International Covenant on Civil and Political Rights (ICCPR) Treaty also made pursuant to Article VI of the United States Constitution, the 1964 Declaration of Helsinki, the 1979 Belmont Report, and the 1947 Nuremberg Code. This funding and research were not confined to the United States but were carried out in other laboratories, including but not limited to BSL4[244] in China and the Ukraine. Agencies purportedly in place to protect the American people conspired to produce these biological viral weapons and circumvented pauses in gain-of-function spending and research; and in so doing, they have not only violated the aforementioned laws, treaties, and statutes but also have committed national and international terrorism as defined previously.

The American people have submitted more than 10,000 indictment letters to the governors and attorneys general in all fifty states (https://10letters.org/) in addition to more than 73,400 signatures[245] on a complaint to the International Criminal Court (ICC)[246] submitted on behalf of the United Kingdom, Slovakia, France, and the Czech Republic to hold the defendants accountable for their criminal actions. The American people demand that those criminally responsible for the research, development, and eventual deployment of this bioweapon be held criminally accountable. The evidence shows that the defendants intentionally and knowingly conspired to obfuscate the discussion on the origin of the SARS-CoV-2 viruses and, in so doing, allowed a biological viral weapon to spread from country to country until it consumed the planet.

The evidence shows communications and efforts to collect viral and human DNA (genetic) data, made possible through the mass collection of nasal samples. These samples collected personal identifying information, in violation of HIPAA,[247] in addition to US and state constitutional citizens' rights. In the process, federal and state governments exceeded the limits previously respected by all nations since the Nazi Germany Nuremberg trials.

The evidence shows the defendants collectively operated in unison to fulfill these efforts and actions. They therefore met the criteria for a series of conspiratorial actions that included not only the development of biological viral weapons but also the distribution of genetic vaccines, which did not statistically demonstrate a reduction in COVID cases or related deaths. These individuals collectively worked together to increase their authority above and beyond US and state constitutional authority, including but not limited to forced vaccination; forced closure of schools, businesses and places of worship; punishment of those who did not follow their demands; coercion of media sources to accept the government edict in violation of the First Amendment[248] and arguably the Fourth Amendment to the US Constitution, including but not limited to the threat to enter homes, businesses, and places of worship, without court-authorized search warrants and coercion to force vaccination and nasal testing of individuals in their homes, businesses, and places of worship.

The evidence shows the defendants, in furtherance of these criminal violations of statutory, constitutional, and treaty law, intentionally and knowingly took steps in an attempt to conceal the evidence through a series of communications and actions designed to obstruct investigations into their criminal actions. The defendants are not the only individuals actively involved in these criminal actions, and it is clear that they have funded and actively engaged in Gain-of-function research in the United States, China, and Ukraine. The evidence also shows the defendants' involvement with the World Economic Forum (WEF) and the World Health Organization (WHO), which have demonstrated an increased intent to usurp US federal and state Cconstitutional authority and US sovereignty.

The consequences of their actions, in addition to clear violation of law and the potential worldwide economic and geopolitical conflicts resulting from their actions, have resulted in the deaths of more than 6.6 million people worldwide, greater than the number of deaths associated with the Nazi concentration camps of the 1940s. In the United States, almost 1.2 million deaths have resulted from COVID, equaling the number of deaths the United States military has sustained[249] since 1776. There have been more than 1.6 million, with estimates up to 2.5 million[250] recognized vaccine adverse events—more than one hundred thousand more injuries than the US military has sustained since 1776.[251] There have been more than 34,516 reported deaths associated with the use of the mRNA and DNA genetic vaccines, more losses than the US military has suffered in all but our five most costliest wars and far exceeding the fifty-three deaths that resulted in the withdrawal of the swine flu vaccine of the 1970s.

A review of the evidence shows that the defendants not only intentionally, knowingly, willfully, and recklessly carried out this gain-of-function biological viral weapon development program but also recognized it was the work of their research and intentionally, knowingly, willfully, and recklessly failed to make people aware of the risks and potential treatments that could have saved these men, women, children, fathers, mothers, teachers, firefighters, first responders, doctors, nurses, and our fellow Americans.

Some of the defendants have clearly perjured themselves, including but not limited to their responses in Senate hearings. Their behavior, admissible under the Federal Rules of Evidence (FRE) Rule 406[252] may be admitted into evidence. This behavior, as defined by the previously noted bioterrorism statues, passed by the US Congress and enacted into law, is by definition, terrorism—national and international. Accordingly, they are criminally responsible for the death and maiming of millions of American citizens and people around the world, far greater than that accomplished by Nazi Germany. The damage to the United States of America, our economy, our political and civil rights, our ability to provide for and protect our families and the people we love has been irreparable and cannot be compensated by any amount of money. We will never recover those lives, moments, relationships, friendships, or opportunities.

Pertinent to this case, each federally elected official takes an oath to uphold and defend the Constitution of the United States of America,

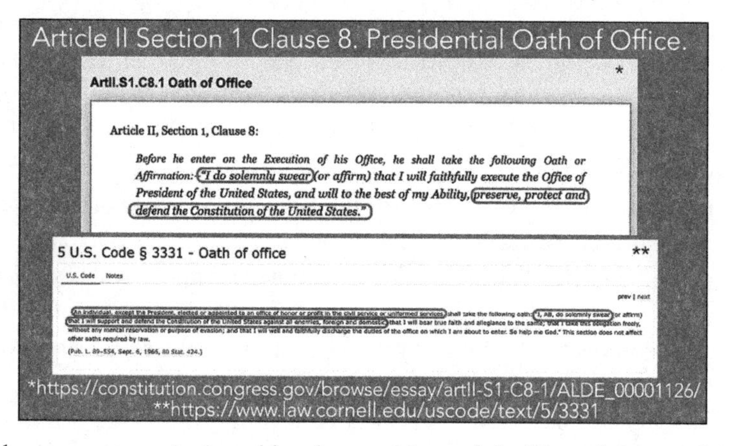

Article II Section 1 Clause 8. Presidential Oath of Office.

ArtII.S1.C8.1 Oath of Office *

Article II, Section 1, Clause 8:

Before he enter on the Execution of his Office, he shall take the following Oath or Affirmation: "I do solemnly swear (or affirm) that I will faithfully execute the Office of President of the United States, and will to the best of my Ability, preserve, protect and defend the Constitution of the United States."

5 U.S. Code § 3331 - Oath of office **

U.S. Code Notes

prev | next

An individual, except the President, elected or appointed to an office of honor or profit in the civil service or uniformed services, shall take the following oath: "I, AB, do solemnly swear (or affirm) that I will support and defend the Constitution of the United States against all enemies, foreign and domestic; that I will bear true faith and allegiance to the same; that I take this obligation freely, without any mental reservation or purpose of evasion; and that I will well and faithfully discharge the duties of the office on which I am about to enter. So help me God." This section does not affect other oaths required by law.

(Pub. L. 89–554, Sept. 6, 1966, 80 Stat. 424.)

*https://constitution.congress.gov/browse/essay/artII-S1-C8-1/ALDE_00001126/
**https://www.law.cornell.edu/uscode/text/5/3331

and, once a treaty is signed by the president of the United States of America (Article II, executive branch) and ratified by Congress (Article I, legislative branch), then, absent changes to a treaty by reservation(s) or understanding(s), it becomes the supreme law of the land, as established by the Supreme Court (Article III, judicial branch).

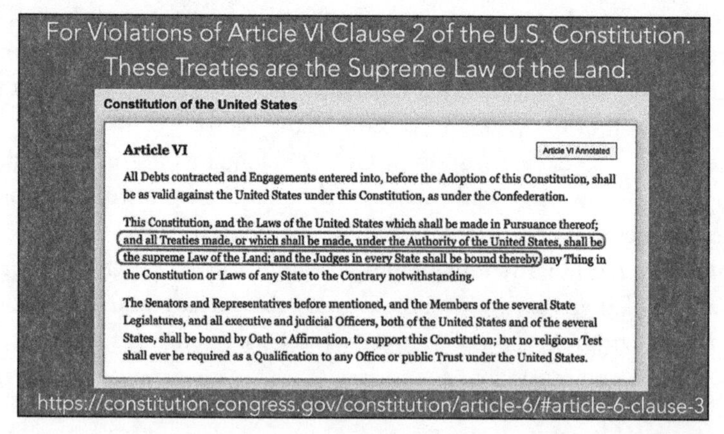

For Violations of Article VI Clause 2 of the U.S. Constitution.
These Treaties are the Supreme Law of the Land.

Constitution of the United States

Article VI Article VI Annotated

All Debts contracted and Engagements entered into, before the Adoption of this Constitution, shall be as valid against the United States under this Constitution, as under the Confederation.

This Constitution, and the Laws of the United States which shall be made in Pursuance thereof; and all Treaties made, or which shall be made, under the Authority of the United States, shall be the supreme Law of the Land; and the Judges in every State shall be bound thereby, any Thing in the Constitution or Laws of any State to the Contrary notwithstanding.

The Senators and Representatives before mentioned, and the Members of the several State Legislatures, and all executive and judicial Officers, both of the United States and of the several States, shall be bound by Oath or Affirmation, to support this Constitution; but no religious Test shall ever be required as a Qualification to any Office or public Trust under the United States.

https://constitution.congress.gov/constitution/article-6/#article-6-clause-3

Any federal or state official having taken a similar oath who then violates a treaty has committed the criminal felony of treason.

18 U.S. Code § 2381 - Treason

U.S. Code

Whoever, owing allegiance to the United States, levies war against them or adheres to their enemies, giving them aid and comfort within the United States or elsewhere, is guilty of treason and shall suffer death, or shall be imprisoned not less than five years and fined under this title but not less than $10,000; and shall be incapable of holding any office under the United States.

(June 25, 1948, ch. 645, 62 Stat. 807; Pub. L. 103–322, title XXXIII, § 330016(2)(J), Sept. 13, 1994, 108 Stat. 2148.)

Accordingly, we call for the immediate indictment, prosecution, and sentencing of the aforementioned defendants, including but not limited to the crimes stated. If found guilty, we call for imposition of life imprisonment or the death penalty of the defendants and their accomplices.

We additionally call for the immediate cessation of any and all genetic research currently being funded or conducted that could be used for gain-of-function, eugenics, or potentially related research or implementation programs, and we call for the immediate establishment of a team of experts to review and recommend further restrictions and oversight to said funding and research, outside the control of the military-industrial complex. Finally, we call for the immediate hearing by the ICC of the cases presented by the United Kingdom, Slovakia, France, and Czech Republic as noted immediately as follows.

By attorneys from **FOUR** countries
UK [Melinda Mayne and Kaira S. McCallum; 143/21],
Slovakia [Peter Weis, Marica Pirosikova,
Erik Schmidt; 133/21],
France [Patrice Lepiller, Raphael Cohen; 271/21], and
The Czech Republic [Tomas Nielsen; 326/21],
calling upon The INTERNATIONAL CRIMINAL COURT (ICC)
to hold the perpetrators criminally accountable for Crimes
Against Humanity.

On 12 August 2021
Affidavits under Oath were submitted by

Kevin W. McCairn, PhD;
Luc Montagnier, PhD; and
Richard M Fleming, PhD, MD, JD,

A **letter of support** was provided by Nazi Concentration
Camp survivors Moshe Brown, Hillel Handler, and Vera
Sharav on 20 September 2021.

EXHIBIT A: Evidence of Gain-of-Function Alteration

EXHIBIT B: Is the Vaccine Efficacy Statistically Significant?

Is there a statistically significant difference between vaccinated and unvaccinated individuals when asking if they later get diagnosed with or die from COVID?

When scientist-physicians want to know if treatments, including the Pfizer, Moderna, and Janssen drug vaccine biologics, work, we must do more than merely look at the numbers. We must statistically compare the results of those treated (vaccinated) with those not treated (unvaccinated) to determine if differences are significant or meaningless. This avoids giving unbeneficial treatments to people.

A variety of statistical methods can be used, and the selection of the correct statistical analysis is determined by the type of research conducted, taking into account what type of numbers we are using. Ordinate numbers are those with units attached (e.g., inches, pounds, milligrams, liter, millimeters, and so forth). Alternatively, when we count or identify something, we use cardinal or nominal numbers to define how many are in that group. For example, a forty-two-year-old Caucasian female and a twenty-seven-year-old Hispanic male and a fifty-nine-year-old Asian female all have the same value—one for each of them.

Cardinal or nominal numbers can be statistically compared using either correlation, which does not provide cause and effect, or chi-square analysis. Chi-square analysis allows the statistical comparison between treatments by asking a fundamental question. If there is no difference (null hypothesis always applied to scientific research) between the two groups, then the expected outcomes (no difference between groups) should match the observed outcomes from the study.

When the EUA documents were used for the statistical analysis of the Pfizer, Moderna, and Janssen drug vaccine biologics and chi-square analysis of the results published in those EUA documents was applied to the Pfizer vaccine as shown in the QR code slide graphics, there was no statistical difference between vaccinated and unvaccinated people diagnosed with having COVID-19. To be statistically different (a benefit for people being vaccinated) the p (probability value) must be less than or equal to five times per hundred people. This is the scientific definition of statistical benefit and is written as "$p < 0.05$." In the graphic the p-value was 0.224418 and is not statistically significant. That is, there is no statistical difference in the number of people diagnosed with COVID who were vaccinated when compared with the nonvaccinated group of people.

When this same approach is taken to the Moderna EUA results, there is no statistical difference between vaccinated and nonvaccinated individuals, with $p = 0.138706$.

When this same approach is used to determine if there is a statistically significant reduction in COVID cases among people vaccinated with the Janssen vaccine, the fourteen-day data shows a statistical benefit with p = 0.020258.

However, two weeks later at twenty-eight days, that benefit was gone with p = 0.138761.

In brief, although there are differences between vaccinated and unvaccinated individuals who were diagnosed as having COVID-19, the differences were not statistically significant.

EXHIBIT C: InflammoThrombotic Response and Disease

[First Diagram] 1994 American Heart Association Presentation;
[Second Diagram] Fleming RM. Chapter 64. The Pathogenesis of Vascular Disease. Textbook of Angiology. John C. Chang Editor, Springer-Verlag New York, NY. 1999, pp. 787-798. doi:10.1007/978-1-4612-1190-7_64.

EXHIBIT D:

InflammoThrombotic, Priogenic, Sudden Cardiac Death, Miscarriages, and Cancer Pathways Promoted by Various Factors, Including Poor Dietary and Lifestyle Practices, SARS-CoV-2 Viruses, and the Genetic Vaccines

EXHIBIT E:

Actual WHO Resolution and 2005 International Health Regulations

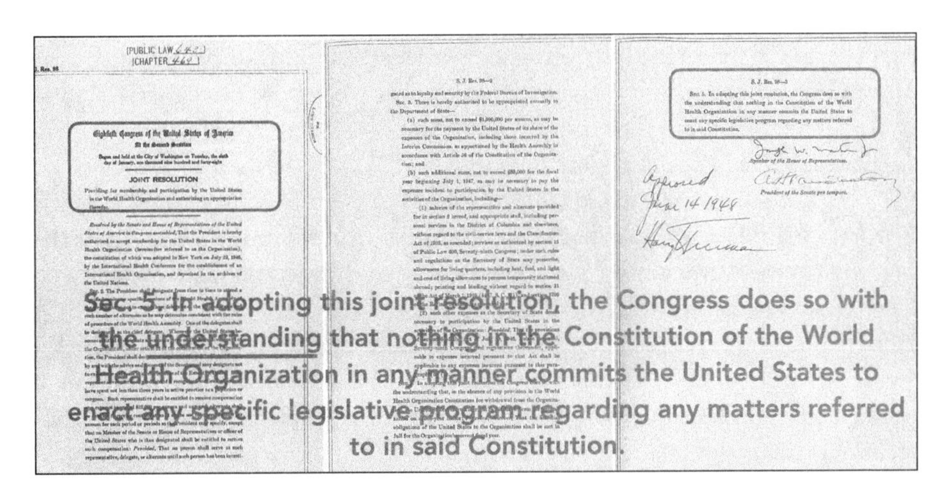

[PUBLIC LAW 643]
[CHAPTER 469]

Eightieth Congress of the United States of America

JOINT RESOLUTION

Sec. 5. In adopting this joint resolution, the Congress does so with the understanding that nothing in the Constitution of the World Health Organization in any manner commits the United States to enact any specific legislative program regarding any matters referred to in said Constitution.

Approved June 14 1948

CHAPTER 7

A Tale of Two Cities

In 1859 Charles Dickens released his now-famous book *A Tale of Two Cities.* He begins his book by describing the cities of London and Paris leading up to the French Revolution—5 May 1789 to 9 November 1799:

> It was the best of times, it was the worst of times, it was the age of wisdom, it was the age of foolishness, it was the epoch of belief, it was the epoch of incredulity, it was the season of Light, it was the season of Darkness, it was the spring of hope, it was the winter of despair, we had everything before us, we had nothing before us, we were all going direct to Heaven, we were all going direct the other way.

Does Dicken's description of the French Revolution sound familiar to you? It should, for the behavior Dickens wrote about centuries ago, including that of people like Maximilien Robespierre, whom we looked at in chapter 3, is the same type we are dealing with today: behaviors that have polarized and divided people more than unified them and which have been employed to obfuscate what happened for the gain of the manipulators.

The last chapter took a sobering look at what a number of people intentionally and knowingly did over many decades. Their behavior demonstrated their character and habits, as they expanded both a biological weapons program and a eugenics-directed gene-changing program.

One of the best ways to hide information is to put it in plain sight. The words of Klaus Schwab and others have demonstrated their commitment to direct the narrative to the world as they see it, the type of world we should all embrace.

"You'll Own Nothing. And You'll Be Happy."[1]

Another method for controlling the narrative, as we saw in the Wargames SPARS, Event 201 and the NTI, was to allow an overload of information that the general public could not sort out. Therefore, it is easy to control the narrative by controlling mainstream media.

Countering the Chaos

To get beyond this chaotic confusion resulting from those who are intentionally misdirecting people and the mental masturbatory efforts of others who desire to be seen as important requires a cold hard objective analysis, especially in the face of criticism. JFK knew that type of analysis would be required if we were to save the country and ourselves.

It is my belief that a jury listening to the case against these criminals, laid out in the last chapter, will not be confused if the testimony provided by the experts recognizes truths on both sides and the lies promulgated by both sides. Such testimony will allow a jury to see that the experts are not biased, are not testifying for attention, or money, or power but are on the witness stand to expose the truth, the whole truth, and nothing but the truth—for *We the People*. Those experts include:[2]

Richard M. Fleming, PhD, MD, JD (Physicist, Nuclear Cardiologist, Attorney)	Andrew G. Huff, PhD (Infectious Disease Epidemiologist, Former VP, EcoHealth Alliance)
Kevin W. McCairn, PhD (Neuroscientist)	Jennifer Bridges, RN (Former Methodist Hospital RN)
Charles Rixey, MA (US Marines WMD Expert)	Unnamed Physician (Medical Doctor, US Military)
Johanna Deinert, MD (Medical Doctor)	Unnamed Rockefeller Foundation Whistleblower

This book must do the same thing. It must present the truth, the whole truth, and nothing but the truth.

Unfortunately, during the last several years, I have had to address misinformation coming from both sides of the narrative. Every time I do, I receive ad hominem attacks, which pale in the face of the truth.[3] To buttress such attacks is not easy. But the failure of these people to address the facts and to then attack someone using ad hominem fallacy[4] arguments repeatedly proves that we are over the target. Their habitual behavior reminds all of us that these people have no

defense for what they have done and continue to do. I'm also reminded of what my mother used to tell me: "If people were attacking me, they were leaving other people alone."

During the winter of 2021 numerous pieces of misinformation came from various sources, resulting in confusion and the polarization of people. Efforts to address some of this misinformation simply resulted in more misinformation and ad hominem attacks. Due to the confusion and harm caused, at the request of the event organizers, I agreed to put together a presentation discussing this "Jurassic Park of Misinformation."

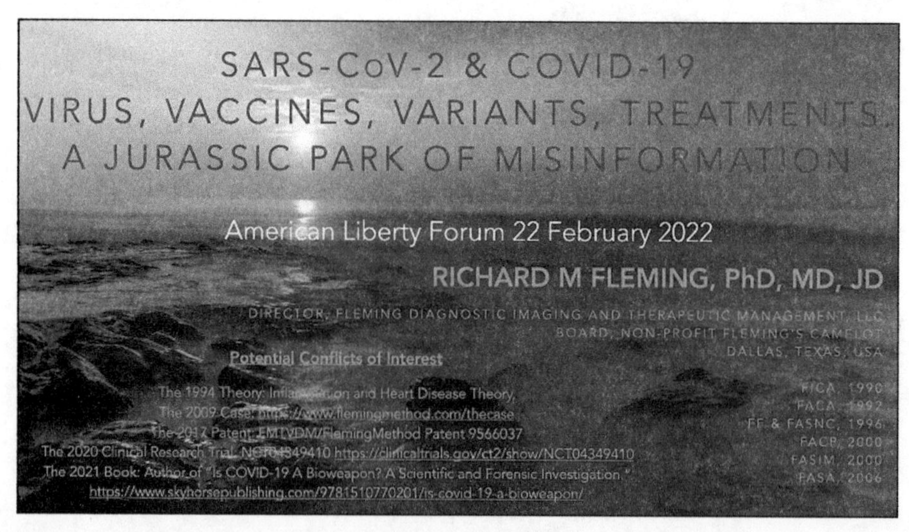

I had the privilege of dedicating that presentation to my friend and colleague Professor Luc Montagnier, who passed on 8 February 2022. Professor Luc and I had known each other for years, and we had been working on the origins of SARS-CoV-2 viruses (COVID), the genetic vaccines, and the effect of certain treatments, including protease inhibitors like Paxlovid, which we saw as problematic given our information on the effect such drugs had on HIV.

A Jurassic Park of Misinformation. What's Real, What Isn't

1. Not everyone who tells you he or she is here for you really is. Some are not telling you the truth. Some have been bought and paid for. Telling a patient about financial conflict of interest is part of providing you informed consent.

When you listen to someone—anyone—I encourage you to keep a certain amount of healthy skepticism. Over time, you will know whether the person has a vested interest in what he or she is telling you. Adopt the heuristic method.

The Heuristic Method
Question everything, everyone—even me.
Demand real proof, not opinions.
Demand to know about any conflicts of interest.
Demand to know who's connected to whom and to what.

As simple as the method may seem, it can be challenging to employ it when in unfamiliar territory with an unfamiliar topic. I think it's fairly safe to say that many people are not that familiar or comfortable with viruses, vaccines, and drug treatments, even today.

Therefore, people look to those they think they can trust for answers. This is true for both sides of the discussions occurring today. If you can see it's a problem for someone on the other side of an argument, please recognize it might also be true for your side of the argument.

Many people profit by promoting certain drugs, genetic vaccines, supplements, or treatments, even if they don't use the word *treatment* or *cure*. There is nothing wrong with people making money for diagnosing or treating someone. However, when the person or the company is misleading you and profiting from that, that, I argue, is wrong. For example, telling people that isotopes used for imaging don't redistribute once injected is wrong in that way.

Another example is when physicians receive money from pharmaceutical companies and then promote the drugs that those companies make. It's particularly telling if the doctor or other health-care provider keeps changing the story about what drugs work and what ones don't. The practitioner might originally tell you drug A is great, only to later say it isn't so great and you should use drug

B. Later, you may be told drug B isn't so great and you should use drug C, and so on.

You might want to check a particular website if you are interested in knowing if your doctor or health-care provider is receiving money from pharmaceutical companies. Curiously, it was put together by Centers for Medicare and Medicaid, which reminds me that even a broken watch is right twice a day.

Just go to the website and enter the name of the person, hospital, or company, and you can see who has been paying that entity money and for what. I know some doctors like myself who have received nothing, whereas other doctors have received millions—I'll let you figure out whom.

The website is https://openpaymentsdata.cms.gov/search The site will look like this:

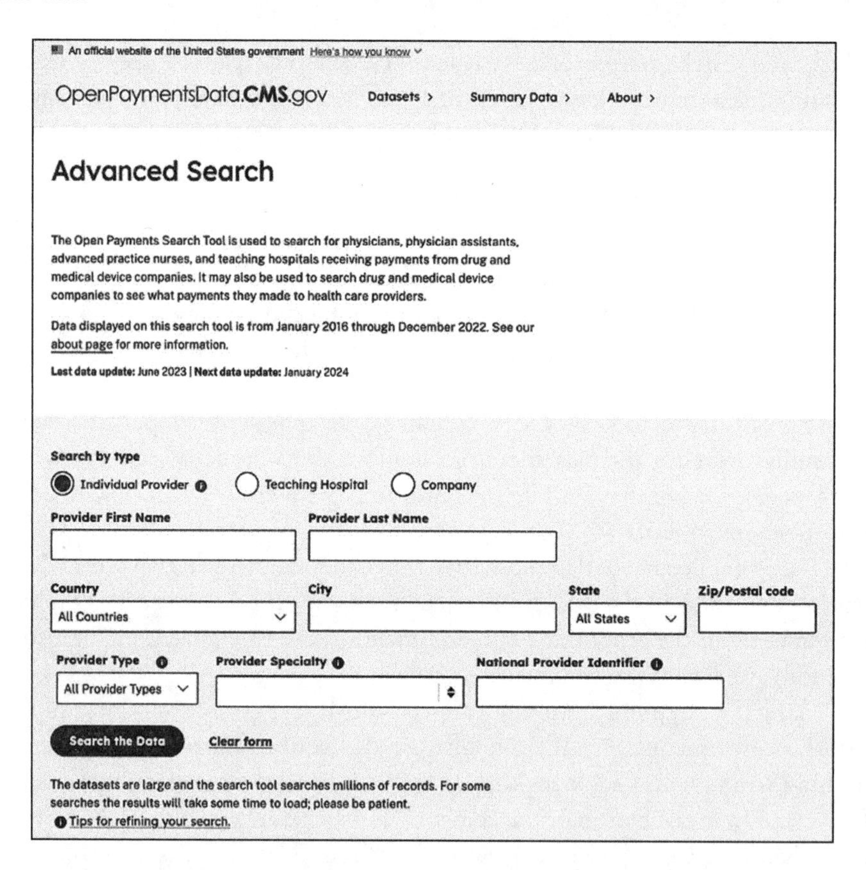

A doctor can't serve two masters. The potential for conflict of interest is too great, and informed consent requires the doctor or health-care provider to be honest with the patient and declare that the drug prescribed is made by the same company paying that individual money.

One such example is shown as follows; the physician has been paid more than $2 million over seven years and is promoting drugs made by AbbVie. One specific drug was intended for treating COVID and is now associated with rebound cases. But it is heavily marketed, so you would immediately think of it if you are diagnosed with COVID.

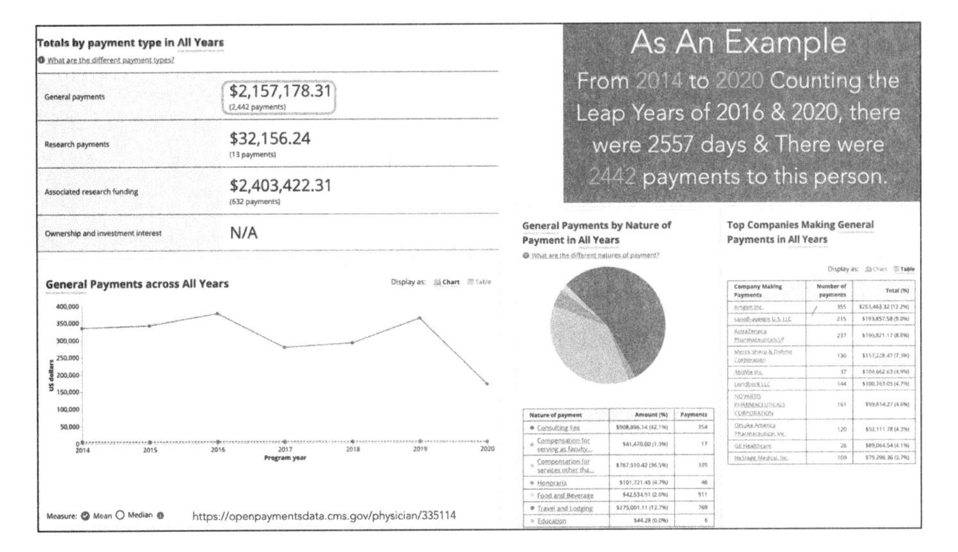

2. Viruses exist, and SARS-CoV-2 has been isolated.

Earlier in this book we discussed the fact that we do not live in the 1850s or even the late 1910s, when the Bohr model of the atom was still being used; or before genetic sequencing and the Sanger Sequencing[5] method, which made it possible to determine the genetic code sequences making up viruses or people; or before electron microscopy made it possible to visualize something as small as a virus.

I understand the temptation for some people to believe that viruses don't exist. If they can convince themselves that viruses aren't real, then there's really nothing to worry about,[6] except for that nasty problem of all the dead and maimed people. It's easier to pretend healthy people suddenly became sick because their body had some type of problem rather than dealing with the fact that a virus passing from one person to another caused the problem.

Viruses are very real, very communicable, and manipulating one into a bioweapon is very illegal. SARS-CoV-2 viruses have been anatomically seen via electron microscopy and genetically defined and entered into the GenBank.[7] Their changes have allowed the tracking of variants that, as we have discussed, appear to have been driven by the use of genetic vaccines, in the same way bacterial resistance to antibiotics occurs with the overuse of antibiotics.

You can say you don't believe in gravity because you can't see it or because no one has been able to successfully "discover" what causes gravity. But if you walk off the top of a twenty-story building, you can only say that for a few seconds—until you experience the life-changing effect of gravity. More on the visualization of SARS-CoV-2 and the genetic sequences in the QR slides.

3. The desire to find a cure doesn't mean you have found one, and just because Big Pharma or the government says you haven't found something that works doesn't mean they are right.

There has been a lot of frustration between members of the medical community and the government, over who treats patients. The FDA, HHS, and CDC do not treat patients; they just make statements. Physicians treat patients. However, just because they do does not always mean they know what they are talking about.

One of the major problems has been the use of in silico (on a computer) models. Everyone has probably had the experience of putting a key in a lock, only to find out it's the wrong key. This is one of the problems with computer modeling and one of the reasons why research and development, which we are going to talk about in the next chapter, can be so expensive. Just because something looks like it should work doesn't mean it will.

In reality, there is more to finding a drug that works (key) for a certain receptor (lock) in a cell that will have the desired effect (opening the door) than just seeing if the key can be put into the lock. As shown in the QR slides, fewer than 10 percent of the drugs expected to work actually do.

Physicians who have little or no research training are not familiar with the steps required to determine if what they think they are seeing is what is really happening. There is a big difference between cause and effect and correlation. Correlation simply means they saw something, but it doesn't explain why.

A classic example of correlation misleading people was published in 2000 by researchers who showed that "storks deliver babies." The correlation was between storks building nests on chimneys and human babies. The correlation was almost perfect, but, I think you would agree, not cause and effect. When analyzed further, it turns out that when Scandinavian and European winters were colder, more storks built nests on the chimneys of homes to keep warm and have their young. At the same time, during colder winters, people tended to stay inside, resulting in greater tendency to perform indoor activities, some of which led to babies.

In addition to seeing something that isn't really there, sometimes doctors fail to see what's right in front of their eyes. A classic example of this was published in 2013, after researchers asked board-certified radiologists, trained to find disease on computed tomography (CT) scans, if they saw "anything" abnormal on the images. As you can see in the last QR link, the blue circles tracked the radiologists' eye movements, showing they looked right at the gorilla images, but only four of the twenty-four reported seeing the gorillas. When the general public was given the same test, nobody saw the gorillas. People see what they expect to see.[8]

Drugs have been approved for SARS-CoV-2 that were not as successful as the powers that be wanted them to be. Other drugs were promoted by physicians without actual scientific, cause-effect proof that what the doctor's thought was working was the result of the drugs. This resulted in conflicts that reflected sadly on the political, legal, and medical systems while causing rifts between physicians and patients.

Just because someone has received a Nobel Prize doesn't mean that work now miraculously solves or applies to everything. Failure of a drug to have any effect on SARS-CoV-2 doesn't mean the Nobel Prize work wasn't valid; it just means the drug doesn't work on everything. This is science, not magic.

William Shockley Jr. was a physicist who received the 1956 Nobel Prize in Physics for his work in helping to develop the transistor in 1947. After his Nobel, in the 1960s and 1970s, Shockley proposed that black people were intellectually inferior to white people. He also believed that intellectually inferior people having children would have a "dysgenic" effect, resulting in the intellectual decline of humanity.[9] Shockley proposed that people with an intelligence quotient (IQ) below 100 should be paid to undergo voluntary sterilization.

Shockley's work on the transistor did not qualify him to be a geneticist. My psychology mentor and graduate advisor, Gordon M. Harrington, PhD, from Harvard and Yale and director of the genetics rat lab I ran for a year as a graduate student, was a recognized psychology and genetics expert. In fact, our lab did work on this very topic of IQ and testing. As part of my graduate thesis in experimental psychology, I was required to discuss and address Shockley's erroneous eugenics ideas.[10]

* * *

The power and influence of Big Pharma and its control over the state and federal governments cannot be overemphasized. It suppresses medications that might work but are no longer patent protected, which would allow less expensive generics to be used in place of brand names. In other instances, misrepresenting the effectiveness of a drug is equally appalling, as it can result in both the misdiagnosis and mistreatment of critical disease and potential loss of life.

The overexuberance about remdesivir (a measured benefit less than random chance) and Paxlovid (multiple reported cases of rebound infections) in treating SARS-CoV-2 and COVID infections are two examples. Another is the issue of sestamibi redistribution, or as the pharmaceutical company would like physicians to believe, the lack of sestamibi redistribution (movement inside the body), which allows Big Pharma to promote the need for two doses of the isotope instead of one[11]—thus, more sales, more profit.

In the absence of being able to measure[12] (quantify) what is happening, it can be impossible to see the critical redistribution (think gorillas on a chest CT) of this isotope and find life-threatening heart disease, like that missed in the case of Tim Russert of *Meet the Press*.[13] Had my warnings about these errors been headed, Mr. Russert might still be alive today. Anyone who tells you something injected into the body (isotopes, vaccines, and such) sticks and doesn't move around is either selling you something, on something, or both.

Not every medical journal has maintained the level of professionalism expected of a scientific journal, as shown in the following publication on 18 February 2020, resulting in my resigning as an external reviewer for *The Lancet*.[14]

Correspondence

Published Online February 18, 2020 https://doi.org/10.1016/S0140-6736(20)30418-9

For the Chinese translation see Online for appendix

Statement in support of the scientists, public health professionals, and medical professionals of China combatting COVID-19

We are public health scientists who have closely followed the emergence of 2019 novel coronavirus disease (COVID-19) and are deeply concerned about its impact on global health and wellbeing. We have watched as the scientists, public health professionals, and medical professionals of China, in particular, have worked diligently and effectively to rapidly identify the pathogen behind this outbreak, put in place significant measures to reduce its impact, and share their results transparently with the global health community. This effort has been remarkable.

We sign this statement in solidarity with all scientists and health professionals in China who continue to save lives and protect global health during the challenge of the COVID-19 outbreak. We are all in this together, with our Chinese counterparts in the forefront, against this new viral threat.

The rapid, open, and transparent sharing of data on this outbreak is now being threatened by rumours and misinformation around its origins. We stand together to strongly condemn conspiracy theories suggesting that COVID-19 does not have a natural origin. Scientists from multiple countries have published and analysed genomes of the causative agent, severe acute respiratory syndrome coronavirus 2 (SARS-CoV-2),[1] and they overwhelmingly conclude that this coronavirus originated in wildlife,[2-10] as have so many other emerging pathogens.[11,12] This is further supported by a letter from the presidents of the US National Academies of Science, Engineering, and Medicine[13] and by the scientific communities they represent. Conspiracy theories do nothing but create fear, rumours, and prejudice that jeopardise our global collaboration in the fight against this virus. We support the call from the Director-General of WHO to promote scientific evidence and unity over misinformation and conjecture.[14] We want you, the science and health professionals of China, to know that we stand with you in your fight against this virus.

We invite others to join us in supporting the scientists, public health professionals, and medical professionals of Wuhan and across China. Stand with our colleagues on the frontline!

We speak in one voice. To add your support for this statement, sign our letter online. LM is editor of ProMED-mail. We declare no competing interests.

To register your support see http://chng.it/SDpTBgKf

Charles Calisher, Dennis Carroll, Rita Colwell, Ronald B Corley, Peter Daszak, Christian Drosten, Luis Enjuanes, Jeremy Farrar, Hume Field, Josie Golding, Alexander Gorbalenya, Bart Haagmans, James M Hughes, William B Karesh, Gerald T Keusch, Sai Kit Lam, Juan Lubroth, John S Mackenzie, Larry Madoff, Jonna Mazet, Peter Palese, Stanley Perlman, Leo Poon, Bernard Roizman, Linda Saif, Kanta Subbarao, Mike Turner

COVID19statement@gmail.com

Colorado State University, Fort Collins, CO, USA (CC); Scowcroft Institute of International Affairs, Texas A&M, College Station, TX, USA (DC); University of Maryland, College Park, MD, USA (RC); NEIDL Institute (RBC), Boston University (GTK), Boston, MA, USA; EcoHealth Alliance, New York, NY, USA (PD); Charité – Universitatsmedizin Berlin, Berlin, Germany (CD); National Center of Biotechnology, Madrid, Spain (LE); The Wellcome Trust, London, UK (JF, JG, MT); School of Veterinary Science, The University of Queensland, Brisbane, QLD, Australia (HF); Leiden University Medical Center, Leiden, Netherlands (AG); Erasmus Medical Center, Rotterdam, Netherlands (BH); Emory University, Atlanta, GA, USA (JMH); World Organization for Animal Health (OIE) Working Group on Wildlife, New York, NY, USA (WBK); University of Malaya, Kuala Lumpur, Malaysia (SKL); Food and Agriculture Organization of the United Nations, Rome, Italy (JL); Curtin University, Perth, WA, Australia (JSM); Massachusetts Medical School, Worcester, MA, USA (LM); University of California at Davis, Davis, CA, USA (JM); Department of Microbiology, Icahn School of Medicine, Mt Sinai Hospital, New York, NY, USA (PP); University of Iowa, Roy J and Lucille A Carver College of Medicine, Iowa City, IA, USA (SP); The University of Hong Kong, Hong Kong (LP); University of Chicago, Chigaco, IL, USA (BR); The Ohio State University, Columbus, OH, USA (LS); and The University of Melbourne, Melboune, VIC, Australia (KS)

1. Gorbalenya A E, Baker SC, Baric RS, et al. Severe acute respiratory syndrome-related coronavirus: the species and its viruses—a statement of the Coronavirus Study Group. bioRxiv 2020; published online Feb 11. DOI:2020.02.07.937862 (preprint).
2. Zhou P, Yang X-L, Wang X-G, et al. A pneumonia outbreak associated with a new coronavirus of probable bat origin. *Nature* 2020; published online Feb 3. DOI:10.1038/s41586-020-2012-7.
3. Lu R, Zhao X, Li J, et al. Genomic characterisation and epidemiology of 2019 novel coronavirus: implications for virus origins and receptor binding. *Lancet* 2020; published online Jan 30. https://doi.org/10.1016/S0140-6736(20)30251-8.
4. Zhu N, Zhang D, Wang W, et al. A novel coronavirus from patients with pneumonia in China, 2019. *NEJM* 2020; published online Jan 24. DOI:10.1056/NEJMoa2001017.
5. Ren L, Wang Y-M, Wu Z-Q, et al. Identification of a novel coronavirus causing severe pneumonia in humans: a descriptive study. *Chin Med J* 2020; published online Feb 11. DOI:10.1097/CM9.0000000000000722.
6. Paraskevis D, Kostaki EG, Magiorkinis G, Panayiotakopoulos G, Tsiodras S. Full-genome evolutionary analysis of the novel corona virus (2019-nCoV) rejects the hypothesis of emergence as a result of a recent recombination event. *Infect Genet Evol* 2020; published online Jan 29. DOI:10.1016/j.meegid.2020.104212.
7. Benvenuto D, Giovanetti M, Ciccozzi A, Spoto S, Angeletti S, Ciccozzi M. The 2019-new coronavirus epidemic: evidence for virus evolution. *J Med Virol* 2020; published online Jan 29. DOI:10.1002/jmv.25688.
8. Wan Y, Shang J, Graham R, Baric RS, Li F. Receptor recognition by novel coronavirus from Wuhan: an analysis based on decade-long structural studies of SARS. *J Virol* 2020; published online Jan 29. DOI:10.1128/JVI.00127-20.
9. US Center for Disease Control and Prevention. Coronavirus disease 2019 (COVID-19) situation summary. Feb 16, 2020. https://www.cdc.gov/coronavirus/2019-nCoV/summary.html (accessed Feb 8, 2020).
10. Andersen KG, Rambaut A, Lipkin WI, Holmes EC, Garry RF. The proximal origin of SARS-CoV-2. Feb 16, 2020. http://virological.org/t/the-proximal-origin-of-sars-cov-2/398. (accessed Feb 17, 2020).
11. Bengis R, Leighton F, Fischer J, Artois M, Morner T, Tate C. The role of wildlife in emerging and re-emerging zoonoses. *Rev Sci Tech* 2004; 23: 497–512.
12. Woolhouse ME, Gowtage-Sequeria S. Host range and emerging and reemerging pathogens. *Emerg Infect Dis* 2005; 11: 1842–47.
13. NASEM. The National Academies of Science Engineering and Medicine of the USA. NAS, NAE, and NAM presidents' letter to the White House Office of Science and Technology Policy. Feb 6, 2020. https://www.nationalacademies.org/includes/NASEM%20Response%20to%20OSTP%20re%20Coronavirus_February%206,%202020.pdf (accessed Feb 7, 2020).

For the SARS-CoV-2 genome analysis see https://www.gisaid.org/epiflu-applications/next-betacov-app/

Submissions should be made via our electronic submission system at http://ees.elsevier.com/thelancet/

The Lancet has since attempted some damage control, going so far as to report on the tendency for the WHO to be overly influenced by WHO-China[15] and the FBI's concern over SARS-CoV-2 possible laboratory origin. The FBI agrees with the laboratory hypothesis with "moderate confidence."[16]

Despite some journals and politicians failing to live up to expected standards, Big Pharma's efforts to control all the journals and politicians had failed, at least as of 2010. In June 2010, despite efforts to tarnish my reputation, the federal government's only peer-reviewed medical journal of the Veterans Affairs (VA), DoD, and Public Health Service (PHS), the *Federal Practitioner*, not only published my research—showing how my patented method correctly found critical heart disease in veterans by measuring the redistribution of sestamibi—but ran the article as the cover story.[17]

Senator Charles Grassley cannot be bought and continues to represent his people. He responded to this publication and another article addressing the government practicing medicine without a license[18] and shared his appreciation for my efforts to improve medical education and patient care, including the care of our military men and women, both active duty and veterans. At the time, Senator Grassley was also the ranking member on the Senate Judiciary Committee.

Reply To:

135 Hart Senate Office Building
Washington, DC 20510-1501
(202) 224-3744
e-mail: grassley.senate.gov/contact.cfm

721 Federal Building
210 Walnut Street
Des Moines, IA 50309-2140
(515) 288-1145

150 1st Avenue NE
Suite 325
Cedar Rapids, IA 52401
(319) 363-6832

United States Senate

CHARLES E. GRASSLEY

WASHINGTON, DC 20510-1501

July 29, 2011

Reply To:

[illegible] Federal Courthouse Building
320 6th Street
Sioux City, IA 51101-1244
(712) 233-1860

210 Waterloo Building
531 Commercial Street
Waterloo, IA 50701-5497
(319) 232-6657

131 West 3rd Street
Suite 180
Davenport, IA 52801-1419
(563) 322-4331

307 Federal Building
8 South 6th Street
Council Bluffs, IA 51501-4204
(712) 322-7103

Dr. Richard M. Fleming

Dear Dr. Fleming:

Thank you for taking the time to contact me with your thoughts and concerns. As your Senator, it's important for me to hear from you.

I appreciate receiving the article that you wrote regarding improving medical care, reducing radiation exposure, and using other ways to diagnose heart disease. I read your article titled Practicing Medicine Without a License. Your article mentioned that the federal government's protocol to conduct a nuclear study of the heart, using a radiotracer, is to use the two injection rest-stress approach. The assertion in your article suggests depending on this method leads to about 100,000 deaths per year due to many diagnoses being missed. The vivid story about the veteran in the article truly drove home your passion regarding this issue. Your attempt to improve medical care as a provider and through education regarding improving testing methods should be commended. Please feel free to contact   if you would like to talk about your thoughts further. He is a member of my staff who handles health care issues and he can be reached at   As a senior member of the Senate Committee on Finance, I will continue to look for legislative vehicles to improve health care services and testing implemented and regulated by the federal government.

By sharing your views with me, Iowans play a vital role in this process. Hearing from you enables me to be a better U.S. Senator, and I very much appreciate the time you took to inform me of your concerns.

Sincerely,

Chuck

Charles E. Grassley
United States Senator

CEG/mh

I assume you get back to Iowa sometimes
"Once an Iowan always an Iowan"
CEG

RANKING MEMBER,
JUDICIARY

Committee Assignments:
AGRICULTURE
BUDGET
FINANCE

CO-CHAIRMAN,
INTERNATIONAL NARCOTICS
CONTROL CAUCUS

PRINTED ON RECYCLED PAPER

Science builds upon science, but efforts by Big Pharma, in addition to the federal and state governments, have adversely affected the fields of science and medicine. The influence of those who stand to profit by controlling the narrative must be constantly monitored and addressed to prevent their lust for money and power from controlling how science and medicine is practiced and to prevent the resulting inherent danger.

4. Getting vaccinated doesn't mean you won't get infected or transmit the virus to someone else.

As a scientist and physician, I had to wonder why and anyone could believe or support the narrative that vaccination prevents infection or transmission of a virus from person to person. When effective, vaccination merely shortens your response time. As we discussed in chapter 2 and shown in appendix D, vaccines are given to people to mimic an actual infection, to expose them to an infectious agent, like a virus.

The person then develops both an innate (T-cell) immune response and an adaptive or humoral (B-cell) antibody response. After the acute infection, or in vaccine exposure, the body produces memory cells of both the T- and B-cell variety. The memory cells are critical; they carry the memory and instructions for how to make the T- and B-cell responses when you actually become infected. In other words, unless you become infected, there is no potential benefit to the vaccine. The benefit is a more rapid response to the infection, with a shorter period required to gear up your immune system to fight the infection.

This does not prevent you from becoming infected, and because you can be, you can then transmit the infection, or the virus, to anyone you come into contact with. You simply should be able to mount your immune response faster.

All the gnashing of teeth and mental gymnastics that went on were completely unfounded scientifically and frankly embarrassing to listen to when physicians and scientists agreed with this. Is it any wonder that the following postings were seen in social media?

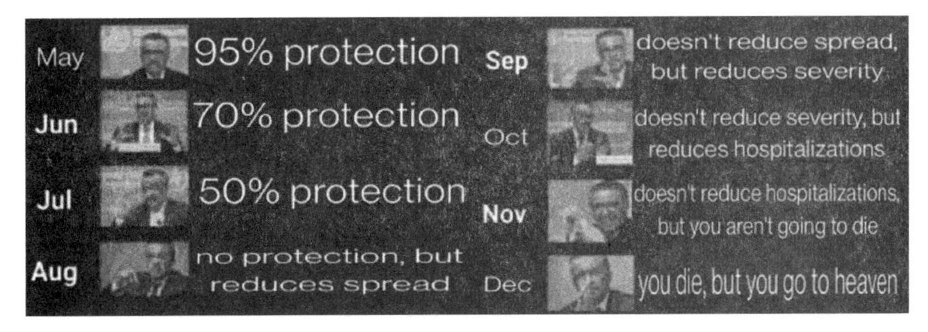

5. The cytokine storm only represents the wheels of the car, not the whole car.

As we have repeatedly discussed throughout this book, when your body is exposed to a pathogen, or infectious agent, such as a virus, your immune system responds. Then both the innate and adaptive or humoral system are activated, and you

have both inflammation and blood clotting (thrombosis), a.k.a. InflammoThrombotic response (ITR). As you can see from the QR 33(5) slides, part of this ITR includes *cytokines* or the cell (cyto) signaling proteins (kines).

The term *cytokine storm* was coined to address problems seen with patients who were receiving cancer treatments (immunotherapy) that proved to be toxic. Physicians were at first unprepared for the symptoms patients were experiencing, when T-cells were taken out of patients and changed to attack the cancer cells. The process of changing the normal T-cell to one that attacked the cancer (antigen) cells is called chimeric antigen receptor T-cell (CAR-T) therapy.

Chimeric changes like this are gain-of-function. When patients received these CAR-T treatments, they experienced nausea, fatigue, fever, heart problems, and problems with infection. The cause of these problems was the massive release of chemicals from the T-cells, which are only part of their immune system. Physicians were armed with a term to explain why patients were feeling these symptoms, but that response is not what happens when patients have an ITR.

You might find it interesting to plug the names of the physicians who use *cytokine storm* into the CMS database and see if they are receiving money from pharmaceutical companies, including those who make CAR-T therapy drugs. The website is https://openpaymentsdata.cms.gov/search.

If you still think the four wheels (cytokine storm) represents the whole car (ITR), I suggest you buy four tires for a spouse, son, or daughter and put the tires in the driveway. Then go inside and tell your loved one you bought that person a car. Let's see if your loved one agrees with what you said you did or if he or she looks at you like you've lost your mind.

6. There is a difference between immunity produced by focusing on part of a virus compared with immunity that targets the entire virus.

A number of people have allergic reactions to penicillin. In many instances they aren't actually allergic to penicillin per se but to the other proteins (antigens) that were present for many years before the process of making penicillin, and other drugs, was improved upon, removing these contaminants.[19]

In reviewing the EUA documents[20] provided to the FDA by Pfizer, Moderna, and Janssen, I see no actual data included in these EUAs for either T-cell or B-cell response, something extremely critical, considering the work of Karikó and Weissman, already discussed, that shows a reduction in interferon response when pseudouridine is used for these genetic vaccines. Interferon is critical for immunity against viruses.

Given the number of mutational changes occurring in the spike protein of the SARS-CoV-2 viruses (in addition to mutations also occurring in the envelope, membrane, and nucleocapsid parts of the virus), the more parts of the virus seen by the immune system, the greater the likelihood that an immune response (either T-cell or B-cell) from prior SARS-CoV-2 virus exposure will protect the body.

Other research has confirmed the findings of Karikó and Weissman, demonstrating measurable impairment of our immunity when reinjected with these genetic vaccines, as shown in the QR 33(6) slides. The research also revealed a blunting of our response to other vaccines, like the influenza vaccine, a reduced ability to make antibodies due to a decreased T-helper 2 cells (Th2), and a reduction in T-cells able to mount an innate immune response.

The repeated use of genetic vaccines focusing only on the spike protein from the originally isolated SARS-CoV-2 Wuhan-Hu-1 viruses has been associated with a shift in our protective viral antibody response of IgG3 to an IgG4 immune tolerant autoimmune response. This shift to IgG4 antibody response is seen with autoimmune diseases (immune tolerance) and cancers. This change in IgG response is associated with cancers being able to escape detection by our immune system (immune tolerance), allowing cancers to grow.

The persistent use of these genetic vaccines has promoted viral selective escape (pressure selection) in the same way that the overuse of antibiotics drives antibiotic-resistant strains of bacteria with life-threatening consequences, such as in the case of methicillin-resistant *Staphylococcus aureus* (MRSA). In the final 33(6) QR code slides, we can see the variant shifts temporally associated with genetic vaccine use, both globally and in the United States. As shown in the QR slides, neither the percent of the population vaccinated nor the addition of boosters overcame the pressure selection of the SARS-CoV-2 variant mutations.

There is simply no scientific data-driven evidence to demonstrate genetic vaccine immunity is superior to immunity acquired by person-to-person contact. In addition, evidence indicates that the continued use of these vaccines promotes pressure selection changes in the virus, leading to more variants. Evidence also indicates that repeated vaccination is resulting in changes in the type of antibody response occurring, which are seen with the development of autoimmune diseases and cancer.

7. What is a Polymerase Chain Reaction (PCR) Test good for?

Necessity, it is often said, is the mother of invention. So it was for Dr. Kary Banks Mullis, a biochemist, when in the 1980s laboratory funding for his work was drying up. In an effort to continue what his laboratory technicians and technologists were doing but soon would be unable to, Mullis patented[21] a new method to amplify genetic sequences in a sample.

As necessary for all patent applications, he described the method and claims for use of the patent. Patents require proof that what a patent is for is real, has been done, and is not a hypothesis or theory. Applicants must prove the usage is new, requiring a review of prior published research by the patent officials issuing the patent. The Founding Fathers considered this so important that it is included in the US Constitution, along with copyrights:

> [The Congress shall have Power . . .] to promote the Progress of Science and useful Arts, by securing for limited Times to Authors and Inventors the exclusive Right to their respective Writings and Discoveries.
>
> Article 1, § 8, Clause 8[22]

An inventor must include enough specifics, so as to avoid the misapplication of the patent for those who are issued a license to use it. These specifics, included in the instance of Mullis's PCR patent, are the information about the number of cycles to be used, including the added benefit obtained by increasing the number of cycles (repeat amplifications) to provide the best signal, and the upper limit of signal benefit, after which artifact (noise) increases without further signal benefit. In electronics and physics, we refer to this as signal-to-noise (S:N) ratio. As Mullis shows in the patent, the cutoff value is twenty cycles, which yields an increase for the specific genetic sequence being sought (the signal) of 1,048,555, as shown in the following table taken from the Mullis patent.

	Number of Double Strands After 0 to n Cycles		
Cycle Number	Template	Long Products	Specific Sequence [S]
0	1	—	—
1	1	1	0
2	1	2	1
3	1	3	4
5	1	5	26
10	1	10	1013
15	1	15	32,752
20	1	20	1,048,555
n	1	n	$(2^n - n - 1)$

There can be no debate over an increase in static (noise) being produced by people carrying out a PCR test using too many cycles, or a loss of signal (detectability) by those using too few cycles. However, that is not the real issue at hand. Mullis's PCR test, used as directed by the patent, correctly identifies the genetic sequences it is looking for, including the spike protein of SARS-CoV-2 viruses.

ANYTIME, ANYPLACE, ANYWHERE, DR. FAUCI

The real issue has been with people stating there was no pandemic due to a virus and, rather, the pandemic was due to a test—the PCR test. Support for this argument came from doctors, many too young or too inexperienced, with few years of clinical practice and no involvement with the HIV-AIDS era of the early 1980s. Many of these physicians said they had never heard of a respiratory virus affecting the kidneys or other parts of the body. These same physicians seem to have no difficulty connecting the genetic vaccines with kidney disease or, again, other parts of the body.

In this book and its QR code slides you have seen multiple slides showing both infected individuals and vaccinated individuals having an InflammoThrombotic response (ITR) producing lung, heart, kidney, liver, pancreas, brain, ovarian, testicular, and other diseases happening throughout the body, including cancers, sudden cardiac death, and miscarriages. These ITRs begin with the SARS-CoV-2 virus or the spike protein produced by the genetic vaccines. When there are other health problems (comorbidities) or a larger amount of the virus or spike protein, particularly when coupled with poor health and immunity problems, ITR injury and death occur. Simply arguing that respiratory viruses don't cause kidney disease ignores what triggers this ITR.

Medically speaking, nothing is worse than seeing a problem, misdiagnosing the cause, and then thinking your treatment will solve the problem. For the pandemic, the cause is not a misapplied PCR test but, rather, the development and release of a dual-use gain-of-function biological viral weapon, which allowed for the enhanced implementation of dual-use gene therapy investigations using the genetic vaccines. Treatment of both requires attention to the ITR and priogenic response resulting from their effect upon the entire human body.[23]

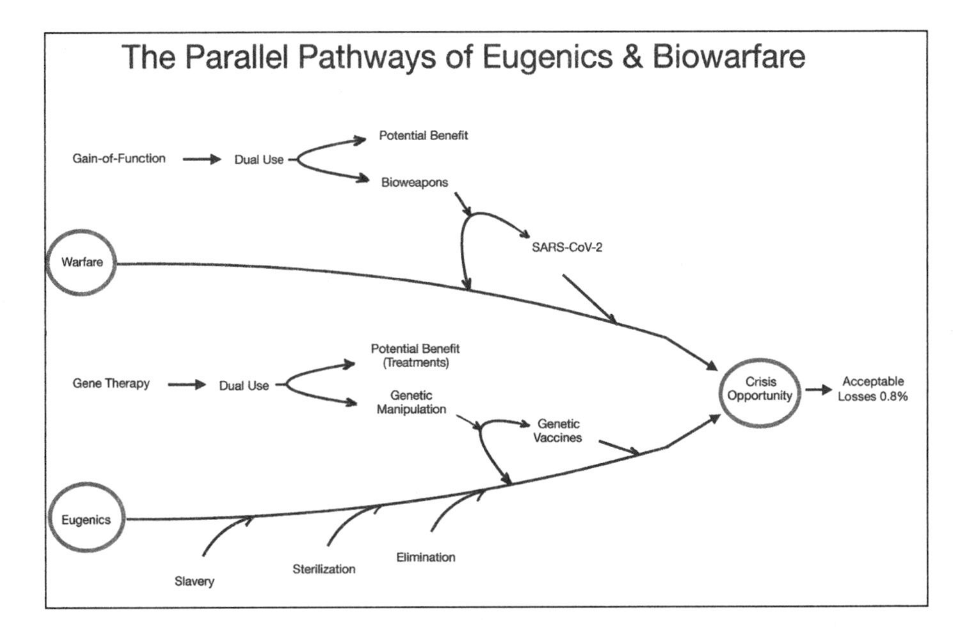

Ignoring the real cause of this real pandemic is to minimize those who have been injured or who have died, and the families and friends left behind in the wake of this egregious crime.

In addition to having patented methods, hopefully to benefit humanity, Dr. Mullis and I have one other thing in common. We have both challenged Dr. Anthony Fauci to debates. With Dr. Mullis passing in 2019,[24] I remain very much alive and ready for our debate. This QR code shows some of our correspondence regarding my offer to debate him on the topic of NIAID funding of biological gain-of-function bioweapons. The offer remains open.

SCAN ME

> Dear Dr. Fauci,
> Now that you are no longer director of the National Institute of Allergy and Infectious Diseases (NIAID), I would like to once again extend to you the opportunity to have an open, honest, public discussion and debate about the origins of SARS-CoV-2 along with your role as NIAID Director in funding Gain-of-Function research, and the consequences of that funding.
>
> On behalf of the almost 1.2 million dead Americans, more than 1.5 million genetic vaccine maimed Americans, the almost 600,000 people who have died from the priogenic and InflammoThrombotic Response (ITR) to the Bioweapon and Genetic Vaccines, the 7 million plus worldwide[25] deaths from COVID, and the 34,600 + genetic vaccine deaths[26] in the United States alone, I think it's time you and I had an open, honest debate about your role in this, moderated by both mainstream and alternative media, with questions coming directly from the general public and people who attend.
>
> Name the time, place, venue and I will join you, America, and the rest of the world for that discussion and debate.
>
> Respectfully,
> Dr. Fleming

8. The SARS-CoV-2 bioweapon and the genetic vaccines are not being used by the Ashkenazi Jews to kill Muslims or African Americans.

The differences that so often divide us once again surface, even in the face of harm to those being accused of doing something malicious. In this instance the accusation exists that somehow the Ashkenazi Jews are trying to exterminate Muslims or, as some people are claiming, trying to eliminate African Americans by targeting the TMPRSS2 receptor, to which SARS-CoV-2 binds.

In neither case is this true. First, although there are differences in the number of TMPRSS2 receptors (covered earlier in the book on how SARS-CoV-2 gets into our cells), everyone, including the Ashkenazi Jews, have TMPRSS2 receptors, so targeting them would also target Jewish individuals.

Second, regarding the misinformation about trying to harm people of Muslim faith, let me point out the obvious: being Muslim is a religion, not a genetic trait, just as being Jewish, Christian (Catholic, Protestant), Hindu, Buddhist, or

Malaysian Frog Worshipping relates to religious or theological beliefs. Such beliefs are not genetic. This attack on Ashkenazi Jews is nothing less than an attempt at ethnic purging, which almost makes the eugenicists look like they are standing on the moral high ground—quite the accomplishment.

In Israel, a country primarily composed of four groups of Jewish individuals having various levels of orthodox to liberal views (just as seen with Christian and Muslim faiths), roughly 73.6 percent of the people are Jewish, followed by 18.1 percent Muslim, 1.9 percent Christian, and 1.6 percent Druze.[27]

If the genetic vaccines were truly developed by Jews to annihilate others, it would seem that the Jewish people in Israel are trying to exterminate themselves, since they embraced the vaccines promoted by the government of Israel, with Israel leading the world in vaccinations. As shown in the 33(8) QR slide set, this came at a cost to the people of Israel, 73.6 percent of whom are Jewish, not Muslim[28] or African American.

The primary genetic vaccine administered in Israel was the Pfizer BioNTech. The founder and CEO of BioNTech is fifty-eight-year-old Uğur Şahin, PhD,[29] a German oncologist and immunologist born in Turkey. He is the recipient of the German Cancer Award, the Order of Merit of the Federal Republic of Germany, and the 2019 Mustafa Prize. The last of those[30] is granted to scholars of the Islamic world. Suggesting that a genetic vaccine administered to most of the individuals in Israel of Jewish faith, yet was developed by an Islamic scholar and recipient of awards from Germany and the Muslim world, was an effort by the Ashkenazi Jews to target and harm Muslims or African Americans, given the fact that the Pfizer BioNTech genetic vaccine wasn't developed by the Ashkenazi Jews but a Muslim, is illogical. What do you think?

The last slide in QR 33(8) shows several of the various parties and businesses involved in the development of several of the genetic vaccines currently available.

9. There are no living creatures or nanotechnology inside these genetic vaccines.[31]

As we saw in chapter 5, much of the purported nanotechnology turned out to be salt crystals. *Nano* means 10^{-9} meters, so I think we can agree that the use of the prefix gives the meaning of something very small. If it's very small, most people don't have microscopes to determine if they are being told the truth or simply being sold snake oil.[32]

There have been reports of parasites and other living creatures in the vials of these various genetic vaccines, or various other changes, simply because of poor technique or damage to the blood or vials of vaccine being examined. Though at least some of these people may have been well-intentioned, their conclusions were not well-founded. In some examples, such as that shown in QR code 33(9), the person reporting on what they thought was a living creature was nothing more than the hair from a plant.

For decades I have taught my students that the first pulse to take in an emergency is your own. It is easy to make mistakes, particularly when attempting to do something you don't normally do. What's important is to understand and correct these mistakes and try not to make them again.

10. Just because someone's a doctor doesn't mean he or she is of the type you want treating you.

As one who has all three designations, I sometimes begin my presentations by explaining that PhDs (doctor of philosophy) solve problems, MDs treat problems, or JDs (attorneys and lawyers) cause problems. Not all PhDs are scientists. A PhD in business or education has advanced training and experience in business or education; this does not make the person a virologist, physicist, attorney, scientist, or physician. It makes the individual someone with an advanced degree in business or education. Remember that when someone is selling you something.

Similarly, not all JDs equally understand patent law. Lawyers who are intellectual property (patents, copyrights, trademarks) attorneys, or those who at least took all of the available courses in the area of intellectual property (IP) law, are better at understanding and handling patent issues. Most people have patent attorneys help write their patent or copyright applications. Some of us, attorneys who have taken all of the available IP classes, elect to file our own patents. When you are reading a patent, you need to notice who developed the patent, who funded it, and what the claims are. Claims are placed at the very end of the patent; they spell out what the patent can do that has not been done before. This is the patent.

Finally, MDs (allopathic[33] doctors), are what you usually think of when you think of a doctor or physician. Unlike other practitioners of health care, including nutritionists, MDs do not attend a specific class studying just food. That information is scattered throughout our medical training in areas such as biochemistry, physiology, pharmacology, endocrinology, internal medicine, surgery, cardiology, obstetrics or gynecology, and family practice. The basics are learned and then applied clinically. When I was a medical student, we had to calculate (by handheld calculator) how many calories, fat (and the various types of fats, including medium-chained triglycerides), carbohydrates, vitamins and minerals, sodium, potassium, chloride, calcium, magnesium, phosphate, and the like were need by our patients and whether the patients would receive that by mouth or intravenously.

Sadly, such information may not be taught as well as it was during the time I was trained, but it is still available and should still be part of the curriculum. If it isn't, the medical colleges need to step up and do their job. Nutritionists and dieticians may be there to help, but they are not there to do our job—prescribe patient care from beginning to end.

Many of the patients I was called to assist during the COVID pandemic were not receiving nutritional support. They also were not receiving necessary IV fluids to prevent dehydration or nutritional support to help with their immune system and body healing. Nor were they receiving the social, psychological, or religious support they needed to help encourage them to recover and return home. Not a single study has ever shown that malnourishment, dehydration, or isolation was a treatment for anything.

All that being said, just because someone uses the term *doctor* as a self-description doesn't mean the person knows anything about medicine, science, or law. You should first ask, "What type of doctor are you?" If he or she isn't the type who was trained and has treated living patients, the doctor just might not be the type you're looking for and need.

In the End, It's Our Character and Behavior that Define Us

When my children were growing up, one of my sons had been hit in the face by a snowball thrown by one of the boys in the neighborhood. His face was swollen. It turned out that the snowball had a rock in it. The boys thought it was funny because they didn't like him, because he was different from them.

All three of my children have demonstrated to me what type of people they are. My daughter used to defend and protect other kids who were harassed because they were different. They were gay or had disabilities, and others picked on them.

My oldest son protected his younger brother when an angry dog jumped a fence and got into our yard during a family gathering. While most people ran to protect themselves, he ran to protect his younger brother. When he was in junior high, I received a call from the principal of his school—he had been in a fight. It turned out, other kids in gym class were picking on an overweight kid and one of the bullies tried to hit this boy. My son stepped in, stopped the attack, and hit the bully trying to hurt the overweight kid. My son threw the second punch, and as most kids can tell you, it's not the kid who throws the first punch who gets caught. I did what my son asked me not to do: I explained to the principal that my son was not at fault and he was protecting someone else. The message to the principal—don't punish my son for doing the right thing. Deal with the bully.

On a Sunday afternoon after watching a movie, my youngest son heard a girl scream in an alleyway. When he went to help her, a man with a knife jumped out from behind garbage dumpsters. My son hit him in the nose and sent him to the hospital. Later the situation seemed like a setup, but at the time he didn't think of that and he went to help the girl.

As a father, I don't know how much prouder I could be of my three children, and yet I am reminded of what my son asked after he was hit in the face with a rock-snowball: "Why can't we all get along?" Good question!

* * *

Although saying we do not all agree is an understatement and although disagreements frequently feel like something out of *A Tale of Two Cities*,[34] I can't help but think of what Gotham used to be like, when it was the shining city on the hill, the beacon of freedom, light, and hope. And I think that perhaps, just perhaps, the people in our city of Gotham, might see what I see:

> I see a beautiful city and a brilliant people rising from this abyss, and, in their struggles to be truly free, in their triumphs and defeats, through long years to come, I see the evil of this time and of the previous time of which this is the natural birth, gradually making expiation for itself and wearing out.[35]

In chapter 6 I showed you how we can bring these criminals to justice and in chapter 5, I showed you how to address the genetic vaccines civil litigation. In the next chapter of this book, we are going to look at what we need to do to restore Gotham to the Shining City on the Hill.

CHAPTER 8

"It's Not Who You Are Underneath, It's What You Do That Defines You"[1]

The significance of carpe diem, or "seize the day," is made clear by hearing the entire Latin phrase: *carpe diem quam minimum credula postero.* That means "seize the day and do not trust in tomorrow." Clearly, the importance of seizing the day is related to the notion that waiting for tomorrow might be too late—tomorrow and the promise of correcting a wrong may never materialize because events may shape the future beyond our control. If we are going to right a wrong, the time to do so is now.

IT'S TIME TO DEFINE WHAT TYPE OF PERSON YOU ARE

In six of the first seven chapters of this book, we have looked at the incredible lust for power and drive for genetic superiority that has plagued much of human history. We have watched people become enslaved, manipulated, isolated, and coerced into extermination, at the bidding of those in power. We have watched the development of biological, chemical, and nuclear weapons. Despite recognizing the horrors and destruction and despite our recognition of needing to control and eliminate the development of these weapons, we have watched countries, including the United States of America, develop and deploy such.

In recognition of the criminal acts committed by Americans and people in other countries, we painstakingly laid out the case in chapter 6 for indicting, prosecuting, and convicting those who committed these crimes, both within and outside of the United States.

Despite warnings from people like Presidents Eisenhower and Kennedy, we failed to correct these wrongs. We allowed the military-industrial complex to continue its growth from MKUltra to the biological weapons development of the current age. The results have yielded devastating human casualties never before seen in the history of humanity. We have given those wanting to advance genetic-manipulation research the very justification needed by them to convince the majority of people that such research development was not only a good thing but a necessity.

To hold these criminals accountable, we will need to carry out the criminal indictment, prosecution, conviction, and sentencing of these very same individual criminals. We will need to hold the courts, judges, and prosecuting attorneys accountable for making this happen and to make certain that this time all the evidence is presented to the American jury for consideration, not just the evidence the judges and prosecutors want to see admitted. This time there will be no pulling the wool over the eyes of the jurors. To do this will take the courage and strength of our combined generations of Americans. But like those who founded this country, we must find that courage and strength once again and act, declaring we will not go quietly into that good night, we will not give up without a fight, we're going to live on.

While you and I are in the process of bringing these criminals to justice, we also need to address the consequential damage to our society, medicine, science, law, and political system. This book has covered the role that lawyers, courts, politicians, physicians, and private organizations have played in contorting medicine into a tool for their own use and power. They did so without the federal or state constitutional authority to do so. In a republic, the power flows from *We the People* to the states where we reside. Whatever power the *People* have given to the states, the states may pass to the federal government. The US Constitution is a compact agreement between the states and federal government. With the states receiving their power from the people. This is what the Founding Fathers clearly established:

Preamble[2] Constitution of the United States of America

> We the People of the United States, in Order to form a more perfect Union, establish Justice, insure domestic Tranquility, provide for the common defence, promote the general Welfare, and secure the Blessings of Liberty to ourselves and our Posterity, do ordain and establish this Constitution for the United States of America.

Within the Preamble to the Constitution, it is *We the People* who establish the Union, for the purpose of justice, security, tranquility, and general benefit for all. And yet we live in a country where justice is obfuscated by judges and lawyers who prevent the truth[3] from being told to the jury, where judges will block the investigation of corrupt lawyers,[4] where the government destroys evidence without notice or consent of the defendants[5] and where federal judges (Article III) enjoy limitless power without limitation in terms—something the Congress (Article I) and president (Article II) do not have. In fact, the amount of time a president, senator, or representative can remain in office is limited—not so for federal judges. Federal judges who determine what is and is not constitutional (the law) are given limitless power for an unlimited period of time by people with limited power and limited time (elected officials). This relationship has resulted in politicians placing judges in positions of power like a brothel-keeper uses a prostitute with similar constitutional results.

Whether we want to admit it or not, we live in a country where money equals power. Money *is* necessary to provide a living for oneself and one's family, and people can make the world better through their wealth. But lust for money and power from individuals and organizations like the Rockefeller Foundation, WEF, WHO, Carnegie Foundation, the military-industrial complex, and the federal and state governments, has produced a world where the rights of the people are abused and the people are subjugated to those with money. We used to call this *slavery*. It still is, but instead of whips and chains, money is used to control your health, food choices, transportation choices. And as we have seen, it has been used to develop biological weapons and promote eugenics through the advancement of gene editing. (Sorry! They're calling it gene *therapy*.)

We live in a country where citizens who want to be heard are *granted* the right to speak if they agree with the narrative selected by those in power. The right to speak is not granted by the government; it was granted by *We the People*.

We also live in a country where the rights of physicians to disagree with narratives mandated by state and federal governments are shackled. When these differences are not discussed transparently (the catchphrase that means we will talk about it if it's in the best interest of those controlling the narrative), contrary views are called misinformation, with threats by governors and presidents to indict those who disagree with the official narrative and state medical boards threaten physicians with revocation of a license to practice, despite physicians having completed their medical training and receiving their medical degrees.

President Ronald Reagan has been quoted to say,

> The nine most terrifying words in the English language are: I'm from the Government, and I'm here to help.

Sadly, it was President Reagan who was responsible for signing 42 U.S.C. §§ 300aa-1 to 300aa-34 into effect. You know it as the National Childhood Vaccine Injury Act (NCVIA),[6] legislation that made it legally impossible to sue vaccine manufacturers for the harm their vaccine might cause. However, it does not protect those companies from SPL suits.

Reagan also signed into effect the National Practitioner Data Bank (NPDB),[7] used by HHS, courts,[8] hospitals, licensing boards, lawyers, and insurance companies to control the privileges of physicians practicing medicine. There is no such control over the courts, hospitals, state or federal governments, licensing boards for lawyers or other professions, or insurance companies.

NPDB controls what hospitals physicians can practice in, which physician your insurance company will allow you to see, where he or she can practice medicine. If a hospital is unhappy with your doctor agreeing to treat you in a particular way, including something you have asked for and your doctor is willing to give you, that hospital may then deny your doctor admitting privileges, putting a black mark on his or her professional reputation for other hospitals, insurance companies, and lawyers who like to sue doctors to see. This is just one example of how the NPDB is used and it has been upheld by a group of judges who are not elected by you and who have no term limits. There is no evidence, no proof, that either the NPDB or the NCVIA has improved the quality of patient care, saved the life of a vaccinated child, or improved the general welfare of the nation.

The law and databank have increased government, court, lawyer, insurance company, hospital, and licensing board control over how people are educated and trained to take care of patients. They have reduced the ability of Americans to make health-care choices, in some instances forcing them to obtain care in other countries, where health-care choices are still possible without interference. The agencies and executive orders issued by governors and the president of the United States have forced people to receive care they do not want, including vaccinations, in direct violation[9] of case law decided by the SCOTUS, *Cruzan v. Director, Missouri Department of Health*, 497 U.S. 261 (1990).

The NCVIA, medical boards, and NPDB have splintered the health-care system between those who will follow the official narrative and those who will ask questions. It was the ability to ask these questions and act on behalf of the patient that led to successful treatment of the 2001 anthrax patients, HIV patients of the 1980s, and more. It was the ability to ask questions and probe for answers that led to FMTVDM[10]—the first method allowing us to measure changes at the tissue level, to find health problems and tailor treatment specific to each patient. Such independent thinking allowed me to honor the dean of my medical college, who told us, on the day we were enrolled as medical students,

> Ninety percent of what we will teach you is wrong. We are teaching you based upon what we have learned. For those of you interested in research, in making medicine better, I encourage you to ask questions and do the research necessary, so that future generations of doctors, will be right more than 10 percent of the time.
>
> John Eckstein, MD
> Former Dean, University of Iowa College of Medicine

Today, the government, courts, lawyers, licensing boards, hospitals, insurance companies, and others, with the help of the NPDB, have told the medical community not to ask questions, not to research, not doubt the narrative. Do what you are told, or face the consequences. That is not the oath I took, and it is not the oath my fellow physicians took. It is time to take back medicine. It does not belong to those groups of people. It is not a business. It belongs to those of us who are physicians. It belongs to those of us who honor our oaths, and our oaths are to our patients—you.

In this chapter we are going to look at areas that need to be addressed and what you, and *We the People* can all do to recover our society and world.

MEDICINE AND THE FEDERAL GOVERNMENT

In December 2008, as we briefly discussed in chapter 4, John D. Podesta[11] of the Obama-Biden transition team asked me and others to convene a town-hall meeting to discuss health care with the American people. The purpose was to see what Americans thought of the upcoming Affordable Care Act (ACA)[12] and to ask questions about health care in general.

When the Obama-Biden transition team asked me to participate, reflecting on the words of President John F. Kennedy, I agreed to help my country. I had hoped that my country and the president elect would listen to what the *People* wanted.

The 29 December 2008 Health Care Forum

During the course of the evening on 29 December 2008, approximately 100 to 125 people from Reno, Nevada, exchanged ideas while I presented slides I had prepared in advance. These slides are in appendix F.

Following the discussion, the notes taken during the evening were typed up and sent to the Obama-Biden transition team, and President Obama, for use in the upcoming congressional and presidential discussions on health care in the United States. I like to think that President Obama read and listened to the *People*, the final legislation known as Obamacare indicates he either didn't or didn't care. I'll let you decide.

Health Care Forum

Health Care Community Discussion Requested by
Obama-Biden Transition Team
December 29, 2008
Grand Sierra Resort, Reno, NV

The forum was led by two physicians, Dr. Richard Fleming (Moderator and Cardiologist) and Dr. Donald VanDyken (President Elect of Washoe Medical Society and Family Medicine). In attendance were about 100–125 people, including (self-identified) physicians, health care professionals (nurses, alternative medicine providers), at least one attorney, a political science professor (Jim Hutter, Iowa State Univ.), a staff member (Katie Pace) of the local Member of Congress (Rep. Dean Heller), a representative of the Nevada ACLU, a representative of a pro-life organization (possibly Right to Life), small business owners (with 40 or fewer employees), some in the insurance industry, and many who did not identify themselves as being other than "ordinary" citizens. The forum began at 7 p.m. and concluded about 10:15 p.m.

The meeting began with statements by the two physicians (Fleming and VanDyken). Dr. Fleming's concerns as moderator focused on (1) CHANGE coming from the people and not the government. In keeping with the words of Theodore Parker and Abraham Lincoln, that since ". . .this government of the people, by the people, for the people. . ." is something most Americans believe in, then it is necessary that such community discussions focus on the people and not the bureaucrats and/or the special interest groups, who up to this point in time have resulted in a Health Care System which is at best dysfunctional despite having some of the best doctors, nurses, technicians, hospitals and other health care providers in the world. (2) That in keeping with this, there be a **moratorium** on all Health Care legislation being considered, proposed or developed until the American people as in all healthy republics or democracies discuss it. (3) That the first Health Care act to be proposed for passage into law, as discussed at this Forum, be an "Access to Emergency Health Care Act - AEHCA" as proposed and modified at this meeting (see below) and finally, that (4) people in Reno, NV have the intelligence, compassion, common sense and determination necessary to establish itself as a Center for Future Health Care Discussions to help lead the way for CHANGE in the Health Care of Americans.

ACCESS TO EMERENCY HEALTH CARE ACT—AEHCA

1. That under emergency, life threatening conditions, and during labor and delivery, all people presenting to emergency departments/rooms/facilities will receive the necessary care required to address and treat their emergency.
2. That they will receive it in a respectful manner consistent with human dignity.
3. That they will receive it in a timely manner.
4. That they will receive it without violating their personal beliefs.
5. That they will receive it independent of their age, race, sex, religious and personal beliefs, or their ability to pay.
6. That insurance (private or government) companies may not exclude, limit, or in any way reduce coverage or payment for such emergency evaluation and treatment, or the treatment plan resulting from such evaluation and treatment; even if it is a pre-existing condition.
7. That non-emergency care is not covered under this act and as such, such individuals may be treated provided:
   a. It does not prohibit the treatment of people with emergency, life threatening conditions,
   b. It is accord with hospital policy, and
   c. Provided it is a condition that the facility has the equipment and appropriately trained personnel to treat.

FREEDOM OF CHOICE ACT—FOCA

1. The act proposes itself as an effort to prevent government "interference" in a private matter.
2. The act adds nothing to "Health Care" for women and as such is not an actual "Health Care" measure. Specifically,
   a. Griswold v. Connecticut, Roe v. Wade and Doe v. Bolton already provide for abortion care prior to fetal viability.
      i. That Roe v. Wade raised the question as to whether fetal viability began at week 24 with discussions of "quickening" indicative of life before that.

ii. That the medical definition of fetal viability is 20–22 weeks and has been so for decades.
iii. That physicians do not require legislation to allow them to follow their Hippocratic Oath which already provides for their response to a woman's life being at risk, and that we do not believe that the government is now planning to practice medicine nor to tell physicians how to practice medicine.
iv. That the 2008 U.S. Department of Health and Human Services "Women's Health and Mortality Chartbook" does not define "Health" of a woman under either condition of:
1. Pregnancy or
2. Abortion

And as such has not provided support or rational for such a statement of "Health" in FOCA.

3. That such an act has not been discussed or approved by the American public, even though it would direct health care dollars away from other Health Care issues, many of which were then discussed.

After about an hour, audience members made their comments and contributions, some of which were as follows.

Many related the types of medical concerns that all of the 2008 presidential candidates heard numerous times at numerous places throughout their campaign travels.

1. High costs of care (e.g., thousands of dollars charged for brief visits to emergency rooms; it's cheaper to fly to India for elective surgery there).
2. Higher costs charged to patients and lower to those covered by insurance (e.g., a $2200 bill reduced by $1600). Specifically, the comments of Kristi Gutierrez,[13] CCS-P, CPC (Director of Billing Compliance, HIPAA Privacy Officer for the University Health Science System, Reno, NV) were read by Dr. Fleming. As noted by Ms. Gutierrez, insurance companies frequently do not fully reimburse hospitals and doctors for their services. In order for the system to financially survive, it must charge those who are uninsured or underinsured higher fees to make up the difference. The result is greater charge for those who can least afford it while insurance companies pocket the difference.
3. High costs of medicines (e.g., 50% of prescriptions are written for US patients but pharmaceutical companies make 90% of their profits from these same people; people go to Mexico to buy their prescriptions). One of Dr. Fleming's proposals was that once a pharmaceutical company initially develops a new drug, that it be reviewed to determine if it will actually provide a new treatment of or if it just an alternative to something which already exists without the promise to improve "health." If there is promise, the drug would undergo further investigation by

the government (it would be less expensive to investigate a drug than to have Americans pay for medications under the current system) with the results published in the medical literature independent of outcome. Currently most research publications only include good outcomes and fail to show that drugs don't work.

4. High costs of insurance (if you can get it). Specifically, the issue of pre-existing conditions was discussed. Failure of a Health Insurance policy to cover a pre-existing condition makes as much sense as having an automobile accident on an interstate while driving 65 mph and having the insurance company tell you they are raising your premium, but will not pay for any accidents you have on an interstate for the next 5 years.
5. Prevention. For every dollar spent on prevention, the system saves $7–10. There is a major distinction between "Health" and "Insurance." The first action taken by the British NHS was to make certain, children received milk, which improved the health of the children, saving money later for health care from malnourished children. "Health" vs. "Insurance."
6. Insurance companies make exorbitant profits (e.g., a $1 billion salary to the CEO of one company; a $22 million building paid for just with the interest received on insurance company deposits, some the result of excessive delays in payment to physicians).

Other comments and proposals included the following.

1. There is no medical need for preadmission screening. If the medical professionals think a person needs to be admitted, they should be and the costs should be paid for. Numerous examples were given of people turned away because of insurance companies not approving patient admission, including the concerns voiced by a mother whose child with Cerebral Palsy could not be taken care of due to insurance issues, even though he had blood in his urine.
2. The review of medical claims should not be done exclusively by insurance company personnel; there should be citizen and medical personnel involved (as a protection against fraud and abuse).
3. Prescription formularies (where different drugs get different insurance treatment) are unreasonable. There should be some regulation on the cost of medications.
4. Medicare pays physicians well in New York state, where doctors prefer Medicare patients to insurance patients, and so poorly in Nevada that doctors have to limit or even refuse to treat Medicare patients, thereby making the problem even worse for those doctors who do treat Medicare patients.

5. Every state has an insurance commissioner of some type; these governmental representatives should be more devoted to protecting citizen/consumers than promoting the interests of the insurance industry in their state.
6. There should be standard provisions of care and coverage in insurance policies so that (a) consumers can reasonably compare prices and (b) add optional coverage that they are willing to pay for; probably these standard provisions should be national instead of state-by-state. There should also be a mechanism to reduce costs paid by consumers when they take an active role in improving their health (e.g., Weight loss, exercising, smoking cessation, etc.).

Considerable discussion happened on the issue of abortion and the proposed FOCA.[14] Approximately 85 percent of those in attendance agreed that FOCA was not a health-related issue but (as established by the wording of FOCA itself) is an individual issue, as is smoking, drinking alcohol, how one drives, and so forth. Given limited funding, equipment, trained physicians and personnel, it is unconscionable for the government to insist that every hospital and clinic provide every medical service to every patient. Contraception and birth control practices should remain between the doctor and patient. There is no supporting evidence for FOCA, and as such it should not be voted on or signed into law by Congress or executive branch of the government, independent of their personal beliefs.

1. Several (as shown by applause a.k.a. ASBA[15]) agreed with the speaker who objected to this topic being considered "Health Care."
2. ASBA, the audience was divided over whether a "baby" was conceived at fertilization.
3. Prof. Hutter noted that the issue of when life began and whether abortion should be legal or illegal was dependent on the fact that the fertilized egg was not independent of any other living creature, residing (for 9 months) in a second person (the mother-to-be); if eggs could be harvested at the beginning of a pregnancy and brought to term wholly without involving an unwilling other person, the nature of the debate would be very different; however, there are two people involved and the issue is not about what happens to just one of them.
4. As a way of summarizing this last concept, Prof. Hutter suggested using a modification of NIMBY (not in my back yard) to illustrate this issue: Not In My Body, thank You!
5. The pro-life lobbyist said that many people and organizations on both sides of this issue had objections to the enactment of FOCA as written; of particular concern were (1) it was retroactive, thereby repealing all previous laws and court decisions on this matter, and (2) it made it

mandatory that physicians and hospitals provide abortions even though they had moral objections to doing so.

The meeting was concluded with participants expressing a desire to have further discussions in Reno and agreeing that the government should bring all health-care bills before the American public before considering them for passage, to avoid special interest groups from controlling what Americans are receiving for health care.

Despite the repeated objections by those in attendance, objections that were sent to President Obama and the transition team, namely that no legislation be passed without returning to the *People* for review and discussion, neither President Obama, Vice President Biden, nor any member of the transition team, including Mr. Podesta, responded to this report or the *People*.

We had reviewed the material and responded, as we were asked. We reminded our elected officials that the *People* were an integral part of any health-care discussion and that it was not up to the president, Congress, or federal government to unilaterally make health care decisions for them.

IMPLEMENTING THE WILL OF THE PEOPLE

To address the general welfare of the people present at this town-hall meeting in Reno and the general welfare of people I have heard from around the country, and after considering changes desperately needed following the deployment of a biological viral weapon, I propose that *We the People* submit to our elected officials, and those running for office, the following American Health-Care Act of 2025. This proposal should be used by anyone running for office representing *We the People*. It should be enacted in 2025—if not before.

The rationale for each of the seven parts of this proposed legislation to be passed by Congress in 2025 will be discussed following the act itself, beginning on page 298.

The American Health-Care Act of 2025—Legislation by and for *We the People*

> We appreciate that American innovation and capitalism is both a driving force in the advancement of health care and medical sciences *and* a hallmark of the United States of America, as reflected by the Founding Fathers, who included in the US Constitution intellectual property protections for the promotion of science and arts and to credit those who make such advancements possible:
>
> > To promote the Progress of Science and useful Arts, by securing for limited Times to Authors and Inventors the exclusive Right to their respective Writings and Discoveries (U.S. Constitution, Article I, Section 8, Clause 8)
>
> And as further subsequently codified in 35 US Title 35, 17 US Title 17, and 15 US Code Chapter 22, this American Health-Care Act of 2025, focuses

on addressing and correcting inconsistencies in patient health care and costs, resulting from (a) instances of corporate greed and corruption, (b) misrepresentation by Big Pharma to the FDA and peoples of the United States, (c) failure of the FDA to hold Big Pharma accountable for these misrepresentations, (d) failure by the FDA to demand full and transparent research data on drugs before approving a drug for use, and (e) the shifting of health-care costs resulting from efforts by those in the field of medicine to be financially viable and attempting to address disparities in reimbursement of costs, while maintaining American innovation and capitalism, we accordingly submit the American Health-Care Act of 2025.

I. **Purpose of the American Health-Care Act (AHCA) of 2025**

The first and primary purpose of the American Health-Care Act of 2025 is to restore oversight and control of any and all state and federal agencies involved in American health care to the citizens of the United States of America.

In recognition that the cost of medical care in the United States of America has significantly increased, in recognition that medical care is best directed by those who have been trained to practice medicine and not those practicing law or politics, and in recognition that the costs of medical care has been driven up by inter alia Big Pharma, insurance companies, and the consequential shifting of medical costs from CMS to other insurance companies and those who are uninsured or underinsured, to compensate for the lower reimbursement paid by CMS, resulting from government overregulation, this legislation is established to:

A. Restore the practice of medicine to those who practice healthcare—specifically physicians, pharmacists, and other health-care providers who are directly involved in the care and treatment of patients and the associated billing of that care.

B. To restore the balance of costs in a free enterprise system of health care, through

1. Transparency of pricing of all healthcare costs including but not limited to:
    a. Physician costs,
    b. Hospital costs,
    c. Prescription and nonprescription drug costs,
    d. Diagnostic-studies costs,
    e. Insurance companies,
    f. Any other costs associated with the diagnosis and treatment of patients.

The act thereby decreases overall health-care costs by promoting capitalistic competition between providers and facilities, allowing patients and not insurance companies to select their health-care providers and facilities—including but not limited to hospitals, doctors, doctors' offices, clinics, and diagnostic and surgical centers. It thereby promotes patient, physician, pharmacist, hospital, clinic, and diagnostic and surgical centers freedom of choice and fair market competition, reducing health-care costs for individuals independent of their insurance status, pharmaceutical and insurance companies, doctors, hospitals, and other health-care providers and facilities.

C. To restore the balance of costs equally across the United States of America, through encouragement of competition of insurance companies for contracts with individuals, doctors, doctors' offices, hospitals, clinics, diagnostic and surgical centers, corporations, unions, and other groups, thereby lowering premiums and out-of-pocket costs for those insured by such insurance companies, in addition to those underinsured and without insurance.

II. Revoking 42 U.S.C. §§ 300aa-1 to 300aa-34, a.k.a. the National Childhood Vaccine Injury Act (NCVIA)[16]

The NCVIA provides blanket immunity for pharmaceutical companies against tort actions or criminal actions depending upon intent.

The function of the FDA is to assure that medications given to Americans are both safe and effective. At the same time, the SCOTUS has said:

> That juries would always find in favor of a child, "even if the defendant manufacturer may have made **as safe a vaccine as anyone reasonably could expect**." [emphasis added]
>
> The dissent's legislative history relies on the following syllogism: A 1986 House Committee Report states that § 300aa–22(b)(1) "sets forth the principle contained in Comment k of Section 402A of the Restatement of Torts (Second)"; in 1986 comment *k* was "commonly understood" to require a case-specific showing that **"no feasible alternative design"** existed; Congress therefore must have intended § 300aa–22(b)(1) to require that showing. The syllogism ignores unhelpful statements in the Report and relies upon a term of art that did not exist in 1986.
>
> H.R. Rep. No. 99–908, pt. 1, p. 25 (1986), U.S. Code Cong. & Admin. News, 1986, pp. 6344, 6366 (hereinafter 1986 Report). *Post,* at 1089–1090. [emphasis added]
>
> Immediately after the language quoted by the dissent, the 1986 Report notes the difficulty **a jury would have in faithfully assessing whether a feasible alternative design exists when an innocent "young child, often badly injured or killed" is the plaintiff.** Eliminating that concern is why the Report's authors "strongly believ[e] that Comment k is appropriate and necessary as the policy for civil actions seeking damages in tort." The dissent's interpretation of § 300aa–22(b)(1) and its version of "the principle in Comment K" adopted by the 1986 Report leave that concern unaddressed. [emphasis added][69]
>
> 1986 Report, at 26, U.S. Code Cong. & Admin. News, 1986, at p. 6367; see *ibid.* ("Even if the defendant manufacturer may have made **as safe a vaccine as anyone reasonably could expect**, a court or jury undoubtedly will find it difficult to rule in favor of the 'innocent' manufacturer if the equally 'innocent' child has to bear the risk of loss with no other possibility of recompense").
>
> *Bruesewitz v. Wyeth LLC*, 562 U.S. 223, 241 (2011)[17][emphasis added]

This case has frequently been mis-stated as meaning the SCOTUS said vaccines have implicit dangers and cannot be made completely safe. This is not what the SCOTUS said. Instead, the SCOTUS decided that rather than erring on the side of protecting an injured child, it would protect the pharmaceutical company, whose product caused injury to the child, removing a jury of the *People* from deciding civil liability.

Congress enacted and the president signed into law the NCVIA, holding vaccine manufacturers nonliable for harm caused by their product. The decision was made to encourage vaccine manufacturers to continue production of vaccines for delivery into the bodies of Americans.

The disconnect between the SCOTUS decision and the FDA responsibility (an agency under the executive branch of the federal government) demonstrates problems within the governmental (Article III and Article II) balance of powers.

As the Preamble of the U.S. Constitution states, it is the obligation of the federal government to provide for the **general welfare of the *People* who have assembled the government**. Accordingly, it is the responsibility of Congress (Article I), to balance the powers and either eliminate the FDA and allow drug manufacturers to have a free for all or revoke the NCVIA that promotes protection of pharmaceutical companies over American children. [emphasis added]

This provision of the AHCA of 2025 effectively revokes the NCVIA, making the SCOTUS decision in *Bruesewitz v. Wyeth LLC*, 562 U.S. 223, 241 (2011) legally irrelevant.

III. Revoking the National Practitioner Data Bank (NPDB)

The practice of medicine is contingent upon successful entry into and graduation from a medical college. Thereafter, physicians must apply to the state medical boards for permission to practice within each state they wish to practice in. This process is known as medical licensing.

The licensing process and the NPDB have been used by those running these boards and HHS to dictate the practice of medicine, to control those practicing medicine, making physicians conform to the approach with which the medical boards, hospitals, federal and state governments, insurance companies, and courts want medicine to be practiced.

Decisions regarding patient care vary from physician to physician, based upon level of training, specialty training, where pysicians were trained, and experiences they have had with different diagnostic and treatment approaches. No one size fits all; yet, the establishment of the NPDB and medical boards—mostly promulgated by lawyers and politicians, not physicians—have been leveraged to require physicians to acquiesce to the opinions of those in power. In short, anyone but the practicing physician trained to practice medicine and surgery makes decisions. This loss of physician autonomy has resulted in less innovation in diagnostic and clinical care of patients, as well as a frustrating loss of options for patients, who for personal reasons may want to forego certain testing and treatment in favor of a different approach.

The NPDB is managed by HHS and collects information on physicians to restrict physician practice, including but not limited to situations where the care provided by the physician may not agree with someone else's approach. Because this someone may be at an insurance company, hospital, licensing board, and the like, the physician is reported to the NPDB, putting a black mark on the physician, damaging his or her professional career and personal life.

The NPDB is also a tool for making certain patients cannot receive alternative care that they may want but the powers that be do not want the patient to have, even though the patient could, if they could afford it, leave the country to obtain that treatment or care.

The primary determinant of what is and isn't accepted has been driven by the Rockefeller Foundation, the WHO, and other agencies, along with the state and federal governments, including many individuals who have never attended or graduated from medical college. This is government and agency overreach, and there is no instance of the NPDB improving the quality of medical care in the United States or saving a life.

The NPDB and the NCVIA are both examples of lawyer and politician overreach into the practice and control of medicine that have been detrimental both to the practice of medicine and to patient choices. Accordingly, *both* the NPDB and the NCVIA should be immediately revoked, with re-evaluation of the power and authority of state licensing boards.

IV. Access to Emergency Health-Care Act—AEHCA

Whereas not everyone agrees about the type of doctor a person would want treatment from for specific health problems, most Americans do agree that in an emergency they would want to be treated in an emergency room. To guarantee that all Americans receive this level of care, the AEHCA is included as a component of the 2025 AHCA.

A. Under emergency, life-threatening conditions, and labor and delivery, all people presenting to emergency rooms or departments, will receive the necessary care required to address and treat their emergency.
B. People will receive this care in a respectful manner consistent with human dignity.
C. They will receive it in a timely manner.
D. They will receive it without violation of their personal beliefs.
E. They will receive it independent of their age, race, sex, religious and personal beliefs, and their ability to pay.
F. Insurance companies (private or government) may not exclude, limit, or in any way reduce coverage or payment for such emergency evaluation and treatment, independent of the treatment plan selected for their evaluation and treatment—even if the visit was for a preexisting condition.
G. Nonemergency care is not covered under this act; as such, individuals seeking that care may be treated, provided:
    a. The delivery of that care does not prohibit the treatment of people with an actual emergency or life-threatening condition,
    b. It is accord with hospital policy, and
    c. The emergency room or department has both the equipment and appropriately trained personnel to treat the condition the person is presenting with.
    d. However, individuals who abuse the use of emergency medical services, including but not limited to ambulances, EMTs, and paramedics, place a strain on the availability of emergency medical services, critical for people

with life-threatening situations. This not only makes it impossible for paramedics and EMTS to be available for legitimate emergency life-saving services but also increases overall health-care costs. Accordingly, when individuals abuse these emergency services, they should be responsible for paying those costs and, in instances of knowingly abusing these emergency services, legal remedies should be considered at the community level.

V. The Investigation, Research and Development, Patenting, FDA Approval Process, Cost of Prescription and Non-prescription Diagnostics and Drugs

Pharmaceutical companies develop and sell medications and diagnostic drugs—including but not limited to radioactive isotopes—after receiving FDA approval and patent protection. The FDA approval and patent protection does not automatically allow pharmaceutical companies the right to arbitrarily fluctuate prices as Enron did with the control of electricity in California during the early 2000s.

Big Pharma has repeatedly justified increasing the costs of newer medications based upon research and development costs associated with the development of such drugs and isotopes. This focus on profit has driven the development of newer medications that can be patented for profit, with encouragement of physicians and patients to abandon older medications that are just as effective and has obfuscated the use of nonprescription medications, including but not limited to vitamins, minerals, lifestyle changes, and preventive medicine.

The solution presented at the aforementioned town-hall meeting is now included as part of The AHCA of 2025. Accordingly, all new diagnostics and drugs proposed, researched, and developed by pharmaceutical companies for the purpose of diagnostic and therapeutic use on people, animals, plants, soil, water, air, or other environments or uses will be evaluated as follows:

A. Members of the FDA, CDC, HHS, or agencies affiliated with these agencies, will be excluded from working in the pharmaceutical industry for a minimum of ten years to avoid any conflicts of interest, with the ultimate responsibility being to the American citizen. Individuals seeking employment within the FDA, CDC, HHS, or agencies affiliated with these agencies, or with the pharmaceutical industry, should be made aware of this ten-year exclusion and agree prior to employment.

B. The FDA shall encourage not only the development and investigation of newer prescription medications but, also the investigation and further development of vitamins, minerals, and alternative diagnostics, medications, and treatments, all of which shall be collectively referred to as "diagnostic" or "drug" for the remainder of this act. The FDA shall evaluate said drugs using the following to determine if a new drug proposed by a pharmaceutical company fits the necessary criteria noted in the sections that follow:

1. All diagnostics and drugs shall be approved by a new committee, entitled the Proposed New Diagnostic and Drug Price Committee (PNDDPC).[18] This committee will consist of volunteer physicians, pharmacists, health-care providers, and members of the general public, who will meet either in person or telephonically—including through Skype, Zoom, and any other secure means considered acceptable by the members of the

committee, given the condition that such means of communication is confidential and restricted to committee members only.

2. The members of this committee must be free of any affiliations with any pharmaceutical company for a minimum of ten years.

C. When a pharmaceutical company has a new diagnostic or drug that may be useful for therapeutic purposes, it must first be presented to the PNDDPC for review and consideration.

D. For a new diagnostic or drug to be considered by the PNDDPC, it must:
1. Be identified for a specific—not hypothetical—purpose.
2. The mechanism or mode of action of the diagnostic or drug must be identified and defined.
    a. If that mechanism or mode of action is identical to medications currently being used for a specific medical purpose, the diagnostic or drug must provide either
        i. An explicit and defined benefit not currently available,
        ii. a reduction in side effects, or
        iii. a reduction in costs.
    b. If the mechanism of action is different from currently applicable medications, the benefit vs. expected risk must be explained.
3. Once cleared for investigation by the PNDDPC, the PNDDPC will determine if there are inter alia nonprofit organizations, universities, institutions, or foundations, with an interest in the proposed diagnostic or drug or health-care problem, which the diagnostic or drug is being developed for use in.
    a. If such entities are identified, these nonprofit organizations, universities, or other institutions or foundations are encouraged to submit their interest in the diagnostic or drug in question, their areas of interest, and contact information, as well as those who would be involved in such diagnostic or drug investigation.
        i. Once identified, further investigation of the diagnostic or drug will be carried out by connecting the not for profit organizations, universities, or other institution with the pharmaceutical company.
        ii. The resulting research will be published, through applicable medical or scientific journals, independent of outcome. If no medical or scientific journal wants to publish the research, it will be available through the PNDDPC's website for the scientific and medical communities and general public.
    b. If no such not for profit organizations, universities, or other institutions or foundations exist, the cost of further research and development will be paid for through funding provided by NIH or other applicable government agencies, with preference given to those diagnostics or drugs and health conditions Congress or the president has expressed a specific interest in. With the same publication requirements outlined in 3(a)(ii) immediately above.

E. Once the research has been completed and published for the new diagnostic or drug, that diagnostic or drug will then be returned to the pharmaceutical company with a report from the PNDDPC, including recommendations for or against FDA approval based upon the following:

1. Recommended approval—the diagnostic or drug offers a benefit greater than risk using a new or alternative pathway of diagnosis or treatment not previously utilized.
2. Recommended approval—the diagnostic or drug offers a greater benefit, or an equal benefit with lower side-effect profile, than currently available diagnostics or drugs using a pathway of treatment already utilized.
3. Denial recommended—the drug either has a greater risk than benefit profile utilizing diagnostic or treatment pathways already in use or offers no alternative diagnostic or treatment mechanism that might prove beneficial to someone not responding through a currently utilized diagnostic or treatment pathway or mechanism.
4. Open recommendation—the data provided does not demonstrate sufficient information to indicate the diagnostic or drug should be denied, nor does it provide sufficient information to recommend the diagnostic or drug.

F. Following the recommendations from the PNDDPC, the diagnostic or drug is returned to the pharmaceutical company.
   1. If the decision was to recommend, the pharmaceutical company must submit the new diagnostic or drug to both the FDA and USPTO. The pharmaceutical company must do so within thirty days of receiving the recommendations from the PNDDPC or forfeit patent rights to the diagnostic or drug.
   2. If the decision by the PNDDPC was to deny, then the diagnostic or drug may not be submitted to the FDA for approval. The pharmaceutical company may submit to the USPTO for diagnostic or drug patent.
   3. If the PNDDPC decision was open recommendation, the FDA may elect to review and consider the diagnostic or drug for use. The pharmaceutical company should submit to the USPTO for diagnostic or drug patent.

G. Once the diagnostic or drug has received both USPTO patent and FDA approval, the pharmaceutical company may market the diagnostic or drug for sale in the United States. The company may only market the diagnostic or drug for use for the purposes approved by the FDA and claims made within the USPTO patent. Off-label use of a diagnostic or drug is a physician determination and should not be marketed by the pharmaceutical company without further USPTO claims and FDA approval.
   1. Under the AHCA of 2025, the pharmaceutical company may not claim research and development (R&D) costs for the R&D of the diagnostic or drug now approved by the FDA and patented by the USPTO. Accordingly,
      a. The price of the diagnostic or drug will consequently be determined by the PNDDPC using the following recommended criteria:
         i. Cost of similar diagnostics or drugs using the same mechanistic pathway,
         ii. The average accepted cost of the diagnostic or drug, if already used, in other countries,[19] and
         iii. Other fair market assessments verified by the PNDDPC.
      b. The set price of a diagnostic or drug determined by the PNDDPC may not be greater than the set market value of the diagnostic or drug determined by the PNDDPC. The cost of the diagnostic or drug sold

by the pharmaceutical company to a wholesale distributor or retail seller of the diagnostic or drug must be the same and cannot be greater for one wholesaler or retailer than other wholesalers or retailers.

i. Pharmacies and pharmacists, in a free market capitalistic system, may compete for sales to patients by selling the diagnostic or drug at a reduced price, as long as the diagnostic or drug is not altered, or, if compounded, falls within 90 percent bioavailability of the original diagnostic or drug, without any increased risks or side effects, as governed by applicable pharmaceutical safety laws.

c. The cost of all prescription diagnostics or drugs must be freely available for review by all physicians, patients, hospitals, insurance companies, and the general public, providing an opportunity for patients, doctors, hospitals, and insurance companies to participate in the decision of where to buy their diagnostic or drug and which diagnostic or drug to use.

H. Pharmaceutical companies must sell their diagnostics and drugs for the same price to all doctors, hospitals, clinics, surgical centers, pharmacies, and pharmacists.[20]

1. Pharmaceutical companies may lower their prices according to free market practices. However, pharmaceutical drug manufacturing companies and wholesale and retail distributors must make available to all doctors, hospitals, clinics, surgical centers, pharmacies, pharmacists, patients, insurance companies, and the general public the differences in prices paid by others, purchasing their diagnostic or drug, thus guaranteeing that everyone is buying the diagnostic or drug for the same price.

VI. Diagnostic Testing and Decision-Making Transparency

A. Physicians will determine what testing and treatment is appropriate for a given patient, not the government or insurance company.

1. In accordance with the Thirteenth Amendment to the US Constitution, physicians and health-care providers are not indentured servants and, as such, cannot be told what to do by inter alia the federal or state government, other agencies, hospitals, insurance companies, diagnostic centers, or clinics.

2. Similarly, in a free society and in a society governed by capitalism (a.k.a. free enterprise) physicians, health-care providers, and health-care facilities cannot be told by inter alia the federal or state governments or insurance companies what the price of their services will be. Patients will determine the fairness of these costs by determining which physicians, health-care providers, and health-care facilities they select for medical care.

a. It is therefore incumbent upon the patients, employers who select insurance coverage, and others who pay for those services to participate in the selection of physicians, health-care providers, and facilities, including but not limited to hospitals, clinics, and diagnostic centers, based upon the quality of care they receive and the cost of that care.

B. Physicians, not insurance companies, lawyers, courts, state, or federal government, should determine what medical testing is appropriate for their patients. In accord with the physicians' Hippocratic or Physician Oath, their responsibility is to their patient and not to the insurance company, Big Pharma, attorneys, courts, or the government. As such, the American Health-Care Act of 2025 eliminates the need for prequalification of testing, which interferes with the practice of medicine and, in turn, eliminates surprise medical bills and costs for the patient.

1. The cost[21] of all diagnostic testing should be freely available for review by all physicians, patients, and insurance companies, providing an opportunity for patients to decide between locations and providers for their diagnostic testing.
2. If a physician orders a diagnostic test, it will be covered by the insurance company. If the patient is uninsured, the cost of a test for the patient will equal the cost of supplies, personnel wages, and equipment costs for the period the patient underwent testing.
    a. This will increase the availability of diagnostic testing while covering the costs of completing such testing.

C. The cost of diagnostic testing, including any and all diagnostic tests used for US citizens, must be consistent across the United States, independent of location. The life of an American citizen in rural America is equal to that of the life of an urban American citizen. The equipment and training required to conduct those tests are not superior in one part of the United States and inferior elsewhere.

D. The function of a physician and other health-care provider is to give the best outcomes for patients under his or her care. If a test is ordered, the patient has the right under this act to determine if and where that testing is done. This encourages the patient to actively participate in health-care decision-making, resulting in a more robust *informed consent* to the satisfaction of the patient, including but not limited to the limitations imposed upon physicians who make treatment decisions in the absence of the information obtained from the diagnostic testing.

VII. Restrictions Placed upon the Federal Government

A. The federal government has been actively involved in efforts to control medicine since 1846, when a group of individuals represented themselves as the American Medical Association (AMA). The focus of this group, which represented only a few physicians and today represents fewer than one in four American doctors, was on limiting who could practice medicine and how. From the relationship between the government and AMA, the federal government passed the Drug Importation Act of 1848.[22]

B. Since 1846 the AMA and federal government have repeatedly determined how medicine would be practiced in the United States, including using the AMA's control of CPT and other billing codes. These codes, though owned by the AMA, are used by CMS to determine what will and will not be paid by Medicare and Medicaid, thus turning the practice of medicine into a business run by the federal government.

C. These codes are also used by the private insurance companies to control the practice of medicine, justifying their decisions as simply following the limits established by the federal government.

D. The US Constitution did not provide explicit powers to Congress to practice medicine. The US Constitution did not provide explicit powers for Congress to go into business or to compete with American citizens, regardless of whether congressional members are professionally or non-professionally educated and trained.

E. The USA Patriot Act provided money to fight terrorism. It has been reported that up to 10 percent of that money was used by the Department of Justice (DOJ) to treat physicians as terrorists, seizing patient medical records and using these AMA billing codes to criminally prosecute physicians for their practice of medicine.[23]
F. The NPDB, state licensing boards, specialty boards, hospitals, insurance companies, malpractice attorneys, courts, and others have used these CPT codes and NPDB reporting, as discussed in Section III (NPDB), to further control the practice of medicine, physicians, and health care.
G. This segment of the AHCA of 2025 effectively expunges any legal action or judicial cases brought against a physician based upon CPT codes.
H. This segment of the AHCA of 2025 officially provides federal legislation limiting the effect of the federal government to the provision of insurance through the Medicare and Medicaid program to US citizens. It officially recognizes, and stipulates on the record, that oversight of medical practice is limited to the free and independent states under the Tenth Amendment to the United States Constitution.
I. This segment of the AHCA of 2025 calls for a specific amendment to the US Constitution stipulating items G, H, and J, restricting the federal government from practicing or being involved in the practice of medicine.
J. The federal government of the United States of America recognizes the *restrictions* and *understanding* placed upon the WHO and IHR, when signed and ratified by the president and Congress. Congress and the president will enter into no further treaties or agreements on health care or the practice of medicine, recognizing that oversight of medicine is a power of the people as established under the Tenth Amendment to the US Constitution. Congress and the president furthermore recognize that the practice of medicine is not a business, is not under federal oversight, and is determined by the physician educated and trained in medicine and surgery.

Discussing Why We Need the American Health-Care Act of 2025

When my children were younger, one of the books I read to them was *If You Give a Mouse a Cookie* by Laura Joffe Numeroff and illustrator Felicia Bond. The book, a bedtime tale of a mouse asking for a cookie, states, "If you give a mouse a cookie, he's going to want a glass of milk, and then—" The mouse tries to avoid going to sleep. (Great book, by the way!)

Congress and its lust for control over American health care and doctors is a lot like the mouse who wanted a cookie. The mouse was simply looking for a way to get what it wanted—to stay up and not go to bed. Both Congress and the president have been increasing their stranglehold on doctors and their practice of medicine. This control isn't constitutional, and it wasn't something doctors asked for. Instead, it was a small group of doctors who wanted to eliminate their competition, as discussed in chapter 4. What these doctors couldn't do on their own they got the federal government to do for them. They restricted the practice of medicine and turned it into a business.

At first glance, one might ask, "Why have a federal law addressing healthcare, when the federal government has overreached into the practice of medicine and healthcare, which belongs with the people within the states, a power that resides with the patient and doctor and no one else?

That's a fair question, and the answer is the same reason you have people apologize for doing something wrong. You need it on the record. You need a recognition of the wrongdoing and for people to see the apology, in addition to the expressed wording of how we are going to specifically move forward, so there isn't any doubt about motives, intent, action.

PURPOSE

The AHCA of 2025 begins with correcting the wrongs that have been done, to officially and legally enter into law the necessary words with action that demonstrates a recognition and correction. The same thing happens when neighborhood bullies are called on the carpet by the parents of a harmed child. The parents (you, me, and all Americans), go to the house of the bullies (Big Pharma, HHS, CDC, AMA, Rockefellers, WHO, and so forth) and talk to those parents (the state and federal governments) responsible for what they have allowed the bully children to do. Parents of the harmed child get an official apology, demand action to address the wrong, and demand that the offensive behavior will not be repeated.

The AHCA of 2025 does just that. It demands an apology, it demands action to address the wrong, and it demands that the action will not be repeated. The last part of the AHCA of 2025 is an expressed stipulation that the people, not the federal government, have the responsibility to address health-care concerns within their free and independent states.

Now let's take a moment to see why each of the individual components has a wrong and the necessary action needed to address these wrongs.

REVOKING THE NATIONAL CHILDHOOD VACCINE INJURY ACT[24] (NCVIA)

The revoking of the NCVIA is not intended to increase pharmaceutical company liability for harm caused by vaccines, but rather to level the liability across the pharmaceutical-industry spectrum.

We would never argue that any other drug company should be given immunity from liability for a drug it made and sold, based upon the idea that the jury would not be able to determine if there was "no feasible alternative design." Nor would we argue that a jury would consequently only find in favor of the plaintiff (the person injured by the drug) independent of the evidence. The SCOTUS ruled just that, in *Bruesewitz v. Wyeth LLC*:

> Immediately after the language quoted by the dissent, the 1986 Report notes the difficulty **a jury would have in faithfully assessing whether a feasible alternative design exists when an innocent "young child, often badly injured or killed" is the plaintiff**.

As a result, the SCOTUS went from being an impartial decision-making body to a biased court, removing the ability of *We the People*, as the jury, to decide on the drug company's liability for harm.

Imagine replacing *vaccine manufacturer* with *doctor*, and *feasible alternative design* with *feasible alternative treatment*. Do you think the SCOTUS would grant protection from liability to doctors practicing medicine? The answer is obviously no, and yet the reasoning would be the same.

We cannot have a law, state or federal, that allows such absolute protections for someone or some company that knowingly is causing harm. The Vaccine Adverse Event Reporting System (VAERS) recognizes harm, which must be addressed and not swept under the carpet.

REVOKING THE NATIONAL PRACTITIONER DATA BANK (NPDB)

The NPDB has become a collection of names of physicians who have practiced medicine in such a way as to differ from what the AMA, federal government, insurance companies, Big Pharma, hospitals, and so forth want physicians to practice. It is being used to control medicine and physicians, preventing physicians from practicing in a manner that they and their patients have chosen, while financially strangulating them. There is no evidence that the NPDB has improved healthcare or patient outcomes. There is no evidence it has removed dangerous physicians and replaced them with better ones. There is no evidence that the NPDB has made medicine safer or more effective or improved medicine or the quality of health care in the United States.

Attorneys, hospitals, insurance companies, big pharma, medical boards and societies, and the government have used the databank to punish and control doctors. Who is behind this, and what are the motives? As we saw in chapter 4, the direction American Medicine took in the late 1800s and early 1900s was determined by a few individuals with a vested interest in controlling medicine. With the efforts of the Rockefeller family, Carnegie family, and the people they funded and put into positions of power, they successfully altered the pathway of diagnostic and treatment decision-making choices toward products they controlled.

The narrative about what is and is not acceptable medicine was, and should remain, based upon outcomes, not Big Pharma, hospitals, profits, insurance companies, or any part of the government. The successful capture and prostitution of medicine by lawyers and government agencies has left many doctors impotent in their practice of medicine, leaving them unable to provide the quality of medical care their patients deserve and personally and professionally dissatisfied[25] with their calling.

Consequently, schools select medical students through controlled standardized testing, including MCAT (Medical College Admission Test). Students can regurgitate answers at the drop of a hat while being unable to interact with patients or solve new problems, like HIV and COVID. That was all too readily apparent during the COVID pandemic. While society asks for diversity,[26] these standardized testing approaches have yielded less diversity, less creativity, less patient-physician satisfaction, and more compliance with the powers that be. Even the American Bar Association has recognized the need to abandon the use of these standardized testing methods and to remove its head from the sand and look for something better, something that works.[27]

All of this has left us with physicians, who are shiny speedboats, in the middle of the desert—an analogy presented by Mr. Brian Wheeler, former paramedic instructor at UCLA.

The result of this failed health-care system, coupled with the poor dietary and lifestyle practices of Americans, is a decline in American longevity[28] and an increase in ITR diseases.[29] Something has been rotting in the practice of medicine and it isn't those people looking for alternative answers.

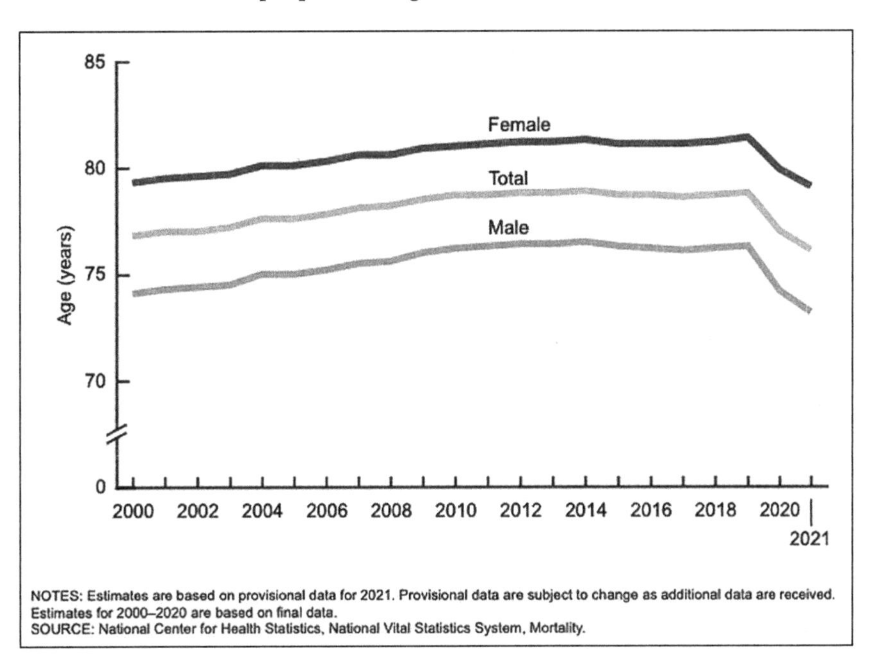

NOTES: Estimates are based on provisional data for 2021. Provisional data are subject to change as additional data are received. Estimates for 2000–2020 are based on final data.
SOURCE: National Center for Health Statistics, National Vital Statistics System, Mortality.

Those who solved the anthrax attacks following 9/11 and saved lives, those of us who grappled with and helped address the diagnosis and treatment of HIV-AIDS and saved lives, were able to do so because we questioned, because we followed the results of our diagnostic and treatment decisions. The development of the InflammoThrombotic response (ITR) disease theory and FMTVDM now makes it possible to determine if the treatment you and your doctor decided upon worked, rather than allowing the government or Big Pharma to make those decisions for you. Like the house arrest of Galileo Galilei, the federal government and Big Pharma are not interested in you seeing the truth and thinking for yourself. They are interested in controlling what you think, just as the Catholic Church was in the time of Galileo. The NPDB, state and federal government, Big Pharma, and those controlling them could potentially return us to the Dark Ages.

It is time to remove this layer of bureaucratic control suffocating the education and practice medicine, and with it you. It is time to revoke the NPDB. It is time to address this wrong.

ACCESS TO EMERENCY HEALTH CARE ACT (AEHCA)

Because President Obama, through the transition team and Mr. Podesta, requested the input by American citizens through the Reno town-hall meeting, it is only fitting that this be included in the health-care act. By and large, the people who attended the town-hall meeting demanded that no change in health care, and certainly nothing that interfered with their selection of physician or options for treatment, be enacted without their approval. The American people thought they should be provided with emergency medical care in life-threatening situations, and their ability to pay for that emergency care should not prevent them from receiving it. They also did not think that people should automatically get to see a doctor or receive care by abusing the emergency services of our country. They believed too many people, "repeat fliers," abuse the medical system, increasing the cost of emergency medical services and threatening the lives of others by taking ambulances with their EMTs and paramedics out of service making them unavailable for real emergencies.

Again, the statement by *We the People* expecting their constitutional authority to be respected and abided by, is accordingly, included in this part of the act.

INVESTIGATION, RESEARCH AND DEVELOPMENT, PATENTING, FDA APPROVAL PROCESS, AND COST OF PRESCRIPTION AND NONPRESCRIPTION DIAGNOSITCS AND DRUGS

Ninety percent of pharmaceutical profits are obtained by the 50 percent of the prescription diagnostics and drugs sold in the United States, demonstrating an unfair burden for Big Pharma profitability coming from the American people. At the same time, the governmental and private-interest control of the practice of medicine, including diagnostics and treatments, has prevented alternative diagnostic and treatment practices—without a demonstrated benefit to the American people.

This provision of the AHCA of 2025 establishes corrective measures for this problem, resulting from lack of citizen oversight of agencies falling under the president of the United States. US citizens must travel to other countries to receive diagnostics and treatments they are currently unable to receive in our country, with no evidence that these diagnostics and treatments adversely affect the health of Americans. Those Americans unable, due to financial or personal circumstances, to travel outside of the United States are limited by the costs imposed for the diagnostics and treatment options available here.

Because the standard of care (SOC), in many instances, is controlled by people in positions of power and by financial influences, including representatives elected to represent *We the People*, not only is there an imbalance of diagnostics and treatment for citizens within the United States, but there is also an unacceptable financial strain upon the people and the US economy.

To address the financial burden and to direct improvements in health care based upon outcomes instead of profits, this component of the AHCA of 2025 stipulates the review of newly proposed diagnostics and drugs. It provides for citizen oversight of the review process of what is being submitted, alternatives to the funding of R&D of such diagnostics and drugs, an expansion of nonprescription drug diagnostics and treatments (many of which have showed improved diagnostic and treatment outcomes, particularly when coupled with currently used diagnostics and treatments)[30] and control of the cost of these diagnostics and drugs currently bankrupting Americans.

DIAGNOSTIC TESTING AND DECISION-MAKING TRANSPARENCY

The purpose of this part of the AHCA of 2025 is to return diagnostic decision-making to the physician and patient, removing it from the control of those not practicing medicine or directly affected by the testing. The interference with physician practice and decision making includes testing decisions. Medicine is not a business. Physicians frequently have to inform patients that they will need to seek approval from the insurance company—which frequently uses CMS to determine what tests are or are not acceptable—leaving both patients and physicians handcuffed by the decisions of those not actively involved in the treatment and care of the patient. This part of the act also calls for equal distribution of costs and the availability of testing for those underinsured or uninsured. It also calls for the recognition that patients should have the right, and therefore the responsibility, of becoming actively involved in the decision-making process, through the use of informed consent, with the recognition that the decision to not undergo testing will limit the ability of the physician to make a clinical decision.

RESTRICTIONS UPON THE FEDERAL GOVERNMENT

The final part of the AHCA of 2025 addresses the federal government's involvement in the education and practice of medicine, which began around 1846 with the aforementioned beginnings of the American Medical Association (AMA), the same group that used dues from its members to invest in cigarette company stock for the AMA's portfolio, while telling Americans they should not smoke—demonstrating a hypocritical model, not Hippocratic Oath.

The specific component of the AHCA of 2025 expressly stipulates that the federal government is restricted to providing insurance through Medicare and Medicaid with passage of this legislation providing a federal law, noting that oversight of medical practice[31] is limited to the people within the free and independent states under the Tenth Amendment to the United States Constitution. Accordingly, a constitutional amendment is called for.

This component of the AHCA of 2025 also clearly defines that the federal government cannot usurp control of medicine from the people and that the practice of medicine is defined by the physician educated and trained as a physician upon receipt of his or her medical or osteopathic diploma. The federal government also acknowledges the *reservations* and *understanding* imposed upon the WHO and IHR agreements, as signed and ratified.

ADDITIONAL AREAS OF CONCERN THAT NEED TO BE ADDRESSED SOONER RATHER THAN LATER

I. THE PRACTICE OF MEDICINE

Although I have argued that the powers that be should stay out of the practice of medicine, it is time we physicians step up and do our jobs. The practice requires (a) taking time to learn what's new so we can provide the best medical care possible, (b) not using others to manipulate our colleagues into practicing medicine how we think medicine should be practiced, and (c) not using our medical practices to sell drugs, vitamins, minerals, or other treatments. If you want to run a business, run a business. If you want to practice medicine, practice medicine. It's a conflict of interest to do both.

The practice of medicine should result from the demonstration of education and proficiency determined by the degree bestowed upon the graduate. Medical boards, the AMA, specialty societies, and such should not tell a physician who has a medical or osteopathic degree how to practice medicine. These same organizations have a vested interest in the outcome, and that vested interest, is a conflict of interest that people need to know about.

Government agencies should not control the practice of medicine anymore. That's what the Nazis did, and everyone paid for it. German doctors during Hitler's reign repeatedly told people they didn't have a choice. When the government tells doctors what to do, to whom, and when, you have a Nazi government. Nazism is a mindset, not a geographic location or a system of government lost to history.

II. SCIENCE, SCIENTIFIC RESEARCH, AND PUBLICATION

Science should be carried out to understand something, scientific truth or how something works. In medicine, scientists seek to understand the cause of diseases, how to find disease, how to treat it, how to improve health, and so forth. Science is not about how to make money or create a patient's medical bill. It's not about Big Pharma. Pharmaceutical companies have the ability to make diagnostics or medications that can help people, including their own families. They do not have the right to destroy data that shows the drugs or diagnostics they have don't work or are harmful. Science teaches us some of our best lessons by showing where we are wrong, so we can improve our understanding, so we can do better than being right only 10 percent of the time.

Scientific journals should not be run by Big Pharma. When science journals, the AMA, and professional associations, boards, or organizations receive funding from those with vested interests in the outcome, including but not limited to the pharmaceutical industry, then those very same scientists and journals are compromised. There is a conflict of interest that prevents the truth from being told, limiting our advancement and potentially causing harm.

Universities, colleges (including medical), researchers, scientists, and physicians must be able to research and carry out patient care free of control. We cannot serve two masters, and you will not be whom we are interested in helping if the people controlling the funding have the control.

There should be no advertisements by pharmaceutical companies in any scientific journal—medical or otherwise. Although nonscientists or nonphysicians may have an opinion, it should be viewed in light of their motives and personal interests. If there is any merit in their opinions, it will undoubtedly be discussed by scientists and physicians. Too much of the discussion about what is and isn't scientific is being driven by people with personal motives, not scientific or medical expertise.

III. ATTORNEYS

Attorneys are educated to practice law, not medicine.[32] Physicians are educated to practice medicine, not law. Medical billers are educated to practice medical billing, not medicine or law. Physicians should not be involved in the regulation of lawyers. Lawyers should not be involved in the regulation of physicians. All lawyers should be removed from any and all organizations, boards, and the like that oversee physicians. Doctors are not making decisions about lawyers.

IV. COURTS AND TERM LIMITS—INCLUDING THE SCOTUS

The focus of providing no term limits to federal judges came from the belief that judges (the court) would be fair and impartial. If the NCVIA and my case[33] has taught us anything, it is that federal judges are not impartial. The impartial court system has become partial. The judges (court) will hide substantive exculpatory evidence[34] from juries, prevent juries from being able to hear evidence, and block any efforts to expose[35] their interference. In essence, the courts have become tools used by Big Pharma and the government for control of *We the People*. They no longer provide the expected justice or protection for the American people. Accordingly, and given the ever-increasing number of years federal judges are living and serving,[36] we need to implement term limits—including for the SCOTUS. Even the federal judges concur—it is time to limit the number of years a federal judge can serve.[37]

* * *

ACTION YOU CAN TAKE—BEGINNING TODAY!

When engaged in research and clinical labs and when I'm treating patients, I begin each week by looking at what we have accomplished and what we want to accomplish. By the end of the week, we meet to discuss what we have accomplished and why and why not.

During the meetings, we have to ask some tough questions to understand where we are, what led us to this place, what has gone right, what went wrong, and what we can do about it. This book has hopefully provided you, the reader, with some much-needed information and raised some serious questions in your mind to motivate you to act:

How did humanity end up where we are today?
Why have we allowed so many people to enslave others?
Why have we allowed so many people to experiment upon others?
What are some of those experiments?
Where has the United States gone wrong?
Is the United States of America somehow more special than other countries?
What can we do about the crimes that have been committed?
What can we do about the overreach of our state and federal Governments?
What is your role and responsibility in restoring America?

As Americans, we have been given the privilege of electing our representatives. For many years this has been taken for granted. We believed that those elected would automatically represent *We the People*, but like the proverbial fox in the henhouse, those in power can be easily enticed by those behind the scenes who are willing to stay there as long as they are ultimately in control.

For at least the last 170 years, the direction the United States has been heading is not where many thought it was. The influence by small groups of individuals, families, foundations, and such has slowly etched their way into controlling the nation. Agreements have been reached, laws have been passed, powers have been abused, weapons (biological, chemical, and nuclear to name a few) have been developed. Those who abused people in other countries have been brought to the United States to continue their work.

This has not gone unnoticed. Presidents Eisenhower and Kennedy had tried to warn us about the military-industrial complex. President Trump had tried to warn us about the WHO and IHR. Members of Congress have tried to make us aware of the illegal use of federal money, our taxpayer money, that has gone to the development of biological weapons.

The country, and the world, has been effectively polarized, just as predicted in every war game I was ever a part of, which you have been learning about, including the SPARS pandemic, Event 201, and the Nuclear Threat Initiative.

You have watched the initiation of war in the Ukraine, where American bioweapon labs have been brought to light. You have seen documents attempting to hide this information, with other documents showing the US involvement in these bioweapon labs. As a result, we have watched a great number of Americans die, as I previously described.

It is one thing not to know. It is quite another to know and not act. In chapter 6, I laid out many details necessary to show what crimes were committed by those involved in the development and resultant deployment of the bioweapons. You have seen the crimes, the individuals, the evidence. It is now time for you to act on behalf of yourself, your family, your friends, and the Americans killed and injured. What can you do?

FIRST, WE ADDRESS THE CRIMES—10LETTERS.ORG

To address the crimes, all you need to do is go to https://www.10Letters.org or go to this QR code to log into the website.

Once there, go to

a. Start Building My Letter.
   a. It will ask for your name and address.
   b. You will want two letters: one for your governor and one for your state attorney general.
   c. You can then click on the box to hide your address if you don't want the persons (your governor and attorney general) to whom you are sending the letters to know your address.
b. Download and print two copies of the letter, one for your governor and one for your state attorney general.
c. The cover letter includes your address labels which you can attach to the envelopes. Mail one envelope with cover letter to your
   a. Your governor, and one to
   b. Your state attorney general.
d. Then click on the link to email your friends and have them send in letters and tell their friends to do the same thing.

You and your friends can do this as many times as you want. Then call your governor and attorney general repeatedly to remind them that you want them to file criminal indictments on these people for the reasons stated in the indictment letter. Your governor and attorney general were elected to represent you. You pay their salaries; they do not pay yours. They work for you!

SECOND, WE ADDRESS THE GOVERNMENT OVERREACH AND DEMAND LEGISLATION BE PASSED TO STOP THIS ABUSE OF THE AMERICAN PEOPLE

This chapter laid out proposed legislation, changes in laws you wanted passed when President Obama took office, laws protecting your rights as a patient and the rights of your doctor to practice without interference, both at the state and federal level. These new laws remove old laws from the books and restores the education and practice of medicine, including preventive medical care, back to the people.

You can either tell your elected representatives about the book, send them a copy of the book, or photocopy the pages beginning with "Medicine and the Federal Government," and send to your governor, state representative and senator, as well as your federal senator and representative. Let them know you want the federal and state governments to stay out of medicine and health care, out from between you and your doctor. These pages explain why we need these elected representatives of *We (You) the People*, to pass The American Health Care Act of 2025 *into law*.

THIRD, WE NEED TO REMIND THEM WHO THIS COUNTRY BELONGS TO—*WE THE PEOPLE!*

If it was important enough for President-elect Obama to ask me to hold a town-hall meeting to get you, the *People,* involved, and you showed up, then it's important enough for the representatives of the *People* to listen to what you and I (*the people*) have to say.

I showed up, the people showed up. It's time our representatives show up.

CHAPTER 9

Are We Next?

As you are nearing the end of this book and contemplating what to do next, I would like you to consider the following information, to encourage you to embrace your next actions with courage and strength. The courage and strength of modern Vikings seeking the truth and embracing challenges with confidence and resilience.

CONSPIRACY THEORY?

We have all heard the term *conspiracy theory*, usually used derisively to describe someone as a *conspiracy theorist* grasping at straws to get attention. The term *conspiracy* is not something thought of in recent years. The legal definition of the word[1] is the following:

> Conspiracy is an agreement between two or more people to commit an illegal act, along with an intent to achieve the agreement's goal. Most U.S. jurisdictions also require an overt act toward furthering the agreement. An overt act is a statutory requirement, not a constitutional one. See Whitfield v. United States, 453 U.S. 209 (2005). The illegal act is the conspiracy's "target offense."

As you can see, conspiracy is not a concept describing a flight of ideas or a hallucination about the actions of someone else, but rather a legally defined term used by prosecuting attorneys dating back to 1864.[2] Simply put, *conspiracy* defines multiple people acting together to carry out crimes, making a conspiracy a crime in and of itself.

IMAGINE IF YOU WILL

You have just been elected the thirty-fourth president of the United States of America. It is 20 January 1953 and you have just been sworn into office. You graduated from West Point military academy in 1915, after which you served in the US Army during both the First and Second World Wars. In 1941 you achieved the rank of brigadier general (one star), and you went on to oversee military operations in North Africa, Sicily, and Normandy. Later in life, you would advise your successor John Fitzgerald Kennedy on the Cuban Bay of Pigs invasion, but before then you would become the military governor for the American occupied region of Germany (1945), the army chief of staff (1945–1948) and the first supreme commander for the North Atlantic Treaty Organization (NATO) from 1951 to 1952.

Having just been sworn into office, you will now see material few people will ever see. From those materials, you will learn that a series of operations, including Operation Paperclip following the defeat of Nazi Germany, which resulted in the smuggling of German scientists and intelligence officers out of Germany to be placed in positions of influence within the United States—including working for and with the US government.

The very people you were trying to kill—the people you were trying to keep from killing your military men and women—you are now sworn as president to protect. You won the election and lost your war[3]. But your war isn't over yet. You would serve two terms as president, but before you leave you know you need to warn the country about the resulting military-industrial complex that had captured the nation. The existential threat to the country is real.

You have long pondered on how to best tell the nation, perhaps because you finally realized you were part of the reason all this was possible. White House memos revealed you had planned to warn the country for years.

May 20, 1959

MEMO FOR RECORD:

The President mentioned in an aside this morning, when I brought up the topic of selective major addresses for the remainder of his term, that he had one speech he would like very much to make.

He hoped, he said, that the Congress might invite him to address them before he left office, at which time he would like to make a 10 minute farewell address to the Congress and the American people.

I think this is a brilliant idea if it can be carried off with a minimum of fanfare and emotionalism, and we should be dropping ideas into a bin, to get ready for this.

It is not uncommon for people in positions of responsibility, who have held their tongue while in service, to deliver a warning message as they leave office, or shortly thereafter. Former CDC head Robert Redfield, MD, for example, stated[4] that he believed SARS-CoV-2 leaked from the Wuhan Institute of Virology laboratory:

> "I'm of the point of view that I still think the most likely etiology of this pathology in Wuhan was from a laboratory. You know, escaped," Redfield told the network's chief medical correspondent, Sanjay Gupta. "Other people don't believe that. That's fine. Science will eventually figure it out."

As a result of this statement, he has received death threats.[5]

> "I was threatened and ostracized because I proposed another hypothesis," he said. "I expected it from politicians. I didn't expect it from science."

Then, as now, the behavior of our leaders has been to warn people while they are still in a position to be heard. On 17 January 1961, three days before President Eisenhower left office, he delivered his farewell address, warning the nation of the dangers of the military-industrial complex that had now infected the United States of America like the plague.

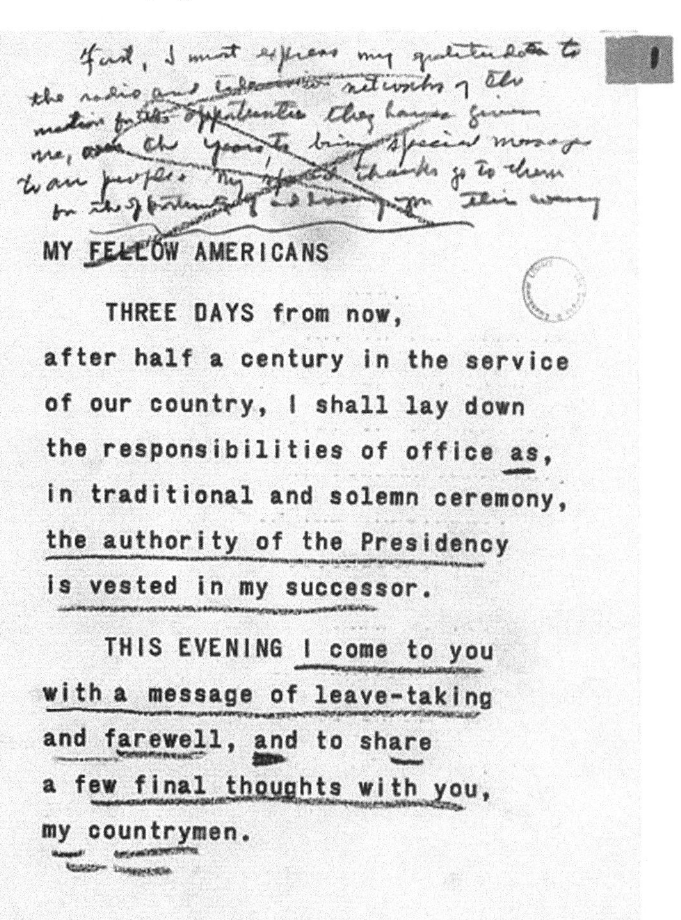

MY FELLOW AMERICANS

THREE DAYS from now,
after half a century in the service
of our country, I shall lay down
the responsibilities of office as,
in traditional and solemn ceremony,
the authority of the Presidency
is vested in my successor.

THIS EVENING I come to you
with a message of leave-taking
and farewell, and to share
a few final thoughts with you,
my countrymen.

Years later, the one and only American president who would speak out against the CIA (the agency that had been involved with Project 63, Operation Paperclip and MKUltra) and the military-industrial complex, John F. Kennedy, would be assassinated in Dallas, Texas. He would have a bullet put in his head, as all evidence indicates, by the very same people he was opposing.[6]

When I was a medical student at the University of Iowa, Richard G. Lynch, MD, the chair for the Department of Pathology[7] who had received significant support from both the Rockefeller Foundation and Abraham Flexner, lectured several times to our medical class. I was also fortunate enough to interact with Dr. Lynch more directly as one of the phlebotomists at the university hospital, a job I held while being a medical student to help pay my way through medical college.

As you can see from the Department of Pathology website and the memorial to Dr. Lynch, medical students at Iowa received a more than healthy dose of immunology training, as it was one of Dr. Lynch's areas of research and expertise. During one of the presentations to our class, Dr. Lynch was explaining how to distinguish suicide from homicide and, if the death was homicide, how to find out who committed the murder. During this presentation, one of my classmates interrupted to ask questions. The student had heard that Dr. Lynch was there when JFK's body was brought in. He asked, "What killed JFK? What happened?"

Dr. Lynch was not a dark-complexioned man, but at that moment, he turned whiter than anyone I had ever seen, and I'm Nordic. Our science auditorium was a steep room designed to hold the 187 students in our class. Dr. Lynch told us to shut off all recording devices (this during the 1981 to 1982 academic year). He walked up the stairs to make certain every door was locked and secured and then went to the main doors on the lecture level and did the same. He turned to us and said that everything we had been told was a lie and we would never talk about this ever again. He said that JFK's brain was no longer inside his body. It had been removed and transported separately. Dr. Lynch told us to look closely at footage of Air Force One and watch how the casket was taken out from the rear of the jet. Years later, documents and documentaries would prove Dr. Lynch was right about the ballistics in JFK's head and about the removal of the brain to prevent others from seeing it.

Now that Dr. Lynch has passed and I am growing older and it has become clear that President Biden is determined not to release the records on JFK's assassination that expose the role of the CIA and operations within Cuba, I am compelled to speak the truth and expose the training of Oswald by the CIA in Louisiana and the CIA's role with him. That information should be known but yet is not. It should be impossible to believe that the CIA would be involved in the assassination of President John F. Kennedy; yet, the origins of the CIA and its involvement in assassinating other world leaders and otherwise removing them from power in favor of those the CIA wanted in power reflects the evolution of the CIA. Having grown from the OSS and the incorporation of Nazis into the government, it altered the trajectory of the United States in the same way Wernher von Braun altered the trajectory of his V2 rocket program, turning it into the Apollo program. Like Eisenhower, Kennedy, Lynch, and Baer,[8] I give you yet another warning about the military-industrial complex.

So, when you ask, if a military-industrial complex is such a threat, why hasn't anyone else done anything about it? Just remember, a Republican and Democratic president both tried to—one as he was leaving office and another before he was assassinated.

Ask yourself why the current president and his predecessor don't want you to know about the sealed records. Ask yourself why the current president refuses to offer security details to protect Robert Kennedy Jr, the Independent Party candidate for president. Ask yourself why there are even records that need to be hidden or destroyed. There are two types of intelligence agencies. The first keeps no records of what it is doing. No records mean no evidence and nothing to leak or be destroyed. The second keeps records because it wants to receive recognition in the future for what it has done, because it can't trust people in the agency and it wants leverage on them, or both. Not trusting the people is a problem within the agency, while wanting recognition is a problem with arrogance. The Nazis had this level of arrogance. They kept extensive records and video archives, all intended to be shared with people under the new world order that they would dominate. If that world order didn't happen, the records would have to be destroyed. Such records are where information leaks come from. Such records are what Edward Snowden and Julian Assange reportedly released or began to release. Such records are why the Nazi war criminals were convicted. In their arrogant lust for glory and recognition, they kept the records that exposed them, which they attempted to destroy as their world was crumbling around them. Records were found. Records sent some to prison and others to be hung. Records, as you have seen in this book, sent thousands of others to the United States, Britain, France and Russia, where they could continue their eugenics and human experimentation programs.

OUR DIFFERENCES MAKE US STRONGER

Throughout the course of human history, some people have tried eradicating others, either by enslaving, sterilizing, or outright killing them. Our differences, our prejudices, even our indifferences, have divided people as sharply as the borders between countries. And while some borders may have become blurred, our differences, indifferences, and prejudices have been used against us by those in power, by people who would try to convince us to eliminate all differences and purify our human genome.

Although our differences can easily be used against us, they can also become our greatest strength. Our genetic differences have kept humanity alive, as with sickle cell anemia, a difference the people carrying out these eugenic and human experimentation programs are currently trying to eradicate as their first great test of changing human DNA, all under the guise of making humanity better. While sickle cell may appear to be a weakness, it prevented the deaths of unknown numbers of people living in areas where malaria was endemic. Should this genetic difference be lost and malaria spread, the eradication of this genetic escape mechanism is only one such example of humanity being placed at risk for extinction.

It has been estimated[9] that 99 percent of all species that have ever existed on our planet are now extinct. Most extinctions apparently are not the result of human activity, although some are. The survival of any species, including ours, is dependent upon the combined interaction of our DNA, environment, behavior, and competitors. We tend to think of ourselves as being the most advanced form of life on the planet, but only people have such a lust for power and control over others that they are willing to enslave or kill anyone who is different.

The pretense used by so many of our elected representatives is that they are looking for inclusion of everyone. This is simply an illusion being sold, particularly during elections, to encourage us to vote for them. Once in office, the elected representatives expect us to go about our lives, allowing them to continue with their plans of deciding what is and is not acceptable for being included. Inclusion is typically determined by whether you agree with what others want—then you're included. If you disagree, then you're not included—all in the name of making humanity better, of course.

This has been true since the beginning of human history. and it is no different today. An effort to resist this approach, to limit this control and coercion by others, is what the Founding Fathers and people who fled to America were trying to accomplish by establishing a form of government *of, for, and by the People*. Yet, as we have seen, even this form of government can become corrupted by those in power. Constraints on those in power do not exist if *We the People* do not demand restraints on those in power and demand our rights. Rights are not given by those in power; rights are inalienable. It does not matter where you think those rights come from. What is important is that you realize they do not come from the government. They do not come from people of wealth or power. They do not come from the WHO, WEF, Rockefellers, Carnegies, federal agencies, the AMA, or anyone else. Some are yours by right and were yours from the moment of your birth. Others you earned—the government did not give them to you. They are not the government's to take away. Only you can allow someone to enslave you.

"It Is Better to Die on Your Feet than to Live on Your Knees"

Throughout the chapters in this book, we have looked at important points in history where people lost control of their lives, and times when people regained control of their lives, often at a price they were willing to pay both for those they held dear and the principles they lived by. But shining cities on the hill are only bright as long as those of us responsible for keeping the light alive do so. It is up to all of us, and future generations, to maintain the principles of these inalienable rights and diversity, not as weaknesses but as necessary strengths and differences that include our physical, mental, spiritual, and political differences. Differences do not merely lend themselves to platitudes about inclusion or involve a free-for-all, devil-may-care chaos, but are the basis for intelligent recognition of the importance of our inalienable uniqueness.

Efforts to exterminate those who are different than us, independent of whether we want to use religion, race, sex, or ancestry as a reason, have no place in a country based upon the principles that everyone has these inalienable rights. A country of such principles does not believe in eliminating these differences but, rather, using them to strengthen humanity. It does not embrace homogenization or eugenics, for eliminating our differences risks the survival of humanity. Nazism is not a lost vestige of history; it is a mindset.

Human DNA, our genetic code, carries with it the untapped potential for life. Unpacked and laid out in a single strand, human DNA is approximately sixty-seven billion miles[10] in length. That equals more than 150,000 round trips to the moon. This tightly wrapped genetic code is present in every cell of our body, except mature red blood cells. As you have seen earlier in this book, these red blood cells are critical to our survival, and, yet, when the genetic vaccines make contact with our red cells, they turn gray and clot, a phenomenon incompatible with life.

In addition to the DNA in the nucleus of our cells, DNA is also in the mitochondria (mtDNA)[11] of our cells. Mitochondria are referred to as the "powerhouses" of our cells and are essential for turning the food we eat, the water we consume, and the oxygen we inhale into the energy that runs our cells, keeping us alive.

Human mtDNA contains two strands, including a heavy strand that has twelve protein subunits for control of energy production (oxidative phosphorylation), in addition to two subunits for ribosomal RNA (rRNA, specifically, 12S and 16S)[12] and fourteen subunits for transfer RNA (tRNA), which brings the amino acids (AAs) to our ribosomes to build proteins. The light strand has one protein subunit for oxidative phosphorylation (energy production) and eight tRNAs. Like nuclear DNA, mtDNA is also involved in the health—or absence of health—in our bodies.

The genetic codes (DNA) in the nucleus of our cells are tightly twisted, wrapped, and folded. Every trait, every characteristic (phenotype) you have, with the exception of a few (e.g., blue eyes or sickle cell anemia), is the result of multiple genes working together—not a single gene.

Errors in one part of your DNA may not be so critical if other areas of your DNA are functioning correctly. You can think of this as your genetic backup system—or multiple sophisticated backup systems, critical for *survival of the*

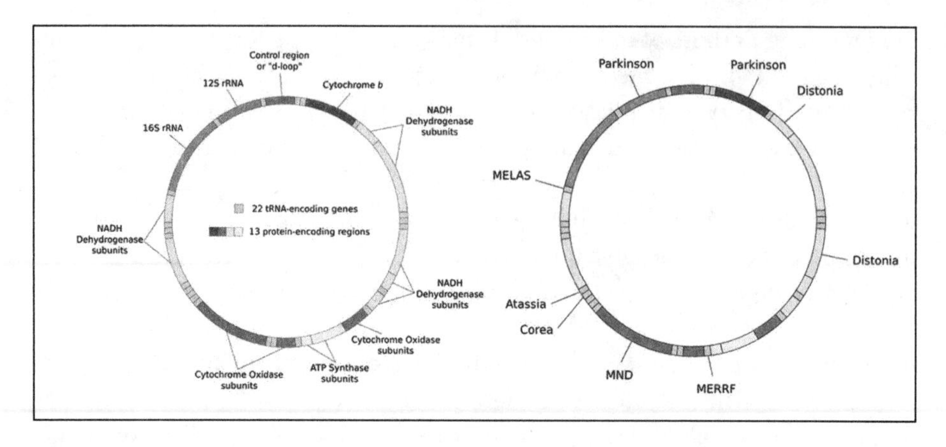

species. The more complex, the more critical the system is for survival, the more backup systems are involved.[13] Failure of more simplistic genetic sequences not necessary for life probably will not result in a problem like death. Failure of more critical life-dependent systems can result in death and are frequently the cause of spontaneous abortions during pregnancy. Simply put, the more critical the failure, the more likely you will die. Playing with genetic codes in one part of your DNA means that similar sequences elsewhere in your genetic code could also be changed, including changes to vital areas of DNA keeping you alive. Manipulating the genetic code and thinking you will only change the area of DNA you want to change, given 67 billion miles of DNA, is, at best, overly optimistic.

This overly simplistic approach of looking at physical (phenotypic) differences in people and trying to separate them from us led to the Greeks killing almost a quarter of their children. It also led to the eugenicists of the mid-1800s selectively trying to eliminate people who were different from them. After all, they saw themselves as genetically superior and these other people as inferior and undesirable. This view of racial, intellectual, physical, geographic, and religious superiority and differences in people led to "master" races trying to eliminate anyone they considered inferior. Britain, Germany, and the United States played a major role in this eugenics movement.

Although Hitler died and Nazi Germany may have technically lost the war, their biological experimentation and eugenics programs are very much alive and well today, which is more than you can say for the people they experimented on. It is only fitting that Hitler's eugenics and biological experimentation programs still exist in the United States and Britain, as these countries gave birth to them. The military-industrial complex that Presidents Eisenhower and Kennedy warned about is nothing but the continuation of those biological weapons and eugenics programs begun in the mid-1800s, now accelerated thanks to the ushering in of the age of genetic vaccines.

PARALLEL PATHWAYS—EUGENICS AND BIOWARFARE

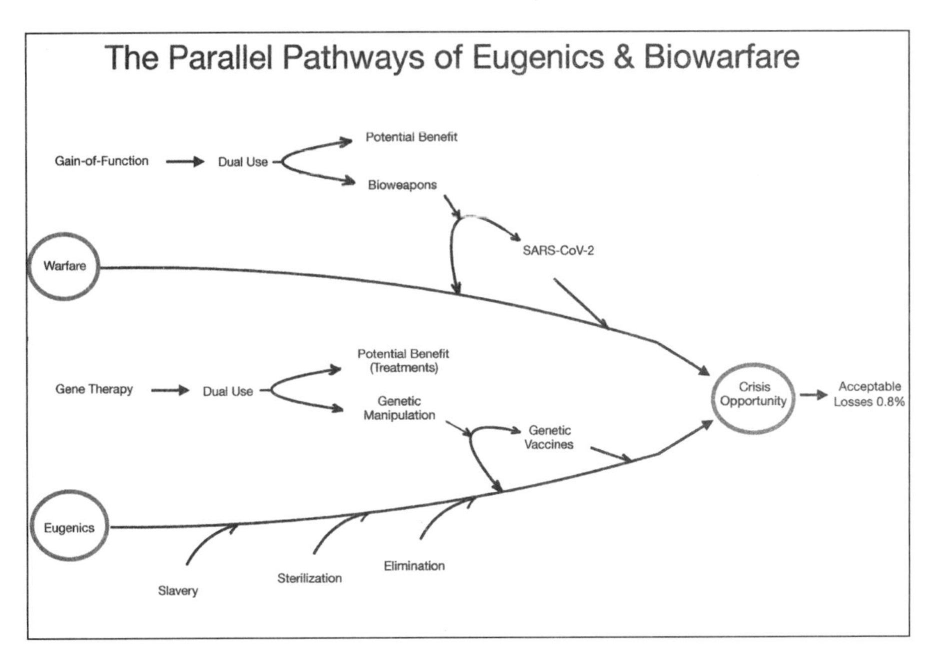

Despite warnings, Americans have continued to elect representatives from both the Republican and Democratic parties, believing that somehow elections determine the trajectory of the United States. Little if any attention was paid to the MIC, which continued moving forward, unrestrained by either political party. Efforts to control it have resulted in serious consequences. Would Americans have acted differently had they known the MIC was developing biological and eugenics weapons? Would Americans have heeded the call to stop the MIC had they recognized what was happening to American medicine and the American physician?

The war-game scenarios I shared with you show just how easily Americans and people around the world can be manipulated by their fears and self-interest, how easily distractable people are when presented with more than one problem, and how willing Americans and others are to surrender their inalienable rights, to turn a blind eye to the problems in front of them and the problems of others. As Göring so unwaveringly pointed out during his trial, it is quite easy to turn one person against another instead of confronting the real enemy, to make people hand over control of their lives instead of taking control of their lives.

As history has repeatedly demonstrated, the only thing necessary for evil to survive is for good men to do nothing. We have seen the virtue, integrity, and valor of those who by necessity came to America following persecution in other countries. We have listened to people from other countries tell us the stories of their persecution in the countries they were born in and how they came to the United States in hope of a better life. We have seen the consequences of Americans refusing to address nefarious actions, including the expansion of biological experiments carried out on members of our military, our children, ourselves. We have seen the consequences of Americans ignoring the development of biological weapons, the usurping of medicine, the manipulation of fellow citizens, justified by people thinking it simply didn't affect them. In some instances, they were told their fellow Americans were placing them and their loved ones at risk. We do not need to return to the America of pre-2020. Rather, we need to establish a new American perspective, one of holding our leaders responsible for their actions, and ourselves responsible for our response. It's time for a change.

We no longer live in a time of luxury, where we can blame others for our inactions. If you want to know what you would have done in Nazi Germany in the 1930s and 1940s, you now know—even if you don't want to admit it. But they are coming—for you, your children, and for those you love and care about. The question is no longer what would you have done in Nazi Germany. It is no longer what you would have done at any other time in history. The only question that matters is, what are you going to do now? Now that you have read this book and your eyes have been opened, the only question is which path are you going to choose now: *extinction* or *life*?

Ask yourself, what would Presidents Eisenhower and Kennedy ask you to do? How will you answer this call at this time in history? Will you be able to look back on these days not with regret, but with a sense of honor and confidence in what you did, the same confidence and honor that Captain Fleming and General George Washington had on Christmas Eve 1776?

CHAPTER 10

Obstruction of Justice

Most people would agree that justice requires a transparent presentation of the facts—the truth. To prevent this truth from being seen in the criminal justice system means you have obstructed justice. By definition[1]:

> Obstruction of justice broadly refers to actions by individuals that **illegally prevent or influence the outcome of a government proceeding**. While the quintessential example of obstruction of justice involves tampering in a judicial proceeding, there are numerous laws on obstruction of justice, covering all branches of government and targeting different kinds of obstruction. Instead of one law, law on obstruction of justice is located in multiple federal and state statutes, given the numerous methods in which obstruction can be carried out.
>
> While varying greatly, each obstruction of justice statute typically requires proof that the defendant 1) knew of a government proceeding and 2) acted with the intent to interfere with the proceeding. Here are a few of the most important federal obstruction of justice statutes:
>
> Obstruction of Jurors and Court Officers 18 U.S.C. § 1503: makes it illegal for someone to "corruptly" or through threats or force influence a juror or officer of the court in carrying out their duties before a judicial proceeding. The punishment for this crime can reach over 20 years imprisonment in the most extreme cases.
>
> Obstructing Witnesses and Evidence 18 U.S.C. § 1512: makes it illegal in any way to harm, threaten, delay, or otherwise influence a witness to an official proceeding, punishable by up to 30 years imprisonment. **The law also makes it a crime to destroy, change, or hide evidence that could be used in an official proceeding.**
>
> Obstruction of Congressional and Agency Proceedings 18 U.S.C. § 1505: makes it illegal for someone to "corruptly" or through threats or force influence any proceeding or inquiry in Congress or by a federal agency. The law also makes it illegal to obstruct documents and other evidence given in response to inquiries under the Antitrust Civil Process Act. The punishment of this crime can lead to up to 5 years imprisonment. [emphasis added]

Other statutes and laws can be relevant to obstructing justice such as bribery or contempt of court. Generally, both officials and judges take seriously and vigorously prosecute attempts by individuals to corruptly interfere with proceedings, regardless of the method.

In the instance of my trial, as explained,[2] not only was evidence of my actual innocence hidden from the jury,[3] but the judge prevented investigation of himself or the attorneys who hid that evidence[4] and later allowed the prosecutor to destroy all evidence when the prosecutor filed a document to destroy the evidence, without either the prosecutor or the court notifying me so I could object to the destruction of the evidence proving my innocence.

Given my prior experience with people who hide and destroy evidence, I warned in the author notes that I have copies of the materials referenced in this book. That includes evidence that the Department of Defense paid $419,511 to Ukrainian biolabs for a report on COVID in November 2019, seven weeks before the world would officially hear about SARS-CoV-2 infections and the deaths in Wuhan, China.

This $419,511 was paid to Black and Veatch Special Projects Corporation, which is connected with Metabiota and USAID Predict. As you saw earlier in this book, USAID funded gain of function biological viruses whose PCR sequences match SARS-CoV-2. According to material published about connections with Hunter Biden[5] and Metabiota, there are concerns that Mr. Biden is connected with these biolabs.

In the final editing of this book and in preparation for a presentation I provided to the Constitutional Sheriffs and Peace Officers Association (CSPOA) in Las Vegas, Nevada on 17 April 2024, I went to the USA Spending website to review the DoD postings for monies paid to Black and Veatch. Not only had the site been wiped of the monies paid for the COVID report, but the DoD now showed conflicting reports on money paid to Black and Veatch after 31 January 2013, with one report showing the HDTRA108D0007 funding contract closed on 31 January 2013, while another DoD grant HDTRA117F0044 was completed on 31 March 2021.

The discrepancies between cessation of funding to Black and Veatch by the DoD, and the statement that the HDTRA108D0007 funding of Black and Veatch was closed on 31 January 2013, when my prior investigations of the DoD funding site clearly showed the use of HDTRA108D0007 to fund the Ukrainian biolabs COVID report, raises serious concerns about the US government obstructing justice by destroying, or attempting to destroy evidence.

As we previously discussed under FRE 406,[6] the habit of the US government is the destruction or hiding of evidence. They did so in my case, we have seen several incidents throughout this book, and we see it yet again with this intentional and knowing removing of evidence and replacement with other material which you will see using the QR code provided here.

> Evidence of a person's habit or an organization's routine practice may be admitted to prove that on a particular occasion the person or organization acted in

accordance with the habit or routine practice. The court may admit this evidence regardless of whether it is corroborated or whether there was an eyewitness.

Summary and Explanation

Federal Rule of Evidence 406 focuses on the admissibility of evidence regarding a person's habit or an organization's routine practice in legal proceedings. This rule is part of the United States Federal Rules of Evidence, which govern the use of evidence in federal courts.

Key points of Rule 406 include:

1. Habit and Routine Practice Defined:
   Habit: Refers to a person's regular response to a specific situation or circumstance. Habitual behavior is consistent and predictable, almost reflexive in nature.
   Routine Practice: Pertains to the regular conduct of an organization or group. It implies a standardized response to repetitive situations.
2. Admissibility for Predictive Purposes:
   The rule allows the introduction of evidence of a habit or routine practice as proof that the conduct of the person or organization on a particular occasion was in accordance with that habit or routine practice.
   This is based on the principle that evidence of regular and consistent responses in similar situations can be a reliable predictor of behavior in a specific instance.
3. Examples of Application:
   For an individual, this might include evidence that a driver always stops at stop signs, to prove they likely stopped at a sign on a specific occasion.
   For an organization, evidence could demonstrate that a company follows strict maintenance protocols, suggesting adherence to these protocols in a specific instance.
4. No Need for Eyewitnesses:
   Unlike other types of evidence, proof of habit or routine practice does not require an eyewitness to the specific instance in question. The consistency of the behavior itself supports the inference of its occurrence.
5. Distinguishing from Character Evidence:
   Rule 406 is distinct from rules about character evidence (like Rules 404 and 405). Habit or routine practice is more specific and predictive than general character traits.

Given the ongoing investigations into the origins of SARS-CoV-2, the violations of treaty law, federal law, state law, and international law, and the ongoing conflict in the Ukraine with investigations into the potential involvement of Hunter Biden, and others, it raises the question: Why would the DoD now intentionally and knowingly obstruct justice by deleting records showing their funding of a COVID report seven weeks before the world heard about the gain of function biological viral weapon? Why would they then add records stating this funding grant stopped in January 2013?

While I never expect to see an expungement of my case or reparations for the harm caused to me and my family by the federal government, I find it unconscionable that the federal government should be allowed to obfuscate the facts, hide the evidence, destroy evidence, make false evidence, and remove the evidence from the USAspending.gov website. While at the same time having the impudence to state at the bottom of the USAspending.gov webpage "building a more transparent government."

If this is the federal governments definition of transparency, they are defining it differently than the rest of us are[7].

> Transparency [noun] "the quality of being done in an open way without secrets"

While those causing death, destruction, famine, hunger and war, continue to prosper through their lies, bioweapons, and eugenics programs, millions of people around the world are dead or maimed as a result of it.

If you are tired of these criminals getting away this, then it's time we stand together and do more than simply talk about it.

> There are of course those who do not want us to speak. I suspect even now, orders are being shouted into telephones, and men with guns will soon be on their way. Why? Because while the truncheon may be used in lieu of conversation, words will always retain their power. Words offer the means to meaning, and for those who will listen, the enunciation of truth. And the truth is, there is something terribly wrong with this country, isn't there?
>
> Cruelty and injustice, intolerance and oppression. And where once you had the freedom to object, to think and speak as you saw fit, you now have censors and systems of surveillance coercing your conformity and soliciting your submission. How did this happen? Who's to blame? Well certainly there are those who are more responsible than others, and they will be held accountable, but again truth be told, if you're looking for the guilty, you need only look into a mirror.
>
> I know why you did it. I know you were afraid. Who wouldn't be? War, terror, disease. There were a myriad of problems which conspired to corrupt your reason and rob you of your common sense. Fear got the best of you, and in your panic you turned to the [government who] promised you order, promised you peace, and all [it] demanded in return was your silent, obedient consent.
>
> Modified from V for Vendetta Broadcast Speech[8]

Sound familiar? If you're tired of the lies, the cover-ups, the blaming of innocent citizens like Julian Assange and Edward Snowden for exposing these wrongs; if you're tired of feeling like your freedoms have been taken away from you and you spend more time worried about what the government is doing to you than for you; if you're tired of feeling manipulated and not represented by the people you elected to serve you in office; if you feel like you're constantly walking in a mine field of lies, bioweapons, and eugenicists; then it's time for you to act and become part of the solution instead of part of the masses that allows these people to do what their Nazi predecessors did in World War II and have been doing here since.

It's time we act together to stop the criminals practicing biowarfare and eugenics. Join me and thousands of others as we call for the criminal indictments of these individuals, for civil litigation caused by the harm resulting from gene therapies carried out without informed consent, and for changes in the laws that have brought us to where we are today. The actions taken by the military-industrial complex (DoD) and its most recent destruction of evidence means we must act if we are to have a future and not become the next endangered species. It's time to give them a message they shall never forget. You have already taken the first step by reading this book and pulling back the veil of truth. Your next step is easier and it begins with 10Letters.org.

THE APPENDIXES

APPENDIX A

Fleming Doctorate Research and Training

APPENDIX B

Severe Acute Respiratory Syndrome Viruses (SARS-CoV-2) Gain-of-Function

(The following slides are taken from my 2020–2024 presentations, including the 2022 Crimes Against Humanity Tour presentations.)

APPENDIX C

Genetic Vaccine Development: InflammoThrombotic Response (ITR), Prion Diseases, Sudden Cardiac Death, and Spontaneous Abortions

(The following slides are taken from my 2020–2024 presentations, including the 2022 Crimes Against Humanity Tour Presentations.)

APPENDIX D

Human Immunity

(The following slides are taken from my 2020–2024 presentations, including the 2022 Crimes Against Humanity Tour Presentations.)

APPENDIX E

Rockefeller Foundation Whistleblower Materials

APPENDIX F

Fleming Presentation 29 December 2008 on the ACA

(Request by Obama-Biden Transition Team for Richard M Fleming, PhD, MD, JD, to hold a Town-Hall Meeting to Discuss the Affordable Health Care Act [ACA] and Concerns Americans Have with Federal Government Involvement in Health Care.)

APPENDIX G

All Images in *Are We the Next Endangered Species?*

Endnotes

Acknowledgments

1 This book does not promote one religious belief over another. You alone must determine what your religious beliefs are, or, perhaps more important, if you think humanity is the pinnacle of evolution. My arguments here are those of a physicist. If you believe this species is the ultimate, then how sad the universe truly is.

2 They gave themselves the name, not me.

3 "The Crimes Against Humanity Indictment Campaign—USA," Crimes Against Humanity Tour USA, n.d., https://crimesagainsthumanitytour.com/.

Author Notes on Why I Am Here in This Battle for Humanity

1 Arthur Herman, *The Viking Heart: How Scandinavians Conquered the World* (Boston: Mariner Books, 2021).

2 Fischer, David Hackett (2006). *Washington's Crossing*. New York: Oxford University Press. ISBN 0–19-518159-X; *George Washington Papers at the Library of Congress*, 1741–1799: Series 3a Varick Transcripts. The Library of Congress. Retrieved 2012–12-21.

3 Certified shorthand reporter.

4 A copy of this deposition under oath can be found at: https://rumble.com/v26m8gg-is-covid-19-a-bioweapon-a-scientific-and-forensic-investigation.html; and https://rumble.com/v1836p4-covid-crimes.html.

5 It's mythology when it's no longer what's accepted by the majority of people; if it's accepted, then we call it science or theology.

Foreword

1 Bruno Canard and Etienne DeCroly, "Wuhan, a New Hiroshima," *Maniere de voir* 179 (October/November 2021):

Chapter 1

1 James B. Donovan and E. Ray Kellog, dirs., *Nazi Concentration Camps*, videocassette, National Archives, NAID 43452, local ID 238.2, ca. 1945, https://catalog.archives.gov/id/43452; and https://en.m.wikipedia.org/wiki/Nazi_Concentration_Camps_(film).

2 Stanford prison experiment funded by the U.S. Office of Naval Research [ONR grant N00014–67-A-0112–0041]. Haney C, Banks C and Zimbardo P.

Interpersonal Dynamics in a Simulated Prison. Intern J Crim and Penology 1973:1, 69–97.

3 Richard M. Fleming, "Event 2021: What if the People you Trust are the People Causing the Problem?", Dallas, Texas, 5 June 2021.

4 USA Patriot Act. Pub. L. 107–56, 115 Stat. 272, 26 October 2001.

5 https://en.wikipedia.org/wiki/Assassination_of_John_F._Kennedy; Foreign and Military Intelligence, Book I, Final Report of the Select Committee to Study Governmental Operations with respect to Intelligence Activities United States Senate. 94th Congress, 2d session. Report No. 94–755. 14 April 1976.

6 Svrcina, "Svrcina—Meet Me on the Battlefield [Official Lyric Video]," 20 May 2016, video, 4:30, https://www.youtube.com/watch?v=GZrddJPGp1I.

Chapter 2

1 Our strengths and weaknesses are frequently the same. The outcome is determined by what you do with these strengths and weaknesses—the how and why.

2 Torah, Gen. 5:27. See the Israel Bible, https://theisraelbible.com/bible/genesis-5/.

3 Gen. 5:27.

4 Exod. 15:26, RSV.

5 Exod. 9:1–35, https://bible.usccb.org/bible/exodus/9.

6 George Homs, "Sneferu. Pharaoh of Egypt (± 2620-± 2547)," Genealogie Online, last modified 22 May 2012, https://www.genealogieonline.nl/en/stamboom-homs/I6000000003645877978.php.

7 Modern Slavery. Countries With the Highest Prevalence of Slavery written by Anna Fleck, https://www.statista.com/chart/30666/estimated-number-of-people-in-modern-slavery-per-1000/.

8 https://www.nationalarchives.gov.uk/help-with-your-research/research-guides/british-transatlantic-slave-trade-records/.

9 When writing the Declaration of Independence, Thomas Jefferson translated the Flemish document *Het Plakkaat van Verlatinge en de Declaration of Independence* into English, forming roughly two-thirds of the Declaration of Independence. He adapted the remaining one-third from the British citizens' complaints against King Charles I and and King Charles II of the 1600s.

10 https://www.monticello.org/thomas-jefferson/jefferson-slavery/thomas-jefferson-and-sally-hemings-a-brief-account/.

11 The Ethics and Religious Liberty Commission of the Southern Baptist Convention written by ERLC staff. 24 June 2022 found at: https://erlc.com/resource-library/articles/a-brief-timeline-of-when-slavery-ended-in-the-united-states/.

12 https://www.loc.gov/classroom-materials/civil-war-the-nation-moves-towards-war-1850-to-1861/

13 Galileo arrives in Rome to face charges of heresy. 13 February 1633. https://www.history.com/this-day-in-history/galileo-in-rome-for-inquisition.

14 Harvard Project Physics. Wiley Periodicals, 1968; The Project Physics Course (Text) Rutherford J, Holton G, Watson F. Holt, Rinehart and Winston © 1970.

15 His theory was the inheritance of acquired characteristics. Michael Bulmer, *Francis Galton: Pioneer of Heredity and Biometry* Johns Hopkins University Press, Baltimore, Maryland © 2003).
16 James J. Nagle, *Heredity and Human Affairs* (C.V. Mosby, 1974).
17 Charles Darwin, *On the Origin of Species by Means of Natural Selection* (London: John Murray, 1859).
18 Herbert Spencer, *The Principles of Biology in Two Volumes*, New York and London, D. Appleton and Company © 1866. Spencer used the term "survival of the species" in 1864, two years before publication of this book.
19 Johann Gregor Mendel, *Versuche über Pflanzenhybriden Verhandlungen des naturforschenden Vereines in Brünn*, Bd. IV für das Jahr, 1865 Abhandlungen, 3–47, http://www.mendelweb.org/Mendel.plain.html. http://www.esp.org/foundations/genetics/classical/gm-65.pdf.
20 Leslie A. Pray, "Discovery of DNA Structure and Function: Watson and Crick," *Nature Education* 1, no. 1 (2008): 100, https://www.nature.com/scitable/topicpage/discovery-of-dna-structure-and-function-watson-397/#.
21 As we now more fully appreciate, most genetic traits are the result of multiple chromosomal factors and not a single gene sequence. R. M. Fleming et al., "Nature vs. Nurture—Will the Species Survive?," *Scholarly Journal of Food and Nutrition* 2019;2(1):174–177, https://doi.org/10.32474/SJFN.2019.02.000127.
22 That is the very focus of the CRISPR, the genetic vaccine program to be discussed later in the book.
23 Francis Galton, "On the Anthropometric Laboratory at the Late International Health Exhibition," *Journal of the Anthropological Institute of Great Britain and Ireland* 14 (January 1885): 205–21, https://doi.org/10.2307/2841978.
24 Francis Galton, *Hereditary Genius: An Inquiry into Its Laws and Consequences* (London: Macmillan, 1869).
25 It is the science of improving stock—in this instance, human stock.
26 Alfred R. Wallace, review of *Hereditary Genius* by Francis Galton, *Nature*, 17 March 1870, 502–3.
27 Following the filing of my case (https://www.flemingmethod.com/about), the Nebraska court demanded that I submit to the collection of my DNA, as was done in all such cases. Clearly, the US federal court system has the same eugenics mentality that existed more than a century ago. Based upon the rationale, I argue that anyone who hides evidence of actual innocence (substantive exculpatory evidence) should be genetically tested.
28 Winston Churchill was a racist warmonger who tried to sterilize the mentally ill. By Zing Tsjeng found at: https://www.vice.com/en/article/ae55v8/winston-churchill-racist-warmonger-sterilize-mentally-ill.
29 The dark truth about Teddy Roosevelt. By Nick Vrchoticky at https://www.grunge.com/247661/the-dark-truth-about-teddy-roosevelt/; https://stljewishlight.org/top-story/takeaways-from-from-episode-1-of-the-u-s-and-the-holocaust/.
30 Margaret Sanger, interview by Mike Wallace, 21 September 1957, *The Mike Wallace Interview*, container 9.1, *The Mike Wallace Interview* Collection, manuscript collection

MS-04387, Harry Ransom Center, Austin, Texas; https://norman.hrc.utexas.edu/fasearch/findingAid.cfm?eadid=01040.

31 American Eugenics Society, 1972 Edited by Margaret Sanger; The Margaret Sanger Papers Project. "Birth Control and Racial Betterment" found at: https://m-sanger.org/items/show/1440; Margaret Sanger, "Birth Control and Racial Betterment," *Speeches and Writings of Margaret Sanger, 1911–1959*, accessed April 2, 2024, https://m-sanger.org/items/show/1440.

32 Gur-Arie, Rachel, Eugenic sterilization in the United States (1922), by Harry H. Laughlin. Embryo Project Encyclopedia (2015–08-12). ISSN: 1940–5030 found at: https://embryo.asu.edu/pages/eugenical-sterilization-united-states-1922-harry-h-laughlin.

33 Philip R. Reilly, "Eugenics and Involuntary Sterilization: 1907–2015," *Annual Review of Genomics and Human Genetics* 16 (2015): 351–68, https://doi.org/10.1146/annurev-genom-090314–024930.

34 "An Act Relating to the Prevention of Criminality, Insanity, Feeble-mindedness, and Epilepsy," chap. 693 in *Wisconsin Session Laws*, no. 559, July 31, 1913 (Madison, WI: Democrat Printing, 1913), https://books.google.com/books?id=rOogAAAAIAAJ&lpg=PA972&ots=X7ha3qA9N2&dq=July%2030%2C%201913%20Wisconsin%20law%20Chapter%20693&pg=PA971#v=onepage&q&f=false

35 https://en.wikipedia.org/wiki/Race_Betterment_Foundation; Engs RC. *The Progressive Era's Health Reform Movement: A Historical Dictionary.* Praeger; 2003; Eugenic sterilizations performed in US through 1932, Human Betterment Foundation, Cold Spring Harbor, Eugenics Record Office, Charles B. Davenport, 1933–1934. Archived at: http://www.eugenicsarchive.org/html/eugenics/static/images/1759.html.

36 https://en.wikipedia.org/wiki/Eugenics_Record_Office; http://www.eugenicsarchive.org/html/eugenics/static/themes/19.html.

37 Edwin Black, "Eugenics and the Nazis—the California Connection," *San Francisco Chronicle*, 9 November 2003, https://www.sfgate.com/opinion/article/Eugenics-and-the-Nazis-the-California-2549771.php.

38 https://en.wikipedia.org/wiki/Eugenics_in_the_United_States; Edwin Black (9 November 2003). "Eugenics and the Nazis—the California connection," *San Francisco Chronicle.* Retrieved 2 February 2017.

39 "Social Problems Have Proven Basis of Heredity; What the Work Done in the Eugenics Record Office at Cold Spring Harbor Has Proved in Scientific Race Investigation," *New York Times*, 12 January 1913, https://www.nytimes.com/1913/01/12/archives/social-problems-have-proven-basis-of-heredity-what-the-work-done-in.html, https://timesmachine.nytimes.com/timesmachine/1913/01/12/100604842.html.

40 H.H. Laughlin, Bulletin No. 10A: Report of the Committee to Study and to Report on the Best Practical Means of Cutting Off the Defective Germ-Plasm in the American Population: I. The Scope of the Committee's Work, 1914. CSHL Archives, Cold Water Spring Harbor Laboratory.; https://archivesspace.cshl.edu/repositories/2/archival_objects/95696.

41 Harry H. Laughlin, *Report on the Best Practical Means of Cutting Off the Defective Germ-Plasm in the American Population: 1. The Scope of the Committee's Work, 1914*, box 17, folder 5, bulletin no. 10A, Eugenics Record Office Collection, Cold Spring Harbor Laboratory Archives, Cold Spring Harbor, NY, https://archivesspace.cshl.edu/repositories/2/archival_objects/95696. The Eugenics Image Archive is a conglomerate of digital items from the following institutions: Cold Spring Harbor Laboratory, American Philosophical Society; Truman State University; Rockefeller Archive Center at Rockefeller University; University of Albany, State University of New York; National Park Service, Statue of Liberty National Monument; University College, London; International Center of Photography; Archiv zur Geschichte der Max Planck Gesellschaft, Berlin-Dahlem; and Special Collections, University of Tennessee, Knoxville.

42 Lutz Kaelber, "Eugenics: Compulsory Sterilization in 50 American States—Virginia" (presentation, Social Science History Association, 2012), https://www.uvm.edu/~lkaelber/eugenics/VA/VA.html#:~:text=Under%20the%20Eugenical%20Sterilization%20Act,84.

43 https://www.flemingmethod.com/thecase; https://www.youtube.com/watch?v=-5Va31X6Rq8

44 https://247wallst.com/special-report/2021/11/20/the-32-states-that-used-to-sterilize-their-citizens/; The 32 States That Used to Sterilize Their Citizens. By Liz Blossom. 24/7 Wall St. editorial team. Published 20 November 2021.

45 Jacobson v. Massachusetts, 197 U.S. 11 (1905), https://supreme.justia.com/cases/federal/us/197/11/.

46 https://en.wikipedia.org/wiki/Pierce_Butler_(judge); Stephen Jay Gould, "Does the Stonless Plum Instruct the Thinking Reed," in *Dinosaur in a Haystack* (1995) p. 287; Thompson, Phillip (February 20, 2005). "Silent Protest: A Catholic Justice Dissents in Buck v. Bell" (PDF). *Catholic Lawyer.* 43 (1): 125–148. Archived from the original (PDF) on January 13, 2013. Retrieved July 24, 2012.

47 https://www.irishcentral.com/roots/history/billy-graham-jfk-catholic; https://www.history.com/news/jfk-catholic-president; and https://archive.nytimes.com/campaignstops.blogs.nytimes.com/2011/12/10/when-a-catholic-terrified-the-heartland/; Billy Graham tried to stop JFK from becoming president because he was a Catholic. By Niall O'Dowd, Irish Central 12 July 2022; The Kennedy Speech that Stoked the Rise of the Christian Right by John Huntington, Politico, 8 March 2020.

48 The phenomenon is noticed today with the mandated genetic vaccines, supported by the belief that they are necessary to prevent more harm to people and society. But that fails to acknowledge that vaccines do not prevent infection or transmission but rather are designed to shorten infection response time. In the end, the EUA documents did not show a statistically significant reduction in COVID cases. https://www.flemingmethod.com/documentation. Documents 2, 3, and 51.

49 https://supreme.justia.com/cases/federal/us/410/113/; and Jerome A. Barron et al., *Constitutional Law: Principles and Policy, Cases and Materials.* 7th ed. (New York: LexisNexis, 2006), 511–528.

50 http://www.eugenicsarchive.org/html/eugenics/essay8text.html; Image Archive on the American Eugenics Movement. Dolan DNA Learning Center, Cold Spring Harbor Laboratory. Eugenics Sterilization Laws, Paul Lombardo, University of Virginia, found at: http://www.eugenicsarchive.org/html/eugenics/essay8text.html.

51 Paul Popenoe, *Applied Eugenics* (New York: Macmillan, 1918).

52 Edwin Black, "Eugenics and the Nazis—the California Connection," *San Francisco Chronicle*, 9 November 2003, https://www.sfgate.com/opinion/article/Eugenics-and-the-Nazis-the-California-2549771.php

53 Arguably, the same treatment happened for many hospitalized COVID patients, who were not given life-sustaining fluids or nutritional support during their hospitalizations. https://www.washingtonpost.com/arts-entertainment/2021/09/08/jimmy-kimmel-hospitals-unvaccinated-ivermectin/; https://www.nbcnews.com/think/opinion/if-covid-vaccine-refusers-are-turned-away-hospitals-doctor-offices-ncna1277475; https://www.beckershospitalreview.com/hospital-physician-relationships/one-physician-s-case-for-refusing-to-treat-unvaccinated-patients-in-person.html.

54 Madison Grant, *The Passing of the Great Race, or The Racial Basis of European History* (New York: Charles Scribner's Sons, 1916).

55 Henry Ford, *The International Jew: The World's Foremost Problem*, The Ford International Weekly, Dearborn, Michigan 1920.

56 https://history.state.gov/milestones/1921–1936/immigration-act; and https://library.missouri.edu/specialcollections/exhibits/show/controlling-heredity/america/immigration; The Immigration Act of 1924 (The Johnson-Reed Act). Department of State U.S.A. Office of the Historian. Milestones: 1921–1936. Found at: https://history.state.gov/milestones/1921–1936/immigration-act.

57 Adolf Hitler, *Mein Kampf* (London: Paternoster Library, 1935), https://archive.org/details/b3001024x.

58 That name is the approved translation of the Gesetz zur Verhütung erbkranken Nachwuchses, enacted on 14 July 1933 (Berlin: Reichsdruckerei, 1935). The Reichsdruckerei gave the official translation of the law into English.

59 His degree was in the Science of Racial Cleansing.

60 The Nuremberg Race Laws, United States Holocaust Memorial Museum, found at: https://encyclopedia.ushmm.org/content/en/article/the-nuremberg-race-laws and https://encyclopedia.ushmm.org/content/en/gallery/the-nuremberg-race-laws

61 Being Jewish is religious; it is *not* a genetic sequence or racial group.

62 https://en.wikipedia.org/wiki/Aktion_T4; Aktion T4, The Nazi Program That Slaughtered 300,000 Disabled People. Written by Richard Stockton, *All That's Interesting* 29 November 2021.

63 SS stood for Schutzstaffel, and, initially, Saal-Schutz.

64 The decoder of the Enigma Nazi computer, Alan Turing, was sentenced to chemical castration in March 1951 for the crime of "gross indecency" (homosexuality). Britannica, s.v. "Alan Turing," last modified 12 February 2024, https://www.britannica.com/biography/Alan-Turing/Computer-designer.

65 Genetic vaccination for the expressed purpose of altering native genes may be beneficial, or it may be used as a tool to modify the genes of those considered inferior. The outcome depends upon the motives and actions of the individuals involved.

66 Arthur Herman, *The Viking Heart: How Scandinavians Conquered the World* (Boston: Mariner Books, 2021), pp386–398.

67 Wernher von Braun, "Construction, Theoretical, and Experimental Solution to the Problem of the Liquid Propellant Rocket" (thesis, Technical University of Berlin, 16 April 1934).

68 Laura Schumm, "What Was Operation Paperclip?," History (website), last modified 4 March 2020, https://www.history.com/news/what-was-operation-paperclip#.

69 https://www.opendemocracy.net/en/countering-radical-right/false-flags-paperclips-and-super-spies-secret-nazi-origins-cia/; Records of the Secretary of Defense (RG 330), Interagency Working Group, The U.S. National Archives and Records Administration found at: https://www.archives.gov/iwg/declassified-records/rg-330-defense-secretary.

70 Brianna Nofil, "The CIA's Appalling Human Experiments with Mind Control," History (website), n.d., https://www.history.com/mkultra-operation-midnight-climax-cia-lsd-experiments.

71 This was called Operation Midnight Climax.

72 https://www.businessinsider.com/how-nazi-scientists-inspired-the-cia-to-use-lsd-2014-2?r=US&IR=T. How Nazi Scientists Taught the CIA To Dose Soviet Spies with LSD. By Michael B Kelly and Jeremy Bender. 13 February 2014.

73 "1942–1975: US Soldiers Experimental Guinea Pigs," Alliance for Human Research Protection, 19 December 2014, https://ahrp.org/1942–1975-u-s-soldiers-experimental-guinea-pigs/.

74 https://ahrp.org/1947–1960-sarin-soman-and-tabun-were-tested-on-soldiers/; 1947–1960: Sarin, Soman and Tabun Were Tested on Soldiers. Alliance for Human Research Protection; Raffi Khatchadourian Primary Sources: Operation Delirium. The New Yorker, 21 December 2012.

75 Biochemical Weapons and Pharmaceutical Rejects Were Tested on U.S. Soldiers, U.S. Army Chemical Research and Development Laboratories Technical Report by James S. Ketchum, et al. July 1964; David S. Martin, Vets feel abandoned after secret drug experiments. CNN Health, 1 March 2012. https://ahrp.org/biochemical-weapons-and-pharmaceutical-rejects-were-tested-on-u-s-solders-2/.

76 https://en.wikipedia.org/wiki/Josef_Mengele; Weindling, Paul (2002). "The Ethical Legacy of Nazi Medical War Crimes: Origins, Human Experiments, and International Justice." In Burley, Justine; Harris, John (eds.). *A Companion to Genethics. Blackwell Companions to Philosophy.* Malden, MA; Oxford: Blackwell. pp. 53–69. doi:10.1002/9780470756423.ch5. ISBN 978–0-631–20698-9; Kubica, Helena (1998) [1994]. "The Crimes of Josef Mengele." In Gutman, Yisrael; Berenbaum, Michael (eds.). *Anatomy of the Auschwitz Death Camp.* Bloomington: Indiana University Press. pp. 317–337. ISBN 978–0-253–20884-2.

77 https://www.washingtonexaminer.com/weekly-standard/the-military-industrial-complex-has-a-more-insidious-relative?utm_source=google&utm_medium

=cpc&utm_campaign=Pmax_USA_High-Intent-Audience-Signals&gad_source=2&gclid=EAIaIQobChMI9orly8aUgwMVcjfUAR0UjAvkEAEYASAAEgJBQPD_BwE; Jeff Bergner The Military-Industrial Complex Has a More Insidious Relative, Washington Examiner, 22 December 2016.

78 Inside the Vaults, "Eisenhower's 'Military-Industrial Complex' Speech Origins and Significance," 19 January 2011, video, 3:16, https://www.youtube.com/watch?v=Gg-jvHynP9Y.

79 Dwight D. Eisenhower, "President Dwight D. Eisenhower's Farewell Address (1961)," transcript, National Archives, Milestone Documents, https://www.archives.gov/milestone-documents/president-dwight-d-eisenhowers-farewell-address.

80 https://www.history.com/topics/cold-war/cuban-missile-crisis; Cuban Missile Crisis—Causes, Timeline & Significance by History.com Editors 4 January 2010.

81 https://www.archives.gov/milestone-documents/president-john-f-kennedys-inaugural-address; Inaugural Address, Kennedy Draft, 01/17/1961; Papers of John F. Kennedy: President's Office Files, 01/20/1961–11/22/1963; John F. Kennedy Library, National Archives and Records Administration.

82 Civil Rights Act of 1964, Pub. L. 88–352, 78 Stat. 241.

83 Executive Order APP #43, issued 29 January 1963; and Richard M. Fleming, Mirror-Mirror: An Honest Look at Our Courts, Big Pharma, Health Care System & Government, Amazon., November 2020 ISBN: 9798565528169.

84 https://news.yahoo.com/newly-released-jfk-documents-point-to-what-the-cia-was-hiding-002728388.html?fr=yhssrp_catchall; Michael Isikoff, Chief Investigative Correspondent, Newly released JFK documents point to what the CIA was hiding. 16 December 2022; Final Report of the Assassination Records Review Board. September 1998.

85 Philip Shenon Yes, the CIA Director Was Par of the JFK Assassination Cover-Up. Politico Magazine. 6 October 2015. https://www.politico.com/magazine/story/2015/10/jfk-assassination-john-mccone-warren-commission-cia-213197/.

86 https://www.nytimes.com/2023/07/16/us/politics/biden-jfk-assassination-papers.html.

87 David Marks, Why is Biden administration denying Robert F. Kennedy Jr. Secret Service Protection? | Opinion. *Miami Herald* 16 December 2023. https://www.msn.com/en-us/news/opinion/why-is-biden-administration-denying-robert-f-kennedy-jr-secret-service-protection-opinion/ar-AA1lBkpA.

88 Erin Blakemore How the Hitler Youth Turned a Generation of Kids into Nazis. History 11 December 2017. https://www.history.com/news/how-the-hitler-youth-turned-a-generation-of-kids-into-nazis.

89 See appendix A for my specific training and research.

90 Federal Policy Governing the Granting of Academic Degrees by Federal Agencies and Institutions, Director, Bureau of the Budget, to Secretary, Health, Education, and Welfare, 23 December 1954.

91 Executive Order APP #43, issued 29 January 1963

92 As part of this doctoral training, participants routinely engaged in 201 scenarios, making decisions about who lived and who didn't based upon the needs of the group for survival.

93 This Day in History: Robert F. Kennedy is fatally shot. History. https://www.history.com/this-day-in-history/bobby-kennedy-is-assassinated.

94 Martin Luther King Jr. Assassination History 28 January 2010. https://www.history.com/topics/black-history/martin-luther-king-jr-assassination.

95 Richard M Nixon Presidential Library and Museum Conversation 700–010 3 April 1972. https://www.nixonlibrary.gov/white-house-tapes/700/conversation-700–010.

96 He was the former chief of Family Planning Division at the Office of Economic Opportunity (OEO). The OEO was responsible for administering most of the War on Poverty programs created by President Lyndon B. Johnson.

97 A similar theme occurred during the SARS-CoV-2 COVID-19 pandemic.

98 Senate Health Subcommittee Hearing, 93rd Cong. (1973) Statement of Dr. Warren Hern, Former Chief of OEO Family Planning Division.

99 Bill Kovach Guidelines Found on Sterilization Discovered in a Warehouse—Physician Disputes O.E.O. 7 July 1973 https://www.nytimes.com/1973/07/07/archives/guidelines-found-on-sterilization-discovered-in-a-warehouse.html.

100 Relf v. Weinberger, 372 F. Supp. 1196 (D.D.C. 1974) at https://www.splcenter.org/sites/default/files/d6_legacy_files/Relf_Original_Complaint.pdf.

101 Relf v. Weinberger, Civil Action 1557–73, United States District Court for the District of Columbia.

102 https://www.theguardian.com/us-news/2020/jun/30/california-prisons-forced-sterilizations-belly-beast; and https://calmatters.org/newsletters/whatmatters/2023/03/forced-sterilization-california/. Shilpa Jindia, Belly of the Beast: California's dark history of forced sterilizations. *The Guardian* 30 June 2020; Lynn LA, More pain for California's forced sterilization patients. Cal Matters. 22 March 2023.

103 COVID means **co**rona**vi**rus **d**isease, first called COVID in 2019; hence, COVID-19. This disease is the result of inflammation and blood clotting, or InflammoThrombotic response to a series of viruses known as severe acute respiratory syndrome (SARS) 2 (a.k.a. SARS-CoV-2). https://www.flemingmethod.com/inflammothromboticresponse-itr-disease. This InflammoThrombotic response (ITR) was first described by me at the 1994 American Heart Association meetings in Dallas, Texas, and later published in 2000 in a cardiology textbook. Richard M. Fleming, "Pathogenesis of Vascular Disease," chap. 64 in *Textbook of Angiology*, edited by J. B. Chang, 787–98 (New York: Springer, 2000), https://doi.org/10.1007/978-1-4612–1190-7_64.

104 The saying is attributed to Winston Churchill. Echoes are in M. F. Weiner, Medical Economics 1976;53(5):227, "Don't Waste a Crisis—Your Patient's or Your Own"; Pat Graham et al., Never let a good crisis go to waste. *State of the Oil & Gas Industry* McKinsey & Company, December 2014).

105 https://oir.nih.gov/sourcebook/ethical-conduct/special-research-considerations/dual-use-research; *Dual Use Research of Concern: Balancing Benefits and Risks.* Hearing Before the Committee on Homeland Security and Governmental Affairs,

United States Senate. (26 April 2012) (statement by Anthony S. Fauci, Director, NIAID).

106 Richard M. Fleming, *Is COVID-19 a Bioweapon? A Scientific and Forensic Investigation* (New York: Skyhorse Publishing, 2021).

107 The Human Genome Project, National Human Genome Research Institute (NIH) at: https://www.genome.gov/human-genome-project.

108 F. Barré-Sinoussi et al., "Isolation of a T-Lymphotropic Retrovirus from a Patient at Risk for Acquired Immune Deficiency Syndrome (AIDS)," *Science* 220 (1983): 868–71, https://history.nih.gov/display/history/discovery+of+hiv.

109 A. Cervantes-Ayalc, R. R. Esparza-Garrido, and M. A. Velázquez-Flores, "Long Interspersed Nuclear Elements 1 (LINE 1): The Chimeric Transcript L1-MET and Its Involvement in Cancer," *Cancer Genetics* 241 (2020): 1–11.

110 That is a family of DNA sequences found in the genome of prokaryotic (bacteria and archaea) organisms.

111 DARPA, "Broad Agency Announcement Preventing Emerging Pathogenic Threats (PREEMPT)," Biological Technologies Office, HR001118S0017, 19 January 2018.

112 Edward U. Condon, edited by Daniel S. Gilmore. Final Report of the Scientific Study of Unidentified Flying Objects Conducted by the University of Colorado Under Contract to the United States Air Force, Board of Regents of the University of Colorado, © 1968). Sturrock, Peter A (1987). "An Analysis of the Condon Report on the Colorado UFO Project." *Journal of Scientific Exploration.* 1 (1): 75. Archived from the original on 2012–07-17.

113 Albert Einstein, *Relativity: The Special and General Theory* (New York: Henry Holt, 1920).

114 I learned and applied that lesson early on.

115 Stanley L. Miller, "A Production of Amino Acids Under Possible Primitive Earth Conditions," *Science* 117, no. 3046 (1953): 528–29.

116 Tunneling is the disappearance of a proton at one site and appearance of the same proton at a different site, resulting in the rearrangement of hydrogen bonds and, in the instance of DNA, the change of base pairs, which produces mutational changes, one nucleoside base at a time.

117 Tularemia is a bacterium spread by ticks, deer flies, and contact with infected animals. Prior to modern antibiotics, the mortality rate was approximately 60 percent.

118 S. I. Trevisanato, "The 'Hittite plaque", and Epidemic of Tularemia and the First Record of Biological Warfare," *Medical Hypothesis* 69, no. 6 (2007): 1371–74; and David P. Clark, Nanette J. Pazdernik, "Biological Warfare: Infectious Disease and Bioterrorism," chap. 22 in *Biotechnology 2nd edition,* Elsevier, 2016).

119 https://www.britannica.com/technology/biological-weapon/Biological-terrorism; and Friedrich Frischknecht, The history of biological warfare. 2003;4:S47-S52. https://www.ncbi.nlm.nih.gov/pmc/articles/PMC1326439/pdf/4-embor849.pdf.

120 This graphic and the material in appendix B, to be discussed in this chapter, are taken directly from Dr. Fleming's slides used from 2019 to the present, including but not limited to the Crimes Against Humanity Tour (https://crimesagainsthu manitytour.com/).

121 Biological terrorism definition at https://www.britannica.com/technology/biological-weapon/Biological-terrorism.
122 Some of the events discussed here are shown in appendix B.
123 R. M. Fleming, D. M. Fleming, and R. Gaede, "Treatment of Hyperlipidemic Patients: Diet versus Drug Therapy" (paper, Council on Arteriosclerosis, 67th Scientific Sessions of the AHA, November 1994(, 101; R. M. Fleming, D. M. Fleming, and R. Gaede, "Improving the Accuracy of Reading Percent Diameter Stenosis from Coronary Arteriograms: A Training Program" (paper, Council on Arteriosclerosis, 67th Scientific Sessions of the AHA, November 1994), 101; and R. M. Fleming, "Arteriosclerosis as Defined by Quantitative Coronary Arteriography" (paper, Council on Arteriosclerosis, 67th Scientific Sessions of the AHA, November 1994), 102.
124 Richard M. Fleming, "Pathogenesis of Vascular Disease," chap. 64 in *Textbook of Angiology*, edited by John B. Chang, 787–98 (New York: Springer, 2000), https://doi.org/10.1007/978–1-4612–1190-7_64.
125 I had tried desperately to avoid the field of infectious diseases, believing that Flemings had contributed sufficiently to the field. I had also tried to avoid getting involved with food and nutrition, again believing Flemings had contributed sufficiently to the field. Little did I know how much my decades of research would inevitably bring me into these fields.
126 Richard M. Fleming, "Pathogenesis of Vascular Disease," chap. 64 in *Textbook of Angiology*, edited by John B. Chang, 787–98 (New York: Springer, 2000), https://doi.org/10.1007/978–1-4612–1190-7_64.
127 2001–2004 DoD and ACS applications.
128 R. M. Fleming, "Determining the Outcome of Risk Factor Modification Using Positron Emission Tomography (PET) Imaging" (presentation, International College of Angiology, 40th Annual World Congress, Lisbon, Portugal, 28 June - 3 July, 1998); R.M. Fleming, L. Boyd, and M. Forster, "Angina Is Caused by Regional Blood Flow Differences—Proof of a Physiologic (Not Anatomic) Narrowing," (presentation, Joint Session of the European Society of Cardiology and the American College of Cardiology, Annual American College of Cardiology Scientific Sessions, Anaheim, CA, 12 March 2000), www.prous.com (for physician training and CME credit, April 2000); R. M. Fleming, L. Boyd, and M. Forster "Unified Theory Approach Reduces Heart Disease and Recovers Viable Myocardium" (presentation, 42nd Annual World Congress, International College of Angiology, San Diego, CA, 29 June 2000); R. M. Fleming, L. B. Boyd, and C. Kubovy, "Myocardial Perfusion Imaging Using High-Dose Dipyridamole Defines Angina. The Difference Between Coronary Artery Disease (CAD) and Coronary Lumen Disease (CLD)" (presentation, 48th Annual Scientific Session of the Society of Nuclear Medicine, Toronto, ON, 28 June 2001); R. M. Fleming, "The Influence of Diet on Homocysteine" (presentation, 4th International Homocysteine Conference, Basel, Switzerland, June 2003); R. M. Fleming, and W. C. Dooley, "Breast Enhanced Scintigraphy Test (B.E.S.T.) Imaging Utilizes Vascularity/Angiogenesis and Mitochondrial Activity to Distinguish Between Normal Breast Tissue, Inflammation and Breast

Cancer" (presentation, 8th International Conference, Vascular Endothelium: Translating Discoveries into Public Health Practice, Crete, Greece, June 2005); and R. M. Fleming, G. M. Harrington, and R. Baqir, "Use of Parabolic Model in Tomographic Diagnosis of Infarction and Stenosis" (presentation, 1st Congress on Controversies in Cardiovascular Diseases: Diagnosis, Treatment and Intervention (C-Care), Berlin, Germany, 4–5 July 2008.

129 Neu 5Ac is N-acetylneuraminic acid; Neu 5Gc is N-glycolylneuraminic acid.

130 See appendix B for publications.

131 See details at https://www.flemingmethod.com/gain-of-function.

132 This is also referred to as SARS, from the 2002 outbreak.

133 These are bioweapons by definition, as it is unconscionable to support an altruistic claim of working with viruses—or other pathogens—that could possibly combine into a new virus, or are extinct, or have been modified with a life-threatening virus to attack people when it could not previously do so.

134 M. Furmanski, "Threatened Pandemics and Laboratory Escapes: Self-Fulfilling Prophecies," *Bulletin of the Atomic Scientists* 31 (March 2014): https://thebulletin.org/2014/03/threatened-pandemics-and-laboratory-escapes-self-fulfilling-prophecies/.

135 Richard H Kruse and Manuel S. Barbeito, A History of the American Biological Safety Association Part II: Safety Conferences 1966–1977. ABSA International at https://absa.org/about/hist02/; https://absa.org/topic/covid19/; and https://absa.org/abj/abj/121704FAHenkel.pdf.

136 Allie Griffin Newly released documents on 2018 coronavirus research proposal reveal scientists' safety concerns over Chinese lab. *New York Post*. 19 December 2023 at https://nypost.com/2023/12/19/news/2018-covid-docs-reveal-scientists-safety-concerns-over-chinese-lab/.

137 "U.S. Government Gain-of-Function Deliberative Process and Research Funding Pause on Selected Gain-of-Function Research Involving Influenza, MERS, and SARS Viruses," 17 October 2014 at https://www.phe.gov/s3/dualuse/documents/gain-of-function.pdf.

138 Scroll down to the Gene Snap program and viral data bases at https://www.flemingmethod.com/gain-of-function.

139 Interferon is a critical part of the innate (first part) human immune defense system to protect from viruses.

140 Documentation of Department of Defense funding to Ukrainian labs for a report on COVID-19 on 12 November 2019 issued through Black and Veatch Special Projects Corporation can be found in Appendix B and at https://www.usaspending.gov/award/CONT_IDV_HDTRA108D0007_9700 On 3 April 2024 site https://www.usaspending.gov/award/CONT_AWD_HDTRA117F0044_9700_HDTRA108D0007_9700 and the original site https://www.usaspending.gov/award/CONT_IDV_HDTRA108D0007_9700 shows no funding after 20 August 2015.

141 "Statement in Support of the Scientists, Public Health Professionals, and Medical Professionals of China Combatting COVID-19," *The Lancet* 395, no. 10226 (7 March 2020: e42–e42, https://doi.org/10.1016/s0140–6736(20)30418–9.

142 Sharon Lerner, Mara Hvistendahl, Maia Hibbett, NIH Documents Provide New Evidence U.S. Funded Gain-of-Function, The Intercept. 9 September 2021 at https://theintercept.com/2021/09/09/covid-origins-gain-of-function-research/.

143 R. M. Fleming "Unmasking CoViD, Part 1," (self-pub., 2020), Kindle; and R. M. Fleming, "Unmasking CoViD, Part 2," (self-pub., 2020), Kindle.

144 Vaccinated.

145 Unvaccinated.

146 "FDA History," US Food & Drug Administration, last modified 29 June 2018, https://www.fda.gov/about-fda/fda-history#:~:text=Although%20it%20was%20not%20known%20by%20its%20present,commerce%20in%20adulterated%20and%20misbranded%20food%20and%20drugs.

147 R. M. Fleming and T. K. Chaudhuri, "Should Big Pharma Be Held Legally Liable for Misrepresentations Made to the FDA?," special issue 1, *Acta Scientific Medical Sciences* (2019): 5–7, https://doi.org/10.31080/ASMS.2019.S01.0003 at https://actascientific.com/ASMS/pdf/ASMS-SI-01–0003.pdf and Richard M. Fleming and Tapan K. Chaudhuri, "FDA Drug Recalls Epidemic. Re: Weight Loss Pill Praised as "Holy Grail" Is Withdrawn from US Market over Cancer Link, *The BMJ* 368 (21 February 2020) at https://www.bmj.com/content/368/bmj.m705/rr.

148 https://www.flemingmethod.com/documentation. Numbers 2, 3, and 51 are the three FDA submitted EUA documents for Pfizer, Moderna and Janssen respectively.

149 This disease (coronavirus disease with its InflammoThrombotic response, or ITR) makes the person sicker than someone merely infected with the virus (SARS-CoV-2).

150 That is the best argument for mass vaccination of everyone every two weeks.

151 Katalin Karikó et al., "Incorporation of Pseudouridine into mRNA Yields Superior Nonimmunogenic Vector with Increased Translational Capacity and Biological Stability," *Molecular Therapy* 16, no. 11 (2008): 1833–40, https://doi.org/10.1038/mt.2008.200. See appendix C.

152 That's the exact opposite of what one would want if the purpose was to make a vaccine that helped people.

153 These pathways also describe in greater detail why both the SARS-CoV-2 viruses and genetic vaccines produce myocardial infarctions (heart attacks), cerebral vascular accidents (strokes), and cancers and are associated with spontaneous abortions (miscarriages) and yield long-term health problems—post-acute sequalae, a.k.a. long COVID. My website at https://www.flemingmethod.com/ includes three presentations that you can listen to.

154 The T-antigens (T for tumor or cancer).

155 Most of the DNA in human chromosomes is not involved with carrying the blueprints for making us. Most of the DNA is involved in making decisions about what should or shouldn't be made. These are the promoter and suppressor genes. Promoter genes are involved in making certain that something is made. Suppressor genes are involved in making certain that something isn't made. LINE-1 elements are responsible for adding genetic information into our DNA—new genetic code.

156 Myocardial infarction is heart-tissue death, frequently referred to by the general public as a heart attack.

157 Cerebral Vascular Accident (CVA), Reversible Ischemic Neurologic Deficit (RIND) and Transient Ischemic Attack (TAI) are interrupted blood flow to the brain with damage (symptoms) lasting greater than 72 hours, 24–72 hours, and less than 24 hours, respectively. Frequently referred to as a stroke by the general public.

158 More technically, it is known as pulmonary (lung) embolus (a moving blood clot), a.k.a. PE. These blood clots frequently form elsewhere in the body and travel to the lungs, blocking blood flow and the ability of the lungs to exchange critical oxygen into the body or remove of acid (carbon dioxide) out of the body.

159 R. M. Fleming et al., "Pfizer, Moderna, and Janssen Vaccine InflammoThrombotic and Prion Effect on Erythrocytes When Added to Human Blood. *Haematology International Journal* 7, no 1 (2023): https://doi.org/10.23880/hij-16000210.

160 Three free presentations discussing this material are available at https://www.fleming method.com/.

161 That research, with published information, has now received the 2023 Nobel Prize in Medicine.

162 See graphics in appendix D.

163 For example, if you develop strep throat, antibodies to the bacteria *streptococcus* may actually attack the valves of your heart, causing rheumatic heart disease. If this happens, you can have multiple heart problems and may even require surgery to replace the valves. Your doctor should therefore prescribe antibiotics in this circumstance to prevent you from making antibodies that could harm your heart.

164 These are diseases in which the immune system attacks the person's body.

165 If you think the incidence of cancer is high, you should know that it is postulated that each person has a new cancer cell develop in their body every day. These cancers are killed by the natural killer (NK) cells of our immune system.

166 https://www.cdc.gov/vaccines/schedules/hcp/imz/child-adolescent.html#notes; and https://www.cdc.gov/vaccines/covid-19/downloads/COVID-19-immunization -schedule-ages-6months-older.pdf.

167 When the energy bond of phosphate is added to a nucleoside base, it is now called a nucleotide (nucleoside + phosphate group).

168 Sanger sequencing is the method used to sequence nucleosides and nucleotides and reconstruct the genetic sequences.

169 Failure of the Operation Warp Speed and genetic vaccine manufacturers to appreciate their pressure selection using this approach can be summed up in the single word BlondStar. See https://www.lsxmag.com/news/blondestar-spoof-of-onstar/.

170 More on this can be seen in the slides provided in appendix B.

171 https://www.britannica.com/biography/Josef-Mengele.

172 Sarah McCammon, A medical ethicist weighs in on how to approach treating unvaccinated people. NPR 23 January 2022 at https://www.npr.org/2022/01/23 /1075168721/a-medical-ethicist-weighs-in-on-how-to-approach-treating-unvaccinated -people.

173 Laura Santhaham, Breakthrough COVID infections show 'the unvaccinated are now putting the vaccinated at risk' PBS News Hour 29 July 2021 at https://www

.pbs.org/newshour/health/breakthrough-covid-infections-show-the-unvaccinated-are-now-putting-the-vaccinated-at-risk.

174 https://www.genome.gov/about-genomics/educational-resources/fact-sheets/human-genome-project.

175 At the center of every cell of your body, except your red blood cells that lack a nucleus once matured and released from your bone marrow, is a special region with.

176 The average human nucleus ranges from 5 to 20 microns, or 0.00019685 inches to 0.000787402 inches.

177 When DNA makes RNA, this is called *transcription*. When RNA makes DNA, this is called *reverse transcription*. When RNA makes protein using your ribosomes (rRNA) this is called *translation*.

178 Histones at https://en.wikipedia.org/wiki/Histone; Mariño-Ramírez L, Kann MG, Shoemaker BA, Landsman D (October 2005). "Histone structure and nucleosome stability" https://www.ncbi.nlm.nih.gov/pmc/articles/PMC1831843/.

179 mRNA is messenger RNA, which carries the DNA message of what is to be made. rRNA is the RNA that makes up the ribosomes, which are the specialized organelles (little organs) in the cytoplasm of cells, where the mRNA and tRNA come together to make proteins (translation). tRNA is the RNA that transfers the amino acids to the ribosome, where the amino acids are attached according to the genetic code brought to the ribosome by the mRNA.

180 Despite all the rhetoric, only the United States has detonated a nuclear weapon upon another country, along with one in the Nevada desert to prove it would work.

181 It's interesting to note that it is the DOE that concurred with the discussions that SARS-CoV-2 was consistent with possible gain-of-function research. That's another example, I fear, of physicists understanding the importance of cautiously investigating the universe, while biologists bound forward into the quagmire of biological ocean algae, unaware of the dangers that accompany their recklessness.

182 Sanger sequencing has made it possible for many laboratories to be able to sequence genetic nucleoside bases. How does Sanger Sequencing Work?—Seq it Out #1. Thermo Fischer Scientific. 17 June 2015 at https://www.youtube.com/watch?v=e2G5zx-OJIw.

183 ORF X interferes with people being able to make the interferon needed to protect against viruses.

184 Human Genome Project Information Archive 1990–2003, U.S. Department of Energy Office of Science, Office of Biological and Environmental Research at https://web.ornl.gov/sci/techresources/Human_Genome/project/hgp.shtml#:~:text=The%20Human%20Genome%20Project%20(HGP,accessible%20for%20further%20biological%20study.

185 US Department of Health and Human Services and US Department of Energy, *Understanding Our Genetic Inheritance. The U.S. Human Genome Project: The First Five Years FY 1991–1995*, NIH Publication No. 90–1590 (Bethesda, MD: National Institutes of Health, April 1990), http:// http://www.esp.org/misc/genome/firstfiveyears.pdf.

186 The SRY gene provides male traits.

187 https://en.m.wikipedia.org/wiki/Human_Genome_Project; Robert Krulwich (2003). *Cracking the Code of Life* (Television Show). PBS.

188 Wrighton, Katharine (February 2021). "Filling in the gaps telomere to telomere." *Nature Milestones: Genomic Sequencing:* S21.

189 "Board of Scientific Counselors," National Human Genome Research Institute, last modified 24 January 2024, https://www.genome.gov/about-nhgri/Institute-Advisors/Scientific-Counselors-board.

190 If we can't sterilize, abort, or euthanize the undesirable individuals in society, we might now be able to simply change them into our image of perfection.

191 CRISPR means clustered regularly interspaced palindromic repeats.

192 Bacteriophages are viruses that feed (-phage) on bacteria.

193 Yes, Virginia, viruses are real. Just because you cannot see them doesn't mean they aren't real, any more than gravity isn't real. The challenge for people who wished viruses weren't real is to get them past the mid 1800s, when eugenics began and people were launching biological weapons at the enemy without any idea of what they were really killing other people with. Remember, Dr. Ignaz Semmelweis tried to get people to wash their hands between patients to reduce mortality. Germs weren't detectable yet, but their effect, like that of gravity, was seen. He was attacked and lost his membership in professional communities and his medical practice, and he died before he was vindicated. Maybe he just billed in a way his attackers didn't approve of?

194 Eric S. Lander, The Heroes of CRISPR. *Cell* 2016;164:18–28. https://www.broadinstitute.org/what-broad/areas-focus/project-spotlight/crispr-timeline;

195 The abbreviation cr stands for CRISPR.

196 Cas = **C**RISPR **as**sociated genes.

197 Lisa Miorin, et al. SARS-CoV-2 Orf6 hijacks Nup98 to block STAT nuclear import and antagonize interferon signaling. 2020;117(45):28344–28354. https://www.flemingmethod.com/_files/ugd/659775_5e6e2c89836a4a7d9895f1a99b6aa492.pdf.

198 Hansjörg Schwertz, et al. Endogenous LINE-1 (Long Interspersed Nuclear Element-1) Reverse Transcriptase Activity in Platelets Controls Translational Events Through RNA–DNA. Arterioscler Thromb Vasc Biol 2018;38:801–815 at https://www.flemingmethod.com/_files/ugd/659775_404687bc70984491bafd5bb8b25be869.pdf; and Jessica S. Wong, et al. Characterization of full-length LINE-1 insertions in 154 Genomes. Genomics 2021;113(6):3804–3810. at https://www.flemingmethod.com/_files/ugd/659775_62ebd28609e84ff790b9ae2e1e446c80.pdf.

199 Elitza Deltcheva et al., "CRISPR RNA Maturation by *Trans*-encoded Small RNA and Host Factor RNase III," *Nature* 471 (2011): 602–7, https://doi.org/10.1038/nature09886.

200 Katie Hunt, How human gene editing is moving on after the CRISPR baby scandal. CNN Health. 9 March 2023 At https://www.cnn.com/2023/03/09/health/genome-editing-crispr-whats-next-scn/index.html.

201 That is gain-of-function.

202 A vector is the way to get into the cell. Currently, that means LNP or adenovirus for the SARS-CoV-2/COVID genetic vaccines.

203 Prions are abnormally folded proteins that lead to further proteins misfolding when they touch. When this happens, the organ involved stops working correctly.

204 Arcady R. Mushegian and Santiago F. Elena, RNAs That Behave Like Prions. mSphere 2020;5:e00520–20 at https://www.flemingmethod.com/_files/ugd/659775_1309c785aba141359d869daa802bc5a5.pdf.

205 Rainer H Müller, et al., 20 years of lipid nanoparticles (SLN and NLC): present state of development and industrial applications. 2011;8(3):207–227 at https://pubmed.ncbi.nlm.nih.gov/21291409/.

206 LNPs assemble based upon orientation of these fat- and water-soluble components, much like oil and water.

207 This is just one of the reasons why different vaccine batches can have significant differences in outcomes.

208 Pieter R. Cullis and Michael J. Hope, Lipid Nanoparticle Systems for Enabling Gene Therapies. *Molecular Therapy* 2017;25(7):1467–1475 at https://www.ncbi.nlm.nih.gov/pmc/articles/PMC5498813/pdf/main.pdf.

209 Sonia Ndeupen, et al., The mRNA-LNP platform's lipid nanoparticle component used in preclinical vaccine studies is highly inflammatory. bioRxiv preprint doi: 10.1101/2021.03.04.430128 at https://www.flemingmethod.com/_files/ugd/659775_b47d90b16e6543b68dcbb98fcb387772.pdf.

210 Human cells (other than our red blood cells, or RBCs) have two sets of DNA: one that results from the combination of our parents' DNA (found inside the nucleus of our cells), and one that we obtained from our mother (mitochondrial DNA).

211 Janssen is Johnson & Johnson.

212 R. M. Fleming et al., "Pfizer, Moderna and Janssen Vaccine InflammoThrombotic and Prion Type Effect on Erythrocytes When Added to Human Blood," *Haematology International Journal* 7, no. 1 (2023): https://doi.org/10.23880/hij-16000210.

213 Erythrocytes (RBCs) are red in color when oxygen is attached to the hemoglobin molecule inside the RBC.

214 Gene therapy. What is biotechnology? at https://www.whatisbiotechnology.org/index.php/science/summary/gene-therapy/; and What is Gene Therapy? U.S. Food and Drug Administration at https://www.fda.gov/vaccines-blood-biologics/cellular-gene-therapy-products/what-gene-therapy.

215 Long Term Follow-Up After Administration of Human Gene Therapy Products. Guidance for Industry. US. Dept HHS, FDA, CBER. January 2020 at https://www.fda.gov/media/113768/download?attachment.

216 These are not necessarily the only reports on gene therapy.

217 Design and Analysis of Shedding Studies for Virus or Bacteria-Based Gene Therapy and Oncolytic Products; Guidance for Industry Availability, Federal Register Volume 80, Number 166. 27 August 2015 at https://www.govinfo.gov/content/pkg/FR-2015–08-27/html/2015–21235.htm.

218 For anyone debating whether shedding is a real phenomenon, you need look no further than the government's own report on the subject. Shedding Is Real (2015).

219 The furin cleavage site is the proline-arginine-arginine-alanine (PRRA) amino acid insertion discussed in *Is COVID-19 a Bioweapon? A Scientific and Forensic Investigation* (New York: Skyhorse Publishing, 2021).

220 Either *replicase* (an enzyme that promotes replication), or *promoter genes* (e.g., sv40) will allow for increased production of the *transgene* (the gene being introduced into the cell, the gene to make the desired protein—e.g., the SARS-C0V-2 spike protein).

221 This was published in the journal *Gene Therapy*.

222 That is the virus Professor Ralph Baric was working with prior to SARS viruses. Methods For Producing Recombinant Coronavirus USPTO Patnet No.: 7279327 B2. 9 October 2007 at https://www.flemingmethod.com/_files/ugd/659775_209bf49687244cbdb41ac978cde2edb3.pdf.

223 A replicon is that which increases the replication of the spike protein.

224 As a journal reviewer, I have reviewed material people are trying to publish, but what I have seen to date does not meet the level of science required for the scientific or medical community to make statements—which could be just as harmful as not.

225 FDA Approves First Gene Therapies to Treat Patients with Sickle Cell Disease. US FDA. 8 December 2023 at https://www.fda.gov/news-events/press-announcements/fda-approves-first-gene-therapies-treat-patients-sickle-cell-disease.

226 Those cells that give rise to all of your erythrocytes (red blood cells) you will ever make for your body.

227 Hemoglobin is that part of your red blood cells responsible for carrying oxygen and carbon dioxide. Hemoglobin (Hb) is made up of 2-alpha (α) chains and 2-beta (β) chains.

228 Selami Demirci et al., "βT87Q-Globin Gene Therapy Reduces Sickle Hemoglobin Production, Allowing for *Ex Vivo* Anti-Sickling Activity in Human Erythroid Cells," *Molecular Therapy: Methods and Clinical Development* 17 (June 2020): 912–21, https://pubmed.ncbi.nlm.nih.gov/32405513/, https://www.cell.com/molecular-therapy-family/methods/pdfExtended/S2329–0501(20)30073–5.

229 To harvest bone marrow, a doctor needs to surgically go into the bone and remove the marrow part.

230 Gen. 1:1.

231 The power is certainly not consistent with the meaning of the term *humane*: "having or showing compassion; civilized or refined behavior").

232 Gen. 1:26 (KJV).

233 Gen. 2:17 (KJV).

234 Just for the record, I am the last person who will shy away from advancing the knowledge of humanity. In the learning and applying of such knowledge, we can advance humanity, defeat disease, and improve who we are. However, how we use that knowledge has been a guiding principle in my life. As one of my mentors used to tell people, "Richard doesn't suffer fools gladly," and I don't suffer the criminality of the government, courts, or would-be dictators any more gladly.

235 Tristan Hughes, Why Was 900 Years of Euroopean History Called "the Dark Ages"? History HIT, 26 October 2022 at https://www.historyhit.com/why-were-the-early-middle-ages-called-the-dark-ages/.

236 Failure to act is an action. Chamberlain showed that when he negotiated with Hitler. Chamberlain and Hitler 1938, The National Archives at https://www.nationalarchives.gov.uk/education/resources/chamberlain-and-hitler/.

237 U.S. and World Population Clock, United States Census Bureau at https://www.census.gov/popclock/.

238 US Coronavirus vaccine tracker. USA Facts at https://usafacts.org/visualizations/covid-vaccine-tracker-states/.

239 Youri Benadjaoud and Mary Kekatos, About 36M American adults have received the updated COVID vaccine: CDC, ABC News. 14 November 2023 at https://abcnews.go.com/Health/36m-american-adults-received-updated-covid-vaccine-cdc/story?id=104874582.

240 I am not antivaccine; I am anti stupid, and as explained, the EUA documents provide no evidence of a desired vaccine outcome. For your information, no vaccine has ever prevented a person from getting infected or spreading the infection. Despite what you may have heard, the "function" of a vaccine, as taught to scientists and allopaths (MDs like myself) is to expose you to something, to make an immune response to that something, and then to make *memory cells* (both T- and B-memory cells). When you become infected, you will respond sooner, hopefully shortening your response time and symptoms. This should shorten your communicability (likelihood of spreading the infection). But it does not prevent you from getting infected or spreading the infection. In fact, you must become infected for a vaccine to provide you with any benefit. Additionally, if the effort that went into justifying the use of these three genetic vaccines is a reflection of the effort that went into the others, then my confidence in any of the vaccines is not scientifically reassured.

241 VAERS Vaccine Adverse Event Reporting System, HHS at https://vaers.hhs.gov/; and VAERS Analysis , Weekly Analysis of the VAERS data at https://vaersanalysis.info/2023/12/02/vaers-summary-for-covid-19-vaccines-through-11–24-2023/

242 It is important to take these numbers at face value rather than mentally masturbating over whether this reflects 1 percent, 10 percent, or all of the vaccine adverse events. In the end, if this is the accepted number, and if this is not enough, then adding one or two zeros behind the number isn't going to change anything.

243 Excess Deaths Associated with COVID-19, CDC, Archived 27 September 2023 at https://www.cdc.gov/nchs/nvss/vsrr/covid19/excess_deaths.htm#dashboard.

244 COVID Data Tracker, CDC at https://covid.cdc.gov/covid-data-tracker/#data tracker-home.

245 That's more deaths than the number of those who died in the Nazi concentration camps.

246 This is identical to the 201 games we played out repeatedly.

247 Remember, I took the worst numbers showing the most harm.

Chapter 3

1 Concupiscence.

2 Gen. 1:26.

3 Exod. 21, שְׁמוֹת; Parashat Mishpatim: פרשת משפטים. [A summary of Moses' discussions with the people of Israel regarding the ethical and ritual laws sealing the covenant.]

4 Exod. 9:1–3 (KJV), https://www.biblegateway.com/passage/?search=Exodus%20 9:1–7&version=KJV.

5 Arthur Herman, *The Viking Heart: How Scandinavians Conquered the World* (Boston: Mariner Books, 2021).

6 K. S. Lai excerpted from the book "Muslim Slave System in Medieval India", Medieval India: Enslavement of Hindus by Arab & Turkish Invaders. *Sanskriti.* at https://www.sanskritimagazine.com/medieval-india-enslavement-hindus-arab -turkish-invaders/.

7 Gen. 4:8.

8 Rabbi Wein, The Garden of Eden. @torah.org found here: https://torah.org /torah-portion/rabbiwein-5775-bereishis/.

9 What were a pharaoh's roles and responsibilities? History Skills at https://www .historyskills.com/classroom/ancient-history/anc-role-of-pharaoh-reading/#:~:text =The%20pharaoh%20was%20considered%20to%20be%20a%20living,meant %20something%20like%20%22order%22%2C%20%22truth%22%20and%2For %20%22justice%22%29%20prevailed.

10 1953: Queen Elizabeth takes coronation oath." BBC News. 2 June 1953 at http: //news.bbc.co.uk/onthisday/hi/dates/stories/june/2/newsid_2654000/2654501.stm and https://en.wikipedia.org/wiki/Coronation_of_Elizabeth_II.

11 Robert G. Ingersoll, The Works of Robert G. Ingersoll, Vol 3 of 12. Dresden Edition Lectures. 9 February 2012. at https://www.gutenberg.org/files/38803/38803-h /38803-h.htm.

12 Graham C. Lilly, Daniel J. Capra, and Stephen A. Saltzburg, *Principles of Evidence*, 5th ed. (St. Paul, MN: West Academic Publishing, 2009).

13 Federal Rules of Evidence 2024 Edition at https://www.rulesofevidence.org /fre/article-iv/rule-406/ and Title 28, Appendix—Rules of Evidence, Rule 406, Authenticated U.S. Government Information, GPO, page 366 at https://www .govinfo.gov/content/pkg/USCODE-2011-title28/pdf/USCODE-2011-title28-app -federalru-dup2-rule406.pdf.

14 Rules of Evidence, Pub. L. 93–595, § 1, 88 Stat. 1932 (2 January 1975), at 356, https://www.govinfo.gov/content/pkg/USCODE-2009-title28/pdf/USCODE -2009-title28-app-federalru-dup2-rule407.pdf; Robert P. Mosteller, McCormick on Evidence, 8th (Practitioner Treatise Series), Thomson Reuters, © 2020.

15 Larry Holzwarth, These 30 Rulers in History Were Hated by All, History Collection, 25 July 2019 at https://historycollection.com/these-30-rulers-in-history -were-hated-by-all/.

16 Senatus populusque Romanus (The senate and people of Rome).

17 *Magna Carta Libertatum* (Great charter of freedoms), 1215. The charter appears to be more of a document discussing the relationship of barons with the monarch.

18 Jeff Ostrowski, Study of the Roman Rite "Missarum Sollemnia". Corpus Christi Watershed, 25 January 2014 and https://en.wikipedia.org/wiki/Mass_in_the_Catholic_Church#:~:text=The%20Mass%20is%20the%20central%20liturgical%20service%20of,and%20become%20the%20body%20and%20blood%20of%20Christ.

19 Walter Scott, *Waverley; or, 'Tis Sixty Years Since* (1814), Chap. 2 and https://en.wikipedia.org/wiki/English_Civil_War.

20 J. Griffiths, "Doctor Thomas Dimsdale, and Smallpox in Russia: The Variolation of the Empress Catherine the Great," Bristol Medico-Chirurgical Journal 99, no. 1 (1984): 14–16.

21 *Death And The Civil War*, PBS, aired 18 September 2012 and at https://www.pbs.org/wgbh/americanexperience/films/death/ and https://www.pbs.org/wgbh/americanexperience/features/death-numbers/.

22 Habeas corpus (Latin for "that you have the body") is the legal protection against being unlawfully imprisoned, under Article I, § 9, Clause © 2 Habeas Corpus. "The Privilege of the Writ of Habeas Corpus shall not be suspended, unless when in Cases of Rebellion or Invasion the public Safety may require it." and at https://constitution.congress.gov/browse/article-1/section-9/clause-2/.

23 Blue mass pills contain mercury (Hg).

24 In 1823, President James Monroe establishes the Monroe Doctrine, which opposed European colonization of the Western world.

25 He was part of the royal family extending from Britain, Germany, Greece, and other nation-states.

26 Arthur Herman, *The Viking Heart: How Scandinavians Conquered the World* (Boston: Mariner Books, 2021), [pp. 378–379, 381, 384–385, 392, 409.].

27 That was the chancellery department's street address.

28 Growing up in this era, I too had a draft number assigned by birth date. My parents, siblings, and I watched the daily body-bag counts on the evening news, then we watched as my brother and I waited to see where our birthdates placed us in the draft lottery. Blake Stillwell, We Are The Mighty, This Is How to See if You Would've Been Drafted for Vietnam, Military.com 19 October 2018 at https://www.military.com/history/how-see-drafted-vietnam.html.

29 The Holy Book of Destiny, Kaivalya International, Redondo Beach, CA © 2011 at http://www.maitreyathefriend.com/The%20Holy%20Book%20of%20Destiny%20-%20Maitreya%20The%20Friend.pdf.

30 Bombing of Hiroshima and Hagasaki, History Channel 18 April 2023 at https://www.history.com/topics/world-war-ii/bombing-of-hiroshima-and-nagasaki; and This Day in History, The first atomic bomb test is successfully exploded. 16 July 1945 at https://www.history.com/this-day-in-history/the-first-atomic-bomb-test-is-successfully-exploded.

31 Edited by Hannah Holbert, The Story of the Atomic Bomb, Ohio State University 2023 at https://ehistory.osu.edu/articles/story-atomic-bomb.

32 "Nevada Test Site north of Las Vegas gets new name: Nevada National Security Site, or N2S2." Fox News. March 20, 2015. At https://en.wikipedia.org/wiki/Nevada_Test_Site; Louis Casiano, US conducts nuclear test in Nevada hours after Russian move to revoke global test ban. FOX News. 19 October 2023 at https://www.msn.com/en-us/news/other/us-conducts-nuclear-test-in-nevada-hours-after-russian-move-to-revoke-global-test-ban/ar-AA1ix4cX.

33 Bridget Judd, Three months before the coronavirus outbreak, researchers simulated a global pandemic. ABC News, 31 January 2020 at https://www.abc.net.au/news/2020–02-01/coronavirus-outbreak-researchers-simulated-severe-pandemic/11906562; and Tabletop Exercise Event 201, Johns Hopkins Bloomberg School of Public Health, Center for Health Security, 18 October 2019 at https://centerforhealthsecurity.org/our-work/tabletop-exercises/event-201-pandemic-tabletop-exercise#scenario.

34 In *Star Trek*, that was known as the Kobayashi Maru.

35 Annie Jacobsen (2014). *Operation Paperclip: The Secret Intelligence Program that Brought Nazi Scientists to America. Little, Brown.* ISBN 978–0-316–22105-4 and https://en.wikipedia.org/wiki/Office_of_Strategic_Services. A good friend of mine (Dottie Forsberg) had been one of the six women responsible for deciphering and decoding messages critical to the success of World War II and later with some of my work. Her work and her husband's work (Dr. Gordon Harrington, my graduate psychology mentor and director of the rat lab I ran during my psychology graduate studies) was critical in military communications.

36 National Archives, President John F. Kennedy's Inaugural Address (1961) at https://www.archives.gov/milestone-documents/president-john-f-kennedys-inaugural-address.

37 Roughly half of the scenarios were carried out underwater or off planet.

38 David Rockefeller, *Memoirs* (New York: Penguin Random House, 28 October 2003),; and David Rockefeller, "Quotable Quote," Goodreads (website), n.d.,https://www.goodreads.com/quotes/10465260-for-more-than-a-century-ideological-extremists-at-either-end.

39 Zbigniew Brezezinski, "Between Two Ages Quotes," Goodreads (website) n.d., https://www.goodreads.com/work/quotes/3269714-between-two-ages.

40 *Revision of the United Nations Charter: Hearings Before the Subcommittee of the Committee on Foreign Relations*, 81st Cong. (1950) (statement of James Warburg, board member of Council on Foreign Relations), https://en.wikisource.org/wiki/James_Warburg_before_the_Subcommittee_on_Revision_of_the_United_Nations_Charter#STATEMENT_OF_JAMES_P._WARBURG_OF_GREENWICH,_CONN. See also https://tinyurl.com/dhvx6uv9.

41 Hillary Clinton, "Hillary Clinton Provides US Foreign Policy Address at the Council of Foreign Relations," 15 July 2009, US Department of State, video, YouTube, 1:09:02, Secretary Clinton Provides U.S. Foreign Policy Address at the Council of Foreign Relations. U.S. Department of State. 18 July 2009 at https://tinyurl.com/5n6v43b6; and https://en.wikipedia.org/wiki/Hillary_Clinton.

42 file:///Users/richardfleming/Downloads/NTI_Paper_BIO-TTX_Final-1.pdf; https://www.nti.org/analysis/articles/strengthening-global-systems-to-prevent-and-respond-to-high-consequence-biological-threats/; and https://www.nti.org/events/report-launch-strengthening-global-systems-to-prevent-and-respond-to-high-consequence-biological-threats/.

43 Casadevall A, Dermody TS, Imperiale MJ, Sandri-Goldin RM, Shenk T. 2015. Dual-use research of concern (DURC) review at American Society for Microbiology journals. mBio 6(4):e01236–15. doi:10.1128/mBio.01236–15. https://www.ncbi.nlm.nih.gov/pmc/articles/PMC4542189/pdf/mBio.01236–15.pdf.

44 The White House, *National Biodefense Strategy and Implementation Plan: For Countering Biological Threats, Enhancing Pandemic Preparedness, and Achieving Global Health Security* (Washington, DC: The White House, October 2022), https://whitehouse.gov/wp-content/uploads/2022/10/National-Biodefense-Strategy-and-Implementation-Plan-Final.pdf.

45 SPARS is the St. Paul Acute Respiratory Syndrome Coronavirus (SPARS-CoV; SPARS), held before the Event 201 scenario. The SPARS Pandemic 2025–2028. A Futuristic Scenario for Public Health Risk Communicators. The Johns Hopkins Center for Health Security. 3027 at https://archive.org/details/spars-pandemic-scenario_202104; https://ia803407.us.archive.org/2/items/spars-pandemic-scenario_202104/spars-pandemic-scenario.pdf; https://asprtracie.hhs.gov/technical-resources/resource/5479/the-spars-pandemic-2025–2028-a-futuristic-scenario-for-public-health-risk-communicators; and https://www.centerforhealthsecurity.org/sites/default/files/2022–12/spars-pandemic-scenario.pdf.

46 Who We Are, Johns Hopkins Bloomberg School of Public Health, Center for Health Security at https://centerforhealthsecurity.org/who-we-are-at-the-center-for-health-security-0.

47 US Const. art. 2, § 2, cl. 2, https://constitution.congress.gov/browse/essay/artII-S2-C2-1-10/ALDE_00012961/.

48 Fleming v. U.S.A., Docket No. 13–17230 (9th Cir. App. 2014).

49 About Treaties, United States Senate at https://www.senate.gov/about/powers-procedures/treaties.htm.

50 Legal Basis for Executive Agreements, Article II, Section 2, Clause 3. Cornell Law School Legal Information Institute at https://www.law.cornell.edu/constitution-conan/article-2/section-2/clause-3/legal-basis-for-executive-agreements.

51 Timothy J. Keeler, et al. Sole Executive Agreements and Their Role in US Law. 12 April 2023 at https://www.mayerbrown.com/en/insights/publications/2023/04/sole-executive-agreements-and-their-role-in-us-law#t4

52 Stephen P. Mulligan, *International Law and Agreements: Their Effect upon US Law* (Washington, DC: Congressional Research Service, last modified 13 July 2023, https://crsreports.congress.gov/product/pdf/RL/RL32528.

53 Mulligan, *International Law and Agreements*, 25. Note that the US definition of *treaty* is significantly narrower than the international definition, which is "any international agreement concluded between states or other entities with international personality . . . if the agreement is intended to have international

legal effect." Frederic L. Kirgis, "International Agreements and U.S. Law," *Insights* 2, no. 5 (27 May 1997): https://www.asil.org/insights/volume/2/issue/5/international-agreements-and-us-law.

54 Fact Sheet: The Biden Administration's Commitment to Global Health. The White House. 2 February 2022 at https://www.whitehouse.gov/briefing-room/statements-releases/2022/02/02/fact-sheet-the-biden-administrations-commitment-to-global-health/.

55 International Health Regulations (2005) Second Edition, WHO Library ISBN 978 92 4 158041 0 at https://iris.who.int/bitstream/handle/10665/43883/9789241580410_eng.pdf?sequence=1

56 Josephine Moulds, How is the World Health Organization funded? World Economic Forum, 15 April 2020 at https://www.weforum.org/agenda/2020/04/who-funds-world-health-organization-un-coronavirus-pandemic-covid-trump/.

57 GAVI The Vaccine Alliance at https://www.gavi.org/.

58 *"The U.S. Government and the World Health Organization". The Henry J. Kaiser Family Foundation. 24 January 2019* at https://www.kff.org/global-health-policy/fact-sheet/the-u-s-government-and-the-world-health-organization/ and https://en.wikipedia.org/wiki/World_Health_Organization#:~:text=The%20WHO%20is%20funded%20primarily%20by%20contributions%20from,approved%20budget%20for%202022%E2%80%932023%20is%20over%20%246.2%20billion. "To propose conventions, agreements and regulations, and make recommendations with respect to international health matters and to perform (article 2 of the Constitution)."

59 Julia Crawford, Does Bill Gates have too much influence in the WHO? Swiss Info Channel 10 May 2021 at https://www.swissinfo.ch/eng/politics/does-bill-gates-have-too-much-influence-in-the-who-/46570526.

60 World Economic Forum and UN Sign Strategic Partnership Framework, Webwire 14 June 2019 at https://www.webwire.com/ViewPressRel.asp?aId=242272 and Alem Tedeneke, Media Manager for World Economic Forum, World Economic Forum and UN Sign Strategic Partnership Framework 13 June 2019 at https://www.weforum.org/press/2019/06/world-economic-forum-and-un-sign-strategic-partnership-framework/

61 My special thanks and appreciation go to Pastor Stephen Broden and members of the Fair Park Bible Fellowship Church for opening their doors for this presentation and to Del Bigtree and *The Highwire* crew for airing this live. https://thehighwire.com/videos/live-from-event-2021-in-dallas-tx/

62 They are criminals, as you will see in chapter 6.

63 Found here as well https://www.amnesty.org.uk/files/2019–01/First%20They%20Came%20by%20Martin%20Niem%C3%B6ller_0.pdf?l6HOtWW1N8umC_ELxnQI6NpaAYbxRCJj=

Chapter 4

1 Exod. 15:26.

2 2 Chron. 16:12 (ESV).

3 The thinking results in so many people electing not to believe in the existence of a God or creator, based upon the misinformation that God is intentionally harming people. Such a vindictive being is certainly not one to be depended upon. But, then again, such is behavior of the world of governments, lowering any creator to the level of man.

4 N. P. Heeßel, "Diagnosis, Divination, and Disease: Towards an Understanding of the Rationale Behind the Babylonian Diagonostic Handbook," in *Magic and Rationality in Ancient Near Eastern and Graeco-Roman Medicine*, edited by Herman F. J. Horstmanshoff, Marten Stol, and C. R. Van Tilburg, vol. 27, Studies in Ancient Medicine, 97–116 (Leiden, The Netherlands: Brill Academic Publishers), https://books.google.com/books?id=p6rejN-iF0IC&pg=PA97#v=onepage&q&f=false 97–116.

5 Herodotus, *The Histories*, Book 2, Chapter 84; English translation by A.D. Godley, Cambridge. Harvard University Press © 1920.

6 We have almost returned to this concept today, with an ever-increasing wealth of knowledge and subspecialization.

7 The classic and revised Hippocratic Oath can be found at "The Hippocratic Oath: The Original and Revised Version," *Practo for Doctors* (blog), 10 March 2015, https://doctors.practo.com/the-hippocratic-oath-the-original-and-revised-version/. The modern version is included in this book further on.

8 The Greeks, while fondly remembered by many as the foundation of democracy, believed in many gods who relieved the Greeks from personal responsibility. The goal was to keep the gods happy to prevent vengeance being taken out on the humans by them. Clearly, the Greeks believed in strokes. The Greeks also believed in sexual orientation of young men from preteen years by older men. The Greeks believed in sacrificing any children who were not perfect, with estimates of up to 25 percent of the children being killed shortly after birth. So, they clearly believed in eugenics.

9 I was a physician services contractor, inspecting military medical facilities in 2007.

10 September 17, 1787: A Republic, If You Can Keep It. Independence National Historical Park, National Park Service at https://www.nps.gov/articles/000/constitutionalconvention-september17.htm.

11 "English Declaration of Rights." The Avalon Project. Yale University; Oliver Joseph Thatcher, ed. (1907). *The library of original sources.* University Research Extension. p. 10. Bill of Rights 1689 December 16 and https://en.wikipedia.org/wiki/Bill_of_Rights_1689.

12 James Madison (1902) *The Writings of James Madison*, vol. 4, *1787: The Journal of the Constitutional Convention, Part II* (edited by G. Hunt), pp. 501–502 and https://en.wikipedia.org/wiki/Signing_of_the_United_States_Constitution.

13 This Day in History, U.S. Constitution Ratified 21 June 1788 at https://www.history.com/this-day-in-history/u-s-constitution-ratified

14 Barry Adamson (2008). *Freedom of Religion, the First Amendment, and the Supreme Court: How the Court Flunked History.* Pelican Publishing. p. 93. ISBN

9781455604586; Kenneth R. Thomas, The Constitution of the United States of America Analysis and Interpretation, Centennial Edition, 112th Congress, 2nd session, Document No. 112–9, U.S. Government Printing Office, Washington 2013 and https://en.wikipedia.org/wiki/United_States_Bill_of_Rights.

15 Jerome A. Barron et al., *Constitutional Law: Principles and Policy, Cases and Materials.* 7th ed. New York: LexisNexis, 2006; *Constitutional Law: Principles and Policy. Cases and Materials* (7th Edition). Barron J.A., et al. Lexis Nexis © 2006; https://www.senate.gov/about/origins-foundations/senate-and-constitution/constitution.htm.

16 Those were established in Articles I, II, and III, respectively.

17 Article I, § 7, clause 2, https://www.law.cornell.edu/constitution-conan/article-1/section-7/clause-2/the-veto-power.

18 Marbury v. Madison, 5. U.S. (1 Cranch) 137 (1803). The SCOTUS asserted that it alone had the ultimate decision-making power over any lower court. The case is fascinating. Prior to Thomas Jefferson being sworn in as president, outgoing President John Adams made several court appointments. One of the deliveries of an appointment (signed at the time by the outgoing US Secretary of State John Marshall) had not been delivered, and Jefferson ordered the delivery *not* to be made. John Marshall moved from secretary of state position to chief justice of the SCOTUS; he ultimately determined that he and the other members of the SCOTUS had the ultimate decision-making power on what was legal or constitutional and what was not. Pursuant to § 13 of the Judiciary Act of 1789, Congress had placed limitations on the SCOTUS, declaring the SCOTUS did not have jurisdiction in the case. The SCOTUS simply struck §13 as being unconstitutional, stating the Constitution did *not* agree with the statute passed by Congress, thus making the SCOTUS the ultimate power holder and stating that if what Congress passed as law disagreed with what the Constitution expressly stated, then the statute was unconstitutional and void.

19 John Marshall, Fourth Chief Justice of the United States; https://en.wikipedia.org/wiki/John_Marshall.

20 Stare decisis means "to stand as decided." Merriam-Webster.com, s.v. "stare decisis," n.d., https://www.merriam-webster.com/dictionary/stare%20decisis.

21 Tenth Amendment U.S. Constitution, Cornell Law at https://www.law.cornell.edu/constitution/tenth_amendment.

22 *Cruzan v. Director, Missouri Department of Health*, 497 U.S. 261 (1990) at https://en.wikipedia.org/wiki/Cruzan_v._Director,_Missouri_Department_of_Health

23 Legislative Attorney (name redacted) Health Care: Constitutional Rights and Legislative Powers, Congressional Research Service, R40846, 9 July 2012 at https://www.everycrsreport.com/files/20120709_R40846_27d29c09b692d83ed3386e207cb0e31b58775fcc.pdf.

24 Mervyn Susser, Zena Stein, *Germ Theory, Infection and Bacteriology.* Eras in Epidemiology: The Evolution of Ideas. New York: Oxford University Press. 2009.

25 Ignaz Semmelweis, Die Aetiologie, der Bergriff und die Prophylaxis des Kindbettfiebers (Etiology, Concept and Prophylaxis of Childbed Fever), C.A. Hartleben's Verlag-Expedition © 1861.

26 Thucydides, *History of the Peloponnesian War*, trans. Richard Crawley (London: J.M. Dent & Sons, 1910), 3.51, 131–32.

27 Including the introduction of Inflammation and Blood Clotting [InflammoThrombotic] diseases, and the quantification of nuclear imaging as discussed here https://www.flemingmethod.com/about.

28 K. C. Carter, "Ignaz Semmelweis, Carl Mayrhofer, and the Rise of Germ Theory," *Medical History*, 29, no. 1 (January 1985): 33–53, https://doi.org/10.1017/S0025727300043738.

29 Double-Slit Experiment Explanation and Equation. Study.com at https://study.com/buy/academy/lesson/double-slit-experiment-explanation-equation.html?src=ppc_bing_nonbrand&rcntxt=aws&crt=&kwd=SEO-PPC-BUY&kwid=dat-2329040505886940:loc-190&agid=1235851302596746&mt=b&device=c&network=o&_campaign=SeoPPC&msclkid=6e9053c08d7c1c14dbc63659ed749c2f.

30 https://microbiologysociety.org/membership/membership-resources/outreach-resources/antibiotics-unearthed/antibiotics-and-antibiotic-resistance/the-history-of-antibiotics.html.

31 Germ theory is the scientific explanation that infectious fungi, bacteria, mycobacteria, viruses, and so forth cause infectious diseases.

32 https://www.khanacademy.org/science/ap-biology/gene-expression-and-regulation/biotechnology/a/dna-sequencing.

33 While there are those who argue that viruses should not be thought of as alive, this approach is somewhat limited ignoring the fact that viruses have genetic codes (RNA or DNA) that when reproduced by entering some other species makes more of the virus. The virus is not making the species it invaded but itself. The killing of the virus is the stopping of the virus from being able to replicate or reproduce itself. If the requirement that something is alive only if it does not require another living organism to survive, then you could argue people are not alive because we require the multitude of bacteria in our gastrointestinal tracts for our survival. Eliminate those bacteria and people cannot survive, so by the same argument, people are not alive. Clearly an erroneous argument.

34 HIV is human immunodeficiency virus.

35 AIDS is acquired immunodeficiency syndrome.

36 Paxlovid is a combination of nirmatrelvir and ritonavir.

37 Kathy Katella, What Is Paxlovid Rebound? 9 Things to Know. Yale Medicine at https://www.yalemedicine.org/news/what-is-paxlovid-rebound-covid-rebound.

38 Arielle Mitropoulos, Fauck says he's taking 2nd course of Paxlovid after experiencing rebound with the antiviral treatment. ABC News. 29 June 2022 at https://abcnews.go.com/US/fauci-taking-2nd-paxlovid-experiencing-rebound-antiviral-treatment/story?id=85922417.

39 https://www.britannica.com/topic/Project-Paperclip.

40 What I find truly fascinating about physicians and their relationship with the AMA and Rockefeller and Carnegie Foundations is the willingness for physicians to cede control of medicine to such a small group of people, even if some of them include physicians. When the AMA was formed, a handful of men arbitrarily assumed control of medicine, citing grievances with other practitioners who treated patients differently. With funding and support from Rockefeller and Carnegie, they seized control of medicine and promoted legislative efforts to control and limit the practice of medicine in the United States. As of 2022 the AMA reported 271,660 members (presuming all are physicians) (https://en.wikipedia.org/wiki/American_Medical_Association) out of 1,077,115 (https://www.statista.com/topics/1244/physicians/#topicOverview), representing only one in four American physicians.

41 Allopathic physicians are medical doctors (MDs). Other types of doctors are not MDs, including DO (osteopath), ND (naturopath), PhD (doctor of philosophy, of which there are many fields: science, social science, education, and so forth), OD (optometist, not to be confused with an ophthalmologist), DDS (dentist), DC (chiropractor), DPM (podiatrist, who deals with feet), and others.

42 What we fail to learn from history, we are destined to repeat. Uniting and Strengthening America by Providing Appropriate Tools Required to Intercept and Obstruct Terrorism (USA-PATRIOT) Act of 2001, Pub. L. No. 107–56, 115 Stat. 272, https://www.history.com/topics/21st-century/patriot-act.

43 Jeffrey B. Roth, Yellow Fever, Greek Fire and Poison Gas—Secret Weapons of the U.S. Civil War, Military History Now, 25 April 2014 at https://militaryhistorynow.com/2014/04/25/yellow-fever-greek-fire-and-poison-gas-secret-weapons-of-the-u-s-civil-war/.

44 117 NJLJ 259, opinion 581 (27 February 1986), https://law.justia.com/cases/new-jersey/advisory-committee-on-professional-ethics/1986/acp581–1.html.

45 CPT® purpose and mission. AMA at https://www.ama-assn.org/about/cpt-editorial-panel/cpt-purpose-mission; and https://www.flemingmethod.com/about.

46 Bob Zebroski, *A Brief History of Pharmacy: Humanity's Search for Wellness* (New York: Routledge, 2016), at https://archive.org/details/briefhistoryofph0000zebr.

47 A History of the FDA and Drug Regulation in the United States found at https://www.fda.gov/media/73549/download; and W. J. Heath, "America's First Drug Regulation Regime: The Rise and Fall of the Import Drug Act of 1848," *Food and Drug Law Journal* 59, no. 1 (2004):169–99.

48 Wesley J. Heath, America's First Drug Regulation Regime: The Rise and Fall of the Import Drug Act of 1848. *Food and Drug Law Journal* 2004;59(1):169–199 at https://www.jstor.org/stable/26660236?read-now=1&seq=3#page_scan_tab_contents.

49 A History of the FDA and Drug Regulation in the United States found at https://www.fda.gov/media/73549/download; and Milestones in U.S. Food and Drug Law, U.S. Food and Drug Administration at https://www.fda.gov/about-fda/fda-history/milestones-us-food-and-drug-law.

50 Ronald Hamowy, "The Early Development of Medical Licensing Laws in the United States, 1875–1900," *Journal of Libertarian Studies*, 3, no. 1 (1979): 93,

https://web.archive.org/web/20190929201236/http:/austrianeconomics.org/sites/default/files/3_1_5_0.pdf.

51 *Transactions of the American Medical Association*, vol. 18 (Philadelphia: Collins Printer, 1867), 30, https://ama.nmtvault.com/jsp/PsImageViewer.jsp?doc_id=6863b9b4-a8b5–4ea0–9e63-ca2ed554e876%2Fama_arch%2FAD200001%2F00000018.

52 James Norment Baker, "Annual Message of the President," in *Transactions of the Medical Association of the State of Alabama, 1916* (Montgomery, AL: Brown Printing Co., 1916), 160.

53 Abraham Flexner, *Medical Education in the United States and Canada: A Report to The Carnegie Foundation for the Advancement of Teaching*, bulletin 4 (Boston: D. B. Updike, Merrymount Press, 1910), http://archive.carnegiefoundation.org/publications/pdfs/elibrary/Carnegie_Flexner_Report.pdf.

54 E. Richard Brown. *Rockefeller Medicine Men: Medicine and Capitalism in America.* University of California Press © 1979.

55 John C. Goodman, and Gerald L. Musgrave, *Patient Power: Solving America's Health Care Crisis* (Washington, DC: Cato Institute, 1992), 142–48. ISBN 978–0-932790–92-7 also available here https://www.book-info.com/isbn/0–932790-92–5.htm

56 https://www.ncbi.nlm.nih.gov/pmc/articles/PMC2567554/pdf/12163926.pdf; and Barbara Barzansky and Norman Gevitz, *Beyond Flexner: Medical Education in the Twentieth Century* (New York: Greenwood Press, 1992).

57 Abraham Flexner, Medical Education in Europe: A Report to the Carnegie Foundation for the Advancement of Teaching, bulletin 6, Tspace, University of Toronto (Boston: D. B. Updike, Merrymount Press, 1912), https://tspace.library.utoronto.ca/handle/1807/33462.

58 Abraham Flexner, "Black Physicians and Black Hospitals" (PDF). p. 24. Archived from the original (PDF) on 2016–10-02, https://web.archive.org/web/20161002104238/http:/medicine.missouri.edu/ophthalmology/uploads/ch06.pdf.

59 The Cabinet. The Biden-Harris Administration. The White House at https://www.whitehouse.gov/administration/cabinet/.

60 Mosher, Frederick C. *American Public Administration: Past, Present, Future.* 2nd ed. Birmingham, Ala.: University of Alabama Press, 1975. ISBN 0–8173-4829–8

61 "Message to Congress on the Reorganization Act." April 25, 1939. John T. Woolley and Gerhard Peters. *The American Presidency Project.* Santa Barbara, Calif.: University of California (hosted), Gerhard Peters (database); Pub. L. 76–16 Statute 561.

62 The hospital was first opened by the District of Columbia in 1855 and was also known as the Government Hospital for the Insane.

63 HHS Historical Highlights, U.S. Department of Health and Human Services, at https://www.hhs.gov/about/historical-highlights/index.html.

64 The previous title of this webpage at the time of the initial writing of this book was "Enhancing the health and well-being of all Americans." During the editing of this book, that title had been removed. It later reappeared at the time of the final

editing. Shakespeare would have such fun writing about the US government and people of our time.

65 Richard M. Fleming, *Is COVID-19 a Bioweapon? A Scientific and Forensic Investigation* (New York: Skyhorse Publishing, 2021).

66 Richard M. Fleming, PhD, MD, JD. The Unified Theory on Vascular Disease and FMTVDM; https://www.flemingmethod.com/about.

67 See chapter 8.

68 See appendix F.

69 Milestones in U.S. Food and Drug Law, U.S. Food and Drug Administration found at https://www.fda.gov/about-fda/fda-history/milestones-us-food-and-drug-law.

70 Regulations: Good Clinical Practice and Clinical Trials, U.S. Food and Drug Administration at https://www.fda.gov/science-research/clinical-trials-and-human-subject-protection/regulations-good-clinical-practice-and-clinical-trials.

71 Research with Children FAQs, U.S. Department of Health and Human Services at https://www.hhs.gov/ohrp/regulations-and-policy/guidance/faq/children-research/index.html; [Name of Code or Regulation], 45 CFR 46, 30 May 1974 at https://www.ecfr.gov/current/title-45/subtitle-A/subchapter-A/part-46; and Best Pharmaceuticals for Children Act of 2002, Public L. No. 107–109 107th Congress, 21 USC 301 (4 January 2002 at https://www.congress.gov/107/plaws/publ109/PLAW-107publ109.pdf.

72 https://www.hhs.gov/ohrp/sites/default/files/the-belmont-report-508c_FINAL.pdf.

73 Food and Drug Administration, organizational chart, March 2024 at https://www.fda.gov/media/171675/download?attachment.

74 https://qz.com/1656529/yet-another-fda-commissioner-joins-the-pharmaceutical-industry.

75 See appendix C.

76 HHS Agencies and Offices. U.S. Department of Heatlh and Human Services. at https://www.hhs.gov/about/agencies/hhs-agencies-and-offices/index.html.

77 "Associated Press, "Epidemic Intelligence Service Is Organized," *News from Frederick, Maryland*, https://www.newspapers.com/newspage/8506893/; and "Promotion and Recruitment Tools," Centers for Disease Control, last reviewed February 29, 2024, https://www.cdc.gov/EIS/Recruitment.html.

78 Richard M. Fleming, The Fleming Method for Tissue and Vascular Differentiation and Metabolism (FMTVDM) Using Same State Single or Sequential Quantification Comparisons, US Patent No. US9566037B2, filed 13 June 2013, and issued 14 February 2017, https://patents.google.com/patent/US9566037B2/en.

79 R. M. Fleming and W. C. Dooley, Breast Enhanced Scintigraphy Test (BEST) Imaging Utilizes Vascularity/Angiogenesis and Mitochondrial Activity to Distinguish Between Normal Breast Tissue, Inflammation and Breast Cancer" (presentation, 8th International Conference, "Vascular Endothelium: Translating Discoveries into Public Health Practice, Crete, Greece, June 2005; and R. M. Fleming et al., "What Effect Do Isocaloric Low-Fat, Low-Carbohydrate and Moderate-Fat Diets Have on Obesity and Inflammatory Coronary Artery Disease?,"

8th International Conference, "Vascular Endothelium: Translating Discoveries into Public Health Practice," Crete, Greece, June 2005.

80 Richard M. Fleming, PhD, MD, JD. The Unified Theory on Vascular Disease and FMTVDM; https://www.flemingmethod.com/about

81 CMS Program History, Centers for Medicare and Medicaid Services at https://www.cms.gov/about-cms/who-we-are/history.

82 The USA PATRIOT Act, Public Law 107–56 enacted 26 October 2001 found at https://irp.fas.org/agency/doj/oig/patriot071703.pdf. After his investigations, I was told by Professor Gordon M. Harrington, my psychology mentor, that a potential 10 percent of the money for the Patriot Act was used by the DOJ to go after doctors and charge them with billing fraud. That would bring more money into the federal government and send a clear message to doctors that the federal government was in charge. The odds are never in your favor; the Government always wins. It has an endless budget and resources that the American people pay for, money we pay in taxes so the government can attack us. https://www.youtube.com/watch?v=mtuBqolFOVs&t=15s; and 16 April 2009, https://www.youtube.com/watch?v=-5Va31X6Rq8.

83 Medicare Advantage Medical Policy Bulletin, R-5, 3:11-cv-00848 at https://www.flemingmethod.com/_files/ugd/659775_c848d96047e840e7ad776c5b0b4d4f01.pdf.

84 Jennifer Trajkovski, BS, RHT, Coding Associate, CPT Information Services for the American Medical at Association https://www.flemingmethod.com/_files/ugd/659775_1736ad14978744edacdbb224a45d9e24.pdf.

85 National Practitioner Data Bank (NPDB), NCSBN at https://www.ncsbn.org/nursing-regulation/discipline/reporting-and-enforcement/npdb.page; and https://www.federalregister.gov/documents/2013/04/05/2013–07521/national-practitioner-data-bank.

86 https://www.flemingmethod.com/documentation.

87 Frank D. Borio et al., "Death Due to Bioterrorism-Related Inhalational Anthrax," *JAMA Express* 286, no. 20 (2001): 2554–59; and J. G. Bartlett, T. V. Inglesby, and L. Borio, "Management of Anthrax. Invited Article Confronting Biological Weapons," *Clinical Infectious Diseases* 35 (2002): 851–58.

88 Some bacteria referred to as endospores have the ability to produce a single endospore containing the genetic material of the bacterium that can endure dormant periods of extreme conditions. These endospores can remain viable for over ten thousand years, and claims of revival of these spores after millions of years have been made. Examples of endospore organisms include anthrax, tetanus, botulism, *Clostridium perfringens* (gas gangrene), and the *Clostridioides difficile* (*C. difficile*) associated with antibiotic-associated pseudomembranous colitis. https://en.wikipedia.org/wiki/Endospore.

89 C. Stroud et al., eds., "A Decision-Aiding Framework," chap. 5 in *Prepositioning Antibiotics for Anthrax* (Washington, DC: National Academies Press, 2011), https://www.ncbi.nlm.nih.gov/books/NBK190041/.

90 National Federation of Independent Business v. Sebelius, 567 U.S. 519 (2012), https://s3.documentcloud.org/documents/5776934/11–393c3a2.pdf; and R. M. Fleming, "Doctors and Patients—Not Insurance Companies—Should Determine the Course of Patient Diagnostics and Treatment," comment on Martha E. Gaines, Austin D. Auleta, and Donald M. Berwick, "Changing the Game of Prior Authorization: The Patient Perspective," *JAMA Network Open* 323, no. 8 (5 February 2020): 705–6, https://jamanetwork.com/journals/jama/fullarticle/2760655;

91 Fleming v. US, Docket No. 13–17230 (9th Cir. App. 2014); R. M. Fleming and T. K. Chaudhuri, "CMS Is Primary Reasons Healthcare Costs Have Increased for Everyone," comment on Steffie Woodhandler and David U. Himmelstein, "The American College of Physicians' Endorsement of Single-Payer Reform: A Sea Change for the Medical Profession," *Annals of Internal Medicine*, supplement, 172, no. S2 (7 February 2020): https://annals.org/aim/fullarticle/2759531/american-college-physiciansendorsement-single-payer-reform-sea-change-medical?searchresult=1; R. M. Fleming et al., Development of Patient-Specific, Patient-Directed, Patient-Responsive Treatment Will Improve Patient Outcomes and Reduce Healthcare Costs in Single-, Double- and Triple-Receptor Status," *JAMA Network Open* (13 February 2020); R. M. Fleming, M. R. Fleming, and T. K. Chaudhuri, "We Can Reduce Out-of-Pocket Costs—but at What Price?," comment on Hiroshi Gotanda et al., "Out-of-Pocket Spending and Financial Burden Among Low Income Adults After Medicaid Expansions in the United States: Quasi-Experimental Difference-in-Difference Study," *BMJ* 368 (February 2020): https://www.bmj.com/content/368/bmj.m40/rapid-responses; Richard M. Fleming, Tapan K. Chaudhuri, and Gordon M. Harrington, "The Statistical Analysis of Data Validity as Acknowledged by ORI. How Three Statisticians Validated Original Data and Exposed Plagiarism by a Public Defender. The Implications for All Research and Article III Courts," *Biomedical Journal of Scientific & Technical Research* 24, no. 3 (2020):: https://doi.org/10.26717/BJSTR.2020.24.004053.; R. M. Fleming, T. K. Chaudhuri, and G. M. Harrington, "Establishing the Irrefutable Statistical Standards Which ORI and Article III and State Courts Must Apply in Determining Data Validity. A Call for Impeachment, Debarment and Presidential Pardon," Journal of Cardiovascular Medicine and Cardiology 7, no. 1 (2020): 6–23, https://www.peertechz.com/articles/JCMC-7–205.php; and R. M. Fleming, M. R. Fleming, and T. K. Chaudhuri, "Efforts to Visually Match 5- and 60-Minute Post-Stress Images Following a Single Injected Dose of Sestamibi Clearly Demonstrate Changes in Sestamibi Distribution: Demonstrating Once and for All Clinical Recognition That Sestamibi Redistributes," Journal of Cardiovascular Medicine and Cardiology 6, no. 2 (2019): 30–35, http://doi.org/10.17352/2455–2976.000087.

92 "Bruesewitz v. Wyeth," *SCOTUS blog*, n.d., https://www.scotusblog.com/case-files/cases/bruesewitz-v-wyeth/.

93 National Childhood Vaccine Injury Act (NCVIA) of 1986, 42 U.S.C. §§ 300aa-1 to 300aa-34.

94 COVID Research; https://www.flemingmethod.com/documentation See documents no. 2, 3, and 51 which are the respective EUA documents filed by Pfizer,

Moderna and Janssen for FDA approval. Also see analysis throughout this book showing no statistical reduction in COVID cases or deaths in the EUA vaccine materials among those vaccinated. R. M. Fleming, M. R. Fleming, and T. K. Chaudhuri, "Establishing Data Validity: Statistically Determining if Data Is Fabricated, Falsified or Plagiarized," *ACTA Scientific Medical Sciences* 3, no. 8 (August 2019): 169–91, https://actascientific.com/ASMS/pdf/ASMS-03–0365.pdf.

95 National Childhood Vaccine Injury Act of 1986—Title I: Vaccines, Subtitle 1: National Vaccine Program, Pub. L. No. 114–255 (1986), https://www.congress.gov/bill/99th-congress/house-bill/5546, https://www.hrsa.gov/sites/default/files/hrsa/vicp/title-xxi-phs-vaccines-1517.pdf.

96 James Myhre and Dennis Sifris, MD, HIV Statistics You Should Know. verywell health, 31 August 2023; https://www.verywellhealth.com/hiv-statistics-5088304?print.

Chapter 5

1 In addition to running one of the top four genetic rat labs in the world for a year as a graduate student, I also conducted research into human responses to data input and how this could be used for the benefit of the person inputting the data.

2 Proclamation on Declaring a National Emergency Concerning the Novel Coronavirus Disease (COVID-19) Outbreak. The White House, 13 March 2020 at https://trumpwhitehouse.archives.gov/presidential-actions/proclamation-declaring-national-emergency-concerning-novel-coronavirus-disease-covid-19-outbreak/.

3 Elizabeth Crisp, Trump Launces 'Operation Warp Speed' Effort to Develop Coronavirus Vaccine by Year's End, Newsweek 15 May 2020 at https://www.newsweek.com/trump-launches-operation-warp-speed-effort-develop-coronavirus-vaccine-years-end-1504458.

4 Katalin Karikó Facts, The Nobel Prize at https://www.nobelprize.org/prizes/medicine/2023/kariko/facts/.

5 Emily and Jackie: Our Most Important Message YET! 18 February 2021 at https://rumble.com/ve0y2x-emily-and-jackie-our-most-important-message-yet.html.

6 Yes, I have saved all the reviewers' and journals' comments. Either the research was considered not important enough or it would raise questions about the vaccine safety. For the record, research is supposed to provide answers and raise questions. That's how science advances.

7 J. L. Diamond, L. C. Levine, and M. S. Madden, *Understanding Torts*, 3rd ed. (New York: LexisNexis, 2008).

8 Having been an attorney (JD) for ten years, I have been looking forward to writing this chapter. I thought I would walk you through the process I would use in litigating such a case against these vaccine manufacturers for their defective product, using our research, as shown. R. M. Fleming et al., "Pfizer, Moderna and Janssen Vaccine InflammoThrombotic and Prion Type Effect on Erythrocytes When Added to Human Blood," *Haematology International Journal* 7, no. 1 (2023): https://doi.org/10.23880/hij-16000210.

9 Evidence is supposed to be presented based upon the Federal Rules of Evidence (FRE).

10 In addition to the published research noted, approximately sixty hours of recordings were made of the research study itself. This is research that the FDA, Pfizer, Moderna, or Janssen—or anyone interested in knowing what the genetic vaccines actually look like under a microscope or what the impact of adding the genetic vaccines directly to human blood was—could have done. What Happens When The Genetic Vaccines Come Into Contact With Human Blood? at https://rumble.com/v2labco-what-happens-when-the-genetic-vaccines-come-into-contact-with-human-blood.html.

11 Lars Noah, *Law, Medicine, and Medical Technology*. 2nd ed., Eagan, MN: Foundation Press, 2007.

12 Courts and legal experts have repeatedly looked at areas of civil litigation and "revised" their thinking on what should or shouldn't be interpreted as a wrong or harm that the defendant is responsible for and what plaintiffs should or shouldn't expect to be able to recover damages for. When the majority of courts (judges) or legal experts hold a certain view or opinion, this is called the "majority view." Over the course of time some experts and judges will decide that change is needed; when they do and provide a different outcome in court than previously expected, this change is considered the "minority view." Over time, many, but not all, minority views become the majority view.

13 SV40 versus Strict Product Liability. This podcast presentation with host Laurie Gagan discusses material looking at the effects of SARS-CoV-2 and the genetic vaccines presented at the 2023 Vitamin C International Consortium Institute (VCICI) Conference.

Chapter 6

1 States with citizen-initiated grand juries. Ballot Pedia at https://ballotpedia.org/States_with_citizen-initiated_grand_juries.

2 Jointly and severally, makes all parties responsible for damages up to the entirety of the amount.

3 A Crime of Conspiracy occurs when two or more people agree to commit an illegal act and take some step toward its completion.

4 Federal Rules of Evidence, Rule 406, focuses on the admissibility of evidence regarding a person or organizations habit of repeated behavior. Unlike Rules 404 and 405, Rule 406 makes such behavior admissible in a court of law as it is more specific and predictive than general character traits.

5 Richard M. Fleming, PhD, MD, JD at https://www.flemingmethod.com/about.

6 Pursuant to several SCOTUS rulings, declarations have no effect upon treaty law, though reservations and understandings modify a treaty as reserved and understood by the US Congress when the treaty is ratified.

7 Trump Declares State of Emergency for COVID-19, National Conference of State Legislatures, 25 March 2020 at https://www.ncsl.org/state-federal/trump-declares-state-of-emergency-for-covid-19.

8 Failure of the citizens of the United States to demand enforcement of our treaties and laws and to hold those individuals criminally accountable for the violations of the BWC Treaty has exposed the United States to criticisms and potential attacks by other countries.

9 Moshe Brown, Hillel Handler, Vera Sharav, Letter in support of the joint 'Request for Investigation' to the ICC from the UK, Slovakia, France and the Czech Republic, 20 September 2021 at https://www.flemingmethod.com/_files/ugd/659775_955f5cb3a13344afb57c53702b9c5c30.pdf.

10 The Belmont Report 18 April 1979 at https://www.hhs.gov/ohrp/sites/default/files/the-belmont-report-508c_FINAL.pdf.

11 The ICCPR Treaty was championed by Eleanor Roosevelt (https://sites.psu.edu/jlia/a-brief-history-of-the-international-covenant-on-civil-and-political-rights/); its purpose was to make certain the atrocities of the Nazis were never repeated again. The ICCPR has had limited use in US courts, due to an error that I am confident time will correct. It was introduced by Ted Cruz, who erroneously referred to a declaration of the ICCPR Treaty as a reservation. The SCOTUS failed to hear the case that would have protected US citizens' rights under the ICCPR Treaty and correct the SCOTUS error. Pursuant to the rule of law, the SCOTUS must correct its error or abandon stare decisis. Fleming v. US, Docket No. 13–17230 (9th Cir. App. 2014).

12 EVENT 2021—The Published Science on SARS-CoV-2 & COVID-19. Dallas, TX 5 June 2021 at https://www.flemingmethod.com/event-2021.

13 Pursuant to tort law, voluntariness requires an understanding and acceptance of the risks involved in a behavior or act—that is, assumption of risk. John L Diamond, Lawrence C. Levine, M. Stuart Madden, *Understanding Torts*, 3rd edition, Lexis Nexis, © 2008; and Assumption of Risk, Cornell Law, Legal Information Institute at https://www.law.cornell.edu/wex/assumption_of_risk#:~:text=Assumption%20of%20risk%20is%20a%20common%20law%20doctrine,themselves%20and%20absolve%20potential%20defendants%20from%20any%20liability.

14 Thank you Del Bigtree and crew of The Highwire at. https://www.flemingmethod.com/live-stream-of-event-2021

15 Frequently referred to as the Johnson & Johnson (J & J) vaccine.

16 Bridges v. Houston Methodist Hospital, 4:21-cv-01774 (S.D. Tex.12 June 2021).

17 18 U.S. Code § 1001—Statements or entries generally. Cornell Law, LII at https://www.law.cornell.edu/uscode/text/18/1001; George Khoury, Esq. What Are the Penalties for Lying to Congress? FindLaw, 21 March 2019 at https://www.findlaw.com/legalblogs/criminal-defense/what-are-the-penalties-for-lying-to-congress/#:~:text=Under%20the%20United%20States%20Code%2C%20title%2018%2C%20section,to%20the%20feds%2C%20is%20guilty%20of%20a%20crime.

18 Also, a part of the 1989 Biological Weapons Anti-Terrorism Act.

19 18 USC 1621: Perjury generally from Title 18-Crimes and Criminal Procedure. House.gov at https://uscode.house.gov/view.xhtml?req=(title:18%20section:1621%20edition:prelim).

20 Course: Civil Procedure, Professor: Arthur R. Miller, Semester: Fall 2013 at https://www.law.nyu.edu/sites/default/files/upload_documents/Miller.CivPro.Fall13%282%29.pdf. Professor Arthur R. Miller was one of my law professors and authored several of our textbooks.

21 Latin phrases are included here to facilitate the process of prosecuting attorneys convening grand juries and submitting the criminal complaint to the applicable courts of law.

22 The jurisdictional responses provided by a few attorneys general might be considered complicit by some in failing the people they took an oath to protect. Some of the responses provided to citizens in their states who submitted requests via the www.10Letters.org website include ad hominem attacks upon me. Personal attacks are not legal arguments addressing the facts in the matter of these defendants and the crimes these defendants have committed. Some respondents argue that the states do not have jurisdiction and this case must be handled in federal court, completely ignoring that the independent and sovereign states composing the United States are where the citizens of the United States have been harmed, injured, /or killed, as a result of these crimes established under the Subject Matter Jurisdiction. They also ignore that the defendants have conducted business within each and every state of the union and are accordingly subject to prosecution in each state.

23 SARS-CoV-2, COVID, GoF, CRISPR and the Genetic Vaccines. Why are People Dying and What Answers Is Science Teaching Us? Houston, TX 18 September 2023 at https://www.flemingmethod.com/; Live EVENT 2021 at https://www.flemingmethod.com/live-stream-of-event-2021.

24 COVID Data Tracker, Centers for Disease Control and Prevention at https://covid.cdc.gov/covid-data-tracker/#datatracker-home.

25 Richard M. Fleming, *Is COVID-19 a Bioweapon? A Scientific and Forensic Investigation* (New York: Skyhorse Publishing, 2021).

26 Richard M. Fleming, "Pathogenesis of Vascular Disease," chap. 64 in *Textbook of Angiology*, edited by John B. Chang, 787–98 (New York: Springer, 2000), https://doi.org/10.1007/978–1-4612–1190-7_64.

27 https://www.flemingmethod.com/documentation Documents ## 2, 3, and 51.

28 American Medical Association, "AMA Code of Medical Ethics," Informed Consent in Research, Opinion 7.1.2., n.d., https://code-medical-ethics.ama-assn.org/sites/default/files/2022–08/7.1.2.pdf.

29 National Institutes of Health, Office of NIH History & Stetten Museum, "The Nuremberg Code," Web Ref: https://tinyurl.com/bdfszzxf and https://history.nih.gov/display/history/Nuremberg%2BCode.

30 Convention on the Prohibition of the Development, Production and Stockpiling of Bacteriological (Biological) and Toxin Weapons and on their Destruction; *"Biological Weapons Convention—UNODA". United Nations Office for Disarmament Affairs* https://en.wikipedia.org/wiki/Biological_Weapons_Convention#:~:text=The%20

BWC%20was%20opened%20for%20signature%20on%2010,Union%2C%20 the%20United%20Kingdom%2C%20and%20the%20United%20States%29.; https://www.state.gov/about-the-biological-weapons-convention/.

31 About the Biological Weapons Convention, U.S. Department of State, 4 August 2022 at https://www.state.gov/about-the-biological-weapons-convention/.

32 Stern, Scott. *Biological Resource Centers: Knowledge Hubs for the Life Sciences*, (Google Books), Brookings Institution Press, 2004, p. 31, (ISBN 0815781490), and discussed at https://en.wikipedia.org/wiki/Biological_Weapons_Anti-Terrorism_Act_of_1989#cite_note-lawtxt-4.

33 SARS-CoV-2 is severe acute respiratory syndrome coronavirus—identified as 2.

34 18 U.S. Code § 175—Prohibitions with respect to biological weapons, Cornell Law, LII at https://www.law.cornell.edu/uscode/text/18/175#:~:text=Whoever%20knowingly%20develops%2C%20produces%2C%20stockpiles%2C%20transfers%2C%20acquires%2C%20retains%2C,life%20or%20any%20term%20of%20years%2C%20or%20both.

35 UN General Assembly, Resolution 2200A (XXI), International Covenant on Civil and Political Rights (16 December 1966), https://www.ohchr.org/en/instruments-mechanisms/instruments/international-covenant-civil-and-political-rights.

36 Fleming v. US, Docket No. 13–17230 (9th Cir. App. 2014). "The ICCPR Treaty is self-executing. Declarations do not change the terms or conditions of a treaty. . . . A declaration is not part of a treaty in the sense of modifying. . . . The treaty is law. The Senate's declaration is not law. . . . The Senate's power under Article II extends only to the making of reservations that require changes to a treaty before the Senate's consent will be efficacious. [emphasis added] INS v. Chadha, 462 US 919, 103 S. Ct. 2764, 77 L. Ed.2d 317 (1983); and Igartua-De La Rosa v. US, 417 F.3d 145, 190–91 (1st Cir. 2005).

37 International Covenant on Civil and Political Rights, General Assembly resolution 2200A (XXII, 16 December 1966 at https://www.ohchr.org/EN/ProfessionalInterest/Pages/CCPR.aspx.

38 The Belmont Report, Ethical Principles and Guidelines for the Protection of Human Subjects of Research, Office of the Secretary, U.S. Department of Heatlh and Human Services, 12 July 1974 at https://www.hhs.gov/ohrp/regulations-and-policy/belmont-report/read-the-belmont-report/index.html.

39 This is a required part of the National Institutes of Health (NIH) training for researchers participating in human subject research, which the author has completed certification in. Mandatory Training, National Institutes of Health at https://hr.nih.gov/training-center/mti/mandatory-training.

40 Laws are established to reflect the code of conduct (morals) unacceptable by members of our society. These fundamental codes of acceptable, and more importantly unacceptable, conduct are the established laws of medicine and in this instance are therefore the laws of our society, written and agreed to by those practicing medicine and science.

41 https://uscode.house.gov/view.xhtml?req=(title:18%20section:1621%20edition:prelim)

42 Further details of gain-of-function funding are in Richard M. Fleming, *Is COVID-19 A Bioweapon? A Scientific and Forensic Investigation* (New York: Skyhorse Publishing, 2021).

43 National Institutes of Health, The NIH Director, *Statement on misinformation about NIH support of specific "gain-of-function" research,* 19 May 2021. Web Ref: https://tinyurl.com/4zwx5yez.

44 National Library of Medicine, National Center for Biotechnology Information, Reverse genetics with a full-length infectious cDNA of severe acute respiratory syndrome coronavirus by Boyd Yount, Kristopher M. Curtis, Elizabeth A. Fritz, Lisa E. Hensley, Peter B. Jahrling, Erik Prentice, Mark R. Denison, Thomas W. Geisbert and Ralph S. Baric. Web Ref: https://tinyurl.com/7xtr5z7w.

45 PCR sequencing shown at https://www.flemingmethod.com/_files/ugd/659775_aa76331579184c83a4271ebccf9bbefe.pdf

46 Boyd Yount et al., "Reverse Genetics with a Full-Length Infectious cDNA of Severe Acute Respiratory Syndrome Coronavirus,"Proceedings *of the National Academy of Sciences of the United States of America* 100, no. 22.

47 Wuze Ren et al., "Difference in Receptor Usage Between Severe Acute Respiratory Syndrome (SARS) Coronavirus and SARS-Like Coronavirus of Bat Origin," Journal of Virology 82 (2008): https://doi.org/10.1128/jvi.01085–07.

48 Ralph S. Baric and Boyd Yount, Directional assembly of large viral genomes and chromosomes, US Patent 6593111B2, filed 21 May 2001, and issued 15 July 2003, https://patents.google.com/patent/US6593111B2/en.

49 National Institutes of Health, "Reverse Genetics with a Coronavirus Infectious Construct," NIH Reporter, n.d., Project No. 5R01GM063228–03, https://reporter.nih.gov/search/besHDSyUtEuweelVgO072A/project-details/6636663. National Institutes of Health, "Studies into the Mechanisms for MHV Replication," NIH Reporter, n.d., Project No. 2R01AI023946–14A1, https://reporter.nih.gov/search/PEMqop0HAUqS9o8RUHU_MQ/project-details/6619188.

50 Sam Husseini, "Peter Daszak's EcoHealth Alliance Has Hidden Almost $40 Million in Pentagon Funding and Militarized Pandemic Science," Independent Science News, 16 December 2020, https://www.independentsciencenews.org/news/peter-daszaks-ecohealth-alliance-has-hidden-almost-40-million-in-pentagon-funding/; and Independent Science News, "Fed. Grants & Contracts," table, https://www.independentsciencenews.org/wp-content/uploads/2020/12/EcoHealth-Funding-as-of-01_10_2020-Fed.-Grants-Contracts.pdf.

51 Fort Detrick is recognized as a principal biowarfare and biodefense facility in the United States. In 2021 Fort Detrick reached out to the author to recruit him as a physicist to work on viral projects funded by NIAID.

52 "Science and Policy Advisors," EcoHealth Alliance, n.d., https://www.ecohealthalliance.org/partners.

53 Steven Salzberg, "Gain-of-Function Experiments at Boston University Create a Deadly New Covid-19 Virus. Who Thought This Was a Good Idea?," *Forbes*, 24 October 2022, https://www.forbes.com/sites/stevensalzberg/2022/10/24/gain-of-function-experiments-at-boston-university-create-a-deadly-new-covid-19-virus

-who-thought-this-was-a-good-idea/?sh=ec592125ca31; and Da-Yuan Chen et al., "Role of Spike in the Pathogenic and Antigenic Behavior of SARS-CoV-2 BA.1 Omicron," preprint, submitted October 14, 2022, https://www.biorxiv.org/content/10.1101/2022.10.13.512134v1.

54 Martin Enserink, "Scientists Brace for Media Storm Around Controversial Flu Studies," *Science*, 23 November 2011, https://www.science.org/content/article/scientists-brace-media-storm-around-controversial-flu-studies.

55 Yang Yang et al., "Receptor Usage and Cell Entry of Bat Coronavirus HKU4 Provide Insight into Bat-to-Human Transmission of MERS Coronavirus," *Proceedings of the National Academy of Sciences of the United States of America* 111, no. 34 2014;111(2014): 12516–21, https://doi.org/10.1073/pnas.1405889111. The project was funded with NIH grants RO1AI089728 and R21AI109094.

56 Yang Yang et al., "Two Mutations Were Critical for Bat-to-Human Transmission of Middle East Respiratory Syndrome Coronavirus," *Journal of Virology* 89, no. 17 (2015): 9199–23, https://doi.org/10.1128/jvi.01279–15.

57 Yang Yang et al., "Two Mutations Were Critical for Bat-to-Human Transmission of Middle East Respiratory Syndrome Coronavirus," *Journal of Virology* 89, no. 17 (2015): 9199–23, https://www.ncbi.nlm.nih.gov/pmc/articles/PMC4524054/.

58 Nature Medicine, *Treatment with interferon-α2b and ribavirin improves outcome in MERS-CoV–infected rhesus macaques* by Darryl Falzarano, Emmie de Wit, Angela L Rasmussen, Friederike Feldmann, Atsushi Okumura, Dana P Scott, Doug Brining, Trenton Bushmaker, Cynthia Martellaro, Laura Baseler, Arndt G Benecke, Michael G Katze, Vincent J Munster & Heinz Feldmann. Published on 8 September 2013. Web Ref: https://tinyurl.com/3dujvwnv.

59 Jocelyn Kaiser and David Malakoff, "U.S. Halts Funding for New Risky Virus Studies, Calls for Voluntary Moratorium," *Science*, 17 October 2014, https://www.science.org/content/article/us-halts-funding-new-risky-virus-studies-calls-voluntary-moratorium; and Jocelyn Kaiser, "Lab Incidents Lead to Safety Crackdown at CDC," *Science*, 11 July 2014, https://www.science.org/content/article/lab-incidents-lead-safety-crackdown-cdc.

60 National Institutes of Health, EcoHealth Alliance projects, n.d., https://tinyurl.com/kk3xyrf6.

61 Fred Guterl, "Dr. Fauci Backed Controversial Wuhan Lab with US Dollars for Risky Coronavirus Research," *Newsweek*, last modified 29 April 2020, https://www.newsweek.com/dr-fauci-backed-controversial-wuhan-lab-millions-us-dollars-risky-coronavirus-research-1500741.

62 Olivia Burke, "Wuhan Wipeout: Covid Cover-Up Fears as China Deletes 300 Wuhan Lab Studies Including All Carried Out by 'Batwoman' Virologist," *The Sun*, 10 January 2021, https://www.thesun.co.uk/news/13701168/covid-cover-up-china-wuhan-studies-batwoman-virologist/.

63 Vineet D. Menachery et al., "Correction: Corrigendum: A SARS-Like Cluster of Circulating Bat Coronaviruses Shows Potential for Human Emergence," *Nature Medicine* 22, no. 4 (6 April 2016): 446, https://doi.org/10.1038/nm0416–446d.

64 National Institutes of Health, Insertion of Furin Protease Cleavage Sites in Membrane Proteins and Uses Thereof, USPTO Patent No. 7,223,390B2, issued 29 May 2007.

65 Vineet Menachery et al., "Correction: Corrigendum: A SARS-Like Cluster of Circulating Bat Coronaviruses Shows Potential for Human Emergence," *Nature Medicine* 22, no. 4 (6 April 2016): 446, https://doi.org/10.1038/nm0416–446d and "Supplementary Information" at https://static-content.springer.com/esm/art%3A10.1038%2Fnm.3985/MediaObjects/41591_2015_BFnm3985_MOESM18_ESM.pdf.

66 "Award Profile: Indefinite Delivery Vehicle," Black & Veatch Special Projects Corp., USA Spending, https://www.usaspending.gov/award/CONT_IDV_HDTRA108D0007_9700. See Chapter Ten for interesting developments on efforts to remove this evidence from the internet.

67 Nathan Wolfe is a virologist involved with WEF, DARPA, and EcoHealth Alliance. Robert Langreth, Finding the Next Epidemic Before It Kills, Forbes, 15 October 2009 at https://www.forbes.com/forbes/2009/1102/opinions-nathan-wolfe-epidemic-ideas-opinions.html?sh=6c5322be1e22 and https://en.wikipedia.org/wiki/Nathan_Wolfe.

68 Bill McCarthy, The facts behind the Russian, right-wing narratives claiming Hunter Biden funded biolabs in Ukraine. PoliticFact, 1 April 2022 at https://www.politifact.com/article/2022/apr/01/facts-behind-russian-right-wing-narratives-claimin/.

69 Meredith Wadman and JonCohen, NIH's axing of bat coronavirus grant a "horrible precedent" and might break rules, critics say. Science, 30 April 2020 at https://www.science.org/content/article/nih-s-axing-bat-coronavirus-grant-horrible-precedent-and-might-break-rules-critics-say.

70 Conor Finnegan, "Trump Admin Pulls NIH Grant for Coronavirus Research over Ties to Wuhan Lab at Heart of Conspiracy Theories," ABC News, 1 May 2020, https://abcnews.go.com/Politics/trump-admin-pulls-nih-grant-coronavirus-research-ties/story?id=70418101; National Institutes of Health, "Understanding Risk of Zoonotic Virus Emergence in EID Hotspots of Southeast Asia," NIH Reporter, n.d., https://reporter.nih.gov/search/Wan71LP3tkimgT8HyEzVow/project-details/9968924.

71 Peter Daszak, "TWiv 615: Peter Daszak of EcoHealth Alliance," interview by Vincent Racaniello, *Virology Blog*, 19 May 2020, video, 35:52, https://virology.ws/2020/05/19/twiv-615-peter-daszak-of-ecohealth-alliance/.

72 Markus Hoffmann, Hannah Kleine-Weber, and Stefan Pöhlmann, "A Multibasic Cleavage Site in the Spike Protein of SARS-CoV-2 Is Essential for Infection of Human Lung Cells," Molecular Cell 78, no. 4 (21 May 2020): 779–84e5, https://doi.org/10.1016/j.molcel.2020.04.022. Refer to "Significance of Spike Protein and Furin Cleavage Site" by Dr. Richard M. Fleming PhD, attached as "Exhibit A."

73 Ralph Baric, "SARS COV2—Identikit di un killer (Identikit of a killer)," interview by Rai, 11 November 2020, video, 9:01, https://www.youtube.com/watch?v=-kt9pVYgqkI; Olivia Burke, "Wuhan Wipeout: Covid Cover-Up Fears as China

Deletes 300 Wuhan Lab Studies Including All Carried Out by 'Batwoman' Virologist," *The Sun*, 10 January 2021, https://www.thesun.co.uk/news/13701168/covid-cover-up-china-wuhan-studies-batwoman-virologist/; "Bat SARS-like Coronavirus RsSHC014, Complete Genome," The Fleming Method, n.d., https://www.flemingmethod.com/_files/ugd/659775_aa76331579184c83a4271ebccf9bbefe.pdf; "SARS Coronavirus MA15 Isolate d3om5, Complete Genome," The Fleming Method, n.d., https://www.flemingmethod.com/_files/ugd/659775_7f8a6ee38f9c4ff085d74cc8c6138f1c.pdf; and "Bat SARS-like Coronavirus Rs3367, Complete Genome," The Fleming Method, n.d., https://www.flemingmethod.com/_files/ugd/659775_3f4d64c2c5b846d9986122bda564da49.pdf. Refer to "Significance of Spike Protein and Furin Cleavage Site" by Dr. Richard M. Fleming PhD, attached as "Exhibit A."

74 "The Future of Cloning Is Smarter and Faster," SnapGene, Dotmatics at https://www.snapgene.com/; and https://www.flemingmethod.com/gain-of-function.

75 Valentin Bruttel, Alex Washburne, and Antonius VanDongen, "Endonuclease Fingerprint Indicates a Synthetic Origin of SARS-CoV-2," preprint, submitted 11 April 2023, https://www.biorxiv.org/content/10.1101/2022.10.18.512756v2.

76 Richard M. Fleming, "Laying Out Evidence That SARS-CoV-2/COVID-19 Is a Gain-of-Function (GoF) Virus That Violates the Biological Weapons Convention (BWC) Treaty. Crimes Against America & Crimes Against Humanity," The Fleming Method, n.d., text and videos, https://www.flemingmethod.com/gain-of-function.

77 Anne Trafton, "Storing Medical Information Below the Skin's Surface," Global MIT, 18 December 2019, https://global.mit.edu/news-stories/storing-medical-information-below-the-skins-surface/.

78 Audrey McNamara, "CDC Confirms First Case of Coronavirus in the United States," CBS News, 21 January 2020, https://www.cbsnews.com/news/coronavirus-centers-for-disease-control-first-case-united-states/.

79 Victor M Corman et al., "Detection of 2019 Novel Coronavirus (2019-nCoV) by Real-Time RT-PCR," *Eurosurveillance* 25, no. 3 (23 January 2020): https://doi.org/10.2807/1560–7917.es.2020.25.3.2000045.

80 US Patent No. 4683202, Process For Amplifying Nucleic Acid Sequences. Issued 28 July 1987 at https://patentimages.storage.googleapis.com/cc/f0/3e/dc51b1fb4af2e6/US4683202.pdf

81 A study funded by the French government in September 2020 demonstrated that "a person receiving a positive PCR test result at a cycle threshold (Ct) of 35 or higher, has a less than 3% possibility of being infected. Concurrently, the probability of a false positive in this case is 97% or higher." [See point 77] This is included in: Daniel P. Oran and Eric J. Topol, Prevalence of Asymptomatic SARS-CoV-2 Infection, *Annals of Internal Medicine* 2020;DOI: 10.7326/M20–3012

82 "NetWorkBench: A Large-Scale Network Analysis, Modeling, and Visualization Tookit for Biomedical, Social Science, and Physics Research," National Science Foundation, grant number IIS-0513650, 12 July 2005, https://www.nsf.gov/awardsearch/showAward?AWD_ID=0513650; Indiana University, "Scientists Assess Risk of Potential Flu Pandemic Spread via Global Airlines," *Science Daily*,

29 January 2007, https://www.sciencedaily.com/releases/2007/01/070128140422.htm.

83 Nur Ibrahim, Did Trump Ban All Travel From China at the Start of the Pandemic? Snopes 22 September 2020 at https://www.snopes.com/fact-check/trump-ban-travel-china-pandemic/.

84 "U.S. Declares Coronavirus a Public Health Emergency, CDC Updates Guidance," American Hospital Association, 31 January 2020, https://www.aha.org/news/headline/2020–01-31-us-declares-coronavirus-public-health-emergency-cdc-updates-guidance.

85 "Naming the Coronavirus Disease (COVID-19) and the Virus That Causes It," World Health Organization, n.d., https://www.who.int/emergencies/diseases/novel-coronavirus-2019/technical-guidance/naming-the-coronavirus-disease-(covid-2019)-and-the-virus-that-causes-it.

86 Charles Calisher et al., "Statement in Support of the Scientists, Public Health Professionals, and Medical Professionals of China Combatting COVID-19," *The Lancet* 395, no. 10226 (7 March 2020): E42–E43, https://doi.org/10.1016/s0140–6736(20)30418–9.

87 Brit McCandless Farmer, "March 2020: Dr. Anthony Fauci Talks with Dr Jon LaPook about COVID-19," 60 Minutes Overtime, CBS News, 8 March 2020, https://www.cbsnews.com/news/preventing-coronavirus-facemask-60-minutes-2020–03-08/; and Anthony Fauci, "Dr. Fauci on Masks (March 8, 2020)," interview by Jon LaPook, *60 Minutes*, March 8, 2020, video, 1:27, https://www.youtube.com/watch?v=jwCynEjFt8E.

88 "READ: Text of Trump's National Emergency Declaration over Coronavirus," CNN, 13 March 2020, https://edition.cnn.com/2020/03/13/politics/trump-national-emergency-proclamation-text/index.html.

89 Libby Cathey, Trump declares national emergency responding to coronavirus: Here's what that means. ABC News, 13 March 2020. https://abcnews.go.com/Politics/trump-declares-national-emergency-responding-coronavirus-heres-means/story?id=69586419.

90 The declaration was further updated on 14 July 2020. Emergency Authoritiese Under the National Emergencies Act, Stafford Act, and Public Health Service Act. Congressional Research Service, 14 July 2020 at https://crsreports.congress.gov/product/pdf/R/R46379.

91 U.S. Department of Health & Human Services, Public Health Emergency, *Waiver or Modification of Requirements Under Section 1135 of the Social Security Act,* 13 March 2020. Web Ref: https://tinyurl.com/6pcm8c4c.

92 World Health Organization, "WHO Director-General's Opening Remarks at the Media Briefing on COVID-19—16 March 2020," Web Ref: originally at but now deleted by those controlling the website https://tinyurl.com/2p8kkt9j.

93 Kristian G. Andersen et al., "The Proximal Origin of SARS-CoV-2," *Nature Medicine* 26 (March 2020): 450–52, https://doi.org/10.1038/s41591–020-0820–9.

94 "Trump Declares State of Emergency for COVID-19," National Conference Of State Legislatures, last modified 25 March 2020, https://www.ncsl.org/state-federal/trump-declares-state-of-emergency-for-covid-19.

95 "2020 CARES Act—FEMA Disaster Relief Fund/Stafford Act Assistance," GovWin (website), 8 June 2020, https://iq.govwin.com/neo/marketAnalysis/view/2020-CARES-Act--FEMA-Disaster-Relief-Fund-Stafford-Act-Assistance/4117?researchTypeId=1&researchMarket=.

96 The question repeatedly asked here and elsewhere is who determines what is factual and what is not? Certainly, a solid scientific debate and discussion is not disinformation. As a scientist-physician-attorney, I understand that transparent discussion is critical to understand and advance scientific knowledge. Alternatively, we could take the position of the Catholic Church when it prosecuted Galileo Galilei as he raised the level of knowledge and understanding, in opposition to the view held by the powers that be. Having lived through the era of HIV and having treated patients then, I can tell you that failure to have such open debate, discussion, and exchange of ideas would have resulted in a situation more crippling than what we experienced.

97 "Trusted News Initiative Announces Plans to Tackle Harmful Coronavirus Disinformation," BBC, 27 March 2020, https://www.bbc.com/mediacentre/latestnews/2020/coronavirus-trusted-news; Bill Pan, "COVID-Narrative Dissenters File Antitrust Action Against Legacy Media over Coordinated Censorship," *Epoch Times*, last modified 12 January 2023, https://www.theepochtimes.com/us/covid-narrative-dissenters-file-antitrust-action-against-legacy-media-over-coordinated-censorship-4977641?src_src=partner&src_cmp=thethinkingconservative; and Children's Health Defense, Robert F. Kennedy Jr., Trialsite, Creative Destruction Media, Erin Elizabeth Finn, Jim Hoft, Ben Tapper, Ben Swann, Joseph Mercola, Ty Bollinger, and Charlene Bollinger v. Washington Post Co., British Broadcasting Corp., Associated Press, and Reuters, Case 2:23-cv-00004-Z (N.D. Tex. 2023), https://childrenshealthdefense.org/wp-content/uploads/TNI-Complaint-1.10.22.pdf.

98 These policies, as you will see in this book, were wholeheartedly backed by the Rockefeller Foundation, which also backed the eugenics movement.

99 In fact, in spite of anything you may have heard, vaccinations do not prevent infection or transmission of disease. The premise of vaccination is to expose an individual to a pathogen, to produce T- and B-memory cells, so that when the person becomes exposed to the pathogen, the time required to mount an immune response is shortened, the goal of which is to reduce symptoms and possible consequences. There is no basis for a belief that a vaccine somehow prevents the person so immunized from transmitting the infectious agent, in this case a virus, to someone else.

100 Jiachuan Wu et al., "Stay-at-Home Orders Across the Country," NBC News, last modified 29 April 2020, https://www.nbcnews.com/health/health-news/here-are-stay-home-orders-across-country-n1168736.

101 "Apple and Google Partner on COVID-19 Contact Tracing Technology," Apple, 10 April 2020, https://www.apple.com/newsroom/2020/04/apple-and-google-partner-on-covid-19-contact-tracing-technology/.

102 "Guidance for Certifying Deaths Due to Coronavirus Disease 2019 (COVID-19)," Vital Statistics Reporting Guidance, last modified February 2023, https://www.cdc.gov/nchs/data/nvss/vsrg/vsrg03–508.pdf.

103 The terms SARS-CoV-2 and COVID or COVID-19 are frequently, incorrectly, used interchangeably. The former is a virus, and the latter is the InflammoThrombotic Response (ITR) to the virus. ITR produces blood clotting, tissue damage, and myriad diseases caused from the combination of inflammation and thrombosis (blood clotting) that occurs in these cases. The resulting CVAs, MIs [myocardial infarction], sudden cardiac death, cancers, miscarriages, and such are therefore the consequence of becoming infected with SARS-CoV-2, or receiving genetic vaccines that produce the SARS-CoV-2 spike protein, responsible for much of the ITR, priogenic disease, and sudden cardiac death. Consequently, it is correct to define these causes of death, related to disturbance in sodium channel flow in conduction system, as being due to SARS-CoV-2 (COVID ITR).

104 Brit McCandless Farmer, "March 2020: Dr. Anthony Fauci Talks with Dr Jon LaPook about COVID-19," 60 Minutes Overtime, CBS News, 8 March 2020, https://www.cbsnews.com/news/preventing-coronavirus-facemask-60-minutes-2020–03-08/; and Anthony Fauci, "Dr. Fauci on Masks (March 8, 2020)," interview by Jon LaPook, *60 Minutes*, March 8, 2020, video, 1:27, https://www.youtube.com/watch?v=jwCynEjFt8E.

105 Scott Jensen, "Dr. Jensen Calls Out 'Ridiculous' CDC Guidelines for Coronavirus-Related Deaths," interview by Laura Ingraham, Fox News, 9 April 2020, https://www.foxnews.com/video/6148398329001.

106 Matt Naham, "Vaccine Expert Files Whistleblower Complaint, Claims Trump Admin Illegally Retaliated Against Him," Law & Crime, 5 May 2020, https://lawandcrime.com/high-profile/vaccine-expert-files-whistleblower-complaint-claims-trump-admin-illegally-retaliated-against-him/; and The Rockefeller Foundation, "Dr. Rick Bright Joins the Rockefeller Foundation to Lead Pandemic Prevention Institute Development," press release, 8 March 2021, https://www.rockefellerfoundation.org/news/dr-rick-bright-joins-the-rockefeller-foundation-to-lead-pandemic-prevention-institute-development/.

107 Dr. Bright is also a member of the Council on Foreign Relations (CFR) and serves as a Senior Fellow at the Foreign Policy Association.

108 Dan Sanchez, "YouTube to Ban Content That Contradicts WHO on COVID-19, Despite the UN Agency's Catastrophic Track Record of Misinformation," Foundation for Economic Education, 23 April 2020, https://fee.org/articles/youtube-to-ban-content-that-contradicts-who-on-covid-19-despite-the-un-agency-s-catastrophic-track-record-of-misinformation/; and Linda Lacina, "YouTube's Susan Wojcicki on the Creator Economy, Competition, and Staying Ahead of Misinformation," 27 May 2022, https://www.weforum.org/agenda/2022/05/davos-2022-mtl-susan-wojcicki-misinformation/.

109 COVID-19 Testing, Reaching, and Contacting Everyone (TRACE) Act, H.R. 6666, 116th Cong. (2020), https://www.congress.gov/bill/116th-congress/house-bill/6666/text.

110 https://www.govtrack.us/congress/bills/116/hr6666/summary

111 Daniel Ortner, "The Wisconsin Supreme Court Stay-at-Home Ruling Defends the Separation of Powers and Individual Liberty," Pacific Legal Foundation, 26 May 2020, https://pacificlegal.org/wisconsin-supreme-court-ruling-separation-of-powers/.

112 US Department of Defense, "Trump Administration Announces Framework and Leadership for 'Operation Warp Speed,'" news release, 15 May 2020, https://www.defense.gov/News/Releases/Release/Article/2310750/trump-administration-announces-framework-and-leadership-for-operation-warp-speed/.

113 Klaus Schwab, "Now Is the Time for a 'Great Reset,'" World Economic Forum, https://www.weforum.org/agenda/2020/06/now-is-the-time-for-a-great-reset/.

114 The Brookings Institution is funded by the Rockefeller Foundation.

115 Zia Khan and John W. McArthur, "Rebuilding Toward the Great Reset: Crisis, COVID-19, and the Sustainable Development Goals," Brookings Institution, 19 June 2020, https://www.brookings.edu/articles/rebuilding-toward-the-great-reset-crisis-covid-19-and-the-sustainable-development-goals/.

116 Felix Richter, "The U.S. Employment-Population Ratio Drops to a Historic Low," World Economic Forum, 6 July 2020, https://www.weforum.org/agenda/2020/07/employment-population-ratio-drops-to-historic-low/.

117 Klaus Schwab and Thierry Malleret, *COVID-19: The Great Reset* (Geneva, Switzerland: World Economic Forum, 2020).

118 World Economic Forum, Agenda Contributors, website search. Web Ref: https://tinyurl.com/35uhaxyh.

119 The Malone Institute, The Pharos Media Foundation, List of U.S. Politicians (not including 2022 YGL (WEF) Graduates. Web Ref: https://tinyurl.com/56y7v5c9. Download the WEF Graduate List, Web Ref: https://tinyurl.com/3ztfmavv; https://en.wikipedia.org/wiki/Young_Global_Leaders

120 David Rockefeller, *Memoirs* (New York: Penguin Random House, 28 October 2003), and David Rockefeller, "Quotable Quote," Goodreads (website), n.d.,https://www.goodreads.com/quotes/10465260-for-more-than-a-century-ideological-extremists-at-either-end.

121 "Collaboration in a Fractured World: Klaus Schwab MC/MPA Speaks at Harvard Kennedy School," Harvard Kennedy School, 2 October 2017, article and video, 1:01:00, https://www.hks.harvard.edu/more/alumni/alumni-stories/collaboration-fractured-world-klaus-schwab-mcmpa-speaks-harvard-kennedy.

122 Ida Auken, "Welcome to 2030: I Own Nothing, Have No Privacy and Life Has Never Been Better," Medium.com, 12 November 2016, https://medium.com/world-economic-forum/welcome-to-2030-i-own-nothing-have-no-privacy-and-life-has-never-been-better-ee2eed62f710. The article was posted on the *Forbes* magazine website on 10 November 2016.

123 Anthony Fauci, "TWiV 641: COVID-19 with Dr. Anthony Fauci," interview by Vincent Racaniello and Rich Condit, Microbe TV, 17 July 2020, video, 36:47, https://www.microbe.tv/twiv/twiv-641/.
124 Donald Trump, "Remarks by President Trump to the 75th Session of the United Nations General Assembly," US Embassy & Consulates in Italy, 22 September 2020, https://it.usembassy.gov/remarks-by-president-trump-to-the-75th-session-of-the-united-nations-general-assembly-september-22–2020/.
125 Rita Jaafar et al., "Correlation Between 3790 Quantitative Polymerase Chain Reaction–Positives Samples and Positive Cell Cultures, Including 1941 Severe Acute Respiratory Syndrome Coronavirus 2 Isolates," *Clinical Infectious Diseases* 72, no. 11 (1 June 2021): e921, https://doi.org/10.1093/cid/ciaa1491.
126 Justin Trudeau, "Coronavirus: Trudeau Tells UN Conference That Pandemic Provided 'Opportunity for a Reset,'" speech, 29 September 2020, video, 6:11, https://tinyurl.com/ypk7axyr; and Brian Lilley, Toronto Sun, "Lilley: Trudeau Has Eyes Set on a Great Reset for Canada," *Toronto Sun*, 16 November 2020, https://torontosun.com/opinion/columnists/lilley-trudeau-has-eyes-set-on-a-great-reset-for-canada.
127 "Separation of Powers: An Overview," National Conference of State Legislatures, last modified 1 May 2021, https://www.ncsl.org/about-state-legislatures/separation-of-powers-an-overview; David Komer, "Michigan Supreme Court Strikes Down Whitmer's Virus Orders; Gov. fires back," Fox2 Detroit, last modified 2 October 2020, https://www.fox2detroit.com/news/michigan-supreme-court-strikes-down-whitmers-virus-orders-gov-fires-back; and Associated Press, "Michigan Gov. Whitmer Headed to Europe, WEF Meeting in Davos," Fox News, 11 January 2023, article and video, 3:33, https://www.foxnews.com/politics/michigan-gov-whitmer-headed-europe-wef-meeting-davos.
128 Martin Kulldorff, Sunetra Gupta, and Jay Bhattacharya, "Great Barrington Declaration," Gbdeclaration.org, 4 October 2020, https://gbdeclaration.org.
129 Steve Anderson, "Vaccines and Related Biological Products Advisory Committee October 22, 2020, Meeting Presentation" (presentation, VRBPAC Meeting), https://www.fda.gov/media/143557/download; and Steve Anderson, "Vaccines and Related Biological Products Advisory Committee—10/22/2020," YouTube, video, 8:50:55 at 2:33:40, https://tinyurl.com/58vdcch7.
130 Tom Shimabukuro, "Vaccines and Related Biological Products Advisory Committee—10/22/2020," YouTube, video, 8:50:55 at 2:06:29, https://tinyurl.com/4wxfbj92.
131 Centers for Disease Control and Prevention, "Share Your COVID-19 Vaccination Experience with V-Safe," YouTube, 29 March 2021, video, 0:31, https://www.youtube.com/watch?v=8e4qJYwtVRg.
132 Tucker Carlson, "Tucker Carlson: The Elites Want COVID-19 Lockdowns to Usher in a 'Great Reset' and That Should Terrify You," Fox News, 16 November 2020, article and video, 10:41, https://www.foxnews.com/opinion/tucker-carlson-coronavirus-pandemic-lockdowns-great-reset.

133 Ethan Yang, "The Lasting Consequences of Lockdowns," American Institute for Economic Research, 26 November 2020, https://www.aier.org/article/the-lasting-consequences-of-lockdowns/.

134 Memorandum by Florida Department of Health, "Mandatory Reporting of COVID-19 Laboratory Test Results: Reporting of Cycle Threshold Values," 3 December 2020, https://www.flhealthsource.gov/files/Laboratory-Reporting-CT-Values-12032020.pdf.

135 Vaccines and Related Biological Products Advisory Committee Meeting, FDA Briefing Document Pfizer-BioNTech COVID-19 Vaccine, Sponsor Pfizer and BioNTech, 10 December 2020 at https://www.flemingmethod.com/_files/ugd/659775_1136b2851e6e48b1886457ab98b4feef.pdf.

136 Vaccines and Related Biological Products Advisory Committee Meeting, FDA Briefing Document Moderna COVID-19 Vaccine, Sponsor ModernaTX, Inc., 17 December 2020 at https://www.flemingmethod.com/_files/ugd/659775_2b26a980a8d44de89cd21c42af406565.pdf.

137 Vaccines and Related Biological Products Advisory Committee Meeting, FDA Briefing Document Janssen Ad26.COV2.S Vaccine for the Prevention of COVID-19, Sponsor Janssen Biotech, Inc., 26 February 2021 at https://www.flemingmethod.com/_files/ugd/659775_ac0cd50998c5405584aefe60ea4a1707.pdf.

138 Robert Cuffe, "Coronavirus Death Rate: What Are the Chances of Dying?," BBC News, 24 March 2020, https://www.bbc.com/news/health-51674743.

139 Health Resources & Services Administration, "National Vaccine Injury Compensation Program Monthly Statistics Report," 1 March 2023, https://www.hrsa.gov/sites/default/files/hrsa/vicp/vicp-stats.pdf.

140 Digital Healthcare Research, *Electronic Support for Public Health—Vaccine Adverse Event Reporting System (ESP:VAERS)* (Rockville, MD: Agency for Healthcare Research and Quality, 29 September 2010), https://digital.ahrq.gov/sites/default/files/docs/publication/r18hs017045-lazarus-final-report-2011.pdf.

141 I accept the CDC numbers at face value without questioning if they represent 1 percent, 10 percent, or 100 percent of the adverse events.

142 Rowan Dean, "'You Will Own Nothing, and You Will Be Happy': Warnings of 'Orwellian' Great Reset," Sky News Australia, 12 December 2020, video, 7:24, https://www.youtube.com/watch?v=NcAO4-o_4Ug.

143 Lena H. Sun and Isaac Stanley-Becker, "Front-Line Essential Workers and Adults 75 and Over Should Be Next to Get the Coronavirus Vaccine, a CDC Advisory Group Says," *Washington Post*, 20 December 2020, https://www.washingtonpost.com/health/2020/12/20/covid-vaccine-front-line-workers/.

144 Stephen M. Hahn and Peter Marks, "FDA Statement on Following the Authorized Dosing Schedules for COVID-19 Vaccines," news release, Food and Drug Administration, 4 January 2021, https://www.fda.gov/news-events/press-announcements/fda-statement-following-authorized-dosing-schedules-covid-19-vaccines.

145 This nomenclature is used in scientific statistical analysis of data. Here, the probability (p) that the differences between the two groups (those who were vaccinated

and those who were not vaccinated) is tested for. When there is no statistically significant difference between the two groups, this is referred to as not statistically significant or 'NS'. In other words, when the results of the Pfizer, Moderna and Janssen COVID vaccines were analyzed, there was no difference between the vaccinated and unvaccinated people, either in terms of the number of people who either were diagnosed with COVID or who died.

146 Nina Kheladze and Elena Stepanova, "The Thomson Reuters Foundation Partners with Sabin Vaccine Institute to Deliver 'Reporting on Immunisation and Vaccination' Journalism Training," Thomson Reuters Foundation, 3 February 2021, https://www.trust.org/i/?id=0b6d2d16–3bb0–4e1f-b7d2-ccbc0335b574.

147 "James C. Smith Elected to Pfizer's Board of Directors," news release, Pfizer, 26 June 2014, https://www.pfizer.com/news/press-release/press-release-detail/james_c_smith_elected_to_pfizer_s_board_of_directors.

148 "COVID-19 Crisis Reporting Hub," Thomson Reuters Foundation, n.d., https://reportinghub.trust.org/#.

149 Megan Redshaw, "Conflict of Interest: Reuters 'Fact Checks' COVID-Related Social Media Posts, but Fails to Disclose Ties to Pfizer, World Economic Forum," The Defender, 11 August 2021, https://childrenshealthdefense.org/defender/reuters-fact-check-covid-social-media-pfizer-world-economic-forum/.

150 Richard M. Fleming and Matthew R. Fleming, "FMTVDM Quantitative Nuclear Imaging Finds Three Treatments for SARS-CoV-2," *Biomedical Journal of Scientific & Technical Research* 33, no. 4 (February 2021): 26041–83, https://doi.org/10.26717/BJSTR.2021.33.005443.

151 Richard M. Fleming, The Fleming Method for Tissue and Vascular Differentiation and Metabolism (FMTVDM) Using Same State Single or Sequential Quantification Comparisons, US Patent No. US9566037B2, filed 13 June 2013, and issued 14 February 2017, https://patentimages.storage.googleapis.com/5a/21/2d/f66558a0b9b861/US9566037.pdf.

152 R. M. Fleming and M. R. Fleming, "Treating SARS-CoV-2 Based upon the Fleming InflammoThrombotic Response (ITR) & Cardiovascular Disease Theory," *Haematology International Journal* 7, no. 1 (January 2023): https://doi.org/10.23880/hij-16000209.

153 Xiaopeng Liang, Oscar Hou In Chou, and Bernard M. Y. Cheung, "The Effects of Human Papillomavirus Infection and Vaccination on Cardiovascular Diseases, NHANES 2003–2016," *American Journal of Medicine* 136, no. 3 (March 2023): 294–301, https://doi.org/10.1016/j.amjmed.2022.09.021.

154 Rohan Khera et al., "Association of COVID-19 Hospitalization Volume and Case Growth at US Hospitals with Patient Outcomes," *American Journal of Medicine* 134, no. 11 (November 2021): 1380–88, https://doi.org/10.1016/j.amjmed.2021.06.034.

155 Rev, Dr. Fauci Testimony Speech Transcript March 18: "Masks Are Not Theatre". Web Ref: https://tinyurl.com/3emaeeey. PBS NewsHour YouTube channel, WATCH: 'Masks are not theater,' Dr. Fauci says. Web Ref: https://tinyurl.com/ycyx5es2.

156 "Dr. Fauci, CDC Director Testify Before Senate on COVID-19 Guidelines Transcript," testimony before Senate, Rev (website), 11 May 2021, https://tinyurl.com/276rf4xx; and Anthony Fauci, "Live: Dr. Fauci and Dr. Walensky testify on efforts to combat COVID-19," Reuters, 11 May 2021, video, 2:43:32 at 55:56, https://tinyurl.com/4kzuvach.

157 Katherine Eban, "The Lab-Leak Theory: Inside the Fight to Uncover COVID-19's Origins," *Vanity Fair*, 3 June 2021, https://www.vanityfair.com/news/2021/06/the-lab-leak-theory-inside-the-fight-to-uncover-covid-19s-origins.

158 ICAN, Informed Consent Action Network, ICAN OBTAINS OVER 3,000 PAGES OF TONY FAUCI'S EMAILS, on 4 June 2021. (Email Ref: ICAN_000103), Web Ref: https://tinyurl.com/43dj8r62. Link to download ICAN Fauci Emails in PDF format uploaded on 03 June 2021: https://tinyurl.com/5cy5asjw.

159 Meghan O'Sullivan is former special assistant to President George W. Bush and professor of International Affairs, well known for her role as deputy national security advisor for Iraq and Afghanistan. She is listed as an agenda contributor on the World Economic Forum (WEF) website and is North American chair at the Trilateral Commission. Trilateral Commission meetings are listed on the past-events section of the commission's website. However, the March13–15 Trilateral Commission Meeting with Dr. Fauci in Washington, DC, is not listed in the past-events section or anywhere else on the Trilateral Commission website.

160 "ICAN Fauci Emails_2021_06_03," MediaFire, 3 June 2021, ICAN_ 000274–000277, https://www.mediafire.com/file/e7wi0lqjd1d5lcx/Combined_Fauci_Emails_in_Chron_Order_OCRd_FINAL.pdf/file. The MediaFile source has a link to download ICAN Fauci Emails in PDF format and may also be accessed via https://tinyurl.com/5cy5asjw.

161 "ICAN Fauci Emails," ICAN_001778; "The Bill & Melinda Gates Foundation," Gavi, the Vaccine Alliance, n.d., https://www.gavi.org/investing-gavi/funding/donor-profiles/bill-melinda-gates-foundation; and ABC News, "By the Numbers: Bill and Melinda Gates Rank as WHO Organization's 2nd Biggest Donor, 15 April 2020, video, 1:06, https://abcnews.go.com/US/video/numbers-bill-melinda-gates-rank-organizations-2nd-biggest-70174755.

162 Such an email communication between Dr. Anthony Fauci; Emilio A. Emini, CEO of Bill & Melinda Gates Medical Research Institute; Rick Bright; and other NIH and NIAID members with a private individual, Bill Gates and his organization (Bill and Melinda Gates Foundation; BMGF), is unusual and allowed an opportunity to exert direct influence on decisions that impacted the civil rights of many Americans during the pandemic. The Bill & Melinda Gates Foundation is a major investor in the global vaccine industry, founding partner of Gavi, the Vaccine Alliance and the second largest donors to the WHO.

163 "ICAN Fauci Emails," ICAN_ 002471–002472.

164 Jay Bhattacharya, "Ingraham: Are Fauci's Days Numbered?," interview by Laura Ingraham, Fox News, 5 June 2021, https://www.foxnews.com/video/6257486745001.

165 Erika Edwards, "Evidence Grows Stronger for Covid Vaccine Link to Heart Issue, CDC Says," NBC News, 10 June 2021, article and video, 2:02, https:

//www.nbcnews.com/health/health-news/evidence-grows-stronger-covid-vaccine-link-heart-issue-cdc-says-n1270339.

166 Richard M. Fleming, "Atherosclerosis: Understanding the Relationship Between Coronary Artery Disease and Stenosis Flow Reserve," chap. 29 in *Textbook of Angiology*, edited by John B. Chang, 381–87 (New York: Springer, 2000); and Richard M. Fleming, "Pathogenesis of Vascular Disease," chap. 64 in *Textbook of Angiology*, edited by John B. Chang, 787–98 (New York: Springer, 2000), https://doi.org/10.1007/978–1-4612–1190-7_64.

167 Rachel Bovard, "Government Dictating What Social-Media Bans Is Tyrannical," *New York Post*, 16 July 2021, https://nypost.com/2021/07/16/government-dictating-what-social-media-bans-is-tyrannical/.

168 "Fauci, Walensky COVID-19 Response Testimony Senate Hearing Transcript July 20," testimony before Senate, Rev (website), 20 July 2021, https://tinyurl.com/5n7ws7hb; and Anthony Fauci, "Watch Live: Fauci, Walensky Testify Before Senate on COVID-19 Response," PBS NewsHour, 20 July 2021, video, 2:20:37 at 50:05, https://tinyurl.com/5yyx7nvj.

169 18 USC § 1001 includes reference to 18 USC § 2331 and perjury under 18 U.S.C. § 1621.

170 The White House, FACT SHEET: President Biden to Announce New Actions to Get More Americans Vaccinated and Slow the Spread of the Delta Variant, 29 July 2021. Web Ref: https://tinyurl.com/55ppeh48.

171 Laura Ly and Holly Yan, "New York City Vaccine Mandate Extends to All City Workers and Includes a New $500 Bonus, Mayor Says," CNN, 20 October 2021, https://edition.cnn.com/2021/10/20/us/nyc-vaccine-mandate-extension/index.html.

172 Emily Crane, "NIH Admits US Funded Gain-of-Function in Wuhan—Despite Fauci Denials," *New York Post*, 21 October 2021, https://nypost.com/2021/10/21/nih-admits-us-funded-gain-of-function-in-wuhan-despite-faucis-repeated-denials/; and Oversight Committee (@GOPoversight), "July 28th NIH says 'no NIAID funding was approved for Gain of Function research at the WIV.' Obviously, they were lied to," with attached letter, Twitter, 20 October 2021, 5:17 p.m., https://twitter.com/GOPoversight/status/1450934193177903105?s=20.

173 Lawrence A. Tabak's letter to Rep. James Comer (R-KY) is verification that Dr. Fauci perjured himself multiple times before Congress.

174 Christina Maxouris, Artemis Moshtaghian, and Ralph Ellis, CNN, "2,300 NYC Firefighters Call Out Sick as Vaccine Mandate Begins, but Mayor Says Public Safety Not Disrupted," CNN, 1 November 2021, https://edition.cnn.com/2021/11/01/us/new-york-city-vaccine-mandate-first-responder-shortage/index.html.

175 Shannon Pettypiece, Heidi Przybyla, and Lauren Egan, "Biden Announces Sweeping Vaccine Mandates Affecting Millions of Workers," NBC News, 9 September 2021, https://www.nbcnews.com/politics/white-house/biden-announce-additional-vaccine-mandates-he-unveils-new-covid-strategy-n1278735.

176 The White House, "Fact Sheet: Biden Administration Announces Details of Two Major Vaccination Policies," news release, 4 November 2021, https://www.white

house.gov/briefing-room/statements-releases/2021/11/04/fact-sheet-biden-administration-announces-details-of-two-major-vaccination-policies/.

177 Natalie O'Neill, "Rand Paul calls on Fauci to Resign over Gain-of-Function Research," *New York Post*, 4 November 2021, https://nypost.com/2021/11/04/rand-paul-calls-on-fauci-to-resign-over-gain-of-function-research/; and *Hearing on the Biden Administration's COVID-19 Response, Before Senate HELP Committee*, 117 Cong. (2021) (statement of Anthony Fauci), video, 8:33, https://www.c-span.org/video/?c4985080/complete-exchange-sen-rand-paul-dr-anthony-fauci.

178 Ariane de Vogue and Rachel Janfraza, "Federal Appeals Court Issues Stay of Biden Administration's Vaccine Mandate for Private Companies," CNN, 6 November 2021, https://edition.cnn.com/2021/11/06/politics/court-blocks-biden-administration-vaccine-rule-larger-employers/index.html.

179 Jacob Pramuk and Spencer Kimball, "Senate Votes to Block Biden Vaccine Mandate, Which Has Already Hit Roadblocks in Court," CNBC, 8 December 2021, https://www.cnbc.com/2021/12/08/biden-vaccine-mandate-senate-votes-to-overturn-osha-rule.html.

180 Tim Hains, "Rand Paul Grills Fauci: Does Government Pay You To Discredit Other Scientists?," RealClearPolitics, 11 January 2022, https://www.realclearpolitics.com/video/2022/01/11/rand_paul_to_fauci_do_you_think_it_is_your_job_to_call_other_scientists_fringe_if_they_disagree.html; and Senate Hearing on Federal Response to COVID-19 Variants, "Heated Exchange Between Sen. Rand Paul and Dr. Anthony Fauci," CSPAN, 11 January 2022, https://www.c-span.org/video/?c4995558/heated-exchange-sen-rand-paul-dr-anthony-fauci.

181 The Intercept, *House Republicans Release Text of Redacted Fauci Emails On COVID Origins* by Maia Hibbett and Ryan Grim on 13 January 2022. Web Ref: https://tinyurl.com/yd56nneu.

182 Devin Dwyer, "Supreme Court Blocks Biden Vaccine-or-Test Mandate for Large Businesses," ABC News, 14 January 2022, article and video, 1:42, https://abcnews.go.com/Health/supreme-court-blocks-biden-vaccine-test-mandate-large/story?id=82179446.

183 Michael Nevradakis, "Pfizer, FDA Lose Bid to Further Delay Release of COVID Vaccine Safety Data," The Defender, 7 February 2022, https://childrenshealthdefense.org/defender/pfizer-fda-lose-bid-delay-release-covid-vaccine-safety-data/.

184 CDC (@CDCgov), "Remember that #COVID19 nose swab test you took?," Twitter, 16 February 2022, 11:15 a.m., https://twitter.com/cdcgov/status/1493982257790570501.

185 Nicolle K. Strand, "Shedding Privacy Along with Our Genetic Material: What Constitutes Adequate Legal Protection Against Surreptitious Genetic Testing?," *AMA Journal of Ethics* 18, no. 3 (March 2016): 264–71, https://doi.org/10.1001/journalofethics.2016.18.3.pfor2–1603.

186 Kile Green et al., "What Tests Could Potentially Be Used for the Screening, Diagnosis and Monitoring of COVID-19 and What Are Their Advantages and Disadvantages?," Centre for Evidence-Based Medicine, 20 April 2020, https://www.cebm.net/covid-19/what-tests-could-potentially-be-used-for-the-screening

-diagnosis-and-monitoring-of-covid-19-and-what-are-their-advantages-and-disadvantages/.

187 Francis Collins, "Francis Collins Talking About Klaus Schwab's 4th Industrial Revolution and Collecting Genomic Data," Free Thinker Fitness, video, 1:10, .

188 The deletion was an intentional but failed act to destroy evidence.

189 Hearing on Russian Invasion of Ukraine Before Senate Foreign Relations Committee, Senator Rubio Questions Undersecretary Nuland over Biolabs in Ukraine, 117th Cong. (8 March 2022), https://www.c-span.org/video/?c5005520/senator-rubio-questions-undersecretary-nuland-biolabs-ukraine; and Tina Redlup, "Biolab Opens in Ukraine," CDC Outreach Journal, no. 818, 18 June 2010, https://media.defense.gov/2019/Aug/02/2002165966/-1/-1/0/CPC%20OUTREACH%20818.PDF.

190 Sharyl Attkisson, "List of Ukraine Biolab Documents Reportedly Removed by US Embassy," SharylAttkisson.com, 11 March 2022, https://sharylattkisson.com/2022/03/list-of-ukraine-biolab-documents-reportedly-removed-by-us-embassy/.

191 As mentioned, at least forty-six biolabs were later identified.

192 Ryan Morgan, "Pentagon Funding Biolabs in Ukraine—'Real Concern' of Pathogen Releases if Russia Attacks," American Military News, 14 March 2022, https://americanmilitarynews.com/2022/03/pentagon-funding-biolabs-in-ukraine-real-concern-of-pathogen-releases-if-russia-attacks/

193 See prior discussion on Pentagon report from Ukrainian Biolabs on COVID seven weeks prior to Wuhan discussion of SARS-CoV-2 in December 2019.

194 Clearly, if these laboratories were illegal for the Soviet Union, they are illegal for Ukraine and the United States.

195 Cosmin Dzsurdzsa, "Trudeau Government Gave $3 Million to WEF and $1.6 Billion to UN in 2021," True North, 10 May 2022, https://tnc.news/2022/05/10/trudeau-government-gave-3-million-to-wef-and-1–6-billion-to-un-in-2021/; and Government of Canada, *The Public Accounts of Canada,* vol. 3, sect. 6, transfer payments, https://www.tpsgc-pwgsc.gc.ca/recgen/cpc-pac/2021/vol3/ds6/index-eng.html.

196 Eric Katz, "Feds' Vaccine Mandate Enforcement Could Be Days Away, but Agencies Are Not Yet Prepping," Government Executive, 31 May 2022, https://www.govexec.com/workforce/2022/05/feds-vaccine-mandate-enforcement-agencies-not-prepping/367566/.

197 Kelly Gooch, *Becker's Hospital Review*, "Lawsuits Still Piling Up over Hospital Vaccine Mandates," 31 May 2022, https://www.beckershospitalreview.com/legal-regulatory-issues/lawsuits-still-piling-up-over-hospital-vaccine-mandates.html.

198 Ian Schwartz, "Rand Paul Grills Fauci: Have You Received Royalties from a Company That You Later Oversaw the Funding Of?," RealClearPolitics, 16 June 2022, article and video, 7:37, https://www.realclearpolitics.com/video/2022/06/16/rand_paul_grills_fauci_have_you_received_royalties_from_a_company_that_you_later_oversaw_the_funding_of.html#!; and *Hearing on the Federal Response to COVID Before Senate HELP Committee*, 117th Cong. (2022), "Heated Exchange Between Sen. Rand Paul and Dr. Anthony Fauci," https://www.c-span.org/video/?c5020101/heated-exchange-sen-rand-paul-dr-anthony-fauci.

199 National Institutes of Health, "Statement by Anthony S. Fauci, MD," news release, 22 August 2022, https://www.nih.gov/news-events/news-releases/statement-anthony-s-fauci-md.

200 Informed Consent Action Network v. Centers for Disease Control and Prevention and Health and Human Services, 1:22-cv-00481-RP doc. 17 (W.D. Tex. 2022), https://icandecide.org/wp-content/uploads/2022/09/019-Agreed-Scheduling-Order.pdf.

201 Ian Schwartz, "Sen. Rand Paul vs Fauci: When We're in Charge, You Will Have to Divulge Where You Get Your Royalties From," RealClearPolitics, 14 September 2022, article and video, 2:20, https://www.realclearpolitics.com/video/2022/09/14/sen_rand_paul_vs_fauci_when_were_in_charge_you_will_have_to_divulge_where_you_get_your_royalties_from.html#!; and *Hearing on Monkeypox Before Senate*, 117th Cong. (2022), "Exchange Between Sen. Rand Paul and Dr. Anthony Fauci at Monkeypox Hearing," https://www.c-span.org/video/?c5031010/exchange-sen-rand-paul-dr-anthony-fauci-monkeypox-hearing.

202 Hannah Knudsen, "Rand Paul Confronts Anthony Fauci: 'You're Not Paying Attention to the Science," Breitbart, 16 September 2022, https://www.breitbart.com/politics/2022/09/16/rand-paul-confronts-fauci-youre-not-paying-attention-science/.

203 Siri & Glimstad, "CDC's Covid-19 Vaccine V-Safe Data Released Pursuant to Court Order," news release, PR Newswire, 3 October 2022, https://www.prnewswire.com/news-releases/cdcs-covid-19-vaccine-v-safe-data-released-pursuant-to-court-order-301639584.html.

204 "Developing Safe and Effective Vaccines," Vaccines for Your Children, Centers for Disease Control and Prevention, last reviewed 21 June 2023, https://www.cdc.gov/vaccines/parents/infographics/journey-of-child-vaccine.html. The CDC has removed this material from the internet as well. FRE Rule 406 (admissible evidence from habit of a party may be entered into a court of law).

205 Sarah Rumpf-Whitten, "Rand Paul Promises to 'Subpoena Every Last Document of Dr. Fauci' in Victory Speech," Fox News, 8 November 2022, https://www.foxnews.com/politics/rand-paul-promises-subpoena-every-document-fauci-victory-speech.

206 Stefania Lugli, "Sarasota Memorial Hospital's COVID Protocols to Be Investigated After Emotional Meeting, *Sarasota Herald-Tribune*, 30 November 2022, https://www.heraldtribune.com/story/news/coronavirus/2022/11/30/sarasota-memorial-hospital-public-newly-elected-board-first-meeting-covid-protocols/10790847002/.

207 *Long Covid*, while popular, is an incorrect term for what should be scientifically and medically referred to as *postacute SARS-CoV-2 sequela*.

208 Greg Iacurci, CNBC, "Long Covid May Be 'the Next Public Health Disaster'—with a $3.7 Trillion Economic Impact Rivaling the Great Recession," CNBC, last modified 9 December 2022, https://www.cnbc.com/2022/11/30/why-long-covid-could-be-the-next-public-health-disaster.html.

209 Jimmy Tobias, "Unredacted NIH E-Mails Show Efforts to Rule Out a Lab Origin of Covid," *The Nation*, 19 January 2023, https://www.thenation.com/article/society/nih-emails-origin-covid-lab-theory/.
210 Discussed in point 134 to follow.
211 Serial passage to change the virus is gain-of-function. It is how gain-of-function was accomplished prior to newer methods.
212 "Understanding How COVID-19 Vaccines Work," Centers for Disease Control and Prevention, last modified 22 September 2023, https://www.cdc.gov/coronavirus/2019-ncov/vaccines/different-vaccines/how-they-work.html.
213 Overview of COVID-19 Vaccines, Centers for Disease Control and Prevention at https://www.cdc.gov/coronavirus/2019-ncov/vaccines/different-vaccines/overview-COVID-19-vaccines.html.
214 Open VAERS at https://openvaers.com/; and https://www.flemingmethod.com/stand-up-be-heard.
215 Kapil Bahl et al., "Preclinical and Clinical Demonstration of Immunogenicity by mRNA Vaccines Against H10N8 and H7N9 Influenza Viruses," *Molecular Therapy* 25, no. 6 (June 2017): 1316–27, https://www.sciencedirect.com/science/article/pii/S1525001617301569.
216 Katalin Karikó et al., "Incorporation of Pseudouridine into mRNA Yields Superior Nonimmunogenic Vector with Increased Translational Capacity and Biological Stability," *Molecular Therapy* 16, no. 11 (2008): 1833–40, https://doi.org/10.1038/mt.2008.200. Refer to Exhibit D.
217 Although the tumor-causing antigens (T- and t-antigen) of SV40 are not present in the sequences found in the genetic vaccines for SARS-CoV-2 (COVID), SV40 produces the promoter sequences that increase replication of the spike protein in human cells. This is discussed in further detail in this book.
218 Paul W. Barone et al., "Viral Contamination in Biologic Manufacture and Implications for Emerging Therapies," *Nature Biotechnology* 38 (May 2020): 563–72, https://doi.org/10.1038/s41587–020-0507–2.
219 The European Medicines Agency, which regulates vaccines in European Union nations, explained that "non-functional" fragments of SV40's DNA sequence are used as "starting material" in producing the vaccine. Philip Marcelo, No evidence that DNA sequence used in Pfizer shot leads to cancer and other health issues. AP News, 20 November 2023 at https://apnews.com/article/fact-check-covid-vaccine-pfizer-sv40-dna-853343189368.
220 CAACB is the Consortium on Adventitious Agent Contamination in Biomanufacturing.
221 CHO is Chinese Hamster Ovary Cell Line.
222 Jose Alfredo Samaniego Castruita et al., "SARS-CoV-2 Spike mRNA Vaccine Sequences Circulate in Blood up to 28 Days After COVID-19 Vaccination," *Acta Pathologica, Microbiologica, et Scandinavica* 131, no. 3 (March 2023): 128–32, https://doi.org/10.1111/apm.13294.
223 Kevin McKernan et al., "Sequencing of Bivalent Moderna and Pfizer mRNA Vaccines Reveals Nanogram to Microgram Quantities of Expression Vector

dsDNA per Dose," preprint, submitted 10 April 2023, https://osf.io/preprints/osf/b9t7m.

224 Yanfei He et al., "Effect of COVID-19 on Male Reproductive System—a Systematic Review," *Frontiers in Endocrinology* 12 (2021): https://doi.org/10.3389/fendo.2021.677701.

225 R. M. Fleming et al., "Pfizer, Moderna and Janssen Vaccine InflammoThrombotic and Prion Type Effect on Erythrocytes When Added to Human Blood," *Haematology International Journal* 7, no. 1 (2023): https://doi.org/10.23880/hij-16000210.

226 Lize M. Grobbelaar et al., "SARS-CoV-2 Spike Protein S1 Induces Fibrin(ogen) Resistant to Fibrinolysis: Implications for Microclot Formation in COVID-19," *Bioscience Reports* 41, no. 8 (27 August 2021): https://doi.org/10.1042/bsr20210611.

227 Yi Zheng et al., "SARS-CoV-2 Spike Protein Causes Blood Coagulation and Thrombosis by Competitive Binding to Heparan Sulfate," pt. B, *International Journal of Biological Macromolecules* 193 (15 December 2021): 1124–29, https://doi.org/10.1016/j.ijbiomac.2021.10.112.

228 R. M. Fleming et al., "Pfizer, Moderna and Janssen Vaccine InflammoThrombotic and Prion Type Effect on Erythrocytes When Added to Human Blood," *Haematology International Journal* 7, no. 1 (2023): https://doi.org/10.23880/hij-16000210.

229 Martina Patone et al., "Risks of Myocarditis, Pericarditis, and Cardiac Arrhythmias Associated with COVID-19 Vaccination or SARS-CoV-2 Infection," *Nature Medicine* 28 (2022): 410–22, https://doi.org/10.1038/s41591–021-01630–0.

230 Lael A. Yonker et al., "Circulating Spike Protein Detected in Post–COVID-19 mRNA Vaccine Myocarditis," *Circulation* 147, no. 11 (14 March 2023): 867–76, https://doi.org/10.1161/circulationaha.122.061025; and Salim Hayek, summary of "Spike Protein Detected in Post-COVID-19 mRNA Vaccine Myocarditis," American College of Cardiology, 17 January 2023, https://www.acc.org/latest-in-cardiology/journal-scans/2023/01/17/18/00/circulating-spike-protein.

231 Sofie Nyström and Per Hammarström, "Amyloidogenesis of SARS-CoV-2 Spike Protein," *Journal of the American Chemical Society* 144, no. 20 (2022): 8945–50, https://doi.org/10.1021/jacs.2c03925.

232 Zhouyi Rong et al., "SARS-CoV-2 Spike Protein Accumulation in the Skull-Meninges-Brain Axis: Potential Implications for Long-Term Neurological Complications in Post-COVID-19," preprint, submitted 5 April 2023, https://doi.org/10.1101/2023.04.04.535604; and "Prion Disease," Medline Plus, n.d., https://medlineplus.gov/genetics/condition/prion-disease/.

233 Joseph Fraiman et al., "Serious Adverse Events of Special Interest Following mRNA COVID-19 Vaccination in Randomized Trials in Adults," *Vaccine* 40, no. 40 (22 September 2022): 5798–805, https://doi.org/10.1016/j.vaccine.2022.08.036.

234 Marco Cosentino and Franca Marino, "The Spike Hypothesis in Vaccine-Induced Adverse Effects: Questions and Answers," *Trends in Molecular Medicine* 28, no.10 (October 2022): 797–99, https://doi.org/10.1016/j.molmed.2022.07.009.

235 Select Subcommittee on the Coronavirus Pandemic majority staff to Select Subcommittee on the Coronavirus Pandemic members, "New Evidence Resulting from the Select Subcommittee's Investigation into the Origins of COVID-19—'The

Proximal Origin of SARS-CoV-2,'"memorandum, 5 March 2023, https://oversight.house.gov/wp-content/uploads/2023/03/2023.03.05-SSCP-Memo-Re.-New-Evidence.Proximal-Origin.pdf.

236 Miranda Devine, "New Emails Show Dr. Anthony Fauci Commissioned Scientific Paper in Feb. 2020 to Disprove Wuhan Lab Leak Theory," *New York Post*, 5 March 2023, https://nypost.com/2023/03/05/new-emails-show-fauci-commissioned-paper-to-disprove-wuhan-lab-leak-theory/.

237 Erik Katz, "The Federal Employee COVID Vaccine Mandate Remains Blocked, After Appeals Court Ruling," Government Executive, 24 March 2023, https://www.govexec.com/management/2023/03/federal-employee-covid-vaccine-mandate-remains-blocked/384413/.

238 Ron Johnson, "Sen. Johnson Offers the No WHO Pandemic Treaty Act as an Amendment to AUMF Repeal," news release, 28 March 2023, https://www.ronjohnson.senate.gov/2023/3/sen-johnson-offers-the-no-who-pandemic-treaty-act-as-an-amendment-to-aumf-repeal; and Ron Johnson, "Senator Johnson on Senate Floor 3.28.23," RJmedia, video, 2:35, https://rumble.com/v2f6vbo-senator-johnson-on-senate-floor-3.28.23.html.

239 Krista Mahr and Adam Cancryn, "CDC Head Resigns, Blindsiding Many Health Officials," Politico, 5 May 2023, https://www.politico.com/news/2023/05/05/cdc-director-rochelle-walensky-leaving-00095583.

240 Monkeypox virus.

241 Malcolm Roberts, "Children Targeted by WHO 'Standards for Sexuality Education in Europe,'" *The Spectator*, 8 May 2023, https://www.spectator.com.au/2023/05/children-targeted-by-who-standards-for-sexuality-education-in-europe/.

242 Emma Farge, "WHO Pandemic Treaty: What Is It and How Will It Save Lives in the Future?," World Economic Forum, 26 May 2023, https://www.weforum.org/agenda/2023/05/who-pandemic-treaty-what-how-work/.

243 Emily Kopp, "Fauci Was Told of NIH Ties to Wuhan Lab's Novel Coronaviruses by January 2020," U.S. Right To Know,, 5 September 2023, https://usrtk.org/covid-19-origins/fauci-nih-wuhan-coronaviruses/.

244 BSL4 stands for biosafety level 4 laboratories. *"Select agents and toxins list". Centers for Disease Control and Prevention / USDA Animal and Plant Health Inspection Service at* .https://en.wikipedia.org/wiki/List_of_biosafety_level_4_organisms.

245 "Investigation and Prosecution of Those Individuals Responsible for Crimes Against Humanity,"Petitions.net, n.d., https://www.petitions.net/investigation_and_prosecution_of_those_individuals_responsible_for_crimes_against_humanity.

246 "Investigation and Prosecution," https://www.petitions.net/investigation_and_prosecution_of_those_individuals_responsible_for_crimes_against_humanity.

247 HIPAA stands for the Health Insurance Portability and Accountability Act of 1996. Health Insurance Portability and Accountability Act of 1996, Public Law 104–191, 21 August 1996 at https://www.cdc.gov/phlp/publications/topic/hipaa.html; principally written by a friend, Dr. William C. Dooley.

248 The Bill of Rights (Amendments 1—10), National Center for Constitutional Studies, 1 January 2018 at https://nccs.net/blogs/americas-founding-documents/bill-of-rights-amendments-1–10.
249 Wayne Fletcher, How many military members have died since 1776? The Gun Zone, 18 March 2024 at https://thegunzone.com/how-many-military-members-have-died-since-1776/.
250 Open VAERS at https://openvaers.com/.
251 United States military casualties of war at https://en.wikipedia.org/wiki/United_States_military_casualties_of_war.
252 Rule 406. Habit; Routine Practice. Cornell Law, LII at https://www.law.cornell.edu/rules/fre/rule_406.

Chapter 7

1 Stewart R, Charles M.B., Page J. A future with no individual ownership is not a happy one: Property theory shows why. *Futures* 2023;152:103209; https://www.sciencedirect.com/science/article/pii/S0016328723001131?via%3Dihub
2 https://crimesagainsthumanitytour.com/
3 Richard M. Fleming, PhD, MD, JD at https://www.flemingmethod.com/about; Kopf and Hansen side bar 2009 at https://www.youtube.com/watch?v=-5Va31X6Rq8; and The Truth, The Whole Truth and Nothing But The Truth! But not in US Federal Court 2009 at https://www.youtube.com/watch?v=GhwgMbIS-e4.
4 Ad Hominem Fallacy | Definition, Facts and Examples at https://study.com/buy/academy/lesson/ad-hominem-fallacy-definition-examples.html?src=ppc_bing_nonbrand&rcntxt=aws&crt=&kwd=SEO-PPC-BUY&kwid=dat-2329040505886940:loc-190&agid=1235851302596746&mt=b&device=c&network=o&_campaign=SeoPPC&msclkid=016e8522070a17709358e0848872d03f
5 How Does Sanger Sequencing Work?—Seq it Out #1 at https://www.youtube.com/watch?v=e2G5zx-OJIw.
6 Understanding Misinformation About Viruses at https://rumble.com/vir8i1-understanding-misinformation-about-viruses.html.
7 Over 500 cataloged publications on viruses, SARS-CoV-2, COVID-19, InflammoThrombotic Response and more at https://www.flemingmethod.com/documentation; and SARS coronavirus MA15 isolate d3om5, complete genome, GenBank: JF292920.1 at https://www.flemingmethod.com/_files/ugd/659775_7f8a6ee38f9c4ff085d74cc8c6138f1c.pdf.
8 This is just one of the reasons I developed FMTVDM to quantify testing, eliminating the errors reported. For instance, someone could think isotopes don't redistribute because Big Pharma told physicians the isotopes don't redistribute. The consequences of this include overradiating people and missing critical heart disease.
9 Boyer, Edward J. (August 14, 1989). "Controversial Nobel Laureate Shockley Dies." *Los Angeles Times* and https://en.wikipedia.org/wiki/William_Shockley.

10 Just in case you wondered, there is more variability in IQ within the races than between them.

11 As a cardiology fellow, part of the research I carried out was on these new technetium isotopes. I published the first SPECT paper on one of these: R. M. Fleming et al., "A Comparison of Technetium-99m Teboroxime Tomography with Automated Quantitative Coronary Arteriography and Thallium-201 Tomographic Imaging," *Journal of the American College of Cardiology* 17, no. 6 (May 1991): 1297–302, https://doi.org/10.1016/s0735–1097(10)80139–1.

12 R. M. Fleming, G. M. Harrington, R. Baqir, S. Jay, Sridevi Challapalli, K. Avery, and J. Green. The Evolution of Nuclear Cardiology Takes Us Back to the Beginning to Develop Today's "New Standard of Care" for Cardiac Imaging: How Quantifying Regional Radioactive Counts at 5 and 60 Minutes Post-Stress Unmasks Hidden Ischemia. *Methodist DeBakey Cardiovascular Journal* (MDCVJ) 2009;5(3):42–48. PMID: 20308963.

13 NBC's Tim Russert dies of heart attack at 58. NBC News 13 June 2008 at https://www.nbcnews.com/id/wbna25145431.

14 Charles Calisher, et al. Statement in support of the scientists, public health professionals, and medical professionals of China combatting COVID-19. *The Lancet* 2020;395:e42-e43 at https://www.thelancet.com/action/showPdf?pii=S0140–6736%2820%2930418–9; and Richard Horton, Offline: The origin story—division deepens. *The Lancet* 2021;398:2221 at https://www.thelancet.com/action/showPdf?pii=S0140–6736%2821%2902833–6.

15 Udani Samaraskera, Investigating the origins of novel pathogens. *The Lancet* 2021;21:1633 at https://www.thelancet.com/action/showPdf?pii=S1473–3099%2821%2900710–6.

16 Editorial, Searching for SARS-CoV-2 origins: confidence versus evidence. *The Lancet* 2023;4:e200 at https://www.thelancet.com/journals/lanmic/article/PIIS2666–5247(23)00074–5/fulltext.

17 R. M. Fleming et al., "Renewed Application of an Old Method Improves Detection of Coronary Ischemia: A Higher Standard of Care," *Federal Practitioner* 27 (2010): 22–31.

18 Richard M. Fleming, "Practicing Medicine Without a License." *New American*, 27 May 2011, https://thenewamerican.com/us/healthcare/practicing-medicine-without-a-license/.

19 That is a problem we have seen return with the rush to put genetic vaccines on the market, including strict product liability issues discussed in chapter 5.

20 See numbers 2, 3, and 51 at https://www.flemingmethod.com/documentation.

21 Kary B. Mullis et al., Process for amplifying, detecting, and/or cloning nucleic acid sequences, USPTO Patent US4683195A, filed 7 February 1986, and issued 28 July 1987, https://patentimages.storage.googleapis.com/ec/14/bf/0a414f77b2d203/US4683195.pdf.

22 Article I, § 8, C 8.1 Overview of Congress's Power Over Intellectual Property. Constitution Annotated. Analysis and Interpretation of the U.S. Constitution at https://constitution.congress.gov/browse/essay/artI-S8-C8–1/ALDE_00013060/.

23 The flaw with this approach to problem solving can also be seen with election integrity and border crossing issues. When the focus is on the symptom (for example election integrity or illegal aliens crossing the border) instead of addressing why elected politicians are so easily influenced by Big Pharma or the military-industrial complex (making how they got elected moot), or why we have people illegally entering the country (when their entry can alter voting outcomes and the economy), we might temporarily feel good about what we are doing, but we are not solving the problem. We are allowing it to become worse.
24 McClain, Dylan Loeb (August 15, 2019). "Kary B. Mullis, 74, Dies; Found a Way to Analyze DNA and Won Nobel." *The New York Times.* And https://en.wikipedia.org/wiki/Kary_Mullis. I suspect Dr. Mullis will most likely attend in his own way.
25 https://www.worldometers.info/coronavirus/
26 https://vaersanalysis.info/2024/01/05/vaers-summary-for-covid-19-vaccines-through-12–29-2023/
27 Religion in Israel at https://en.wikipedia.org/wiki/Religion_in_Israel.
28 Muslims follow the Islamic faith.
29 McClain, Dylan Loeb (August 15, 2019). "Kary B. Mullis, 74, Dies; Found a Way to Analyze DNA and Won Nobel." *The New York Times.* and https://en.wikipedia.org/wiki/U%C4%9Fur_%C5%9Eahin.
30 Mustafa Prize at https://en.wikipedia.org/wiki/Mustafa_Prize
31 What's Really in the Vaccines? at https://rumble.com/v261ge6-whats-really-in-the-vaccines.html
32 Snake Venom Misinformation at https://rumble.com/v196rya-snake-venom-misinformation.html.
33 There are also DOs (osteopathic doctors).
34 A Call for Unity; A Call for Action at https://rumble.com/v22941q-a-call-for-unity-a-call-for-action.html.
35 Charles Dickens, *A Tale of Two Cities. A Story of the French Revolution.* eBook licensed under Gutenberg. Released 1 January 1994 at https://www.gutenberg.org/cache/epub/98/pg98-images.html and discussed at https://en.wikiquote.org/wiki/A_Tale_of_Two_Cities.

Chapter 8

1 The quote is from *Batman Begins.*
2 Constitution of the United States, The Preamble, Constitution Annotated, Analysis and Interpretation of the U.S. Constitution at https://constitution.congress.gov/constitution/preamble/.
3 Letter from Dr. Gordon Harrington regarding his background, expertise and analysis of Hansen plagiarism of Fleming data at https://www.flemingmethod.com/_files/ugd/659775_1e0f048572374d4bb99bd17dc96c569e.pdf; and https://www.youtube.com/watch?v=-5Va31X6Rq8; Kopf and Hansen sidebar April 2009 at https://www.youtube.com/watch?v=GhwgMbIS-e4.

4 Dennis G. Carlson, Counsel for Discipline of the Nebraska Supreme Court, notification of investigation blockage by Kopf 18 May 2012 at https://www.flemingmethod.com/_files/ugd/659775_723225d0a00547018be12e456fe4353d.pdf.

5 Notice of Intent to Destroy Exhibits with Filing of false certification of service statement with the court 5 May 2015 at https://www.flemingmethod.com/_files/ugd/659775_d4a678c5377c4929863f67abce08d864.pdf

6 Robert Pear, Reagan Signs Bill on Drug Exports and Payment for Vaccine Injuries, The New York Times, 15 November 1986 at https://www.nytimes.com/1986/11/15/us/reagan-signs-bill-on-drug-exports-and-payment-for-vaccine-injuries.html; H.R.5546—National Childhood Vaccine Injury Act of 1986, Congress.gov at https://www.congress.gov/bill/99th-congress/house-bill/5546 and https://en.wikipedia.org/wiki/National_Childhood_Vaccine_Injury_Act.

7 NPDB National Practitioner Data Bank, U.S. Department of Health and Human Services at https://www.npdb.hrsa.gov/topNavigation/timeline.jsp; and National Practitioner Data Bank at https://en.wikipedia.org/wiki/National_Practitioner_Data_Bank.

8 Sharon C. Peters and Eric Werner, SCOTUS Lets National Practitioner Data Bank Safeguards Stand, 3 November 2023 at https://www.healthlawadvisor.com/scotus-lets-national-practitioner-data-bank-safeguards-stand.

9 *Cruzan v. Director, Missouri Department of Health*, 497 U.S. 261 (1990) at https://supreme.justia.com/cases/federal/us/497/261/.

10 FlemingMethod [Quantitative Nuclear Imaging] at https://www.flemingmethod.com/copy-of-fleming-method.

11 John Podesta at https://en.wikipedia.org/wiki/John_Podesta.

12 The ACA is also known as Obamacare.

13 Ms. Gutierrez discussed the importance of improving diagnostic testing. She also submitted evidence on 23 March 2006 that the billing codes I had used for combination planar and SPECT imaging were CORRECT. Shown at https://www.flemingmethod.com/_files/ugd/659775_cdf30f98c0ff4c67ad019d238d2da4b7.pdf. This evidence was not presented at trial by Mr. Hansen, or the court, thus hiding substantive exculpatory evidence from the jury.

14 Freedom of Choice Act.

15 ASBA is a pro-life group with the state of Nevada.

16 The National Vaccine Injury Act of 1986; https://en.wikipedia.org/wiki/National_Childhood_Vaccine_Injury_Act.

17 Bruesewitz v. Wyeth LLC, 562 U.S. 223 at https://casetext.com/case/bruesewitz-v-wyeth-llc.

18 Members of the Proposed New Diagnostic and Drug Price Committee (PNDDPC) must have no vested interests in any pharmaceutical company, including but not limited to employment, stock, and participation in pharmaceutical companies for a minimum of ten years.

19 Currently, 50 percent of all pharmaceutical sales occur within the United States, accounting for 90 percent of the profit. Americans should not be providing a disproportionate share of the burden supporting pharmaceutical companies, many of

which are headquartered outside of the United States. Accordingly, the price of any medication sold in the United States may not be greater than that sold outside the United States based upon the exchange rate.

20 Currently, health-care costs are irregularly distributed, with larger cities and larger health-care facilities paying less for medications, diagnostic drugs, medical supplies, and equipment than smaller metropolitan or rural areas, medical practices, and so forth. This produces increased costs for those in smaller practices or communities, while larger communities and facilities are frequently allowed to bill (while paying less) more for the same current procedural terminology (CPT) billing codes.

21 Transparency in pricing applies not only to diagnostic testing but also to physician costs, physician office costs, hospital costs, emergency transportation by ambulance, prescription and nonprescription drug prices, and all costs related to the health care provided to US citizens under this act.

22 A History of the FDA and Drug Regulation in the United States, found at https://www.fda.gov/media/73549/download. W. J. Heath, "America's First Drug Regulation Regime: The Rise and Fall of the Import Drug Act of 1848," *Food and Drug Law Journal* 59 no. 1 (2004): 169–99.

23 J. Lenzer, "Doctors Outraged at Patriot Act's Potential to Seize Medical Records," *BMJ* 332 (2006): 69,https://doi.org/10.1136/bmj.332.7533.69.

24 National Childhood Vaccine Injury Act (NCVIA) of 1986 (42 U.S.C. §§ 300aa-1 to 300aa-34) discussed at https://en.wikipedia.org/wiki/National_Childhood_Vaccine_Injury_Act.

25 Measuring and addressing physician burnout, The American Medical Association website, 3 May 2023 at https://www.ama-assn.org/practice-management/physician-health/measuring-and-addressing-physician-burnout#:~:text=Physician%20burnout%20rates%20spike%20to%2063%25%20in%202021,symptoms%20of%20burnout%2C%20up%20from%2038%25%20in%202020.

26 The type of diversity we need isn't based upon color or sex or religion; it's based upon different approaches to thinking through and solving problems.

27 Tyler Harvey, Removing the MCAT Could Improve Diversity in Medicine, *Newsweek*, 24 January 2023 at https://www.newsweek.com/removing-mcat-could-improve-diversity-medicine-opinion-1775471#:~:text=An%20article%20published%20in%20JAMA,the%20MCAT%20as%20a%20requirement.

28 Robert H. Shmerling, Why life expectancy in the US is falling. Harvard Health Publishing, 20 October 2022 at https://www.health.harvard.edu/blog/why-life-expectancy-in-the-us-is-falling-202210202835.

29 InflammoThrombotic Response at https://www.flemingmethod.com/inflammothromboticresponse-itr-disease.

30 SARS-CoV-2, COVID, GoF, CRISPR and the Genetic Vaccines. Why are People Dying and What Answers is Science Teaching Us? Houston, TX 18 September 2023 at https://www.flemingmethod.com/.

31 Richard M. Fleming, "Practicing Medicine Without a License." *New American*, 27 May2011,https://thenewamerican.com/us/healthcare/practicing-medicine-without-a-license/.

32 R. M. Fleming and T. K. Chaudhuri, "American Healthcare Is at the Forefront of the 2020 Presidential Election," comment on Donald M. Berwick, "Understanding the American Healthcare Reform Debate," *BMJ* 357 (2017): https://www.bmj.com/content/357/bmj.j2718/rapid-responses.

33 Richard M Fleming, PhD, MD, JD, at https://www.flemingmethod.com/about.

34 The Case For Richard M Fleming, PhD, MD, JD, The high cost of inventing disruptive technology. At https://www.flemingmethod.com/thecase.

35 Setting the stage, at https://www.flemingmethod.com/more.

36 List of justices of the Supreme Court of the United States at https://en.wikipedia.org/wiki/List_of_justices_of_the_Supreme_Court_of_the_United_States

37 Anna-Leigh Firth, The majority of judges polled oppose lifetime appointments. The National Judicial College, 11 November 2020 at https://www.judges.org/news-and-info/the-majority-of-judges-polled-oppose-lifetime-appointments/.

Chapter 9

1 Conspiracy, Cornell Law, LII at https://www.law.cornell.edu/wex/conspiracy.

2 Mike Caulfield, The first use of the term "conspiracy theory" is much earlier—and more interesting—than historians have thought. Hapgood, 24 December 2018 at https://hapgood.us/2018/12/24/the-first-use-of-the-term-conspiracy-theory-is-much-earlier-and-more-interesting-than-historians-have-thought/#:~:text=Thus%2C%20a%20search%20of%20the%20database%20America%E2%80%99s%20Historical,theory%20%281874%29%20blackmail%20theory%20%281874%29%20abduction%20theory%20%281875%29.

3 Election integrity is about more than simply counting ballots correctly. It's about making certain that those elected to office serve the people who elected them and not the interests of those with money and power. Integrity is the quality of being honest and having strong moral principles. Even if the votes are correctly counted, there is no election integrity when those elected are compromised once elected; no longer serving those they were elected to represent.

4 Fmr. CDC head Redfield reinforces belief that Covid 'more likely' spread from lab leak. NBX News, 8 March 2023 at https://www.nbcnews.com/now/video/fmr-cdc-head-reinforces-belief-that-covid-more-likely-spread-from-lab-leak-164714565651.

5 Alexander Hutzler, Ex-CDC Director Robert Redfield Says He Got Death Threats for Saying He Thought COVID Leaked From China Lab 3 June 2021 at https://www.newsweek.com/ex-cdc-director-robert-redfield-says-he-got-death-threats-saying-he-thought-covid-leaked-china-lab-1597382.

6 Winston Cutshall, Producer, JFK Declassified: Tracking Oswald, 2017 at https://www.imdb.com/title/tt6830318/fullcredits?ref_=ttpl_ql_dt_1; Olivia B. Waxman, Former CIA Operative Argues Lee Harvey Oswald's Cuba Connections Went Deep, *TIME*, 25 April 2017 aat https://time.com/4753349/oswald-kennedy-declassified-documentary/; and Winston Cutshall, Producer, JFK Declassified: Tracking Oswald, 2017 at https://www.youtube.com/channel/UCk7zxuQSWg9dPc0FalNAfcA/videos.

7 Dr. Fred Stamler, Winston Cutshall, Producer, JFK Declassified: Tracking Oswald, 2017 at Winston Cutshall, Producer, JFK Declassified: Tracking Oswald, 2017 at Department of Pathology, Departmental History, The Territory of Iowa, University of Iowa Health Care, Carver College of Medicine, The Medical Bulletin from 6 April 1940 at https://medicine.uiowa.edu/pathology/about-us/departmental-history #rockefeller; and Mark E. Sobel, In Memoriam, Richard (Dick) Gregory Lynch, M.D., 1934–2009, The American Journal of Pathology, 2010;177(1):2–3 at https: //ajp.amjpathol.org/action/showPdf?pii=S0002–9440%2810%2960055–7.

8 *JFK: The Ultimate Conspiracy*, directed by Matt Salmon (2020; New York: 1091 Pictures), https://pluto.tv/en/on-demand/movies/62bf84e287b6c600133a88ad. The movie features former CIA agent Bob Baer.

9 Ashley Hamer, 99 Percent Of The Earth's Species Are Extinct—But That's Not The Worst Of It Nature and Wildlife Discovery, Discovery 1 August 2019 at https: //www.discovery.com/nature/99-Percent-Of-The-Earths-Species-Are-Extinct.

10 There are 1,609.34 meters in a mile; therefore, our DNA is roughly 120 trillion meters long.; During our discussions on genetics and DNA, I recalled that the amount of DNA in a human is quite substantial, and if the strands were linked end to end, they would cover a distance of "X". I forgot how long human DNA was, and my class is now curious themselves. Can you tell us how long human DNA is? UCSB Science line, 2002–06-07 at http://scienceline.ucsb.edu/getkey .php?key=144.

11 mtDNA = mitochondrial DNA.

12 S = subunit. R. M. Fleming "Unmasking CoViD, Part 1," (self-pub., 2020), Kindle; and R. M. Fleming, "Unmasking CoViD, Part 2," (self-pub., 2020), Kindle.

13 R. M. Fleming et al., "Nature vs. Nurture—Will the Species Survive?," *Journal of Food and Nutrition Sciences* 2, no. 1 (2019): 174–77, https://doi.org/10.32474/ SJFN.2019.2.000128.; Thomas Peter Bennett, et al., *The Physical Basis of Life* Del Mar, California: CRM Books, 1972; James H. Moller, et al., *Congenital Heart Disease*, Kalamazoo, Michigan, Upjohn, © 1971; K. L. Moore, "Clinically Oriented Embryology," in *The Developing Human*, 2nd ed. Philadelphia, London, Toronto: W. B. Saunders, © 1977; J. J. Moeling, *The Laureates' Anthology, New York City, New York, USA* : Scientific American, © 1990; and J. G. Jorgensen, *Biology and Culture in Modern Perspective*, San Francisco, USA: W. H. Freeman, 1956).

Chapter 10

1 Obstruction of Justice, Cornell Law, LII at https://www.law.cornell.edu/wex /obstruction_of_justice.

2 Richard M. Fleming, PhD, MD, JD, FlemingMethod at https://www.fleming method.com/about.

3 The Truth, The Whole Truth and Nothing But The Truth! But not in US Federal Court, April 2009 at https://www.youtube.com/watch?v=GhwgMbIS-e4; The Case For Richard M Fleming, PhD, MD, JD. The high cost of inventing disruptive technology. at https://www.flemingmethod.com/thecase.

4 Judge block investigation of attorney hiding of evidence of innocence from the jury at https://www.flemingmethod.com/_files/ugd/659775_723225d0a00547018be12e456fe4353d.pdf.

5 Bill McCarthy, The facts behind the Russian, right-wing narratives claiming Hunter Biden funded biolabs in Ukraine. Politifact. 1 April 2022. At https://www.politifact.com/article/2022/apr/01/facts-behind-russian-right-wing-narratives-claimin/.

6 Federal Rules of Evidence, Rule 406 - Habit; Routine Practice, 2024 Edition at https://www.rulesofevidence.org/fre/article-iv/rule-406/#:~:text=Federal%20Rule%20of%20Evidence%20406%20allows%20for%20the,or%20organization%20likely%20acted%20in%20a%20particular%20situation.

7 Transparency, Cambridge Dictionary at https://dictionary.cambridge.org/dictionary/english/transparency.

8 The Wachowskis—*V For Vendetta*—Broadcast Speech, 1 January 2006 at https://genius.com/The-wachowskis-v-for-vendetta-broadcast-speech-annotated.

Index

A

Aktion T4: 21, 76.
AMA: 83, 109–115, 120, 122, 123, 149, 150, 156, 157, 293–296, 299–301, 310.

B

Belmont: 116, 141, 149, 161, 162, 248.
Bioweapon: 1, 26, 27–29, 32, 34, 36, 78, 82, 86, 98, 123, 127, 136, 140, 156, 163, 218, 219, 248, 259, 270, 303, 318.
Blood Clotting—*see InflammoThrombotic Response* (ITR).
BWC Treaty: 3, 30, 32, 141, 142–145, 147, 149, 158, 159, 178, 248.

C

Carnegie: 13, 15, 82, 109, 112–115, 123, 277, 296, 310.
Catholic(s): 10, 16, 17, 74, 270, 298.
Christian: 5, 69, 97, 270, 271.
China: 34, 73, 85, 147, 168, 170–172, 175–177, 180, 183, 187, 188, 200–202, 204, 208, 210, 211, 215, 225, 226, 236, 239, 248, 249, 263, 307, 316.
CIA: 22, 23, 25, 45, 78, 79, 308, 309.
Clinton, Hillary: 81.
CMS: 120, 210, 211, 266, 285, 293, 299.
Collins: 140, 162, 184, 205, 212–215, 218, 227, 236, 237, 240.
Congress: 10, 19, 88–92, 102–106, 114, 130, 145, 147, 152, 154, 158–162, 207, 208, 210, 225, 246, 248, 250, 276, 283, 284, 286, 287, 290, 293, 294, 303.
Conspiracy: 140, 151, 174, 177, 182, 215, 305.
Constitution: 17, 87, 89–92, 96, 101–105, 114, 125, 189, 248–250, 268, 276, 284, 287, 292–294, 300.
COVID-19: Corona Virus Disease—2019 is The InflammoThrombotic Disease caused by SARS-CoV-2. *see ITR and SARS-CoV-2.*
CRISPR: 2, 28, 29, 56–60, 63–65, 118, 201.

D

DARPA: 36, 59.
Daszak: 36–38, 140, 164, 166, 168, 169, 172, 174, 176, 192, 201, 202, 247.
Deinert: ix, 255.
DNA: 2, 11, 28, 29, 49, 53–55, 57–61, 64, 65, 73, 108, 122, 129,

163, 173, 217, 218, 231, 232, 248, 249, 310–312.
Dual-Use: 28, 81, 86, 126, 138, 269.

E

EcoHealth: 36–38, 115, 164–169, 174, 192, 201–203, 210, 247, 248.
Event 201: 78, 79, 81, 82, 86, 95, 254, 303.
EUAs: 46, 47, 60, 117, 123, 128, 129, 150, 156, 190, 193, 194, 196, 252, 266.
Eugenics: 3, 10–15, 17–22, 25, 27–29, 38, 53, 71, 96, 100, 106, 108, 109, 113, 118, 126, 251, 254, 261, 277, 309–313, 318, 319.
Existential: xxi, xxii, 306.

F

Farrar: 82, 176, 184, 205, 214, 227, 228, 236, 238, 239, 240.
Fauci: 38, 140, 167, 176, 177, 181, 182, 184, 187, 199, 200–205, 207, 208, 210–214, 222–227, 236–240, 247, 248, 269, 270.
FBI: 25, 263.
FCS (FURIN): 33, 34, 62, 167, 169, 170, 172, 214, 215, 227, 238.
Fort Detrick: 22, 26, 119, 140, 165, 172.
FDA: 45, 46, 60, 63, 82, 84, 85, 100, 110, 114, 115–118, 122, 123, 128, 129, 140, 189, 190, 192–196, 206, 216, 223, 225, 231, 260, 266, 285–287, 289–291, 298.
Fleming: xxii, 170, 172–174, 193, 198, 206, 207, 232, 234, 255, 270, 279, 281, 322, 327.
Flexner: 112, 113, 115, 308.
FMTVDM: 118, 198, 278, 297.

G

Gain-of-Function: 1, 2, 28, 29, 32, 34, 36–38, 52, 53, 58, 60, 62, 63, 81, 82, 86, 95, 115, 118, 119, 126, 155, 156, 160, 162–166, 168–172, 174, 178, 192, 200–202, 207, 208, 210–212, 215, 219, 227, 247–249, 251, 252, 266, 269, 270, 323.
Galileo: 10, 127, 298.
Genetic(s): 1, 5, 10, 11, 13, 15, 19, 20, 22, 28, 29, 34, 53–59, 60, 61, 63–66, 68, 76, 96, 108, 109, 115, 164, 167, 169, 172, 173, 176, 207, 215, 217, 218, 232, 237, 238, 248, 251, 259–261, 267, 268, 271, 275, 309, 310, 311, 312.
Genetic Vaccine(s): 2, 45–48, 50–53, 59, 60, 62, 67, 68, 95, 116, 117, 123, 125, 128–134, 136, 137, 150, 155, 156, 160, 186, 190, 192–194, 218, 228, 232, 249, 253, 257, 259, 266, 267, 269–271, 274, 311, 312, 324.

H

Helsinki Declaration: 141, 149, 162, 248.
HHS: 32, 34, 55, 62, 84, 85, 86, 100, 114, 115, 118, 120, 123, 128, 129, 149, 162, 164, 176, 182, 201, 260, 277, 281, 287, 289, 295.
Hitler: 3, 5, 11, 19–21, 25, 71, 76, 78, 101, 108, 113, 125, 151, 300, 312.

HIV: 28, 34, 59, 62, 108, 118, 120, 124, 127, 164, 170, 178, 229, 243, 257, 269, 278, 297.
Huff: ix, 255

I
ICAN: 203–205, 223, 225.
ICCPR Treaty: 141, 149, 161, 248.
IHR: 87, 90–92, 94, 242, 253, 294, 300, 303.
Informed Consent: 27, 116, 133, 134, 141, 149, 150, 151, 156, 157, 161, 203, 217, 218, 223, 257, 258, 293, 299, 319.
Inflammation—*see InflammoThrombotic Response (ITR).*
InflammoThrombotic Response (ITR): 32–34, 47, 60, 67, 118, 119, 129, 131–133, 155, 156, 176, 198, 199, 206, 207, 228, 231–233, 253, 266, 269, 270, 297, 324.
Islam: 73, 110, 271.

J
Jews: 9, 18, 20, 69, 70, 74, 76, 108, 270, 271.
Johnson, Ron: 240.

L
Luc Montagnier (Prof.): 28, 108, 147, 257.

M
Mandate(s): 94, 95, 106, 186, 192, 209, 210, 212, 215, 221, 222, 240, 242.
McCairn: ix, xxi, 147, 255.
Medicine: 13, 46, 65, 66, 83, 95, 97–100, 106, 107, 110, 111–113, 115, 119–125, 127, 129, 137, 141, 150, 156, 161, 162, 182, 184, 189, 264, 265, 272, 273, 276–278, 279, 281, 285, 287–289, 293–301, 304, 313, 314.
Mengele: 22, 52, 96.
Military Industrial Complex (MIC): 1, 22, 25, 27, 29, 77 78, 79, 95, 121, 142, 251, 275, 277, 302, 306–309, 312, 313, 319.
MKUltra: 22, 275, 308.
mRNA: 2, 46–48, 54, 59, 60, 122, 129, 196, 198, 206, 228–231, 233, 235, 236, 249.
Muslim(s): 69, 74, 270, 271.

N
Nazi: 3, 15, 19–22, 25, 38, 76–78, 96, 99, 109, 113, 125, 138, 149, 248–250, 300, 306, 309, 311, 312, 314, 318.
NCVIA: 129, 277, 278, 286–288, 295–296, 301.
NIAID: 26, 62, 163, 166–168, 172, 174, 202, 204, 205, 211, 213, 223, 226, 247, 248, 270.
NIH: 34, 37, 58, 85, 162–168, 172, 174, 200–205, 207, 208, 210, 211, 212, 215, 218, 222, 223, 227, 235, 236, 240, 247, 290.
NPDB: 120, 277, 278, 287–288, 294, 296–298.
Nuremberg: 2, 20, 22, 140, 151, 157, 158, 248.
NTI: 81, 82, 86, 87, 95, 254, 303.

O
Operation Paperclip: 22, 77, 108, 306, 308.
Over Reach: 119, 156, 157, 226, 288, 295, 302, 304.

P

Perjury: 151, 153–156, 162, 168.
Polymerase Chain Reaction (PCR): 34, 37, 163, 172, 174–176, 187, 188, 193, 217, 218, 231, 267–269, 316.
President: 17, 21, 22, 25, 26, 75–79, 89, 90, 95, 104, 105, 113, 114, 120, 128, 146, 158, 160, 161, 163, 169, 176, 177, 178, 181, 183, 186–188, 205, 208–210, 212, 221, 239–241, 246, 275, 277, 279, 284, 298, 302, 304, 307, 308, 312, 314.
President Biden: 25, 91, 113, 115, 169, 179, 184, 188, 208–210, 212, 215, 221, 241, 279, 284, 308.
President Bush (George): 160.
President Bush (W): 163.
President Carter: 161.
President Eisenhower: 22, 23, 77, 79, 95, 275, 302, 307, 309, 312, 314.
President Ford: 158.
President Johnson: 26, 76, 120.
President Kennedy: 3, 17, 23, 25, 26, 78, 79, 95, 275, 279, 302, 306, 308, 309, 312, 314.
President Lincoln: 71, 75, 280.
President Nixon: 22, 26, 163.
President Obama: 36, 37, 115, 166, 201, 219, 279, 284, 298, 304.
President Truman: 90, 246.
President Trump: 128, 146, 169, 176–178, 181, 183, 187, 201–203, 239, 242, 302.
President Washington: 101, 314.
Prion: 47, 59, 67, 155, 233–235, 324.

R

Rand Paul: 199–201, 207, 208, 211–213, 222–225.
Redfield: 241, 306, 307.
Rixey: ix, xxii, 255.
Rockefeller: 13, 15, 30, 56, 58, 79, 80, 83, 108, 109, 112–115, 118, 123, 182, 186, 203, 255, 277, 288, 296, 308, 310, 326.
Russia: 9, 19, 21–22, 38, 74, 75, 108, 219, 220, 309.

S

SARS-CoV-2: 2, 26, 28, 32, 34, 37, 38, 47, 48, 50, 51, 53, 58, 62, 86, 98, 107, 108, 117, 119, 122–124, 127, 128, 155, 156, 160, 163, 165, 169–178, 187, 188, 192, 193, 198, 199, 201, 207, 214, 215, 227, 228, 230–234, 236–240, 242, 247, 248, 253, 257, 259–261, 263, 266–270, 306, 316, 317, 323.
Schwab, Klaus: 80, 184–188, 218, 254.
SCOTUS: 15, 17, 88, 102, 105, 106, 122, 123, 129, 130, 278, 286, 287, 295, 296, 301, 302.
SPARS: 83–86, 95, 254, 303.
Spike Protein: 33, 34, 36, 37, 45, 47, 48, 50, 51, 62, 63, 155, 160, 165–167, 169, 170, 178, 200, 210, 214, 228, 231–234, 266–269.
SPL (Strict Product Liability): 47, 60, 129–130, 132, 134–137, 277.

T

Thrombosis—*see InflammoThrombotic Response (ITR).*

U

Ukraine: 38, 169, 174, 218–220, 248, 249, 303, 317.
U.N.: 82, 83, 87, 94, 114, 177, 187, 188, 220, 221, 242–244, 246, 247

V

Vaccine(s): 2, 45–53, 59, 60, 62, 65, 67, 68, 84–86, 95, 116, 117, 122, 123, 125, 128–135, 137, 150, 155, 156, 160, 171, 175, 180, 183, 186, 187, 189–199, 203, 205, 206, 209, 210, 212, 215–218, 221–226, 228–236, 241, 249, 252, 253, 257, 259, 265–267, 269–271, 274, 277, 286, 287, 295, 296, 311.
VAERS: 67, 117, 190, 194, 228, 296.

W

Wargames: 78, 95, 254, 313.
WEF: 79, 80, 83, 93, 94, 183–189, 194, 197, 198, 218, 220, 221, 247, 249, 277, 310.
We The People: 66, 94, 139, 141, 255, 276, 277, 279, 284, 296, 298, 299, 301, 302, 304, 310.
WHO: 68, 82–84, 87, 90–95, 100, 162, 172, 176, 177, 183, 187, 196, 205, 207, 235, 238, 240–247, 249, 253, 263, 277, 288, 294, 295, 300, 303.
WIV: 34, 36, 140, 164, 167, 168, 170, 172, 174, 200, 202, 203, 207, 212, 214, 227, 247, 248, 306.